Sustaining the Earth
An Integrated Approach
SEVENTH EDITION

G. TYLER MILLER, JR.

President, Earth Education and Research

THOMSON

BROOKS/COLE

Australia • Canada • Mexico • Singapore • Spain • United Kingdom • United States

THOMSON

BROOKS/COLE

Publisher: *Jack Carey*
Developmental Editor: *Scott Spoolman*
Production Project Manager: *Andy Marinkovich*
Technology Project Manager: *Keli Sato Amann*
Senior Assistant Editor: *Suzannah Alexander*
Editorial Assistants: *Jana Davis, Chris Ziemba*
Marketing Manager: *Ann Caven*
Marketing Assistant/Associate: *Leyla Jowza*
Advertising Project Manager: *Nathaniel Bergson-Michelson*
Print/Media Buyer: *Karen Hunt*

Production Management, Copyediting, and Composition: *Thompson Steele, Inc.*
Interior Illustration: *Precision Graphics; Sarah Woodward; Darwin and Vally Hennings; Tasa Graphic Arts, Inc.; Alexander Teshin Associates; John and Judith Walker; Rachel Ciemma; Victor Royer, Electronic Publishing Services, Inc.; J/B Woosley Associates; Kerry Wong, ScEYEnce Studios.*
Cover Image: *Mountain Lake (Eagle Cap Wilderness Area, Wallowa Mountains, NE Oregon). Photograph by David Jensen.*
Text and Cover Printer: *Edwards Brothers, Incorporated*

Printed in the United States of America

1 2 3 4 5 6 7 08 07 06 05 04

For more information about our products, contact us at:
Thomson Learning Academic Resource Center
1-800-423-0563
For permission to use material from this text, contact us by:
Phone: 1-800-730-2214
Fax: 1-800-730-2215
Web: http://www.thomsonrights.com

ExamView® and *ExamView Pro®* are registered trademarks of FSCreations, Inc. Windows is a registered trademark of the Microsoft Corporation used herein under license. Macintosh and Power Macintosh are registered trademarks of Apple Computer, Inc. Used herein under license.

Library of Congress Control Number: 2004110310

ISBN 0-534-49672-5

Brooks/Cole-Thomson Learning
511 Forest Lodge Road
Pacific Grove, CA 93950
USA

Asia
Thomson Learning
5 Shenton Way #01-01
UIC Building
Singapore 068808

Australia
Nelson Thomson Learning
102 Dodds Street
South Melbourne, Victoria 3205
Australia

Canada
Nelson Thomson Learning
1120 Birchmount Road
Toronto, Ontario M1K 5G4
Canada

Europe/Middle East/Africa
Thomson Learning
High Holborn House
50/51 Bedford Row
London WC1R 4LR
United Kingdom

Latin America
Thomson Learning
Seneca, 53
Colonia Polanco
11560 Mexico D.F.
Mexico

Spain
Paraninfo Thomson Learning
Calle/Magallanes, 25
28015 Madrid, Spain

BRIEF CONTENTS

DETAILED CONTENTS

PREFACE
FOR INSTRUCTORS

What Are the Major Features of This Book? Science-Based, Solutions-Oriented, Integrated, Flexible, and Balanced

This book is designed for introductory courses on environmental science. It is an *interdisciplinary* study of how nature works and how things in nature are interconnected. This book is *science-based, solutions-oriented, flexible,* and *balanced.* It is divided into five major parts (see Brief Contents, p. iii).

A *flexible format* allows instructors to use almost any course outline. I suggest that instructors use Chapter 1 to provide an overview of environmental problems and solutions and Chapters 2 through 5 to provide a base of scientific principles and concepts. The remaining chapters—6 through 14—can be used in virtually any order or omitted as desired. In addition, sections within chapters can be omitted or rearranged to meet instructor needs.

This book has 48 Case Studies (see topics in bold-faced type in the detailed table of contents, pp. v–viii). They provide a more in-depth look at specific environmental problems and their possible solutions. Twenty-one *Guest Essays* are available on the website for this book (see list of contributors on page xvi).

This book is an integrated study of environmental problems, connections, and solutions. The seven integrative themes are *biodiversity and natural capital (ecosystem services), sustainability, pollution prevention and waste reduction, population and exponential growth, energy and energy efficiency, solutions to environmental problems,* and *the importance of individuals working together to bring about environmental change.*

Balance is achieved by discussing different sides of environmental issues (for example, see the discussion of the advantages and disadvantages of reducing birth rates on p. 84) and using 43 *trade-offs diagrams* to summarize the *advantages* and *disadvantages* of various technologies and proposed solutions to environmental problems and *good* and *bad news* about progress on a particular environmental problem. Examples of such diagrams are Figure 1-4 (p. 7) on good and bad news about economic development; Figure 6-10 (p. 109) on clear-cutting forests; Figure 8-12 (p. 158) on genetically modified food and crops; Figure 10-11 (p. 210) on coal;

Figure 10-37 (p. 233) on hydrogen; and Figure 13-11 (p. 299) on sanitary landfills.

This book is *focused on visual learning.* It has 315 illustrations, including 43 new and 90 improved figures for this edition. These illustrations are designed to present complex ideas in understandable ways and to relate learning to the real world.

This book is *short* (14 chapters covered in 327 pages) and *economical.* We have lowered the book's cost by reducing its size, printing it in black and white, and using a soft cover.

The *website* for this book at www.biology.brooks cole.com/miller7 contains study aids and many ideas for further reading and research. For each chapter there is a summary, review questions, flash cards for key terms and concepts, a multiple-choice practice quiz, interesting Internet sites, references, and a guide for accessing thousands of InfoTrac® College Edition articles from more than 700 periodicals.

This book is *current and accurate.* I have consulted more than 10,000 research sources in the professional literature and about the same number of Internet sites. I have also benefited from the more than 250 experts and teachers who have provided detailed reviews of this and my other three books in this field.

Instructors who want a book covering this material with a different emphasis and length can use one of my three other books written for various types of environmental science courses: *Living in the Environment,* fourteenth edition (642 pages, Brooks/Cole, 2005) *Environmental Science,* tenth edition (538 pages, Brooks/Cole, 2004), and *Essentials of Ecology,* third edition (272 pages, Brooks/Cole, 2005).

Why Have I Changed the Order of the Chapters? Satisfying Text Users

There are hundreds of different ways to organize the chapters in this book to fit the needs of different instructors. That is why one of the key features of this book is the *flexibility* in changing the order of chapters once the basic science has been covered.

Over the years, I have used different chapter orders usually in response to feedback from instructors.

In this edition, I have decided to put the two chapters on biodiversity (Chapters 6 and 7) after the chapters in Part II on Science and Ecological Principles for two reasons.

First, the loss and degradation of biodiversity is one of the most serious environmental problems. *Second,* the majority of instructors teaching this course prefer that I put all of the biology chapters (ecology and biodiversity, Chapter 3–7) together.

There is no best chapter order, but this one makes sense to most text users, and I plan to stick with this order from now on. Because of the book's flexibility, anyone who wants a different order can easily do so.

What Are the Major Changes in the Seventh Edition? The Most Significant Revision Ever

This is the most significant revision since the first edition. Major changes include the following:

- Chapter order changed to put all of the biology (ecology and biodiversity, Chapters 3–7) together

- Entire book rewritten with help from a developmental editor and three in-depth reviews.

- 43 *Trade-Offs diagrams,* 48 *Case Studies,* and 11 *What Can You Do?* diagrams

- Simpler, more learning-oriented writing style

- Improved learning system. Each subsection begins with a question. Then brief answers to the question are given by a short headline and by a one-sentence summary of the subsection. Here is an example:

What Is Biodiversity? Variety is the Spice of Life

A vital renewable resource is the biodiversity found in the earth's variety of genes, species, ecosystems, and ecosystem processes.

Text of the subsection then follows.

- Important new topics: affluenza, future implications of genetic engineering, foundation species, reconciliation ecology, General Motors' fuel-cell car of the future, phytoremediation, and expanded treatment of stewardship environmental worldview.

- Extensive updating. It is essentially impossible for readers to detect the more than 5,000 updates in this book based on information published in 2002, 2003, and 2004. But there are over 200 references throughout the book that specifically cite these years and 4,800 additional updates that are not cited by these years.

What In-Text Study Aids Are Provided? Tried and True Tools Plus Some New Ones

Each chapter begins with a few general questions to reveal how it is organized and what students will be learning. When a new term is introduced and defined, it is printed in boldface type. A glossary of all key terms is located at the end of the book.

Questions are used as titles for all subsections so readers know the focus of the material that follows. In effect, this is a built-in set of learning objectives. A brief attention-grabbing headline follows each subsection question to give students a general idea of the content of the subsection. This is followed by a single sentence that summarizes the key material in each subsection. It provides readers with a built-in running summary of the material (see example at left). I believe this is a more effective learning tool than providing an end-of-chapter summary.

Each chapter ends with a set of *Critical Thinking* questions. The website for the book also contains a set of *Review Questions* covering *all* of the material in the chapter that can be used as a study guide for students. Some instructors download this off the website and give students a list of the particular questions they expect their students to answer.

What Internet and Online Study Aids Are Available?

Qualified users of this textbook have free access to the Brooks/Cole Biology and Environmental Science Resource Center. Access the online resource material for this book by logging on at

http://biology.brookscole.com/miller7

At this website you will find the following material for each chapter:

- Flash Cards, which allow you to test your mastery of the Terms and Concepts to Remember for each chapter

- Environmental Quizzes, which provide a multiple-choice practice quiz

- Student Guide to InfoTrac® College Edition, which will lead you to Critical Thinking Projects that use InfoTrac College Edition as a research tool

- References, which lists the major books and articles consulted in writing this chapter

- A brief What You Can Do list addressing key environmental problems

- Hypercontents, which takes you to an extensive list of websites with news, research, and images related to individual sections of the chapter

Qualified adopters of this textbook also have free access to WebTutor Toolbox on WebCT and Blackboard at:

http://e.thomsonlearning.com

It provides access to a full array of study tools, including flashcards (with audio), practice quizzes, online tutorials, and web links.

Teachers and students using new copies of this textbook also have free and unlimited access to InfoTrac College Edition. This fully searchable online library gives users access to complete environmental articles from several hundred periodicals dating back over the past 25 years.

Other student learning tools include:

- An interactive CD-ROM, *Student CD with Environmental ScienceNow,* that integrates concept summaries and almost 100 engaging animations and interactions based on figures from the text with flashcards and quizzing on the web.

- *Essential Study Skills for Science Students* by Daniel D. Chiras. This book includes chapters on developing good study habits, sharpening memory, getting the most out of lectures, labs, and reading assignments, improving test-taking abilities, and becoming a critical thinker. Your instructor can have this book bundled FREE with your textbook.

- *Laboratory Manual* by C. Lee Rockett and Kenneth J. Van Dellen. This manual includes a variety of laboratory exercises, workbook exercises, and projects that require a minimum of sophisticated equipment.

The following supplementary materials are available to instructors adopting this book:

- *Multimedia Manager.* This CD-ROM, free to adopters, allows you to create custom lectures using over 2,000 pieces of high-resolution artwork, images, and QuickTime movies from the CD and the web, assemble database files, and create Microsoft PowerPoint lectures using text slides and figures from the textbook. This program's editing tools allow use of slides from other lectures, modification or removal of figure labels and leaders, insertion of your own slides, saving slides as JPEG images, and preparation of lectures for use on the Web.

- *Transparency Masters and Acetates.* Includes 100 color acetates of line art and nearly 450 black-and-white master sheets of key diagrams for making overhead transparencies. Free to adopters.

- *CNN™ Today Videos.* These videos, updated annually, contain short clips of news stories about environmental news. Adopters can receive 5 CDs or video tapes containing more than 120, 2- to 5-minute, video clips shot over the past few years. Each year after, adopters will receive a new free CD or video tape of new clips.

- Two videos, *In the Shadow of the Shuttle: Protecting Endangered Species,* and *Costa Rica: Science in the Rainforest,* are available to adopters.

- *Instructor's Manual with Test Bank.* Free to adopters.

- *ExamView.* Allows you to easily create and customize tests, see them on the screen exactly as they will print, and print them out.

Help Me Improve This Book

Let me know how you think this book can be improved; if you find any errors, bias, or confusing explanations, please e-mail them to me at

mtg89@hotmail.com

Most errors can be corrected in subsequent printings of this edition rather than waiting for a new edition.

Acknowledgments

I wish to thank the many students and teachers who responded so favorably to the 6 previous editions of *Sustaining the Earth,* the 13 editions of *Living in the Environment,* and the 10 editions of *Environmental Science,* and who corrected errors and offered many helpful suggestions for improvement. I am also deeply indebted to the more than 250 reviewers, who pointed out errors and suggested many important improvements in the various editions of these three books. Any errors and deficiencies left are mine.

The members of the talented production team, listed on the copyright page, have made vital contributions as well. My thanks also go to production editors Andy Marinkovich at Wadsworth and Andrea Fincke at Thompson Steele, copy editor Anita Wagner, developmental editor Scott Spoolman, page layout artist Bonnie Van Slyke, Brooks/Cole's hard-working sales staff, and Keli Amann, Chris Evers, Steve Bolinger, Joy Westburg and other members of the talented team who developed the multimedia, website, and advertising materials associated with this book

I also thank C. Lee Rockett and Kenneth J. Van Dellen for developing the *Laboratory Manual* to accompany this book; Jane Heinze-Fry for her work on concept mapping, *Environmental Articles, Critical Thinking and the Environment: A Beginner's Guide,* and *Booklet;* and Paul Blanchard for his excellent work on the *Instructor's Manual.*

My deepest thanks go to Jack Carey, biology publisher at Brooks/Cole, for his encouragement, help,

38 years of friendship, and superb reviewing system. It helps immensely to work with the best and most experienced editor in college textbook publishing.

I dedicate this book to the earth that sustains us and to Kathleen Paul Miller, my wife and research associate.

G. Tyler Miller, Jr.

Guest Essayists

Guest Essays by the following authors are available online at the website for this book: **M. Kat Anderson,** ethnoecologist with the National Plant Center of the USDA's Natural Resource Conservation Service; **Lester R. Brown,** president, Earth Policy Institute; **Alberto Ruz Buenfil,** environmental activist, writer, and performer; **Robert D. Bullard,** professor of sociology and director of the Environmental Justice Resource Center at Clark Atlanta University; **Michael Cain,** Bowdoin College; **Herman E. Daly,** senior research scholar at the school of Public Affairs, University of Maryland; **Lois Marie Gibbs,** director, Center for Health, Environment, and Justice; **Garrett Hardin,** professor emeritus (now deceased) of human ecology, University of California, Santa Barbara; **Paul G. Hawken,** environmental author and business leader; **Jane Heinze-Fry,** author, teacher, and consultant in environmental education; **Amory B. Lovins,** energy policy consultant and director of research, Rocky Mountain Institute; **Lester W. Milbrath,** director of the research program in environment and society, State University of New York, Buffalo; **Peter Montague,** director, Environmental Research Foundation; **Norman Myers,** tropical ecologist and consultant in environment and development; **David W. Orr,** professor of environmental studies, Oberlin College; **Noel Perrin,** adjunct professor of environmental studies, Dartmouth College; **John Pichtel,** Ball State University; **David Pimentel,** professor of entomology, Cornell University; **Andrew C. Revkin,** environmental author and environmental reporter for the *New York Times;* and **Nancy Wicks,** ecopioneer and director of Round Mountain Organics; **Donald Worster,** environmental historian and professor of American history, University of Kansas.

INTRODUCTION: LEARNING SKILLS

Students who can begin early in their lives to think of things as connected, even if they revise their views every year, have begun the life of learning.

MARK VAN DOREN

Why Is It Important to Study Environmental Science? Learning How the Earth Works

Environmental science may be the most important course you will ever take.

Welcome to *environmental science*—an *interdisciplinary* study of how nature works and how things in nature are interconnected. This book is an integrated and science-based study of environmental problems, connections, and solutions.

Environmental issues affect every part of your life and are an important part of the news stories presented on television and in newspapers and magazines. Thus, the concepts, information, and issues discussed in this book and the course you are taking should be useful to you now and in the future.

Understandably, I am biased. But *I strongly believe that environmental science is the single most important course in your education.* What could be more important than learning how the earth works, how we are affecting its life support system, and how we can reduce our environmental impact?

We live in an incredibly challenging era. There is a growing awareness that during this century we need to make a new cultural transition in which we learn how to live more sustainably by not degrading our life-support system.

I hope this book and the course you are taking will help you learn about the exciting challenges we face. More important, I hope it will stimulate you to become involved in this change in the way we view and treat the earth that sustains other life, all economies, and us.

How Did I Become Involved with Environmental Problems? Individuals Matter

I became involved in environmental science and education after hearing a scientific lecture in 1966 by Dean Cowie.

In 1966, I heard Dean Cowie, a physicist with the U.S. Geological Survey, give a lecture on the problems of population growth and pollution. Afterward I went to him and said, "If even a fraction of what you have said is true, I will feel ethically obligated to give up my re-

search on the corrosion of metals and devote the rest of my life to research and education on environmental problems and solutions. Frankly, I do not want to change my life, and I am going into the literature to try to show that your statements are either untrue or grossly distorted."

After six months of study, I was convinced of the seriousness of these and other environmental problems. Since then, I have been studying, teaching, and writing about them. This book summarizes what I have learned in more than three decades of trying to understand environmental principles, problems, connections, and solutions.

How Can You Improve Your Study and Learning Skills? Becoming an Efficient and Effective Learner

Learning how to learn is life's most important skill.

Maximizing your ability to learn should be one of your most important lifetime educational goals. This involves continually trying to improve your study and learning skills.

This has a number of payoffs. You can learn more and do this more efficiently. You will also have more time to pursue other interests besides studying without feeling guilty about always being behind. It can also help you get better grades and live a more fruitful and rewarding life.

Here are some general study and learning skills.

Get organized. Becoming more efficient at studying gives you more time for other interests.

Make daily to-do lists in writing. Put items in order of importance, focus on the most important tasks, and assign a time to work on these items. Because life is full of uncertainties, you will be lucky to accomplish half of the items on your daily list. Shift your schedule as needed to accomplish the most important items. Otherwise, you fall behind and become increasingly frustrated.

Set up a study routine in a distraction-free environment. Develop a written daily study schedule and stick to it. Study in a quiet, well-lighted space. Work sitting at a desk or table—not lying down on a couch or bed. Take breaks every hour or so. During each break take several deep breaths and move around to help you stay more alert and focused.

Avoid procrastination—putting work off until another time. Do not fall behind on your reading and other assignments. Accomplish this by setting aside a particular time for studying each day and making it a part of your daily routine. The website for this book has a list of review questions and a multiple-choice practice quiz.

Do not eat dessert first. Otherwise, you may never get to the main meal (studying). When you have accomplished your study goals then reward yourself with play (dessert).

Make hills out of mountains. It is psychologically difficult to climb a mountain such as reading an entire book, reading a chapter in a book, writing a paper, or cramming to study for a test. Instead, break such large tasks (mountains) down into a series of small tasks (hills). Each day read a few pages of a book or chapter, write a few paragraphs of a paper, and review what you have studied and learned. As Henry Ford put it, "Nothing is particularly hard if you divide it into small jobs."

Look at the big picture first. Get an overview of an assigned reading by looking at the main headings or chapter outline. In this textbook, I provide a list of the main questions that are the focus of each chapter.

Ask and answer questions as you read. For example, what is the main point of this section or paragraph? To help you do this I start each subsection with a question that the material is designed to answer. Then to help you further, I follow this question with a brief headline to give you an idea of how the question will be answered. Then I provide a one-sentence summary of the key material in the subsection. This provides you with a running summary of the text. I find this type of summary to be more helpful than providing a harder-to-digest summary at the end of each chapter. My goal is to present the material in more manageable bites. You can also use the one-sentence summaries as a way to review what you have learned. Putting them all together gives you a summary of the chapter.

Focus on key terms. Use the glossary in your textbook to look up the meaning of terms or words you do not understand. Make flash cards for learning key terms and concepts and review them frequently. This book shows all key terms in **boldfaced** type and lesser but still important terms in *italicized* type. Flash cards for testing your mastery of key terms for each chapter are available on the website for this book.

Interact with what you read. I do this by marking key sentences and paragraphs with a highlighter or pen. I put an asterisk in the margin next to an idea I think is important and double asterisks next to an idea I think is especially important. I write comments in the margins, such as *Beautiful, Confusing, Misleading,* or *Wrong.* I fold down the top corner of pages with highlighted passages and the top and bottom corners of especially important pages. This way, I can flip through a chapter or book and quickly review the key ideas. The website for this book has a list of review questions and a multiple-choice practice quiz.

Use the audio version of this book for study and review. You can listen to the chapters in this book while walking, traveling, sitting in your room, or working out. You can use the pin code enclosed in every new copy of this book to download book chapters free from the Web to any MP3 player.

Review to reinforce learning. Before each class review the material you learned in the previous class and read the assigned material. Review, fill in, and organize your notes as soon as possible after each class.

Become a better note taker. Do not try to take down everything your instructor says. Instead, take down main points and key facts using your own shorthand system. Fill in and organize your notes after class.

Write out answers to questions to focus and reinforce learning. Answer questions at the end of each chapter or those assigned to you and the review questions on the website for each chapter. Do this in writing as if you were turning them in for a grade. Save your answers for review and preparation for tests.

Use the buddy system. Study with a friend or become a member of a study group to compare notes, review material, and prepare for tests. Explaining something to someone else is a great way to focus your thoughts and reinforce your learning. If available, attend review sessions offered by instructors or teaching assistants.

Learn your instructor's test style. Does your instructor emphasize multiple-choice, fill-in-the-blank, true-or-false, factual, thought, or essay questions? How much of the test will come from the textbook and how much from lecture material? Adapt your learning and studying methods to this style. You may disagree with this style and feel that it does not adequately reflect what you know. But the reality is that your instructor is in charge.

Become a better test taker. Avoid cramming. Eat well and get plenty of sleep before a test. Arrive on time or early. Calm yourself and increase your oxygen intake by taking several deep breaths. Do this about every 10–15 minutes. Look over the test and answer the questions you know well first. Then work on the harder ones. Use the process of elimination to narrow down the choices for multiple-choice questions. Getting it down to two choices gives you a 50% chance of guessing the right answer. For essay questions, organize your thoughts before you start writing. If you have no idea what a question means, make an educated guess. You might get some partial credit and avoid a zero. Another strategy for getting some credit is to show your knowledge and reasoning by writing: "If this question means so and so, then my answer is _____."

Develop an optimistic outlook. Try to be a "glass is half-full" rather than a "glass is half-empty" person. Pessimism, fear, anxiety, and excessive worrying (especially over things you have no control over) are destructive and lead to inaction. Try to keep your energizing feelings of optimism slightly ahead of your immobilizing feelings of pessimism. Then you will always be moving forward.

Take time to enjoy life. Every day take time to laugh and enjoy nature, beauty, and friendship. Becoming an effective and efficient learner is the best way to do this without getting behind and living under a cloud of guilt and anxiety.

How Can You Improve Your Critical Thinking Skills? Detecting Baloney

Learning how to think critically is a skill you will need throughout your life.

Every day we are exposed to a sea of information, ideas, and opinions. How do we know what to believe and why? Do the claims seem reasonable or exaggerated?

Critical thinking involves developing skills to help you analyze and evaluate the validity of information and ideas you are exposed to and to make decisions. Critical thinking skills help you decide rationally what to believe or what to do. This involves examining information and conclusions or beliefs in terms of the evidence and chain of logical reasoning that supports them. Critical thinking helps you distinguish between facts and opinions, evaluate evidence and arguments, take and defend an informed position on issues, integrate information and see relationships, and apply your knowledge to dealing with new and different problems. Here are some basic skills for learning how to think more critically.

Question everything and everybody. Be skeptical, as any good scientist is. Do not believe everything you hear or read, including the content of this textbook. Evaluate all information you receive. Seek other sources and opinions. As Albert Einstein put it, "The important thing is not to stop questioning."

Do not believe everything you read on the Internet. The Internet is a wonderful and easily accessible source of information. It is also a useful way to find alternative information and opinions on almost any

subject or issue—much of it not available in the mainstream media and scholarly articles. However, because the Internet is so open, anyone can write anything they want with no editorial control or peer evaluation—the method in which scientific or other experts in an area review and comment on an article before it is accepted for publication in a scholarly journal. As a result, evaluating information on the Internet is one of the best ways to put into practice the principles of critical thinking discussed in this preface. Use and enjoy the Internet, but think critically and proceed with caution.

Identify and evaluate your personal biases and beliefs. Each of us has biases and beliefs taught to us by sources such as parents, teachers, friends, role models, and experience. What are your basic beliefs and biases? Where did they come from? What basic assumptions are they based on? How sure are you that your beliefs and assumptions are right and why? According to William James, "A great many people think they are thinking when they are merely rearranging their prejudices."

Be open-minded, flexible, and humble. Be open to considering different points of view, suspend judgment until you gather more evidence, and be capable of changing your mind. Recognize that there may be a number of useful and acceptable solutions to a problem and that very few issues are black or white. There are usually valid points on both (or many) sides of an issue. One way to evaluate divergent views is to get into another person's head or walk in his or her shoes. How do they see or view the world? What are their basic assumptions and beliefs? Is their position logically consistent with their assumptions and beliefs? And be humble about what you know. According to Will Durant, "Education is a progressive discovery of our own ignorance."

Evaluate how the information related to an issue was obtained. Are the statements made based on first-hand knowledge or research or on hearsay? Are unnamed sources used? Is the information based on reproducible and widely accepted scientific studies (*sound* or *consensus science*, p. 17) or preliminary scientific results that may be valid but need further testing (*frontier science*, p. 17)? Is the information based on a few isolated stories or experiences (*anecdotal information*) instead of carefully controlled studies? Is it based on unsubstantiated and widely doubted scientific information or beliefs (*junk science* or *pseudoscience*)? You need to know how to detect junk science, as discussed on p. 17.

Question the evidence and conclusions presented. What are the conclusions or claims? What evidence is presented to support them? Does the evidence support them? Is there a need to gather further evidence to test the conclusions? Are there other, more reasonable conclusions?

Try to identify and assess the assumptions and beliefs of those presenting evidence and drawing conclusions. What is their expertise in this area? Do they have any unstated assumptions, beliefs, biases, or values? Do they have a personal agenda? Can they benefit financially or politically from acceptance of their evidence and conclusions? Would investigators with different basic assumptions or beliefs take the same data and come to different conclusions?

Do the arguments used involve common logical fallacies or debating tricks? Here are six of many examples. *First,* attack the presenter of an argument rather than the argument itself. *Second,* appeal to emotion rather than facts and logic. *Third,* claim that if one piece of evidence or one conclusion is false, then all other pieces of evidence and conclusions are false. *Fourth,* say that a conclusion is false because it has not been scientifically proven (scientists can never prove anything absolutely, but they can establish degrees of reliability, as discussed on p. 17). *Fifth,* inject irrelevant or misleading information to divert attention from important points. *Sixth,* present only either/or alternatives when there may be a number of options.

Become a seeker of wisdom, not a vessel of information. Develop a written list of principles, concepts, and rules to serve as guidelines in evaluating evidence and claims and making decisions. Continually evaluate and modify this list on the basis of experience. Many people believe that the main goal of education is to learn as much as you can by concentrating on gathering more and more information—much of it useless or misleading. I believe that the primary goal is to know as little as possible. This is done by learning how to sift through mountains of facts and ideas to find the few *nuggets of wisdom* that are the most useful in understanding the world and in making decisions. This takes a firm commitment to learning how to think logically and critically, and continually flushing less valuable and thought-clogging information from our minds. This book is full of facts and numbers, but they are useful only to the extent that they lead to an understanding of key and useful ideas, scientific laws, concepts, principles, and connections. A major goal of the study of environmental science is to find out how nature works and sustains itself (*environmental wisdom*) and to use *principles of environmental wisdom* to help make our human societies and economies more sustainable and thus more beneficial and enjoyable. As Sandra Carey put it, "Never

mistake knowledge for wisdom. One helps you make a living; the other helps you make a life."

How Have I Attempted to Achieve Balance? Evaluate Trade-Offs

There are no simple answers to the environmental problems we face.

There are always *trade-offs* involved in making and implementing environmental decisions. My challenge is to give a fair and balanced presentation of different viewpoints, advantages and disadvantages of various technologies and proposed solutions to environmental problems, and good and bad news about environmental problems without injecting personal bias.

Studying a subject as important as environmental science and ending up with no conclusions, opinions, and beliefs means that both the teacher and student have failed. However, such conclusions should be based on using critical thinking to evaluate different ideas and to understand the trade-offs involved.

How Can You Improve This Book? Keep Those Ideas Coming

I welcome your help in improving this book.

Researching and writing a book that covers and connects ideas in such a wide variety of disciplines is a challenging and exciting task. Almost every day I learn about some new connection in nature.

In a book this complex, there are bound to be some errors—some typographical mistakes that slip through and some statements that you might question based on your knowledge and research. My goal is to provide you with an interesting, accurate, balanced, and challenging book that furthers your understanding of this vital subject. I have also attempted to provide balance on presenting various sides of key environmental issues.

I invite you to contact me and point out any remaining bias, correct any errors you find, and suggest ways to improve this book. Over decades of teaching some of my best teachers have been students taking my classes and reading my textbooks. Please e-mail your suggestions to me at

mtg89@hotmail.com

Now start your journey into this fascinating and important study of how the earth works and how we can leave the planet in at least as good a shape as we found it. Have fun.

Study nature, love nature, stay close to nature. It will never fail you.

Frank Lloyd Wright

1 ENVIRONMENTAL PROBLEMS, THEIR CAUSES, AND SUSTAINABILITY

Alone in space, alone in its life-supporting systems, powered by inconceivable energies, mediating them to us through the most delicate adjustments, wayward, unlikely, unpredictable, but nourishing, enlivening, and enriching in the largest degree—is this not a precious home for all of us? Is it not worth our love?

BARBARA WARD AND RENÉ DUBOS

1-1 LIVING MORE SUSTAINABLY

What Is Environmental Science? Defining Some Basic Terms

Environmental science is a study of how the earth works, how we interact with the earth, and how to deal with environmental problems.

Environment is everything that affects a living organism (any unique form of life). **Ecology** is a biological science that studies the relationships between living organisms and their environment.

This textbook is an introduction to **environmental science,** an interdisciplinary study that uses information from the physical sciences and social sciences to learn how the earth works, how we interact with the earth, and how to deal with environmental problems.

Environmentalism is a social movement dedicated to protecting the earth's life support systems for us and other species.

What Keeps Us Alive? The Sun and the Earth's Natural Capital

All life and economies depend on energy from the sun (solar capital) and the earth's resources and ecological services (natural capital).

Our existence, lifestyles, and economies depend completely on the sun and the earth, a blue and white island in the black void of space. To economists, *capital* is wealth used to sustain a business and to generate more wealth.

By analogy, we can think of energy from the sun as **solar capital. Solar energy** includes direct sunlight and indirect forms of renewable solar energy such as *wind power, hydropower* (energy from flowing water), and *biomass* (direct solar energy converted to chemical energy and stored in biological sources of energy such as wood).

Similarly, we can think of the planet's air, water, soil, wildlife, forest, rangeland, fishery, mineral, and energy resources and the processes of natural purification, recycling, and pest control as **natural resources,** or **natural capital** (Figure 1-1). See the Guest Essay on this topic by Paul Hawken on the website for this chapter.

This priceless natural capital that nature provides at no cost to us plus the natural biological income it supplies can sustain the planet and our economies indefinitely as long as we do not deplete then. Examples of *biological income* are renewable supplies of wood, fish, grassland for grazing, and underground water for drinking and irrigation.

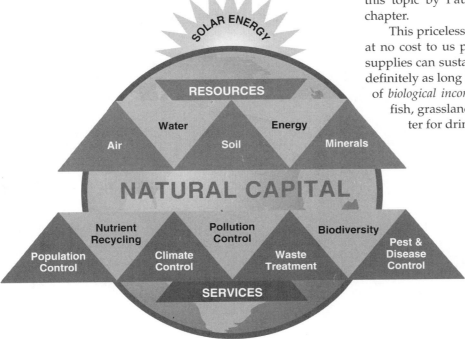

Figure 1-1 The earth's **natural capital.** Energy from the sun (solar capital) and the earth's natural capital provide resources (top) and ecological services (bottom) that support and sustain the earth's life and economies. Wedges from this diagram will be used near the titles of various chapters to indicate the components of natural capital that are the primary focus of such chapters. This diagram also appears in color on the back cover of this book.

What Is an Environmentally Sustainable Society? One That Preserves Natural Capital and Lives off Its Income

An environmentally sustainable society meets the basic resource needs of its people indefinitely without degrading or depleting the natural capital that supplies these resources.

An **environmentally sustainable society** meets the current needs of its people for food, clean water, clean air, shelter, and other basic resources without compromising the ability of future generations to meet their needs. *Living sustainably* means living off natural income replenished by soils, plants, air, and water and not depleting or degrading the earth's natural capital that supplies this income.

Imagine you win $1 million in a lottery. Invest this capital at 10% interest per year, and you will have a sustainable annual income of $100,000 without depleting your capital. If you spend $200,000 a year, your $1 million will be gone early in the 7th year and even if you spend only $110,000 a year, you will be bankrupt early in the 18th year.

The lesson here is an old one: *Protect your capital and live off the income it provides.* Deplete, waste, or squander your capital, and you move from a sustainable to an unsustainable lifestyle.

The same lesson applies to the earth's natural capital. According to many environmentalists and leading scientists, we are living unsustainably by wasting, depleting, and degrading the earth's natural capital at an accelerating rate.

Some people disagree. They contend that environmentalists have exaggerated the seriousness of population, resource, and environmental problems. They also believe we can overcome these problems by human ingenuity, economic growth, and technological advances.

1-2 POPULATION GROWTH, ECONOMIC GROWTH, AND ECONOMIC DEVELOPMENT

How Rapidly Is the Human Population Growing? Pretty Fast

The rate at which world's population is growing has slowed down but is still growing pretty rapidly.

During the past 10,000 years the human population has grown from several million to 6.4 billion (Figure 1-2). The current annual growth rate of the world's population is 1.25%. This does not seem like a very fast rate but it added about 80 million people to the world's population in 2004, an average increase of 219,000 people a day, or 9,900 an hour. At this rate it takes only about 3 days to add the 651,000 Americans killed in battle in all U.S. wars and only about 1.6 years to add the 129 million people killed in all wars fought in the past 200 years!

What Is the Difference between Economic Growth and Economic Development? More Stuff and Better Living Standards

Economic growth provides people with more goods and services, and economic development uses economic growth to improve living standards.

Economic growth is an increase in the capacity of a country to provide people with goods and services. Accomplishing this increase requires population growth (more producers and consumers), more production and consumption per person, or both.

Economic growth is usually measured by the percentage change in a country's **gross domestic product (GDP):** the annual market value of all goods and ser-

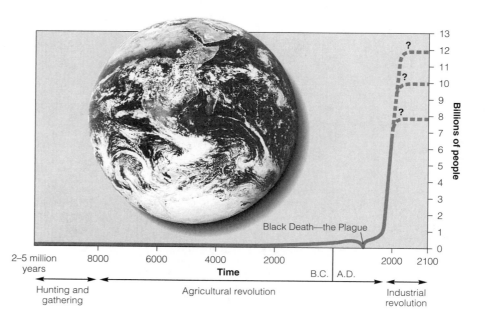

Figure 1-2 The *J*-shaped curve of past world population growth, with projections to 2100. The current world population of 6.4 billion people is projected to grow to 8–12 billion people sometime this century. (This figure is not to scale.) (World Bank and United Nations; photo courtesy of NASA)

vices produced by all firms and organizations, foreign and domestic, operating within a country. Changes in a country's standard of living are measured by **per capita GDP:** the GDP divided by the total population at midyear.

Economic development is the improvement of living standards by economic growth. The United Nations (UN) classifies the world's countries as economically developed or developing based primarily on their degree of industrialization and their per capita GDP.

The **developed countries** (with 1.2 billion people) include the United States, Canada, Japan, Australia, New Zealand, and the countries of Europe. Most are highly industrialized and have a high average per capita GDP. All other nations (with 5.2 billion people) are classified as **developing countries,** most of them in Africa, Asia, and Latin America. Some are *middle-income, moderately developed countries* and others are *low-income countries.*

Figure 1-3 compares some key characteristics of developed and developing countries. About 97% of the projected increase in the world's population is expected to take place in developing countries.

Figure 1-4 summarizes some of the benefits (*good news*) and harm (*bad news*) caused mostly by economic development. It shows effects of the wide and increasing gap between the world's haves and have-nots.

Percent of World's

Population
19
81

Population growth
0.1
1.5

Wealth and income
85
15

Resource use
88
12

Pollution and waste
75
25

■ Developed countries □ Developing countries

Figure 1-3 Comparison of developed and developing countries. (Data from United Nations and the World Bank)

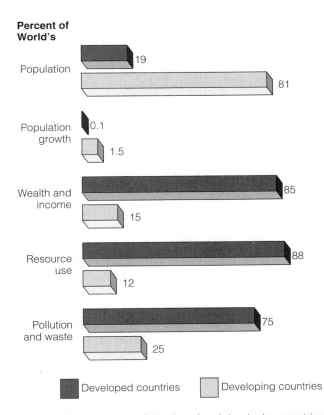

Trade-Offs

Economic Development

Good News

Global life expectancy doubled since 1950

Infant mortality cut in half since 1955

Food production ahead of population growth since 1978

Air and water pollution down in most developed countries since 1970

Number of people living in poverty dropped 6% since 1990

Bad News

Life expectancy 11 years less in developing countries than in developed countries

Infant mortality rate in developing countries over 8 times higher than in developed countries

Harmful environmental effects of agriculture may limit future food production

Air and water pollution levels in most developing countries too high

Half of world's people trying to live on less than $3 (U.S.) per day

Figure 1-4 Trade-offs: good and bad news about economic development. Pick the single piece of good news and bad news that you believe are the most important. (Data from United Nations and World Health Organization)

1-3 RESOURCES

What Is a Resource? Things We Need or Want

We obtain resources from the environment to meet our needs and wants.

From a human standpoint, a **resource** is anything obtained from the environment to meet human needs and wants. Examples include food, water, shelter, manufactured goods, transportation, communication, and recreation. On our short human time scale, we classify the material resources we get from the environment as *perpetual, renewable,* or *nonrenewable.*

Some resources, such as solar energy, fresh air, wind, fresh surface water, fertile soil, and wild edible plants, are directly available for use. Other resources, such as petroleum (oil), iron, groundwater (water found underground), and modern crops, are not directly available. They become useful to us only with some effort and technological ingenuity. For example, petroleum was a mysterious fluid until we learned how to find and extract it, and to refine it into gasoline, heating oil, and other products that we could sell at affordable prices.

What Are Perpetual and Renewable Resources? Resources That Can Last

Resources renewed by natural processes are sustainable if we do not use them faster than they are replenished.

Solar energy is called a **perpetual resource** because on a human time scale it is renewed continuously. It is expected to last at least 6 billion years as the sun completes its life cycle.

On a human time scale, a **renewable resource** can be replenished fairly rapidly (hours to several decades) through natural processes as long as it is not used up faster than it is replaced. Examples of renewable resources are forests, grasslands, wild animals, fresh water, fresh air, and fertile soil.

Renewable resources can be depleted or degraded. The highest rate at which a renewable resource can be used *indefinitely* without reducing its available supply is called **sustainable yield.**

When we exceed a renewable resource's natural replacement rate, the available supply begins to shrink, a process known as **environmental degradation.** Examples include urbanization of productive land, excessive topsoil erosion, pollution, deforestation (temporary or permanent removal of large expanses of forest for agriculture or other uses), groundwater depletion, overgrazing of grasslands by livestock, and reduction in the earth's forms of wildlife (biodiversity) by elimination of habitats and species.

Case Study: The Tragedy of the Commons— Degrading Free Renewable Resources

Renewable resources that are freely available to everyone can be degraded.

One cause of environmental degradation of renewable resources is the overuse of **common-property** or **free-access resources.** No individual owns these, and they are available to all users at little or no charge.

Examples include clean air, the open ocean and its fish, migratory birds, wildlife species, gases of the lower atmosphere, and space.

In 1968, biologist Garrett Hardin (1915–2003) called the degradation of renewable free-access resources the **tragedy of the commons.** It happens because each user reasons, "If I do not use this resource, someone else will. The little bit I use or pollute is not enough to matter, and such resources are renewable."

With only a few users, this logic works. But the cumulative effect of many people trying to exploit a free-access resource eventually exhausts or ruins it. Then no one can benefit from it, and that is the tragedy.

One solution is to use free-access resources at rates well below their estimated sustainable yields by reducing population, regulating access to the resources, or both. Some communities have established rules and traditions to regulate and share their access to common-property resources such as ocean fisheries, grazing lands, and forests. Governments have also enacted laws and international treaties to regulate access to commonly owned resources such as forests, national parks, rangelands, and fisheries in coastal waters.

Another solution is to *convert free-access resources to private ownership.* The reasoning is that if you own something, you are more likely to protect your investment.

This sounds good, but private ownership is not always the answer. One problem is that private owners do not always protect natural resources they own when this conflicts with protecting their financial capital or increasing their profits. For example, some private forest owners can make more money by clearcutting the timber, selling the degraded land, and investing their profits in other timberlands or businesses. A second problem is that this approach is not practical for global common resources—such as the atmosphere, the open ocean, most wildlife species, and migratory birds—that cannot be divided up and converted to private property.

What Is Our Ecological Footprint? Our Growing Environmental Impact

Supplying each person with renewable resources and absorbing the wastes from such resource use creates a large ecological footprint or environmental impact.

The **per capita ecological footprint** is the amount of biologically productive land and water needed to supply each person with the renewable resources they use and to absorb the wastes from such resource use (Figure 1-5, left). It measures the average environmental impact of individuals in different countries and areas.

Bad news. Humanity's ecological footprint per person exceeds the earth's biological capacity to replenish renewable resources and absorb waste by about 15%. (Figure 1-5, right) If these estimates are correct, *it will take the resources of 1.15 planet earths to indefinitely support our current use of renewable resources!* You can estimate your ecological footprint by going to the website www.redefiningprogress.org/. Also, see the Guest Essay by Michael Cain on the website for this chapter.

According to William Rees and Mathis Wackernagel, developers of the ecological footprint concept, it would take the land area of about *four more planet*

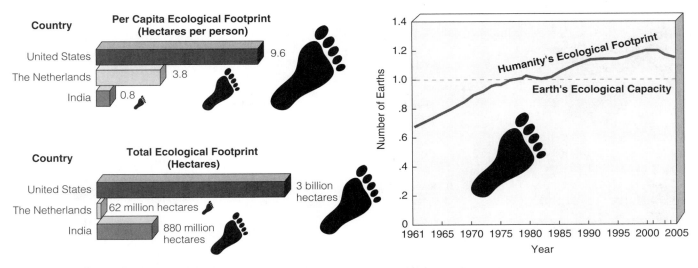

Figure 1-5 Natural capital use and degradation: relative *ecological footprints* of the United States, the Netherlands, and India (left). The *ecological footprint* is a measure of the biologically productive areas of the world needed for producing the renewable resources required per person and absorbing or breaking down the wastes produced by such resource use. Currently, humanity's average ecological footprint per person is 15% higher than the earth's biological capacity per person (right). (William Rees and Mathis Wackernagel, Redefining Progress)

earths for the rest of the world to reach U.S. levels of consumption with existing technology. See the Guest Essay on the website of this chapter by Norman Myers about the growing ecological footprint of China.

What Are Nonrenewable Resources? Resources We Can Deplete

Nonrenewable resources can be economically depleted to the point where it costs too much to obtain what is left.

Nonrenewable resources exist in a fixed quantity or stock in the earth's crust. On a time scale of millions to billions of years, geological processes can renew such resources. But on the much shorter human time scale of hundreds to thousands of years, these resources can be depleted much faster than they are formed.

Exhaustible resources include *energy resources* (such as coal, oil, and natural gas that cannot be recycled), *metallic mineral resources* (such as iron, copper, and aluminum that can be recycled), and *nonmetallic mineral resources* (such as salt, clay, sand, and phosphates that usually are difficult or too costly to recycle).

We never completely exhaust a nonrenewable mineral resource but it becomes *economically depleted* when the costs of extracting and using what is left exceed its economic value. At that point, we have six choices: try to find more, recycle or reuse existing supplies (except for nonrenewable energy resources, which cannot be recycled or reused), waste less, use less, try to develop a substitute, or wait millions of years for more to be produced.

Most published estimates of the supply of a given nonrenewable resource refer to **reserves**: known deposits from which a usable mineral can be extracted profitably at current prices. Reserves can increase when new deposits are found or when higher prices or improved mining technology make it profitable to extract deposits that previously were too expensive to extract.

Depletion time is the time it takes to use up a certain proportion—usually 80%—of the reserves of a mineral at a given rate of use. Extracting the remaining 20% of a resource usually costs more than it is worth. The shortest depletion time assumes no recycling or reuse and no increase in reserves. A longer depletion time assumes that recycling will stretch existing reserves and better mining technology, higher prices, and new discoveries will increase reserves. An even longer depletion time assumes that new discoveries will further expand reserves and that recycling, reuse, and reduced consumption will extend supplies. Finding a substitute for a resource dictates a new set of depletion curves for the new resource.

Some environmentalists and resource experts believe exhaustion of most nonrenewable resources is not a serious environmental problem in the near future. Instead, future supplies of widely used nonrenewable resources may be limited by the damage that their extraction, processing, and conversion to products do to the environment in the form of energy use, land disturbance, soil erosion, water pollution, and air pollution (Figure 1-6, p. 10).

Some nonrenewable material resources, such as copper and aluminum, can be recycled or reused to

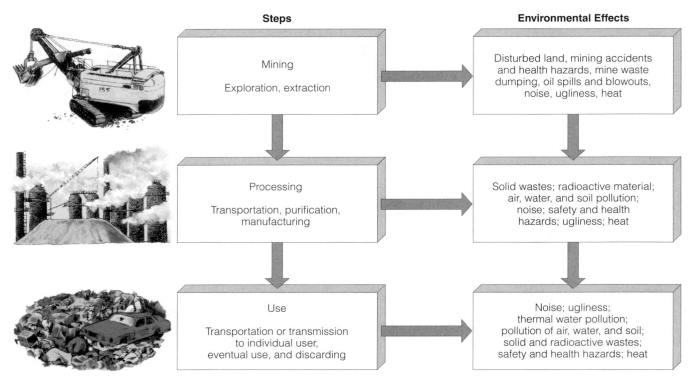

Steps		Environmental Effects
Mining Exploration, extraction	→	Disturbed land, mining accidents and health hazards, mine waste dumping, oil spills and blowouts, noise, ugliness, heat
Processing Transportation, purification, manufacturing	→	Solid wastes; radioactive material; air, water, and soil pollution; noise; safety and health hazards; ugliness; heat
Use Transportation or transmission to individual user, eventual use, and discarding	→	Noise; ugliness; thermal water pollution; pollution of air, water, and soil; solid and radioactive wastes; safety and health hazards; heat

Figure 1-6 Natural capital degradation: some harmful environmental effects of extracting, processing, and using nonrenewable mineral and energy resources. The energy used to carry out each step causes additional pollution and environmental degradation.

extend supplies. **Recycling** involves collecting waste materials, processing them into new materials, and selling these new products. For example, discarded aluminum cans can be crushed and melted to make new aluminum cans or other aluminum items that consumers can buy. Recycling means nothing if we do not close the loop by buying products that are made from or contain recycled materials. **Reuse** is using a again in the same form. For example, glass bottles can be collected, washed, and refilled many times.

Recycling nonrenewable metallic resources takes much less energy, water, and other resources and produces much less pollution and environmental degradation than exploiting virgin metallic resources. Reusing such resources takes even less energy and other resources, and produces less pollution and environmental degradation than recycling.

1-4 POLLUTION

Where Do Pollutants Come From and What Are Their Harmful Effects? Threats to Health and Survival

Pollutants are chemicals found at high enough levels in the environment to cause harm to people or other organisms.

Pollution is the presence of chemicals at high enough levels in air, water, soil, or food to threaten the health, survival, or activities of humans or other organisms. Pollutants can enter the environment naturally (for example, from volcanic eruptions) or through human activities (for example, from burning coal). Most pollution from human activities occurs in or near urban and industrial areas, where pollutants are concentrated. Industrialized agriculture is also a major source of pollution.

Some pollutants contaminate the areas where they are produced and some are carried by wind or flowing water to other areas. Pollution does not respect the neat territorial political lines we draw on maps.

The pollutants we produce come from two types of sources. **Point sources** of pollutants are single, identifiable sources. Examples are the smokestack of a coal-burning power plant, the drainpipe of a factory, and the exhaust pipe of an automobile. **Nonpoint sources** of pollutants are dispersed and often difficult to identify. Examples are pesticides sprayed into the air or blown by the wind into the atmosphere, and runoff of fertilizers and pesticides from farmlands, golf courses, and suburban lawns and gardens into streams and lakes. It is much easier and cheaper to identify and control pollution from point sources than from widely dispersed nonpoint sources.

Pollutants can have three types of unwanted effects. *First,* they can disrupt or degrade life-support systems for humans and other species. *Second,* they can damage wildlife, human health, and property. *Third,* they can be nuisances such as noise and unpleasant smells, tastes, and sights.

Solutions: What Can We Do about Pollution? Prevention Pays

We can try to clean up pollutants in the environment or prevent them from entering the environment.

We use two basic approaches to deal with pollution. One is **pollution prevention,** or **input pollution control,** which reduces or eliminates the production of pollutants. The other is **pollution cleanup,** or **output pollution control,** which involves cleaning up or diluting pollutants after they have been produced.

Environmentalists have identified some problems with relying primarily on pollution cleanup. One is that it is only a temporary bandage as long as population and consumption levels grow without corresponding improvements in pollution control technology. For example, adding catalytic converters to car exhaust systems has reduced air pollution. But increases in the number of cars and in the distance each travels have reduced the effectiveness of this approach.

Also, cleanup often removes a pollutant from one part of the environment only to cause pollution in another. For example, we can collect garbage, but the garbage is then *burned* (perhaps causing air pollution and leaving toxic ash that must be put somewhere), *dumped* into streams, lakes, and oceans (perhaps causing water pollution), or *buried* (perhaps causing soil and groundwater pollution).

In addition, once pollutants have entered and become dispersed into the environment at harmful levels, it usually costs too much to reduce them to acceptable levels.

Both pollution prevention (front-of-the-pipe) and pollution cleanup (end-of-the-pipe) solutions are needed. But environmentalists and some economists urge us to put more emphasis on prevention because it works better and because it is cheaper than cleanup. As Benjamin Franklin observed long ago, "An ounce of prevention is worth a pound of cure."

1-5 ENVIRONMENTAL PROBLEMS: CAUSES AND CONNECTIONS

What Are Key Environmental Problems and Their Basic Causes? The Big Five

The major causes of environmental problems are population growth, wasteful resource use, poverty, poor environmental accounting, and ecological ignorance.

We face a number of interconnected environmental and resource problems, as listed in Figure 1-7. Most of the harmful environmental problems we face today are unintended results of activities designed to increase the quality of human life.

The first step in dealing with the environmental problems we face is to identify their underlying causes, listed in Figure 1-8. Three of these causes are rapid population growth, poverty, and excessive and wasteful use of resources. A fourth is failure to include the harmful environmental costs of items in their market prices, discussed in Chapter 14. The fifth, inadequate understanding of how the earth works, is discussed throughout this book.

What Is the Relationship between Poverty and Environmental Problems? Being Poor Is Bad for People and the Earth.

Poverty is a major threat to human health and the environment.

Many of the world's poor do not have access to the basic necessities for a healthy, productive, and decent

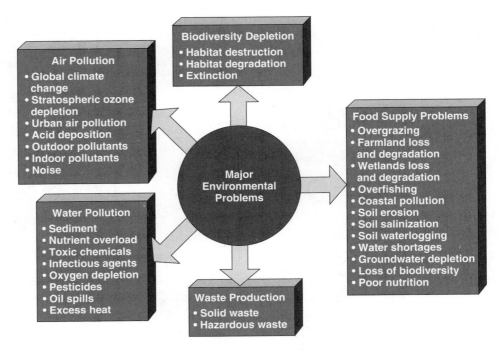

Figure 1-7 Natural capital degradation: major environmental and resource problems.

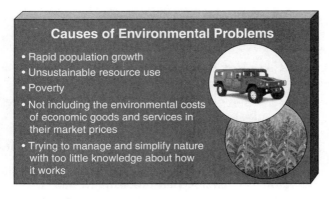

Figure 1-8 Environmentalists have identified five basic causes of the environmental problems we face.

life, as listed in Figure 1-9 Their daily lives are focused on getting enough food, water, and fuel (for cooking and heat) to survive. Desperate for land to grow enough food, many of the world's poor people deplete and degrade forests, soil, grasslands, and wildlife, for short-term survival. They do not have the luxury of worrying about long-term environmental quality or sustainability.

Poverty also affects population growth. Poor people often have many children as a form of economic security. Their children help them grow food, gather fuel (mostly wood and dung), haul drinking water, tend livestock, work, and beg in the streets. The children also help their parents survive in their old age before they die, typically in their 50s in the poorest countries.

Many of the world's desperately poor die from four preventable health problems. One is *malnutrition* from a lack of protein and other nutrients needed for good health. The second is increased susceptibility to

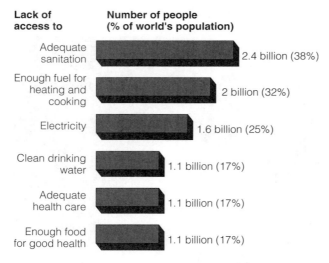

Figure 1-9 Natural capital degradation: some harmful effects of poverty. Which two of these effects do you believe are the most harmful? (Data from United Nations, World Bank, and World Health Organization)

normally nonfatal infectious diseases, such as diarrhea and measles, because of their weakened condition from malnutrition. A third factor is lack of access to clean drinking water. A fourth factor is severe respiratory disease and premature death from inhaling indoor air pollutants produced by burning wood or coal for heat and cooking over open fires or on poorly vented stoves.

According to the World Health Organization, these four factors cause premature death for at least 7 million of the poor a year. *This premature death of about 19,200 human beings per day is equivalent to 48 jumbo jet planes, each carrying 400 passengers, accidentally crashing every day with no survivors.* Two-thirds of those dying are children under age 5. This ongoing human tragedy is rarely covered in the daily news.

What Is the Relationship between Resource Consumption and Environmental Problems? Affluenza

Many consumers in developed countries have become addicted to buying more and more stuff in their search for fulfillment and happiness.

Affluenza (af-loo-en-zuh) is a term used to describe the unsustainable addiction to overconsumption and materialism exhibited in the lifestyles of affluent consumers in the United States and other developed countries. It is based on the assumption that buying more and more things can, should, and does buy happiness.

Most people infected with this contagious *shop-till-you-drop virus* have some telltale symptoms. They feel overworked, have high levels of debt and bankruptcy, suffer from increasing stress and anxiety, have declining health, and feel unfulfilled in their quest to accumulate more and more stuff. As humorist Will Rogers said, "Too many people spend money they haven't earned to buy things they don't want, to impress people they don't like." For some, shopping until you drop means shopping until you go bankrupt. Between 1998 and 2001, more Americans declared bankruptcy than graduated from college.

Globalization and global advertising are spreading the virus throughout most of the world. Affluenza has an enormous environmental impact. It takes about 27 tractor-trailer loads of resources per year to support one American. This amounts to 7.9 billion truckloads a year to support the U.S. population. Stretched end-to-end, these trucks would more than reach the sun!

What can we do about affluenza? The first step for addicts is to admit they have a problem. Then they begin steps to kick their addiction by going on a stuff diet. For example, before buying anything a person with the affluenza addiction should ask: Do I really need this or merely want it? Can I buy it secondhand (reuse)? Can I borrow it from a friend or relative? Another withdrawal strategy: Do not hang out with other addicts. Shopaholics should avoid malls as much as they can.

How Can Affluence Help Increase Environmental Quality? Another Side of the Story

Affluent countries have more money for improving environmental quality.

Some analysts point out that affluence can lead people to become more concerned about environmental quality. And it provides money for developing technologies to reduce pollution, environmental degradation, and resource waste. This explains why most of the important environmental progress made since 1970 has taken place in developed countries.

In the United States, the air is cleaner, drinking water is purer, most rivers and lakes are cleaner, and the food supply is more abundant and safer than in 1970. Also, the country's total forested area is larger than it was in 1900 and most energy and material resources are used more efficiently. Similar advances have been made in most other affluent countries. Affluence financed these improvements in environmental quality.

How Are Environmental Problems and Their Causes Connected? Exploring Connections

Environmental quality is affected by interactions between population size, resource consumption, and technology.

Once we have identified environmental problems and their root causes, the next step is to understand how they are connected to one another. The three-factor model in Figure 1-10 is a starting point.

According to this simple model, the environmental impact (**I**) of population on a given area depends on three key factors: the number of people (**P**), average resource use per person (affluence, **A**), and the beneficial and harmful environmental effects of the technologies (**T**) used to provide and consume each unit of resource and to control or prevent the resulting pollution and environmental degradation.

In developing countries, population size and the resulting degradation of renewable resources as the poor struggle to stay alive tend to be the key factors in total environmental impact (Figure 1-10, top). In such countries, per capita resource use is low.

In developed countries, high rates of per capita resource use and the resulting high levels of pollution and environmental degradation per person usually are the key factors determining overall environmental impact (Figure 1-10, bottom) and a country's ecological footprint per person (Figure 1-5). For example, the average U.S. citizen consumes about 35 times as much as the average citizen of India and 100 times as much as the average person in the world's poorest countries. *Thus, poor parents in a developing country would need 70*

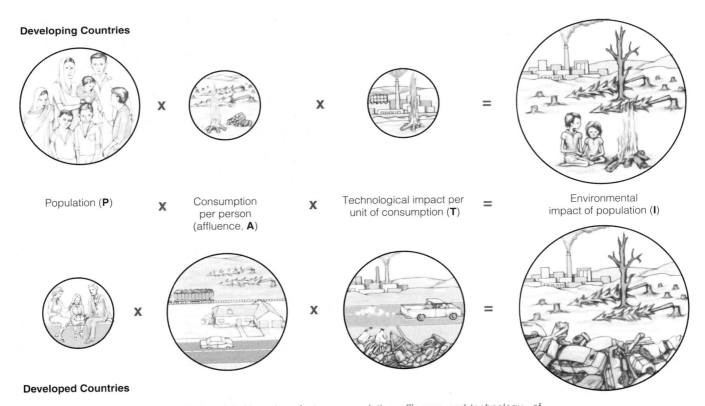

Developing Countries

Population (**P**) ✕ Consumption per person (affluence, **A**) ✕ Technological impact per unit of consumption (**T**) = Environmental impact of population (**I**)

Developed Countries

Figure 1-10 Connections: simplified model of how three factors—*population, affluence,* and *technology*—affect the environmental impact of population in developing countries (top) and developed countries (bottom).

to 200 children to have the same lifetime resource consumption as 2 children in a typical U.S. family.

Some forms of technology, such as polluting factories and motor vehicles and energy-wasting devices, increase environmental impact by raising the T factor in the equation. But other technologies, such as pollution control, solar cells, and energy-saving devices, lower environmental impact by decreasing the T factor. In other words, some forms of technology are *environmentally harmful* and some are *environmentally beneficial*.

The three-factor model in Figure 1-10 can help us understand how key environmental problems and some of their causes are connected. It can also guide us in seeking solutions. However, these problems involve a number of poorly understood political, economic, and technological interactions not included in this simplified model.

1-6 IS OUR PRESENT COURSE SUSTAINABLE?

Are Things Getting Better or Worse? The Answer Is Both.

There is good and bad environmental news.

Experts disagree about how serious our population and environmental problems are and what we should do about them. Some analysts believe human ingenuity, technological advances, and economic growth will allow us to clean up pollution to acceptable levels, find substitutes for any resources that become scarce, and keep expanding the earth's ability to support more humans, as we have done in the past. They accuse many scientists and environmentalists of exaggerating the seriousness of the problems we face and failing to appreciate the progress we have made in improving quality of life and protecting the environment.

Environmentalists and many leading scientists disagree with this view. They cite evidence that we are degrading and disrupting much of the earth's life-support systems for us and other species at an accelerating rate. They are greatly encouraged by the progress we have made in increasing average life expectancy, reducing infant mortality, increasing food supplies, and reducing many forms of pollution—especially in developed countries. But they point out that we need to use the earth in a way that is more sustainable for present and future human generations and other species that support us and other forms of life.

The most useful answer to the question of whether things are getting better or worse is *both*. Some things are getting better and some worse. Our challenge is not to get trapped into confusion and inaction by listening primarily to either of two groups of people. One group consists of *technological optimists*.

They tend to overstate the situation by telling us to be happy and not to worry, because technological innovations and conventional economic growth and development will lead to a wonderworld for everyone.

The second group consists of *environmental pessimists* who overstate the problems to the point where our environmental situation seems hopeless. According to the noted conservationist Aldo Leopold, "I have no hope for a conservation based on fear."

What Is Environmentally Sustainable Economic Development? Rewarding Environmentally Beneficial Activities

An environmentally sustainable economy rewards environmentally beneficial and sustainable activities and discourages environmentally harmful and unsustainable activities.

During this century, many analysts call for us to put much greater emphasis on **environmentally sustainable economic development**. Figure 1-11 lists some of the shifts involved in implementing an *environmental,* or *sustainability, revolution* during this century based on this concept. Study this figure carefully.

This type of development uses economic rewards (government subsidies and tax breaks) to *encourage* environmentally beneficial and more sustainable forms of economic growth, and economic penalties (government

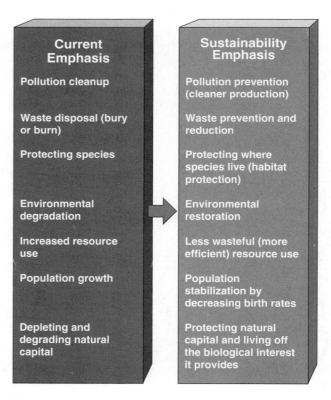

Figure 1-11 Solutions: some shifts involved in the *environmental* or *sustainability revolution.*

taxes and regulations) to *discourage* environmentally harmful and unsustainable forms of economic growth.

According to environmental and health scientists, the most serious environmental problems we face are *poverty, malnutrition, unsafe drinking water, smoking, air pollution, infectious diseases (AIDS, TB, malaria, and hepatitis B), water shortages, climate change, and loss and degradation of biodiversity.* The poor in developing countries bear the brunt of most of these serious problems.

Throughout this book I try to give you a balanced view of good and bad environmental news. Try not to be overwhelmed or immobilized by the bad environmental news, because there is also some *great environmental news.* We have made immense progress in improving the human condition and dealing with many environmental problems. We are learning a great deal about how nature works and sustains itself. And we have numerous scientific, technological, and economic solutions available to deal with the environmental problems we face, as you will learn in this book.

The challenge is to make creative use of our economic and political systems to implement such solutions. One key is to recognize that most economic and political change comes about as a result of individual actions and individuals acting together to bring about change by grassroots action from the bottom up. Social scientists suggest it takes only about 5–10% of the population of a country or of the world to bring about major social change. Anthropologist Margaret Mead summarized our potential for change: "Never doubt that a small group of thoughtful, committed citizens can change the world. Indeed, it is the only thing that ever has."

We live in exciting times during what might be called a *hinge of cultural history.* Indeed, if I had to pick a time to live, it would be the next 50 years as we face the challenge of developing more environmentally sustainable societies.

What's the use of a house if you don't have a decent planet to put it on?

HENRY DAVID THOREAU

CRITICAL THINKING

1. List **(a)** three forms of economic growth you believe are environmentally unsustainable and **(b)** three forms you believe are environmentally sustainable.

2. Give three examples of how you cause environmental degradation as a result of the tragedy of the commons.

3. Explain why you agree or disagree with the following propositions:
   a. Stabilizing population is not desirable because without more consumers, economic growth would stop.

   b. The world will never run out of resources because we can use technology to find substitutes and to help us reduce resource waste.

4. See if the affluenza bug infects you by indicating whether you agree or disagree with the following statements.
   a. I am willing to work at a job I despise so I can buy lots of stuff.
   b. When I am feeling down, I like to go shopping to make myself feel better.
   c. I would rather be shopping right now.
   d. I owe more than a $1,000 on my credit cards.
   e. I usually make only the minimum payment on my monthly credit card bills.
   f. I am running out of room to store my stuff.

If you agree with two of these statements, you are infected with affluenza. If you agree with more than two, you have a serious case of affluenza. Compare your answers with those of your classmates and discuss the effects of the results on the environment and your feelings of happiness.

5. When you read that at least 19,200 human beings die prematurely each day (13 per minute) from preventable malnutrition and infectious disease, do you **(a)** doubt whether it is true, **(b)** not want to think about it, **(c)** feel hopeless, **(d)** feel sad, **(e)** feel guilty, or **(f)** want to do something about this problem?

6. How do you feel when you read that the average American consumes about 35 times more resources than the average citizen of India and human activities are projected to make the earth's climate warmer: **(a)** skeptical about their accuracy, **(b)** indifferent, **(c)** sad, **(d)** helpless, **(e)** guilty, **(f)** concerned, or **(g)** outraged? Which of these feelings help perpetuate such problems, and which can help alleviate these problems?

7. Make a list of the resources you truly need. Then make another list of the resources you use each day only because you want them. Finally, make a third list of resources you want and hope to use in the future. Compare your lists with those compiled by other members of your class, and relate the overall result to the tragedy of the commons and affluenza.

LEARNING ONLINE

The website for this book contains study aids and many ideas for further reading and research. They include a chapter summary, review questions for the entire chapter, flash cards for key terms and concepts, a multiple-choice practice quiz, interesting Internet sites, references, and a guide for accessing thousands of InfoTrac® College Edition articles. Log on to

http://biology.brookscole.com/miller7

Then click on the Chapter-by-Chapter area, choose Chapter 1, and select a learning resource.

SCIENCE, MATTER, ENERGY, AND ECOSYSTEMS: CONNECTIONS IN NATURE

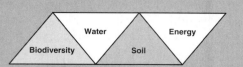

When we try to pick out anything by itself, we find it hitched to everything else in the universe.

JOHN MUIR

2-1 THE NATURE OF SCIENCE

What Is Science, and What Do Scientists Do? Searching for Order in Nature

Scientists collect data, form hypotheses, and develop theories, models, and laws about how nature works.

Science is an attempt to discover order in the natural world and use that knowledge to describe what is likely to happen in nature. Figure 2-1 summarizes the scientific process. Trace the pathways in this figure.

The first thing scientists do is ask a question or identify a problem to be investigated. Then they collect **scientific data,** or facts, by making observations and measurements. They often conduct **experiments** to study some phenomenon under known conditions. The resulting scientific data or facts must be confirmed by repeated observations and measurements, ideally by several different investigators.

The primary goal of science is not the data or facts themselves. Instead, science seeks new ideas, principles, or models that connect and explain certain scientific data and descriptions of what is likely to happen in nature. Scientists working on a particular problem try to come up with a variety of possible explanations, or **scientific hypotheses,** of what they (or other scientists) observe in nature. A scientific hypothesis is an unconfirmed explanation of an observed phenomenon that can be tested by further research.

Two important features of the scientific process are *reproducibility* and *peer review* of results by other scientists. Peers, or scientists working in the same field, check for reproducibility by repeating and checking out one another's work to see if the data can be reproduced and whether proposed hypotheses are reasonable. *Peer review* happens when scientists openly publish details of the methods they used, the results of their experiments, and the reasoning behind their hypotheses for other scientists to examine and criticize.

If repeated observations and measurements support a particular hypothesis or a group of related hypotheses, it becomes a **scientific theory.** In other words, a *scientific theory* is a verified, credible, and widely accepted scientific hypothesis or a related group of scientific hypotheses.

To scientists, *scientific theories are not to be taken lightly.* They are not guesses, speculations, or suggestions. Instead, they are useful explanations of processes or natural phenomena that have a high degree of certainty because they are supported by extensive evidence.

Nonscientists often use the word *theory* incorrectly when they mean to refer to a *scientific hypothesis,* a tentative explanation or educated guess that needs further evaluation. The statement, "Oh, that's just a theory," made in everyday conversation, implies a lack of knowledge and careful testing—the opposite of the scientific meaning of the word.

Another important result of science is a **scientific, or natural, law:** a description of what we find happening in nature over and over in the same way. For example, after making thousands of observations and measurements over many decades, scientists formulated the *second law of thermodynamics.* Simply stated, this law says that heat always flows spontaneously from

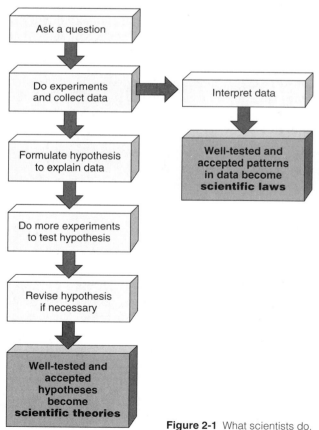

Figure 2-1 What scientists do.

hot to cold—something you learned the first time you touched a hot object.

A scientific law is no better than the accuracy of the observations or measurements upon which it is based. But if the data are accurate, a scientific law cannot be broken.

How Do Scientists Learn about Nature? Follow Many Paths

There are many scientific methods.

We often hear about *the* scientific method. In reality, many **scientific methods** exist: they are ways scientists gather data and formulate and test scientific hypotheses, theories, and laws.

Here is an example of applying the scientific method to an everyday situation:

> *Observation:* You switch on your trusty flashlight and nothing happens.
> *Question:* Why did the light not come on?
> *Hypothesis:* Maybe the batteries are dead.
> *Test the Hypothesis:* Put in new batteries and switch on the flashlight.
> *Result:* Flashlight still does not work.
> *New Hypothesis:* Maybe the bulb is burned out.
> *Experiment:* Replace bulb with a new bulb.
> *Result:* Flashlight works when switched on.
> *Conclusion:* Second hypothesis is verified.

Situations in nature are usually much more complicated than this. Many *variables* or *factors* influence most processes or parts of nature scientists seek to understand. Ideally, scientists conduct a *controlled experiment* to isolate and study the effect of a single variable. To do such a *single-variable analysis*, scientists set up two groups. One is an *experimental group* in which the chosen variable is changed in a known way. The other is a *control group* in which the chosen variable is not changed. If the experiment is designed properly, any difference between the two groups should result from a variable that was changed in the experimental group.

A basic problem is that many of the phenomena environmental scientists investigate involve a huge number of interacting variables. This limitation is sometimes overcome by using *multivariable analysis*—running mathematical models on high-speed computers to analyze the interactions of many variables without having to carry out traditional controlled experiments.

How Valid Are the Results of Science? Very Reliable But Not Perfect

Scientists try to establish that a particular theory or law has a very high probability of being true.

Scientists can do two major things. *First,* they disprove things. *Second,* they establish that a particular theory or

law has a very high probability or degree of certainty of being true. However, like scholars in any field, scientists cannot prove their theories and laws are *absolutely* true.

Although it may be extremely low, some degree of uncertainty is always involved in any scientific theory, model, or law. Most scientists rarely say something like, "Cigarettes cause lung cancer." Rather, the statement might be phrased, "There is overwhelming evidence from thousands of studies that indicate a significant relationship between cigarette smoking and lung cancer."

What Is the Difference between Frontier Science and Sound Science? Preliminary and Well-Tested Results

Scientific results fall into two categories: those that have not been confirmed (frontier science) and those that have been well tested and widely accepted (sound science).

News reports often focus on two things: new so-called scientific breakthroughs, and disputes between scientists over the validity of preliminary and untested data and hypotheses. These preliminary results, called **frontier science,** are often controversial because they have not been widely tested and accepted. At the frontier stage, it is normal and healthy for reputable scientists to disagree about the meaning and accuracy of data and the validity of various hypotheses.

By contrast, **sound** science, or **consensus science,** consists of data, theories, and laws that are widely accepted by scientists who are considered experts in the field involved. The results of sound science are based on a self-correcting process of open peer review. One way to find out what scientists generally agree on is to seek out reports by scientific bodies such as the U.S. National Academy of Sciences and the British Royal Society, which attempt to summarize consensus among experts in key areas of science.

What Is Junk Science and How Can We Detect It? Look Out for Baloney

Junk science is untested ideas presented as sound science.

Junk science consists of scientific results or hypotheses presented as sound science but not having undergone the rigors of the peer-review process. Some scientists, politicians, and other analysts label as junk science any science that does not support or further their particular agenda. Here are some critical thinking questions you can use to uncover junk science.

- How reliable are the sources making a particular claim? Do they have a hidden agenda? Are they experts in this field? What is their source of funding?

- Do the conclusions follow logically from the observations?

- Has the claim been verified by impartial peer review?

- How does the claim compare with the consensus view of experts in this field?

2-2 MATTER AND ENERGY

What Types of Matter Do We Find in Nature? Getting to the Bottom of Things

Matter exists in chemical forms as elements and compounds.

Matter is anything that has mass (the amount of material in an object) and takes up space. Scientists classify matter as existing in various levels of organization (Figure 2-2). There are two chemical forms of matter. One is **elements**: the distinctive building blocks of matter that make up every material substance. The other consists of **compounds**: two or more different elements held together in fixed proportions by attractive forces called *chemical bonds*.

To simplify things, chemists represent each element by a one- or two-letter symbol. Examples used in this book are hydrogen (H), carbon (C), oxygen (O), nitrogen (N), phosphorus (P), sulfur (S), chlorine (Cl), fluorine (F), bromine (Br), sodium (Na), calcium (Ca), lead (Pb), mercury (Hg), arsenic (As), and uranium (U).

What Are Nature's Building Blocks? Matter's Bricks

Atoms, ions, and molecules are the building blocks of matter.

If you had a supermicroscope capable of looking at individual elements and compounds, you could see they are made up of three types of building blocks. The first is an **atom**: the smallest unit of matter that exhibits the characteristics of an element.

The second is an **ion**: an electrically charged atom or combination of atoms. Examples encountered in this book include *positive* hydrogen ions (H^+), sodium ions (Na^+), calcium ions (Ca^{2+}), and ammonium ions (NH_4^+) and *negative* chloride ions (Cl^-), nitrate ions (NO_2^-), sulfate ions (SO_4^{2-}), and phosphate ions (PO_4^{3-}).

A third building block is a **compound**: a substance containing atoms or ions of more than one element that are held together by chemical bonds. Chemists use a shorthand **chemical formula** to show the number of atoms (or ions) of each type in a compound. The formula contains the symbols for each of the elements present and uses subscripts to represent the number of atoms or ions of each element in the compound's basic structural unit. Examples of compounds and their formulas encountered in this book are water (H_2O, read as "H-two-O"), oxygen (O_2), ozone (O_3), nitrogen (N_2), nitrous oxide (N_2O), nitric oxide (NO), hydrogen sulfide (H_2S), carbon monoxide (CO), carbon dioxide (CO_2), nitrogen dioxide (NO_2), sulfur dioxide (SO_2), ammonia (NH_3), sulfuric acid (H_2SO_4), nitric acid (HNO_3), methane (CH_4), and glucose ($C_6H_{12}O_6$).

Table sugar, vitamins, plastics, aspirin, penicillin, and most of the chemicals in your body are **organic compounds** that contain at least two carbon atoms combined with each other and with atoms of one or more other elements such as hydrogen, oxygen, nitrogen, sulfur, phosphorus, chlorine, and fluorine. One exception, methane (CH_4) has only one carbon atom. All other compounds are called **inorganic compounds.**

The millions of known organic (carbon-based) compounds include the following:

- *Hydrocarbons:* compounds of carbon and hydrogen atoms. An example is methane (CH_4), the main component of natural gas, and the simplest organic compound.

- *Chlorinated hydrocarbons:* compounds of carbon, hydrogen, and chlorine atoms. An example is the insecticide DDT ($C_{14}H_9Cl_5$).

- *Simple carbohydrates* (simple sugars): certain types of compounds of carbon, hydrogen, and oxygen atoms. An example is glucose ($C_6H_{12}O_6$), which most plants and animals break down in their cells to obtain energy.

Larger and more complex organic compounds, called *polymers*, consist of a number of basic structural or molecular units (*monomers*) linked by chemical bonds, somewhat like cars linked in a freight train. The three major types of organic polymers are *complex carbohydrates* consisting of two or more monomers of simple sugars (such as glucose) linked together, *proteins* formed by linking together monomers of amino acids, and *nucleic acids* (such as DNA and RNA) made by linked sequences of monomers called nucleotides.

Genes consist of specific sequences of nucleotides in a DNA molecule. These coded units of genetic information about specific traits are passed on from parents to offspring during reproduction. **Chromosomes** are combinations of genes that make up a single DNA molecule, together with a number of proteins. Each chromosome typically contains thousands of genes. Genetic information coded in your chromosomal DNA is what makes you different from an oak leaf, an alligator, or a flea, and from your parents. The relationships of genetic material to cells are depicted in Figure 2-3 (p. 20).

What Are Atoms Made Of? Looking Inside

Each atom has a tiny nucleus containing protons, and in most cases neutrons, and one or more electrons whizzing around somewhere outside the nucleus.

If you increased the magnification of your supermicroscope, you would find that each different type of atom

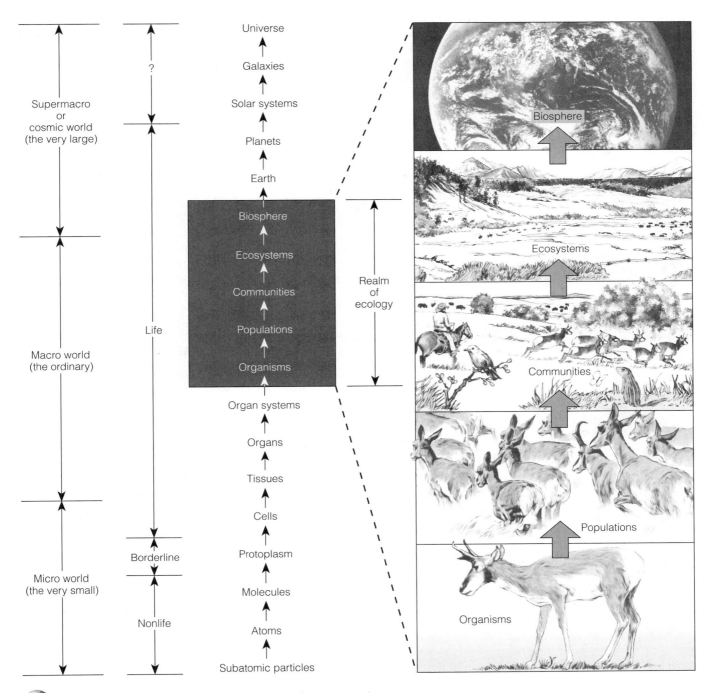

Figure 2-2 Natural capital: levels of organization of matter in nature. Note the five levels that ecology focuses on.

contains a certain number of *subatomic particles.* There are three types of these atomic building blocks: positively charged **protons** (p), uncharged **neutrons** (n), and negatively charged **electrons** (e).

Each atom consists of an extremely small center, or **nucleus.** It contains one or more protons, and in most cases neutrons, and one or more electrons in rapid motion somewhere outside the nucleus. Atoms are incredibly small. More than 3 million hydrogen atoms could sit side by side on the period at the end of this sentence. Now that is tiny.

Each atom has an equal number of positively charged protons inside its nucleus and negatively charged electrons whirling around outside its nucleus. Because these electrical charges cancel one another, *the atom as a whole has no net electrical charge.*

Each element has its own specific **atomic number,** equal to the number of protons in the nucleus of each

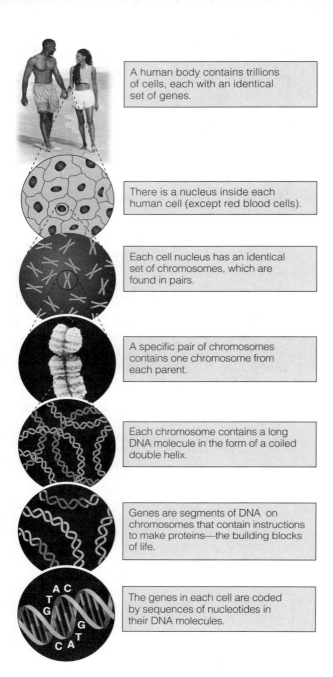

A human body contains trillions of cells, each with an identical set of genes.

There is a nucleus inside each human cell (except red blood cells).

Each cell nucleus has an identical set of chromosomes, which are found in pairs.

A specific pair of chromosomes contains one chromosome from each parent.

Each chromosome contains a long DNA molecule in the form of a coiled double helix.

Genes are segments of DNA on chromosomes that contain instructions to make proteins—the building blocks of life.

The genes in each cell are coded by sequences of nucleotides in their DNA molecules.

Figure 2-3 Relationships among cells, nuclei, chromosomes, DNA, and genes.

of its atoms. The simplest element, hydrogen (H), has only 1 proton in its nucleus, so its atomic number is 1. Carbon (C), with 6 protons, has an atomic number of 6, whereas uranium (U), a much larger atom, has 92 protons and an atomic number of 92.

Because electrons have so little mass compared with the mass of a proton or a neutron, *most of an atom's mass is concentrated in its nucleus.* The mass of an atom is described in terms of its **mass number:** the total number of neutrons and protons in its nucleus. For

example, a hydrogen atom with 1 proton and no neutrons in its nucleus has a mass number of 1, and an atom of uranium with 92 protons and 143 neutrons in its nucleus has a mass number of 235 (92 + 143 = 235).

All atoms of an element have the same number of protons in their nuclei. But they may have different numbers of uncharged neutrons in their nuclei, and thus may have different mass numbers. Various forms of an element having the same atomic number but a different mass number are called **isotopes** of that element. Scientists identify isotopes by attaching their mass numbers to the name or symbol of the element. For example, hydrogen has three isotopes: hydrogen-1 (H-1, with 1 proton and no neutrons in its nucleus), hydrogen-2 (H-2, common name *deuterium,* with 1 proton and 1 neutron in its nucleus), and hydrogen-3 (H-3, common name *tritium,* with 1 proton and 2 neutrons in its nucleus). A natural sample of an element contains a mixture of its isotopes in a fixed proportion or percentage abundance by weight.

What Is Matter Quality? Matter Usefulness

Matter can be classified as having high or low quality depending on how useful it is to us as a resource.

Matter quality is a measure of how useful a form of matter is to us as a resource, based on its availability and concentration, as shown in Figure 2-4. **High-quality matter** is concentrated, usually is found near the earth's surface, and has great potential for use as a matter resource. **Low-quality matter** is dilute, often is deep underground or dispersed in the ocean or the atmosphere, and usually has little potential for use as a material resource.

An aluminum can is a more concentrated, higher-quality form of aluminum than aluminum ore containing the same amount of aluminum. That is why it takes less energy, water, and money to recycle an aluminum can than to make a new can from aluminum ore.

Material efficiency, or **resource productivity,** is the total amount of material needed to produce each unit of goods or services. Business expert Paul Hawken and physicist Amory Lovins contend that resource productivity in developed countries could be improved by 75–90% within two decades using existing technologies.

What Is Energy? Doing Work and Transferring Heat

Energy is the work needed to move matter and the heat that flows from hot to cooler samples of matter.

Energy is the capacity to do work and transfer heat. Work is performed when an object such as a grain of sand, this book, or a giant boulder is moved over some distance. Work, or matter movement, also is needed to boil water or to burn natural gas to heat a house or

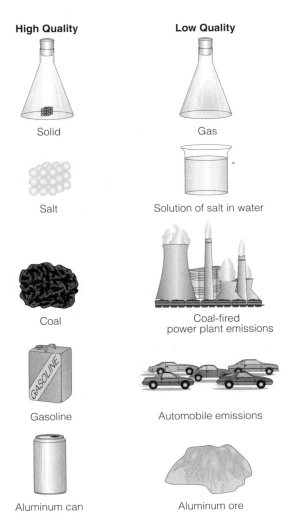

High Quality / **Low Quality**

Solid / Gas

Salt / Solution of salt in water

Coal / Coal-fired power plant emissions

Gasoline / Automobile emissions

Aluminum can / Aluminum ore

Figure 2-4 Examples of differences in matter quality. High-quality matter (left-hand column) is fairly easy to extract and is concentrated; low-quality matter (right-hand column) is more difficult to extract and is more dispersed than high-quality matter.

cook food. Energy is also the heat that flows automatically from a hot object to a cold object when they come in contact.

There are two major types of energy. One is **kinetic energy,** possessed by matter because of the matter's mass and its speed or velocity. Examples of this energy in motion are wind (a moving mass of air), flowing streams, heat flowing from a body at a high temperature to one at a lower temperature, and electricity (flowing electrons).

The second type is **potential energy,** which is stored and potentially available for use. Examples of this stored energy are this book held in your hand, an unlit match, still water behind a dam, the chemical energy stored in gasoline molecules, and the nuclear energy stored in the nuclei of atoms.

Potential energy can be changed to kinetic energy. Drop this book on your foot and the book's potential

energy when you held it changes into kinetic energy. When you burn gasoline in a car engine, the potential energy stored in the chemical bonds of its molecules changes into heat, light, and mechanical (kinetic) energy that propel the car.

What Is Electromagnetic Radiation? Waves of Energy

Some energy travels in waves at the speed of light.

Another type of energy is **electromagnetic radiation.** It is energy traveling in the form of a *wave* as a result of the changes in electric and magnetic fields.

There are many different forms of electromagnetic radiation, each with a different *wavelength* (distance between successive peaks or troughs in the wave) and *energy content,* as shown in Figure 2-5 (p. 22). Such radiation travels through space at the speed of light, which is about 300,000 kilometers per second (186,000 miles per second). That is fast.

What Is Energy Quality? Energy Usefulness

Energy can be classified as having high or low quality depending on how useful it is to us as a resource.

Energy quality is a measure of an energy source's ability to do useful work. **High-quality energy** is concentrated and can perform much useful work. Examples are electricity, the chemical energy stored in coal and gasoline, concentrated sunlight, and the nuclei of uranium-235, used as fuel in nuclear power plants.

By contrast, **low-quality energy** is dispersed and has little ability to do useful work. An example is heat dispersed in the moving molecules of a large amount of matter (such as the atmosphere or a large body of water) so that its temperature is low.

For example, the total amount of heat stored in the Atlantic Ocean is greater than the amount of high-quality chemical energy stored in all the oil deposits of Saudi Arabia. Yet the ocean's heat is so widely dispersed, it cannot be used to move things or to heat things to high temperatures.

What Is the Difference between a Physical and a Chemical Change?

Matter can change from one physical form to another or change its chemical composition.

When a sample of matter undergoes a **physical change,** its chemical composition does not change. A piece of aluminum foil cut into small pieces is still aluminum foil. When solid water (ice) is melted or liquid water is boiled, none of the H_2O molecules involved are altered; instead, the molecules are organized in different spatial (physical) patterns.

In a **chemical change,** or **chemical reaction,** there is a change in the chemical composition of the elements

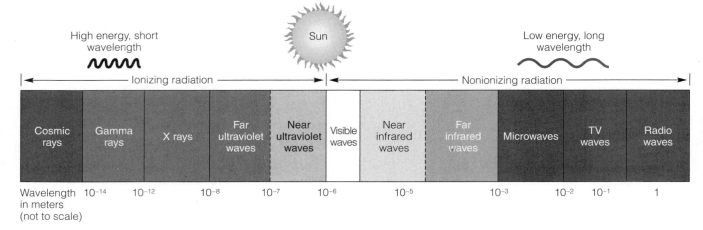

	Ionizing radiation					Nonionizing radiation					
Cosmic rays	Gamma rays	X rays	Far ultraviolet waves	Near ultraviolet waves	Visible waves	Near infrared waves	Far infrared waves	Microwaves	TV waves	Radio waves

Wavelength in meters (not to scale)

10^{-14} 10^{-12} 10^{-8} 10^{-7} 10^{-6} 10^{-5} 10^{-3} 10^{-2} 10^{-1} 1

Figure 2-5 The *electromagnetic spectrum:* the range of electromagnetic waves, which differ in wavelength (distance between successive peaks or troughs) and energy content.

or compounds. Chemists use shorthand chemical equations to represent what happens in a chemical reaction. For example, when coal burns completely, the solid carbon (C) it contains combines with oxygen gas (O_2) from the atmosphere to form the gaseous compound carbon dioxide (CO_2). We can represent this chemical reaction in shorthand form as $C + O_2 \longrightarrow CO_2 + energy$.

Energy is given off in this reaction, making coal a useful fuel. The reaction also shows how the complete burning of coal (or any of the carbon-containing compounds in wood, natural gas, oil, and gasoline) gives off carbon dioxide gas. This is a key gas that helps warm the lower atmosphere (troposphere).

The Law of Conservation of Matter: Why Is There No "Away"?

When a physical or chemical change occurs, no atoms are created or destroyed.

We may change various elements and compounds from one physical or chemical form to another, but in no physical and chemical change can we create or destroy any of the atoms involved. All we can do is rearrange them into different spatial patterns (physical changes) or different combinations (chemical changes). This statement, based on many thousands of measurements, is known as the **law of conservation of matter.**

The law of conservation of matter means there is no "away" as in "to throw away." *Everything we think we have thrown away is still here with us in one form or another.* We can make the environment cleaner and convert some potentially harmful chemicals into less harmful physical or chemical forms. But *the law of conservation of matter means we will always face the problem of what to do with some quantity of wastes and pollutants.*

What Is the First Law of Thermodynamics? You Cannot Get Something for Nothing.

In a physical or chemical change, we can change energy from one form to another but we can never create or destroy any of the energy involved.

Scientists have observed energy being changed from one form to another in millions of physical and chemical changes. But they have never been able to detect the creation or destruction of any energy (except in nuclear changes). The results of their experiments have been summarized in the **law of conservation of energy,** also known as the **first law of thermodynamics:** *In all physical and chemical changes, energy is neither created nor destroyed, but it may be converted from one form to another.*

This scientific law tells us that when one form of energy is converted to another form in any physical or chemical change, *energy input always equals energy output.* No matter how hard we try or how clever we are, we cannot get more energy out of a system than we put in; in other words, *we cannot get something for nothing in terms of energy quantity.* This is one of Mother Nature's basic rules that we have to live with.

What Is the Second Law of Thermodynamics? You Cannot Even Break Even.

Whenever energy is changed from one form to another we always end up with less usable energy than we started with.

Because the first law of thermodynamics states that energy can be neither created nor destroyed, we may be tempted to think there will always be enough energy. Yet if we fill a car's tank with gasoline and drive around or use a flashlight battery until it is dead, something has been lost. If it is not energy, what is it? The answer is *energy quality,* the amount of energy available that can perform useful work.

Countless experiments have shown that when energy is changed from one form to another, a decrease in energy quality always occurs. The results of these experiments have been summarized in what is called the **second law of thermodynamics:** *When energy is changed from one form to another, some of the useful energy is always degraded to lower-quality, more dispersed, less useful energy.* This degraded energy usually takes the form of heat given off at a low temperature to the surroundings (environment). There it is dispersed by the random motion of air or water molecules and becomes even less useful as a resource.

In other words, *we cannot even break even in terms of energy quality because energy always goes from a more useful to a less useful form when energy is changed from one form to another.* No one has ever found a violation of this fundamental scientific law. It is another one of Mother Nature's rules that we have to live with.

Consider three examples of the second law of thermodynamics in action. *First,* when a car is driven, only about 20–25% of the high-quality chemical energy available in its gasoline fuel is converted into mechanical energy (to propel the vehicle) and electrical energy (to run its electrical systems). The remaining 75–80% is degraded to low-quality heat that is released into the environment and eventually lost into space.

Thus, most of the money you spend for gasoline is not used to get you anywhere.

Second, when electrical energy flows through filament wires in an incandescent light bulb, it is changed into about 5% useful light and 95% low-quality heat that flows into the environment. In other words, this so-called *light bulb* is really a *heat bulb. Good news.* Scientists have developed compact fluorescent bulbs that are four times more efficient, and even more efficient bulbs are on the way. Do you use compact fluorescent bulbs?

Third, in living systems, solar energy is converted into chemical energy (food molecules) and then into mechanical energy (moving, thinking, and living). During each of these conversions, high-quality energy is degraded and flows into the environment as low-quality heat (Figure 2-6). Trace the flows and energy conversions in this diagram.

The second law of thermodynamics also means *we can never recycle or reuse high-quality energy to perform useful work.* Once the concentrated energy in a serving of food, a liter of gasoline, a lump of coal, or a chunk of uranium is released, it is degraded to low-quality heat that is dispersed into the environment.

Energy efficiency, or **energy productivity,** is a measure of how much useful work is accomplished by a particular input of energy into a system. *Good news.* There is plenty of room for improving energy efficiency. Scientists estimate that only about 16% of the energy used in the United States ends up performing

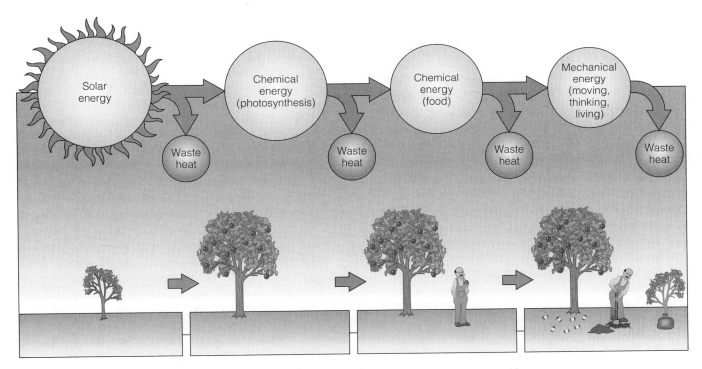

Figure 2-6 The second law of thermodynamics in action in living systems. Each time energy is changed from one form to another, some of the initial input of high-quality energy is degraded, usually to low-quality heat that is dispersed into the environment.

useful work. The remaining 84% is either unavoidably wasted because of the second law of thermodynamics (41%) or unnecessarily wasted (43%).

2-3 EARTH'S LIFE-SUPPORT SYSTEMS: FROM ORGANISMS TO THE BIOSPHERE

💿 What Is Ecology? Understanding Connections

Ecology is a study of connections in nature.

Ecology (from the Greek words *oikos,* "house" or "place to live," and *logos,* "study of") is the study of how organisms interact with one another and with their non-living environment. In effect, it is a study of *connections in nature*—the house for the earth's life. Ecologists focus on trying to understand the interactions among organisms, populations, communities, ecosystems, and the biosphere (Figure 2-2).

An **organism** is any form of life. The **cell** is the basic unit of life in organisms. Organisms may consist of a single cell (bacteria, for instance) or many cells. Look in the mirror. What you see is about 10 trillion cells divided into about 200 different types.

Organisms can be classified into **species,** groups of organisms that resemble one another in appearance, behavior, chemistry, and genetic makeup. Organisms that reproduce sexually by combining cells from both parents are classified as members of the same species if, under natural conditions, they can potentially breed with one another and produce live, fertile offspring.

How many species are on the earth? We do not know. Estimates range from 3.6 million to 100 million—most of them microorganisms too small to be seen with the naked eye. A best guess is that we share the planet with 10–14 million other species. So far biologists have identified and named about 1.4 million species, most of them insects (Case Study, right).

Case Study: What Species Rule the World? Small Matters!

Multitudes of tiny microbes such as bacteria, protozoa, fungi, and yeast help keep us alive.

They are everywhere and there are trillions of them. Billions are found inside your body, on your body, in a handful of soil, and in a cup of river water.

These mostly invisible rulers of the earth are *microbes,* a catchall term for many thousands of species of bacteria, protozoa, fungi, and yeasts—most too small to be seen with the naked eye.

Microbes do not get the respect they deserve. Most of us think of them as threats to our health in the form of infectious bacteria or "germs," fungi that cause athlete's foot and other skin diseases, and protozoa that cause diseases such as malaria. But these harmful microbes are in the minority.

You are alive because of multitudes of microbes toiling away mostly out of sight. You can thank microbes for providing you with food by converting nitrogen gas in the atmosphere into forms that plants can take up from the soil as nutrients that keep you alive and well. Microbes also help produce foods such as bread, cheese, yogurt, vinegar, tofu, soy sauce, beer, and wine. You should also thank bacteria and fungi in the soil that decompose organic wastes into nutrients that can be taken up by plants that we and most other animals eat. Without these wee creatures we would be up to our eyeballs in waste matter.

Microbes, especially bacteria, help purify the water you drink by breaking down wastes. Bacteria in your intestinal tract break down the food you eat. Some microbes in your nose prevent harmful bacteria from reaching your lungs. Others are the source of disease-fighting antibiotics, including penicillin, erythromycin, and streptomycin. Genetic engineers are developing microbes that can extract metals from ores, break down various pollutants, and help clean up toxic waste sites.

Some microbes help control plant diseases and populations of insect species that attack our food crops. Relying more on these microbes for pest control can reduce the use of potentially harmful chemical pesticides.

Case Study: Have You Thanked the Insects Today?

Insects play vital roles in sustaining life on the earth by pollinating plants that provide food for us and other animals and eating insects we classify as pests.

Insects have a bad reputation. We classify many insect species as *pests* because they compete with us for food, spread human diseases such as malaria, and invade our lawns, gardens, and houses. Some people have "bugitis," fear all insects, and think the only good bug is a dead bug. This view fails to recognize the vital roles insects play in helping sustain life on earth.

A large proportion of the earth's plant species depend on insects to pollinate their flowers. In turn, we and other land-dwelling animals depend on plants for food, either by eating them or by consuming animals that eat them. Without pollinating insects, very few fruits and vegetables would be available for us.

Insects that eat other insects help control the populations of at least half the species of insects we call pests. This free pest-control service is an important part of the natural capital that helps sustain us.

Insects have been around for at least 400 million years and are phenomenally successful forms of life. Some insects can reproduce at an astounding rate. For example, a single female housefly and her offspring can theoretically produce about 5.6 trillion flies in only one year!

Insects can also rapidly evolve new genetic traits, such as resistance to pesticides. They also have an exceptional ability to evolve into new species when faced with new environmental conditions, and they are very resistant to extinction. The environmental lesson is that although insects can thrive without newcomers such as us, we and most other land organisms would perish without them.

What Are Populations, Communities, and Ecosystems? Some Interactions in Nature

Members of a species interact in groups called populations; populations of different species living and interacting in an area form a community; and a community interacting with its physical environment of matter and energy is known as an ecosystem.

A **population** is a group of interacting individuals of the same species occupying a specific area. Examples are all sunfish in a pond, white oak trees in a forest, and people in a country. In most natural populations, individuals vary slightly in their genetic makeup, which is why they do not all look or act alike. This is called a population's **genetic diversity.**

The place where a population (or an individual organism) normally lives is its **habitat.** It may be as large as an ocean or as small the intestine of a termite.

A **community,** or **biological community,** consists of all the populations of the different species living and interacting in an area. It is a complex and interacting network of plants, animals, and microorganisms.

An **ecosystem** is a community of different species interacting with one another and with their physical environment of matter and energy. Ecosystems can range in size from a puddle of water to a stream, a patch of woods, an entire forest, or a desert. Ecosystems can be natural or artificial (human created). Examples of artificial ecosystems are crop fields, farm ponds, and reservoirs. All of the earth's ecosystems together make up what we call the **biosphere.**

What Are the Major Parts of the Earth's Life-Support Systems? The Spheres of Life

The earth is made up of interconnected spherical layers that contain air, water, soil, minerals, and life.

We can think of the earth as being made up of several spherical layers, as diagrammed in Figure 2-7. Study this figure carefully. The **atmosphere** is a thin envelope or membrane of air around the planet. Its inner layer, the **troposphere,** extends only about 17 kilometers (11 miles) above sea level. It contains most of the planet's air, mostly nitrogen (78%) and oxygen (21%).

The next layer, stretching 17–48 kilometers (11–30 miles) above the earth's surface, is the **stratosphere.** Its lower portion contains enough ozone (O_3) to filter out most of the sun's harmful ultraviolet radiation. This allows life to exist on land and in the surface layers of bodies of water.

The **hydrosphere** consists of the earth's water. It is found as *liquid water* (both surface and underground), *ice* (polar ice, icebergs, and ice in frozen soil layers called *permafrost*), and *water vapor* in the atmosphere. The **lithosphere** is the earth's crust and upper mantle; the crust contains nonrenewable fossil fuels and minerals we use as well as renewable soil chemicals (nutrients) needed for plant life.

All parts of the biosphere are interconnected. If the earth were an apple, the biosphere would be no thicker than the apple's skin. *The goal of ecology is to understand the interactions in this thin, life-supporting global membrane of air, water, soil, and organisms.*

What Sustains Life on Earth? Sun, Cycles, and Gravity

Solar energy, the cycling of matter, and gravity sustain the earth's life.

Life on the earth depends on three interconnected factors shown in Figure 2-8 (p. 26).

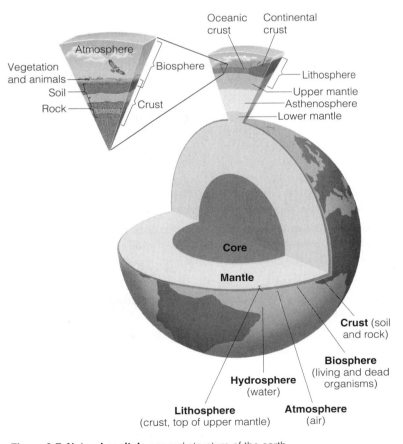

Figure 2-7 Natural capital: general structure of the earth.

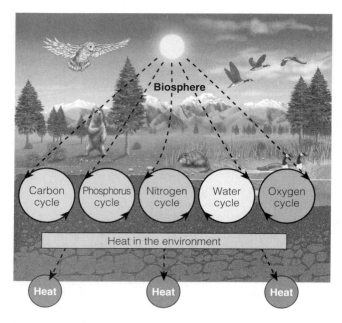

Figure 2-8 Natural capital: life on the earth depends on the *one-way flow of energy* (dashed lines) from the sun through the biosphere, the *cycling of crucial elements* (solid lines around circles), and *gravity,* which keeps atmospheric gases from escaping into space and draws chemicals downward in the matter cycles. This simplified model depicts only a few of the many cycling elements.

- The *one-way flow of high-quality energy* from the sun through materials and living things in their feeding interactions, into the environment as low-quality energy (mostly heat dispersed into air or water molecules at a low temperature), and eventually back into space as heat. No round-trips are allowed because energy cannot be recycled.

- The *cycling of matter* (the atoms, ions, or molecules needed for survival by living organisms) through parts of the biosphere. Because the earth is closed to significant inputs of matter from space, the earth's essentially fixed supply of nutrients must be recycled again and again for life to continue. All nutrient trips in ecosystems are round-trips.

- *Gravity,* which allows the planet to hold on to its atmosphere and causes the downward movement of chemicals in the matter cycles.

What Happens to Solar Energy Reaching the Earth? Just Passing Through

Solar energy flowing through the biosphere warms the atmosphere, evaporates and recycles water, generates winds, and supports plant growth.

About one-billionth of the sun's output of energy reaches the earth—a tiny sphere in the vastness of space—in the form of electromagnetic waves—mostly as visible light (Figure 2-5). Much of this energy is either reflected away or absorbed by chemicals in its atmosphere (Figure 2-9). Trace the flows in this diagram.

Most of the solar radiation making it through the atmosphere hits the earth and is degraded into longer wavelength infrared radiation. This infrared radiation interacts with so-called *greenhouse gases* (such as water vapor, carbon dioxide, methane, nitrous oxide, and ozone) in the troposphere. As this infrared radiation intersects with these gaseous molecules, it increases their kinetic energy, helping warm the troposphere and the earth's surface. Without this **natural greenhouse effect** (Figure 2-10), the earth would be too cold for life as we know it to exist and you would not be around to read this book.

Why Is the Earth So Favorable for Life? The Just Right Planet

The earth's temperature range, distance from the sun, and size result in conditions that are favorable for life as we know it.

Life on the earth as we know it needs a certain temperature range: Venus is much too hot and Mars is much too cold, but the earth is *just right*. Otherwise, you would not be reading these words.

Life as we know it depends on the liquid water that dominates the earth's surface. Temperature is crucial because most life on the earth needs average temperatures between the freezing and boiling points of water.

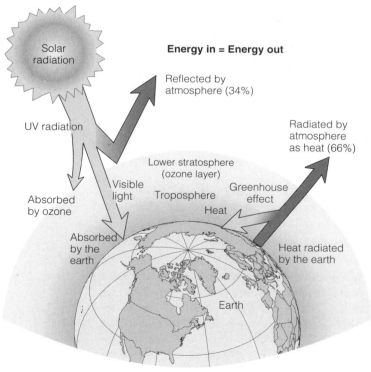

Figure 2-9 Solar capital: flow of energy to and from the earth.

(a) Rays of sunlight penetrate the lower atmosphere and warm the earth's surface.

(b) The earth's surface absorbs much of the incoming solar radiation and degrades it to longer-wavelength infrared (IR) radiation, which rises into the lower atmosphere. Some of this IR radiation escapes into space as heat and some is absorbed by molecules of greenhouse gases and emitted as even longer wavelength IR radiation, which warms the lower atmosphere.

(c) As concentrations of greenhouse gases rise, their molecules absorb and emit more infrared radiation, which adds more heat to the lower atmosphere.

Figure 2-10 Natural capital: the *natural greenhouse effect*. Without the atmospheric warming provided by this natural effect, the earth would be a cold and mostly lifeless planet. According to the widely accepted greenhouse theory, when concentrations of greenhouse gases in the troposphere rise, the average temperature of the troposphere rises. (Modified by permission from Cecie Starr, *Biology: Concepts and Applications*, 4th ed., Pacific Grove, Calif.: Brooks/Cole, 2000)

The earth's orbit is the right distance from the sun to provide these conditions. If the earth were much closer, it would be too hot—like Venus—for water vapor to condense to form rain. If it were much farther away, its surface would be so cold—like Mars—that its water would exist only as ice. The earth also spins; if it did not, the side facing the sun would be too hot and the other side too cold for water-based life to exist.

The earth is also the right size; that is, it has enough gravitational mass to keep its iron and nickel core molten and to keep the light gaseous molecules (such as N_2, O_2, CO_2, and H_2O) in its atmosphere from flying off into space.

On a time scale of millions of years, the earth is enormously resilient and adaptive. During the 3.7 billion years since life arose, the average surface temperature of the earth has remained within the narrow range of 10–20° C (50–68° F), even with a 30–40% increase in the sun's energy output. In short, this remarkable planet that we live on is just right for life as we know it.

What Are the Major Components of Ecosystems? Matter, Energy, Life

Ecosystems consist of nonliving (abiotic) and nonliving (biotic) components.

Two types of components make up the biosphere and its ecosystems. One type, called **abiotic,** consists of nonliving components such as water, air, nutrients, and solar energy. The other type, called **biotic,** consists of biological components—plants, animals, and microbes.

Figures 2-11 (p. 28) and 2-12 (p. 28) are greatly simplified diagrams of some of the biotic and abiotic components in a freshwater aquatic ecosystem and a terrestrial ecosystem. Look carefully at these components and how they are connected to one another through the consumption habits of organisms.

Each population in an ecosystem has a **range of tolerance** to variations in its physical and chemical environment (Figure 2-13, p. 29). Individuals within a population may also have slightly different tolerance ranges for temperature or other factors because of small differences in genetic makeup, health, and age. For example, a trout population may do best within a narrow band of temperatures (*optimum level or range*), but a few individuals can survive above and below that band. But if the water becomes too hot or too cold, none of the trout can survive.

These observations are summarized in the **law of tolerance:** *The existence, abundance, and distribution of a species in an ecosystem are determined by whether the levels of one or more physical or chemical factors fall within the range tolerated by that species.* A species may have a wide range of tolerance to some factors and a narrow range of tolerance to others. Most organisms are least tolerant during juvenile or reproductive stages of their life cycles. Highly tolerant species can live in a variety of habitats with widely different conditions.

A variety of factors can affect the number of organisms in a population. However, sometimes one factor, known as a **limiting factor,** is more important in regulating population growth than other factors. This ecological principle, related to the law of tolerance, is

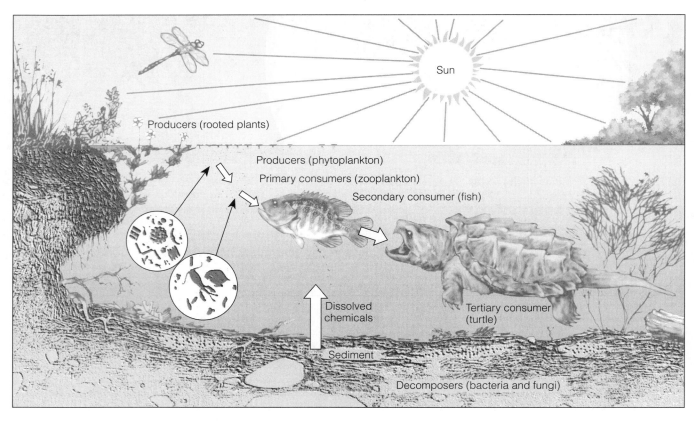

Figure 2-11 Major components of a freshwater ecosystem.

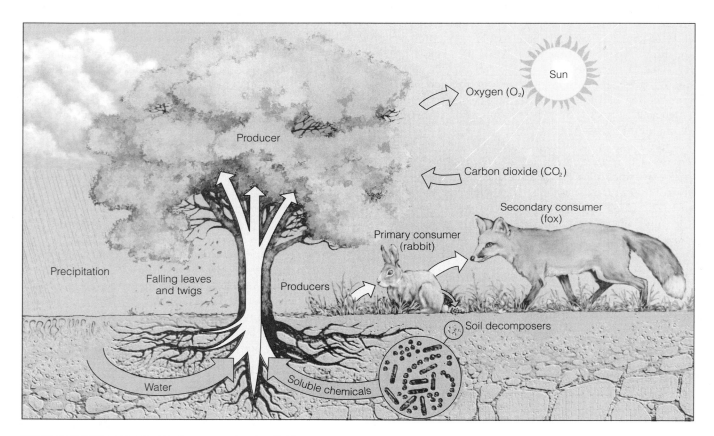

Figure 2-12 Major components of an ecosystem in a field.

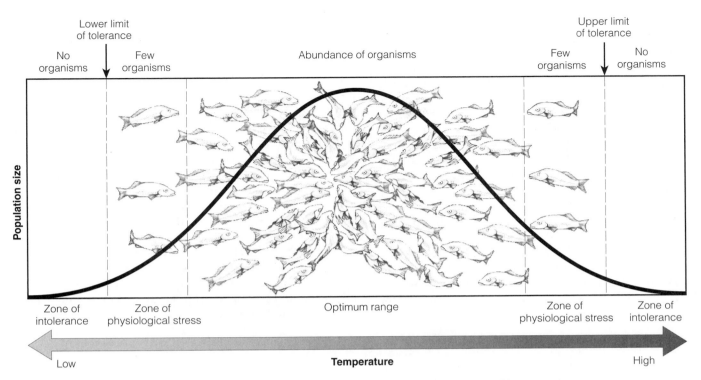

Figure 2-13 Range of tolerance for a population of organisms, such as fish, to an abiotic environmental factor—in this case, temperature.

called the **limiting factor principle:** *Too much or too little of any abiotic factor can limit or prevent growth of a population, even if all other factors are at or near the optimum range of tolerance.*

On land, precipitation often is the limiting factor. Lack of water in a desert limits plant growth. Soil nutrients also can act as a limiting factor on land. Suppose a farmer plants corn in phosphorus-poor soil. Even if water, nitrogen, potassium, and other nutrients are at optimum levels, the corn will stop growing when it uses up the available phosphorus.

Too much of an abiotic factor can also be limiting. For example, too much water or too much fertilizer can kill plants, a common mistake of many beginning gardeners.

Important limiting factors for aquatic life zones include temperature, sunlight, nutrient availability, and **dissolved oxygen (DO) content**—the amount of oxygen gas dissolved in a given volume of water at a particular temperature and pressure. Another limiting factor in aquatic life zones is **salinity**—the amounts of various inorganic minerals or salts dissolved in a given volume of water.

What Are the Major Biological Components of Ecosystems? Producers and Consumers

Some organisms in ecosystems produce food and others consume food.

The earth's organisms either produce or consume food. **Producers,** sometimes called **autotrophs** (self-feeders), make their own food from compounds obtained from their environment.

On land, most producers are green plants. In freshwater and marine life zones, algae and plants are the major producers near shorelines. In open water, the dominant producers are *phytoplankton*—mostly microscopic organisms that float or drift in the water.

Most producers capture sunlight to make complex compounds (such as glucose, $C_6H_{12}O_6$) by **photosynthesis.** Although hundreds of chemical changes take place during photosynthesis, the overall reaction can be summarized as follows:

carbon dioxide + water + **solar energy** ⟶ glucose + oxygen

All other organisms in an ecosystem are **consumers,** or **heterotrophs** ("other-feeders") that get their energy and nutrients they need by feeding on other organisms or their remains. **Decomposers** (mostly certain types of bacteria and fungi) are specialized consumers that recycle organic matter in ecosystems. They do this by breaking down (*biodegrading*) dead organic material or **detritus** ("di-TRI-tus," meaning "debris") to get nutrients. This releases the resulting simpler inorganic compounds into the soil and water, where producers can take them up as nutrients.

Some consumers, called **omnivores,** play dual roles by feeding on both plants and animals. Examples are pigs, rats, foxes, bears, cockroaches, and humans. When you had lunch today were you an herbivore, a carnivore, or an omnivore?

Detritivores consist of detritus feeders and decomposers that feed on detritus. Hordes of these waste eaters and degraders can transform a fallen tree trunk into a powder and finally into simple inorganic molecules that plants can absorb as nutrients (Figure 2-14). *In natural ecosystems, there is little or no waste.* One organism's wastes serve as resources for others, as the nutrients that make life possible are recycled again and again.

Producers, consumers, and decomposers use the chemical energy stored in glucose and other organic compounds to fuel their life processes. In most cells this energy is released by **aerobic respiration,** which uses oxygen to convert organic nutrients back into carbon dioxide and water. The net effect of the hundreds of steps in this complex process is represented by the following reaction:

$$\text{glucose} + \text{oxygen} \longrightarrow \text{carbon dioxide} + \text{water} + \textbf{energy}$$

Although the detailed steps differ, the net chemical change for aerobic respiration is the opposite of that for photosynthesis.

The survival of any individual organism depends on the *flow of matter and energy* through its body. However, an ecosystem as a whole survives primarily through a combination of *matter recycling* (rather than one-way flow) and *one-way energy flow* (Figure 2-15). Figure 2-15 sums up most of what is going on in our planetary home.

Decomposers complete the cycle of matter by breaking down detritus into inorganic nutrients that can be reused by producers. These waste eaters and nutrient recyclers provide us this crucial ecological service and never send us a bill. Without decomposers, the entire world would be knee-deep in plant litter, dead animal bodies, animal wastes, and garbage, and most life as we know it would no longer exist. Have you thanked a decomposer today?

What Is Biodiversity? Variety Is the Spice of Life.

A vital renewable resource is the biodiversity found in the earth's variety of genes, species, ecosystems, and ecosystem processes.

Biological diversity, or **biodiversity,** is one of the earth's most important renewable resources. Kinds of biodiversity include the following:

- **Genetic diversity:** the variety of genetic material within a species or a population.

- **Species diversity:** the number of species present in different habitats.

- **Ecological diversity:** the variety of terrestrial and aquatic ecosystems found in an area or on the earth.

- **Functional diversity:** the biological and chemical processes such as energy flow and matter cycling needed for the survival of species, communities, and ecosystems (Figure 2-15).

The earth's biodiversity is the biological wealth or capital that helps keep us alive and supports our economies. It supplies us with food, wood, fibers, energy, raw materials, industrial chemicals, and medicines—all of which pour hundreds of billions of dollars into the world economy each year. It also helps preserve the quality of the air and water, maintain the fertility of soils, dispose of wastes, and control populations of pests that attack crops and forests.

Biodiversity is a renewable resource as long we live off the biological income it provides instead of the natural capital that supplies this income. Understanding, protecting, and sustaining biodiversity is a major theme of ecology and of this book. Biodiversity is our lifeline!

2-4 ENERGY FLOW IN ECOSYSTEMS

What Are Food Chains and Food Webs? Maps of Ecological Interdependence

Food chains and webs show how eaters, the eaten, and the decomposed are connected to one another in an ecosystem.

All organisms, whether dead or alive, are potential sources of food for other organisms. A caterpillar eats a leaf, a robin eats the caterpillar, and a hawk eats the robin. Decomposers consume the leaf, caterpillar, robin, and hawk after they die. As a result, *there is little waste in natural ecosystems.*

The sequence of organisms, each of which is a source of food for the next, is called a **food chain.** It determines how energy and nutrients move from one organism to another through the ecosystem (Figure 2-16, p. 32).

Ecologists assign each organism in an ecosystem to a feeding level, also called a **trophic level,** depending on whether it is a producer or a consumer and on what it eats or decomposes. Producers belong to the first trophic level, primary consumers to the second trophic level, secondary consumers to the third, and so on. Detritivores and decomposers process detritus from all trophic levels.

Real ecosystems are more complex than this. Most consumers feed on more than one type of organism, and most organisms are eaten by more than one type of consumer. Because most species participate in several different food chains, the organisms in most ecosystems form a complex network of interconnected food chains called a **food web** (Figure 2-17, p. 33). Trace the

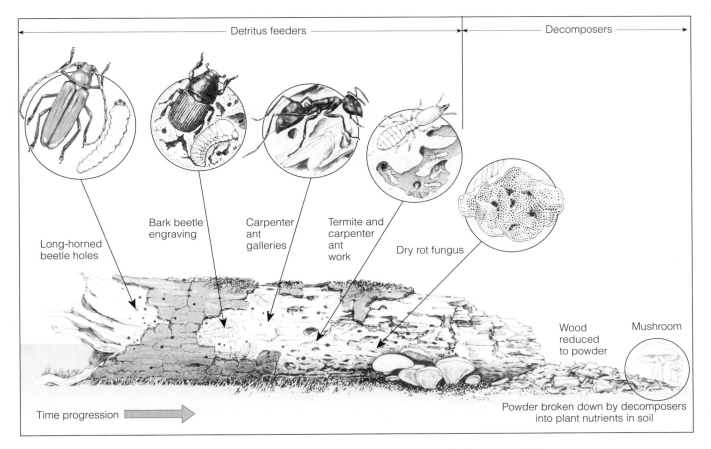

Long-horned
beetle holes

Bark beetle
engraving

Carpenter
ant
galleries

Termite and
carpenter
ant
work

Dry rot fungus

Wood
reduced
to powder

Mushroom

Time progression

Powder broken down by decomposers
into plant nutrients in soil

Figure 2-14 Natural capital: some detritivores, called *detritus feeders,* directly consume tiny fragments of this log. Other detritivores, called *decomposers* (mostly fungi and bacteria), digest complex organic chemicals in fragments of the log into simpler inorganic nutrients. These nutrients can be used again by producers if they are not washed away or otherwise removed from the system.

flows of matter and energy within this web. Trophic levels can be assigned in food webs just as in food chains. A food web shows how eaters, the eaten, and the decomposed are connected to one another. It is a map of life's interdependence.

How Can We Represent the Energy Flow in an Ecosystem? Think Pyramid

There is a decrease in the amount of energy available to each succeeding organism in a food chain or web.

Each trophic level in a food chain or web contains a certain amount of **biomass,** the dry weight of all organic matter contained in its organisms. In a food chain or web, chemical energy stored in biomass is transferred from one trophic level to another.

The percentage of usable energy transferred as biomass from one trophic level to the next is called **ecological efficiency.** It ranges from 2% to 40% (that is, a loss of 60–98%) depending on the types of species and the ecosystem involved, but 10% is typical.

Assuming 10% ecological efficiency (90% loss) at each trophic transfer, if green plants in an area manage

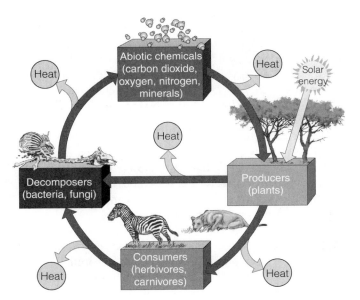

Figure 2-15 Natural capital: the main structural components of an ecosystem (energy, chemicals, and organisms). Matter recycling and the flow of energy from the sun, through organisms, and then into the environment as low-quality heat, link these components.

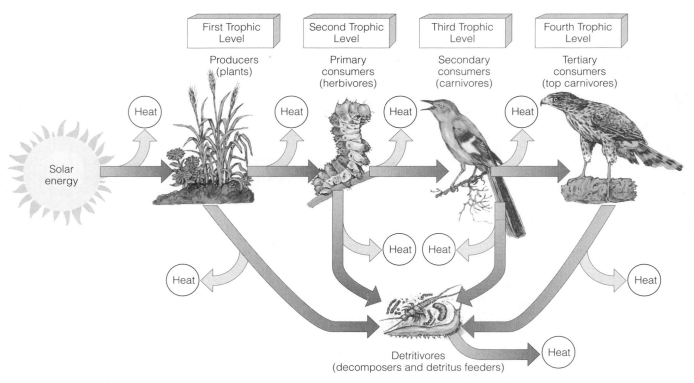

First Trophic Level
Second Trophic Level
Third Trophic Level
Fourth Trophic Level

Producers (plants)
Primary consumers (herbivores)
Secondary consumers (carnivores)
Tertiary consumers (top carnivores)

Solar energy

Heat

Detritivores (decomposers and detritus feeders)

Figure 2-16 A *food chain.* The arrows show how chemical energy in food flows through various *trophic levels* in energy transfers; most of the energy is degraded to heat, in accordance with the second law of thermodynamics. Food chains rarely have more than four trophic levels. Can you figure out why?

to capture 10,000 units of energy from the sun, then only about 1,000 units of energy will be available to support herbivores and only about 100 units to support carnivores.

The more trophic levels or steps in a food chain or web, the greater the cumulative loss of usable energy as energy flows through the various trophic levels. The **pyramid of energy flow** in Figure 2-18 (p. 34) illustrates this energy loss for a simple food chain, assuming a 90% energy loss with each transfer. How does this diagram help explain why there are not very many tigers in the world?

Energy flow pyramids explain why the earth can support more people if they eat at lower trophic levels by consuming grains, vegetables, and fruits directly rather than passing such crops through another trophic level and eating grain eaters as cattle.

The large loss in energy between successive trophic levels also explains why food chains and webs rarely have more than four or five trophic levels. In most cases, too little energy is left after four or five transfers to support organisms feeding at these high trophic levels. This explains why there are so few top carnivores such as eagles, hawks, tigers, and white sharks. It also explains why such species usually are the first to suffer when the ecosystems that support them are disrupted, and why these species

are so vulnerable to extinction. Do you think we are on this list?

How Fast Can Producers Produce Biomass? A Very Important Rate

Different ecosystems use solar energy to produce and use biomass at different rates.

The *rate* at which an ecosystem's producers convert solar energy into chemical energy as biomass is the ecosystem's **gross primary productivity (GPP).** However, to stay alive, grow, and reproduce, an ecosystem's producers must use some of the biomass they produce for their own respiration. **Net primary productivity (NPP),** is the *rate* at which producers use photosynthesis to store energy *minus* the *rate* at which they use some of this stored energy through aerobic respiration, as shown in Figure 2-19 (p. 34). In other words, NPP = GPP − R, where R is energy used in respiration. NPP is a measure of how fast producers can provide the food needed by other organisms (consumers) in an ecosystem.

Various ecosystems and life zones differ in their NPP, as graphed in Figure 2-20 (p. 34). Looking at this graph, what are nature's three most productive and three least productive systems? Despite its low net primary productivity, there is so much open ocean

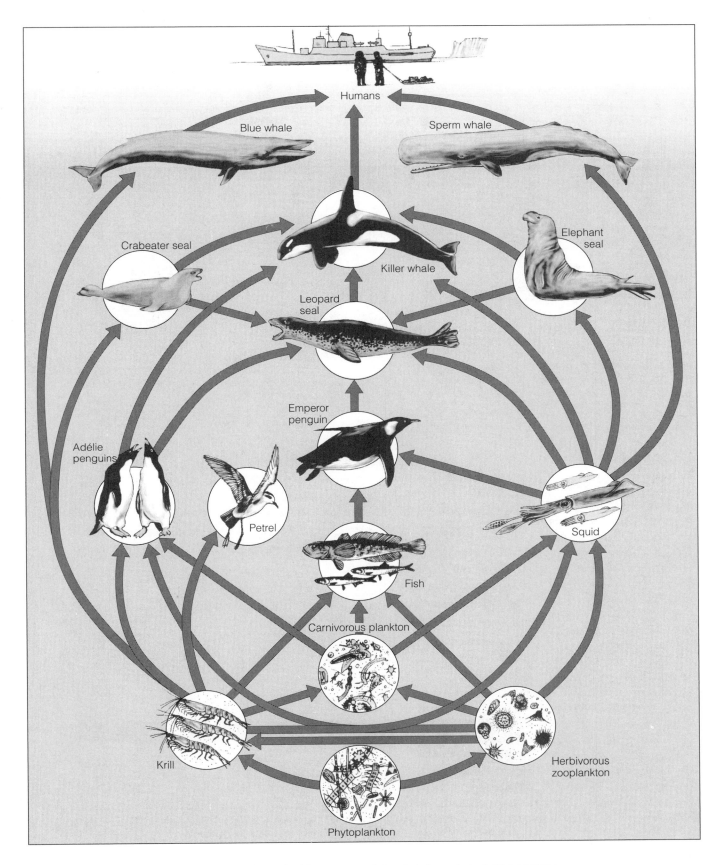

Figure 2-17 Greatly simplified *food web* in the Antarctic. Many more participants in the web, including an array of decomposer organisms, are not depicted here.

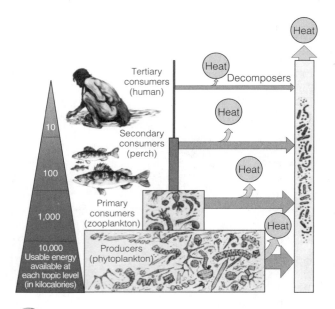

Figure 2-18 Generalized *pyramid of energy flow* showing the decrease in usable energy available at each succeeding trophic level in a food chain or web. In nature, ecological efficiency varies from 2% to 40%, with 10% efficiency being common. This model assumes a 10% ecological efficiency (90% loss in usable energy to the environment, in the form of low-quality heat) with each transfer from one trophic level to another.

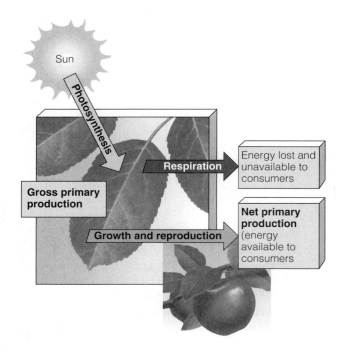

Figure 2-19 Distinction between gross primary productivity and net primary productivity. A plant uses some of its gross primary productivity to survive through respiration. The remaining energy is available to consumers.

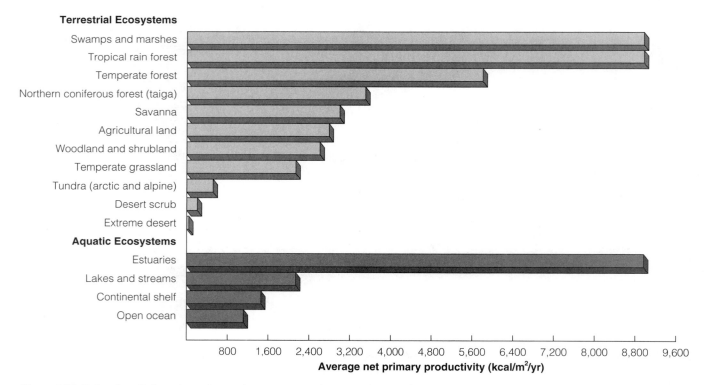

Figure 2-20 Natural capital: estimated annual average *net primary productivity (NPP)* per unit of area in major life zones and ecosystems, expressed as kilocalories of energy produced per square meter per year (kcal/m²/yr). (Data from R. H. Whittaker, *Communities and Ecosystems,* 2nd ed., New York: Macmillan, 1975)

that it produces more of the earth's NPP per year than any of the other ecosystems and life zones shown in Figure 2-20.

As we have seen, producers are the source of all food in an ecosystem. Only the biomass represented by NPP is available as food for consumers and they use only a portion of this. Thus, *the planet's NPP ultimately limits the number of consumers (including humans) that can survive on the earth.* This is an important lesson from nature.

Peter Vitousek, Stuart Rojstaczer, and other ecologists estimate that humans now use, waste, or destroy about 27% of the earth's total potential NPP and 10–55% of the NPP of the planet's terrestrial ecosystems.

They contend that this is the main reason why we are crowding out or eliminating the habitats and food supplies of a growing number of other species. What might happen to us and to other consumer species as the human population grows over the next 40–50 years and per capita consumption of resources such as food, timber, and grassland rises sharply? This is an important question.

2-5 SOILS

What Is Soil and Why Is It Important? The Base of Life on Land

Soil is a slowly renewed resource that provides most of the nutrients needed for plant growth and also helps purify water.

Soil is a thin covering over most land that is a complex mixture of eroded rock, mineral nutrients, decaying organic matter, water, air, and billions of living organisms, most of them microscopic decomposers (Figure 2-21). Soil is a renewable resource but it is renewed very slowly. Depending mostly on climate, the formation of just 1 centimeter (0.4 inch) can take from 15 years to hundreds of years.

Soil is the base of life on land because it provides most of the nutrients needed for plant growth. Indeed, you are mostly soil nutrients imported into your body by the food you eat. Soil is also the earth's primary filter that cleanses water as it passes through. It also helps decompose and recycle biodegradable wastes

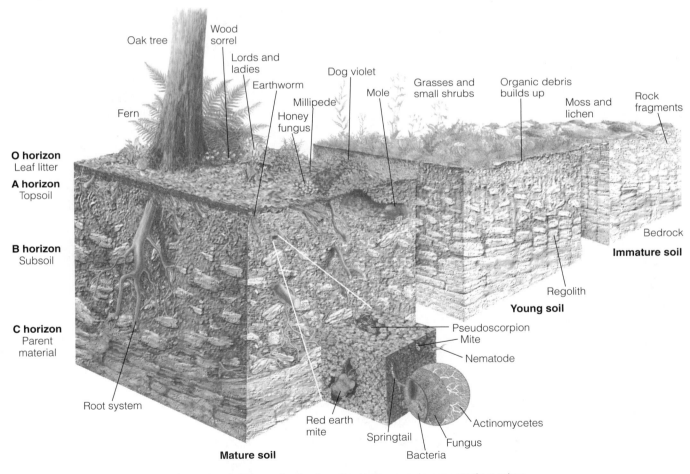

Figure 2-21 Natural capital: soil formation and generalized soil profile. Horizons, or layers, vary in number, composition, and thickness, depending on the type of soil. Soil is the base of life that provides the food you need to stay alive and healthy. (Used by permission of Macmillan Publishing Company from Derek Elsom, *Earth*, New York: Macmillan, 1992. Copyright © 1992 by Marshall Editions Developments Limited)

and is major component of the earth's water recycling and water storage processes.

Yet since the beginning of agriculture, human activities have led to rapid soil erosion, which can convert this renewable resource into a nonrenewable resource. Entire civilizations have collapsed because they mismanaged the topsoil that supported their populations.

What Major Layers Are Found in Mature Soils? Layers Count

Most soils developed over a long time consist of several layers containing different materials.

Mature soils, or soils that have developed over a long time, are arranged in a series of horizontal layers called **soil horizons,** each with a distinct texture and composition that varies with different types of soils. A cross-sectional view of the horizons in a soil is called a **soil profile.** Most mature soils have at least three of the possible horizons (Figure 2-21). Think of them as floors in the building of life underneath your feet.

The top layer is the *surface litter layer,* or *O horizon.* It consists mostly of freshly fallen undecomposed or partially decomposed leaves, twigs, crop wastes, animal waste, fungi, and other organic materials. Normally, it is brown or black.

The *topsoil layer,* or *A horizon,* is a porous mixture of partially decomposed organic matter, called **humus,** and some inorganic mineral particles. It is usually darker and looser than deeper layers. A fertile soil that produces high crop yields has a thick topsoil layer with lots of humus. This helps topsoil hold water and nutrients taken up by plant roots.

The roots of most plants and most of a soil's organic matter are concentrated in a soil's two upper layers. As long as vegetation anchors these layers, soil stores water and releases it in a nourishing trickle.

The two top layers of most well developed soils teem with bacteria, fungi, earthworms, and small insects that interact in complex food webs. Trace the flows and interactions in the food web in Figure 2-22. Bacteria and other decomposer microorganisms found

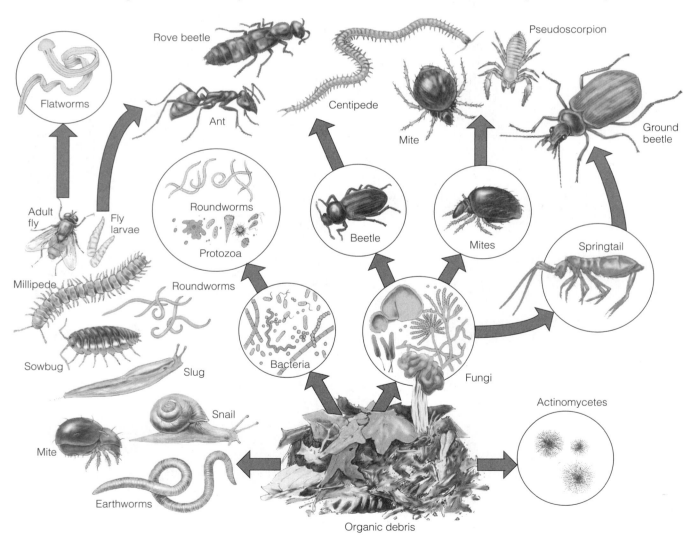

Figure 2-22 Natural capital: greatly simplified food web of living organisms found in soil.

by the billions in every handful of topsoil break down some of its complex organic compounds into simpler inorganic compounds soluble in water. Soil moisture carrying these dissolved nutrients is drawn up by the roots of plants and transported through stems and into leaves as part of the earth's chemical cycling processes.

The color of its topsoil tells us a lot about how useful a soil is for growing crops. Dark-brown or black topsoil is nitrogen-rich and high in organic matter. Gray, bright yellow, or red topsoils are low in organic matter and need nitrogen enrichment to support most crops. Pick up a handful of soil and look at its color. What did you learn?

The spaces, or pores, between the solid organic and inorganic particles in the upper and lower soil layers contain varying amounts of air (mostly nitrogen and oxygen gas) and water. Plant roots need the oxygen for cellular respiration.

Some of the precipitation that reaches the soil percolates through the soil layers and occupies many of the soil's open spaces or pores. This downward movement of water through soil is called **infiltration.** As the water seeps down, it dissolves various minerals and organic matter in upper layers and carries them to lower layers in a process called **leaching.**

Most of the world's crops are grown on soils exposed when grasslands and deciduous forests are cleared. Five important soil types, each with a distinct profile, are shown in Figure 2-23.

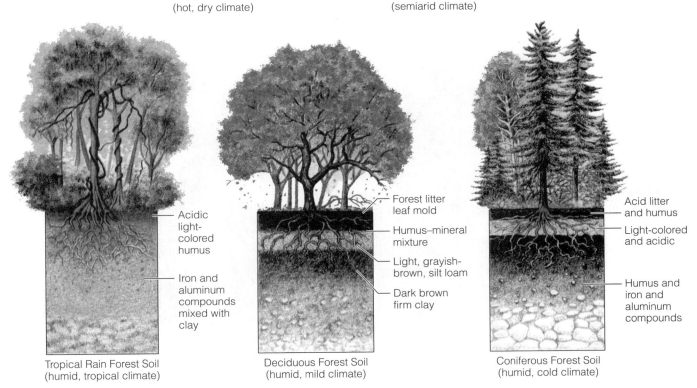

Mosaic of closely packed pebbles, boulders

Weak humus–mineral mixture

Dry, brown to reddish-brown with variable accumulations of clay, calcium carbonate, and soluble salts

Desert Soil (hot, dry climate)

Alkaline, dark, and rich in humus

Clay, calcium compounds

Grassland Soil (semiarid climate)

Acidic light-colored humus

Iron and aluminum compounds mixed with clay

Tropical Rain Forest Soil (humid, tropical climate)

Forest litter leaf mold

Humus–mineral mixture

Light, grayish-brown, silt loam

Dark brown firm clay

Deciduous Forest Soil (humid, mild climate)

Acid litter and humus

Light-colored and acidic

Humus and iron and aluminum compounds

Coniferous Forest Soil (humid, cold climate)

Figure 2-23 Natural capital: soil profiles of the principal soil types typically found in five types of terrestrial ecosystems.

Soils vary in their content of *clay* (very fine particles), *silt* (fine particles), *sand* (medium-size particles), and *gravel* (coarse to very coarse particles). The relative amounts of the different sizes and types of these mineral particles determine **soil texture.**

To get an idea of a soil's texture, take a small amount of topsoil, moisten it, and rub it between your fingers and thumb. A gritty feel means it contains a lot of sand. A sticky feel means a high clay content, and you should be able to roll it into a clump. Silt-laden soil feels smooth, like flour. A loam topsoil, best suited for plant growth, has a texture between these extremes—a crumbly, spongy feeling—with many of its particles clumped loosely together.

A numerical scale of **pH** values is used to compare the acidity and alkalinity in water solutions (Figure 2-24). The pH of a soil influences the uptake of soil nutrients by plants.

2-6 MATTER CYCLING IN ECOSYSTEMS

What Are Biogeochemical Cycles? Going in Circles

Global cycles recycle nutrients through the earth's air, land, water, and living organisms and, in the process, connect past, present, and future forms of life.

All organisms are interconnected by vast global recycling systems known as **nutrient cycles,** or **biogeochemical cycles** (literally, life–earth–chemical cycles). In these cycles, nutrient atoms, ions, and molecules that organisms need to live, grow, and reproduce are continuously cycled between air, water, soil, rock, and living organisms. These cycles, driven directly or indirectly by incoming solar energy and gravity, include the carbon, oxygen, nitrogen, phosphorus, and hydrologic (water) cycles (Figure 2-8).

The earth's chemical cycles also connect past, present, and future forms of life. Some of the carbon atoms in your skin may once have been part of a leaf, a dinosaur's skin, or a layer of limestone rock. Your grandmother, Plato, or a hunter–gatherer who lived 25,000 years ago may have inhaled some of the oxygen molecules you just inhaled.

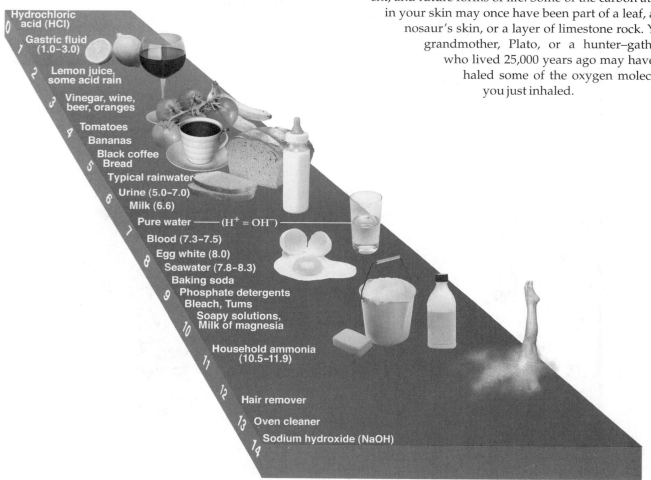

0 Hydrochloric acid (HCl)
1 Gastric fluid (1.0–3.0)
2 Lemon juice, some acid rain
3 Vinegar, wine, beer, oranges
4 Tomatoes
   Bananas
   Black coffee
5 Bread
   Typical rainwater
   Urine (5.0–7.0)
6 Milk (6.6)
7 Pure water —— ($H^+ = OH^-$) ——
   Blood (7.3–7.5)
8 Egg white (8.0)
   Seawater (7.8–8.3)
   Baking soda
9 Phosphate detergents
   Bleach, Tums
   Soapy solutions,
10 Milk of magnesia
   Household ammonia
11 (10.5–11.9)
12 Hair remover
13 Oven cleaner
14 Sodium hydroxide (NaOH)

Figure 2-24 The pH scale, used to measure acidity and alkalinity of water solutions. Values shown are approximate. A solution with a pH less than 7 is acidic, a neutral solution has a pH of 7, and one with a pH greater than 7 is basic. Each whole-number drop in pH represents a tenfold increase in acidity. (Modified by permission from Cecie Starr, *Biology: Concepts and Applications*, 4th ed., Pacific Grove, Calif.: Brooks/Cole, 2000)

How Is Water Cycled in the Biosphere? The Water Cycle

A vast global cycle collects, purifies, distributes, and recycles the earth's fixed supply of water.

The **hydrologic cycle,** or **water cycle,** which collects, purifies, and distributes the earth's fixed supply of water, is shown in Figure 2-25. Trace the flows and paths in this diagram.

Solar energy evaporates water found on the earth's surface into the atmosphere. Some of this water returns to the earth as rain or snow, passes through living organisms, flows into bodies of water, and eventually is evaporated again to continue the cycle.

Some of the fresh water returning to the earth's surface as precipitation in this cycle becomes locked in glaciers. But most precipitation falling on terrestrial ecosystems becomes *surface runoff*. This water flows into streams and lakes, which eventually carry water back to the oceans, where it can be evaporated to cycle again.

Besides replenishing streams and lakes, surface runoff also causes soil erosion, which moves soil and weathered rock fragments from one place to another. Water is thus the primary sculptor of the earth's landscape. Because water dissolves many nutrient compounds, it is a major medium for transporting nutrients within and between ecosystems.

Throughout the hydrologic cycle, many natural processes act to purify water. Evaporation and subsequent precipitation act as a natural distillation process that removes impurities dissolved in water. Water flowing above ground through streams and lakes and below ground in *aquifers* is naturally filtered and purified by chemical and biological processes. Thus, *the hydrologic cycle also can be viewed as a cycle of natural renewal of water quality.*

How Are Human Activities Affecting the Water Cycle? Messing with Nature

We alter the water cycle by withdrawing large amounts of fresh water, clearing vegetation, eroding soils, and polluting surface and underground water.

During the past 100 years, we have been intervening in the earth's current water cycle in three major ways. *First,* we withdraw large quantities of fresh water from streams, lakes, and underground sources.

Second, we clear vegetation from land for agriculture, mining, road and building construction, and other activities and sometimes cover the land with buildings, concrete, or asphalt. This increases runoff, reduces infiltration that recharges groundwater supplies, increases the risk of flooding, and accelerates soil erosion and landslides. We also increase flooding by destroying wetlands, which act like sponges to absorb

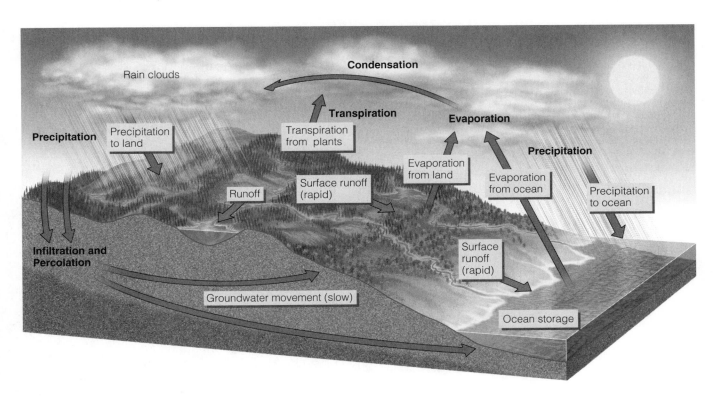

 Figure 2-25 Natural capital: simplified model of the *hydrologic cycle.*

and hold overflows of water. *Third,* we modify water quality by adding nutrients (such as phosphates and nitrates found in fertilizers) and other pollutants.

How Is Carbon Cycled in the Biosphere? Carbon Dioxide in Action

Carbon, the basic building block element of organic compounds, recycles through the earth's air, water, soil, and living organisms.

Carbon, the basic building block of the carbohydrates, fats, proteins, DNA, and other organic compounds necessary for life, is circulated through the biosphere by the **carbon cycle,** as shown in Figure 2-26. Trace the flows and paths in this diagram.

The carbon cycle is based on carbon dioxide (CO_2) gas, which makes up 0.038% of the volume of the troposphere and is also dissolved in water. Carbon dioxide is a key component of nature's thermostat. If the carbon cycle removes too much CO_2 from the atmosphere, the atmosphere will cool; if the cycle generates too much, the atmosphere will get warmer. Thus, even slight changes in the carbon cycle can affect climate and ultimately the types of life that can exist on various parts of the planet.

Terrestrial producers remove CO_2 from the atmosphere, and aquatic producers remove it from the water. They then use photosynthesis to convert CO_2 into complex carbohydrates such as glucose ($C_6H_{12}O_6$).

The cells in oxygen-consuming producers, consumers, and decomposers then carry out aerobic respiration. This process breaks down glucose and other complex organic compounds and converts the carbon back to CO_2 in the atmosphere or water for reuse by producers. This linkage between *photosynthesis* in producers and *aerobic respiration* in producers, consumers, and decomposers circulates carbon in the biosphere and is a major part of the global carbon cycle. Oxygen and hydrogen, the other elements in carbohydrates, cycle almost in step with carbon.

Some carbon atoms take a long time to recycle. Over millions of years, buried deposits of dead plant matter and bacteria are compressed between layers of sediment, where they form carbon-containing *fossil fuels* such as coal and oil (Figure 2-26). This carbon is not released to the atmosphere as CO_2 for recycling until these fuels are extracted and burned, or until long-term geological processes expose these deposits to air. In only a few hundred years, we have extracted and burned fossil fuels that took millions of years to form.

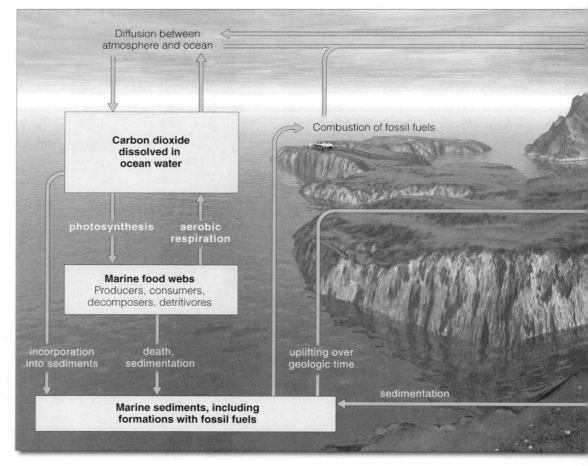

Figure 2-26
Natural capital: simplified model of the global *carbon cycle* that helps keep you alive. The left portion shows the movement of carbon through marine ecosystems, and the right portion shows its movement through terrestrial ecosystems. Carbon reservoirs are shown as boxes; processes that change one form of carbon to another are shown in unboxed print. (From Cecie Starr, *Biology: Concepts and Applications,* 4th ed., Brooks/Cole, [Wadsworth], © 2000)

This is why fossil fuels are nonrenewable resources on a human time scale.

How Are Human Activities Affecting the Carbon Cycle? Messing with Nature's Thermostat

Burning fossil fuels and clearing photosynthesizing vegetation faster than it is replaced can increase the average temperature of the atmosphere by adding excess carbon dioxide to the atmosphere.

Since 1800, and especially since 1950, we have been intervening in the earth's carbon cycle in two ways that add carbon dioxide to the atmosphere: *First,* in some areas we clear trees and other plants that absorb CO_2 through photosynthesis faster than they can grow back. *Second,* we add large amounts of CO_2 by burning fossil fuels and wood.

Computer models of the earth's climate systems suggest that increased concentrations of atmospheric CO_2 and other gases we are adding to the troposphere could enhance the planet's *natural greenhouse effect* that helps warm the lower atmosphere (troposphere) and the earth's surface (Figure 2-10). The resulting *global warming* could disrupt global food production and wildlife habitats, alter temperature and precipitation patterns, and raise the average sea level in various parts of the world.

How Is Nitrogen Cycled in the Biosphere? Bacteria in Action

Different types of bacteria help recycle nitrogen through the earth's air, water, soil, and living organisms.

Nitrogen is the atmosphere's most abundant element, with chemically unreactive nitrogen gas (N_2) making up 78% of the volume of the troposphere. However, N_2 cannot be absorbed and used directly as a nutrient by multicellular plants or animals.

Fortunately, atmospheric electrical discharges in the form of lightning and certain types of bacteria in aquatic systems, in the soil, and in the roots of some plants can convert N_2 into compounds useful as nutrients for plants and animals as part of the **nitrogen cycle,** depicted in Figure 2-27 (p. 42). Trace the flows and paths in this diagram.

The nitrogen cycle consists of several major steps. In *nitrogen fixation,* specialized bacteria convert gaseous nitrogen (N_2) to ammonia (NH_3) that can be used by plants.

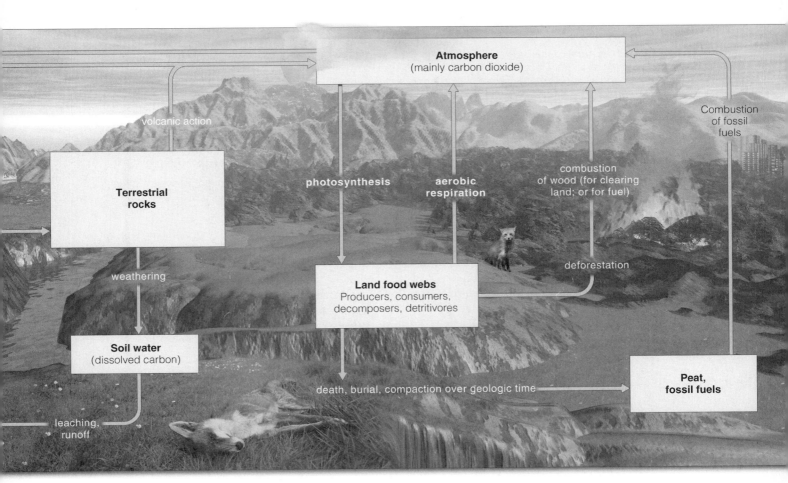

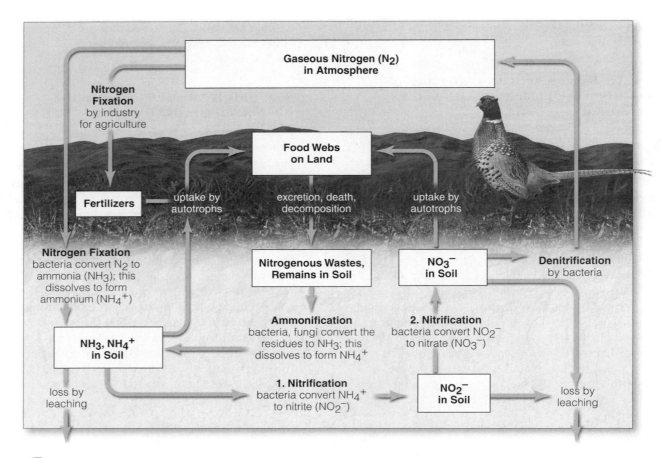

Figure 2-27 Natural capital: simplified model of the *nitrogen cycle* in a terrestrial ecosystem. Nitrogen reservoirs are shown as boxes; processes changing one form of nitrogen to another are shown in circles. This cycle helps keep you alive. (Adapted from Cecie Starr and Ralph Taggart, *Biology: The Unity and Diversity of Life*, 9th ed., Wadsworth, © 2001)

Ammonia not taken up by plants may undergo *nitrification*. In this process, specialized aerobic bacteria convert most of the ammonia in soil to *nitrite ions* (NO_2^-), which are toxic to plants, and *nitrate ions* (NO_3^-), which are easily taken up by plants as a nutrient. Animals, in turn, get their nitrogen by eating plants or plant-eating animals.

Plants and animals return nitrogen-rich organic compounds to the environment as wastes, cast-off particles, and dead bodies. In the *ammonification* step, vast armies of specialized decomposer bacteria convert this detritus into simpler nitrogen-containing inorganic compounds such as ammonia and water-soluble salts containing ammonium ions (NH_2^+).

Nitrogen leaves the soil in the *denitrification* step in which other specialized bacteria in waterlogged soil and in the bottom sediments of lakes, oceans, swamps, and bogs convert NH_3 and NH_4^+ back into nitrite and nitrate ions and then into nitrogen gas (N_2) and nitrous oxide gas (N_2O). These gases are released to the troposphere to begin the cycle again.

How Are Human Activities Affecting the Nitrogen Cycle? Altering Nature

We add large amounts of various nitrogen-containing compounds to the earth's air and water and remove nitrogen from the soil.

We intervene in the nitrogen cycle in several ways. *First,* we add large amounts of nitric oxide (NO) into the troposphere when we burn any fuel. In the atmosphere, this gas can be converted to nitrogen dioxide gas (NO_2) and nitric acid (HNO_3), which can return to the earth's surface as damaging *acid deposition,* commonly called *acid rain*—more on this in Chapter 11.

Second, we add nitrous oxide (N_2O) to the troposphere through the action of anaerobic bacteria on livestock wastes and commercial inorganic fertilizers applied to the soil. This gas can warm the troposphere and deplete ozone in the stratosphere.

Third, we release large quantities of nitrogen stored in soils and plants as gaseous compounds into the troposphere through destruction of forests, grasslands, and wetlands.

Fourth, we upset aquatic ecosystems by adding excess nitrates in agricultural runoff and discharges from municipal sewage systems—more on this in Chapter 9.

Fifth, we remove nitrogen from topsoil when we harvest nitrogen-rich crops, irrigate crops, and burn or clear grasslands and forests before planting crops.

How Is Phosphorus Cycled in the Biosphere? Slow Cycling without Using the Atmosphere

Phosphorus cycles fairly slowly through the earth's water, soil, and living organisms.

Phosphorus circulates through water, the earth's crust, and living organisms in the **phosphorus cycle,** depicted in Figure 2-28. Bacteria are less important here than in the nitrogen cycle. Trace the flows and paths in this diagram. Unlike carbon and nitrogen, very little phosphorus circulates in the atmosphere because soil conditions do not allow bacteria to convert chemical forms of phosphorus to gaseous forms of phosphorus. The phosphorus cycle is slow, and on a short human time scale much phosphorus flows one way from the land to the oceans.

Phosphorous typically is found as phosphate salts containing phosphate ions (PO_4^{3-}) in terrestrial rock formations and ocean bottom sediments. Phosphate can be lost from the cycle for long periods when it washes from the land into streams and rivers and is carried to the ocean. There it can be deposited as sediment on the sea floor and remain for millions of years. Someday geological uplift processes can expose these seafloor deposits from which phosphate can be eroded to start the cyclical process again.

Because most soils contain little phosphate, it is often the *limiting factor* for plant growth on land unless phosphorus (as phosphate salts mined from the earth) is applied to the soil as a fertilizer. Phosphorus also limits the growth of producer populations in many freshwater streams and lakes because phosphate salts are only slightly soluble in water.

How Are Human Activities Affecting the Phosphorus Cycle? More Messing with Nature

We remove large amounts of phosphate from the earth to make fertilizer, reduce phosphorus in tropical soils by clearing forests, and add excess phosphates to aquatic systems.

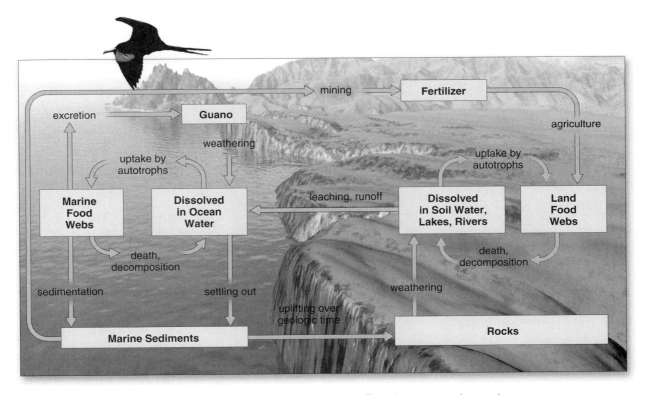

Figure 2-28 Natural capital: simplified model of the *phosphorus cycle.* Phosphorus reservoirs are shown as boxes; processes that change one form of phosphorus to another are shown in unboxed print. This cycle helps keep you alive. (From Cecie Starr and Ralph Taggart, *Biology: The Unity and Diversity of Life,* 9th ed., Wadsworth © 2001)

We intervene in the earth's phosphorus cycle in three ways. *First,* we mine large quantities of phosphate rock to make commercial inorganic fertilizers and detergents. *Second,* we reduce the available phosphate in tropical soils when we cut down tropical forests.

Third, we disrupt aquatic systems with phosphates from runoff of animal wastes and fertilizers and discharges from sewage treatment systems—more on this in Chapter 9.

What Is the Rock Cycle? Really Slow Recycling

The earth's three types of rock are transformed and recycled by a rock cycle that is the planet's slowest cyclical process.

Rock is any material that makes up a large, natural, continuous part of the earth's crust. Based on the way it forms, rock is placed in three broad classes. One is **igneous rock,** formed below or on the earth's surface when molten rock material (magma) wells up from the earth's upper mantle or deep crust, cools, and hardens into rock. Examples are granite (formed underground) and lava rock (formed above ground when molten lava cools and hardens).

A second type is **sedimentary rock,** formed from sediment when existing rocks are weathered and eroded into small pieces and carried by wind and water to be deposited in a lake or river. As deposited layers from weathering and erosion become buried and compacted, the resulting pressure causes their particles to bond together to form sedimentary rocks such as sandstone and shale.

The third type is **metamorphic rock,** produced when an existing rock is subjected to high temperatures (which may cause it to melt partially), high pressures, chemically active fluids, or a combination of these agents. Examples are anthracite (a form of coal), slate, and marble.

Rocks are constantly exposed to various physical and chemical conditions that can change them over time. The interaction of processes that change rocks from one type to another is called the **rock cycle** (Figure 2-29).

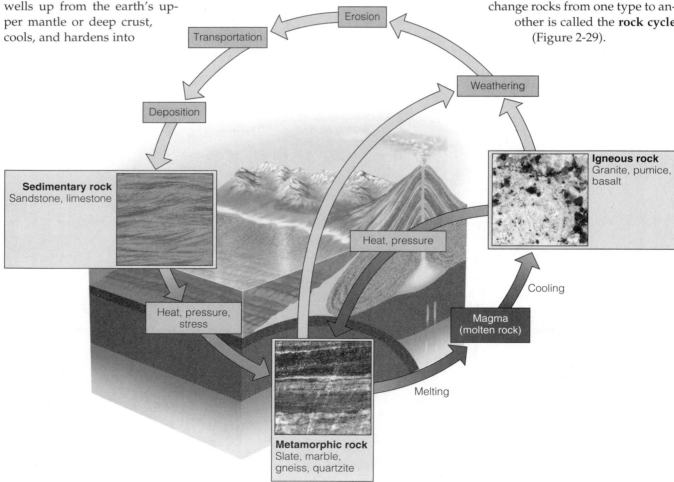

Figure 2-29 Natural capital: the *rock cycle,* the slowest of the earth's cyclic processes. The earth's materials are recycled over millions of years by three processes: *melting, erosion,* and *metamorphism,* which produce *igneous, sedimentary,* and *metamorphic* rocks. Rock of any of the three classes can be converted to rock of either of the other two classes or can even be recycled within its own class.

The rock cycle recycles the earth's three types of rocks over millions of years and is the slowest of the earth's cyclic processes. It concentrates the planet's nonrenewable mineral resources on which we depend. Without the incredibly slow rock cycle, you would not exist.

All things come from earth, and to earth they all return.

MENANDER (342–290 B.C.)

CRITICAL THINKING

1. Respond to the following statements:
   a. Scientists have not absolutely proven that anyone has ever died from smoking cigarettes.
   b. The greenhouse theory—that certain gases (such as water vapor and carbon dioxide) warm the troposphere—is not a reliable idea because it is only a scientific theory.

2. See whether you can find an advertisement or article describing some aspect of science in which **(a)** the concept of scientific proof is misused, **(b)** the term *theory* is used when it should have been *hypothesis,* and **(c)** a sound or consensus scientific finding is dismissed or downplayed because it is "only a theory."

3. A tree grows and increases its mass. Explain why this is not a violation of the law of conservation of matter.

4. If there is no "away," why is the world not filled with waste matter?

5. Use the second law of thermodynamics to explain why a barrel of oil can be used only once as a fuel.

6. **(a)** A bumper sticker asks, "Have you thanked a green plant today?" Give two reasons for appreciating a green plant. **(b)** Trace the sources of the materials that make up the bumper sticker, and decide whether the sticker itself is a sound application of the slogan. **(c)** Explain how decomposers help keep you alive.

7. Using the second law of thermodynamics, explain why **(a)** there is such a sharp decrease in usable energy as energy flows through a food chain or web and **(b)** many poor people in developing countries live on a mostly vegetarian diet.

8. What would happen to an ecosystem if **(a)** all its decomposers and detritus feeders were eliminated or **(b)** all its producers were eliminated?

LEARNING ONLINE

The website for this book contains study aids and many ideas for further reading and research. They include a chapter summary, review questions for the entire chapter, flash cards for key terms and concepts, a multiple-choice practice quiz, interesting Internet sites, references, and a guide for accessing thousands of InfoTrac® College Edition articles Log on to

http://biology.brookscole.com/miller7

Then click on the Chapter-by-Chapter area, choose Chapter 2, and select a learning resource.

3 EVOLUTION AND BIODIVERSITY

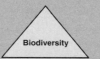

In looking at nature . . . never forget that every single organic being around us may be said to be striving to increase its numbers.

CHARLES DARWIN, 1859

3-1 EVOLUTION AND ADAPTATION

What Is Evolution? Changing Genes

Evolution is the change in a population's genetic makeup over time.

According to scientific evidence, populations of organisms adapt to changes in environmental conditions through **biological evolution,** or **evolution.** It is the change in a population's genetic makeup through successive generations. Note that *populations, not individuals, evolve by becoming genetically different.*

According to the **theory of evolution,** all species descended from earlier, ancestral species. In other words, life comes from life. This widely accepted scientific theory explains how life has changed over the past 3.7 billion years (Figure 3-1) and why life is so diverse today. Study this figure carefully. Religious and other groups offer other explanations, but this is the accepted scientific explanation.

Biologists use the term **microevolution** to describe the small genetic changes that occur in a population. They use the term **macroevolution** to describe long-term, large-scale evolutionary changes through which new species form from ancestral species and other species are lost through extinction.

How Does Microevolution Work? Changes in the Gene Pool

A population's gene pool changes over time when beneficial changes or mutations in its DNA molecules are passed on to offspring.

A population's **gene pool** consists of all of the genes (Figure 2-3, p. 20) in its individuals. *Microevolution* is a change in a population's gene pool over time.

The first step in microevolution is the development of *genetic variability* in a population. Genetic variability in a population originates through **mutations:** random changes in the structure or number of DNA molecules in a cell that can be inherited by offspring. Mutations can occur in two major ways. One is by ex-posure of DNA to external agents such as radioactivity, X rays, and natural and human-made chemicals (called *mutagens*). The other is the result of random mistakes that sometimes occur in coded genetic instructions when DNA molecules are copied each time a cell divides and whenever an organism reproduces.

Mutations can occur in any cells, but only those in reproductive cells are passed on to offspring. Some mutations are harmless but most are lethal. *Every so often, a mutation is beneficial.* The result is new genetic traits that give that individual and its offspring better chances for survival and reproduction under existing environmental conditions or when such conditions change.

Mutations are random and unpredictable, are a source of totally new genetic raw material, and are rare events. Thus, life is a genetic shuffle. In addition to mutations, another source of evolutionary change is the trading of genes between bacteria.

What Role Does Natural Selection Play in Microevolution? Backing Reproductive Winners

Some members of a population may have genetic traits that enhance their ability to survive and produce offspring with such traits.

Natural selection occurs when some individuals of a population have genetically based traits that increase their chances of survival and their ability to produce offspring with the same traits. Three conditions are necessary for evolution of a population by natural selection to occur.

First, there must be *genetic variability* for a trait in a population. *Second,* the trait must be *heritable,* meaning it can be passed from one generation to another. *Third,* the trait must somehow lead to **differential reproduction.** This means it must enable individuals with the trait to leave more offspring than other members of the population.

An **adaptation,** or **adaptive trait,** is any heritable trait that enables organisms to survive and reproduce better under prevailing environmental conditions. *Structural adaptations* include *coloration* (allowing more individuals to hide from predators or to sneak up on prey), *mimicry* (looking like a poisonous or dangerous species), *protective cover* (shell, thick skin, bark, or thorns), and *gripping mechanisms* (hands with opposable thumbs). *Physiological adaptations* include the ability to

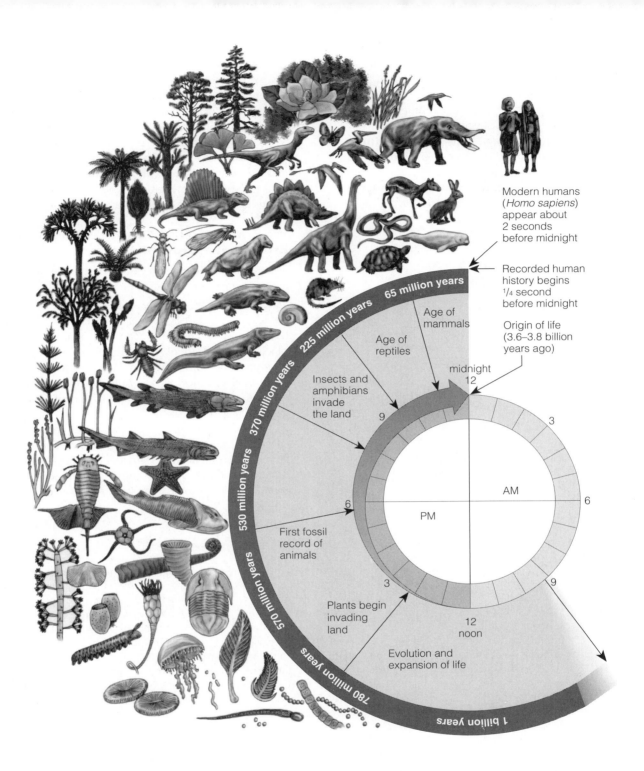

Figure 3-1 Natural capital: greatly simplified overview of the biological evolution of life on the earth, which was preceded by about 1 billion years of chemical evolution. Evidence indicates that microorganisms (mostly bacteria) that lived in water dominated the early span of biological evolution on the earth, between about 3.7 billion and 1 billion years ago. Plants and animals evolved first in the seas. Fossil and recent DNA evidence suggests that plants began invading the land about 780 million years ago, and animals began invading the land about 370 million years ago. Humans arrived on the scene only a very short time ago. We have been around for less than an eye blink of the earth's roughly 3.7-billion-year history of biological evolution. Although we are newcomers, we have taken over much of the planet and now have the power to cause the premature extinction of many of the other species that travel with us as passengers on the earth as it hurtles through space.

hibernate during cold weather and give off chemicals that poison or repel prey. *Behavioral adaptations* include the ability to fly to a warmer climate during winter and various interactions between species.

When faced with a change in environmental conditions, a population of a species has three possibilities: adapt to the new conditions through natural selection, migrate (if possible) to an area with more favorable conditions, or become extinct.

The process of microevolution can be summarized as follows: *Genes mutate, individuals are selected, and populations evolve.*

3-2 ECOLOGICAL NICHES AND ADAPTATION

What Is an Ecological Niche? How Species Coexist

Each species in an ecosystem has a specific role or way of life.

If asked what role a certain species such as an alligator plays in an ecosystem, an ecologist would describe its **ecological niche,** or simply **niche** (pronounced "nitch"). It is a species' way of life or role in an ecosystem and includes everything that affects its survival and reproduction.

A species' ecological niche includes the *adaptations* or *adaptive traits* its members have acquired through evolution. It also includes its range of tolerance for various physical and chemical conditions, such as temperature, the types and amounts of resources it uses, how it interacts with other living and nonliving components of the ecosystems in which it is found, and the role it plays in the energy flow and matter cycling in an ecosystem (Figure 2-15, p. 31).

The ecological niche of a species is different from its **habitat,** the physical location, where its populations and individuals live. Ecologists often say that a niche is like a species' occupation, whereas habitat is like its address.

What Are Generalist and Specialist Species? Broad and Narrow Niches

Some species have broad ecological roles and others have narrower or more specialized roles.

Scientists use the niches of species to classify them broadly as *generalists* or *specialists*. **Generalist species** have broad niches. They can live in many different places, eat a variety of foods, and tolerate a wide range of environmental conditions. Flies, cockroaches (Spotlight, below), mice, rats, white-tailed deer, raccoons, coyotes, copperheads, channel catfish, and humans are generalist species.

For example, *tiger salamanders* are specialists because they can breed only in fishless ponds where their larvae will not be eaten. Threatened *red-cockaded woodpeckers* carve nest holes almost exclusively in old (at least 75 years) longleaf pines. China's highly endan-

---

Spotlight Cockroaches: Nature's Ultimate Survivors

SPOTLIGHT

Cockroaches, the bugs many people love to hate, have been around for about 350 million years and are one of the great success stories of evolution. They are so successful because they are *generalists.*

The earth's 4,000 cockroach species can eat almost anything, including algae, dead insects, fingernail clippings, salts in tennis shoes, electrical cords, glue, paper, and soap. They can also live and breed almost anywhere except in polar regions.

Some cockroach species can go for months without food, survive for a month on a drop of water from a dishrag, and withstand massive doses of radiation. One

species can survive being frozen for 48 hours.

They usually can evade their predators and a human foot in hot pursuit because most species have antennae that can detect minute movements of air, vibration sensors in their knee joints, and rapid response times (faster than you can blink). Some even have wings.

They also have high reproductive rates. In only a year, a single Asian cockroach (especially prevalent in Florida) and its young can add about 10 million new cockroaches to the world. Their high reproductive rate also helps them quickly develop genetic resistance to almost any poison we throw at them.

Most cockroaches also sample food before it enters their mouths

and learn to shun foul-tasting poisons. They also clean up after themselves by eating their own dead and, if food is scarce enough, their living.

Only about 25 species of cockroach live in homes. However, such species can carry viruses and bacteria that cause diseases such as hepatitis, polio, typhoid fever, plague, and salmonella. They can also cause people to have allergic reactions ranging from watery eyes to severe wheezing. About 60% of Americans suffering from asthma are allergic to live or dead cockroaches.

Critical Thinking

If you could, would you exterminate all cockroach species? What might be some ecological consequences of doing this?

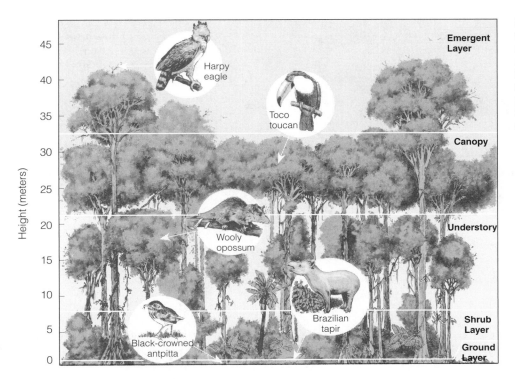

Height (meters)

45
40
35
30
25
20
15
10
5
0

Harpy eagle

Toco toucan

Wooly opossum

Brazilian tapir

Black-crowned antpitta

Emergent Layer

Canopy

Understory

Shrub Layer

Ground Layer

Figure 3-2 Natural capital: stratification of specialized plant and animal niches in various layers of a *tropical rain forest*. The presence of these specialized niches enables species to avoid or minimize competition for resources and results in the coexistence of a great variety of species (biodiversity).

gered *giant pandas* feed almost exclusively on various types of bamboo.

In a tropical rain forest, an incredibly diverse array of species survives by occupying specialized ecological niches in various distinct layers of vegetation exposed to different levels of light (Figure 3-2). Study this figure carefully. The widespread clearing and degradation of such forests is dooming millions of such specialized species to premature extinction.

Is it better to be a generalist than a specialist? It depends. When environmental conditions are fairly constant, as in a tropical rain forest, specialists have an advantage because they have fewer competitors. But under rapidly changing environmental conditions, the generalist usually is better off than the specialist.

What Limits Adaptation? Life Is a Genetic Dice Game.

A population's ability to adapt to new environmental conditions is limited by its gene pool and how fast it can reproduce.

Will adaptations in the not to distant future allow our skin to become more resistant to the harmful effects of ultraviolet radiation, our lungs to cope with air pollutants, and our liver to better detoxify pollutants?

The answer is *no* because of the two limits to adaptations in nature: *First,* a change in environmental conditions can lead to adaptation only for genetic traits already present in the gene pool of a population. You must have genetic dice to play the genetic dice game.

Second, even if a beneficial heritable trait is present in a population, the population's ability to adapt may be limited by its reproductive capacity. Populations of genetically diverse species that reproduce quickly—such as weeds, mosquitoes, rats, bacteria, or cockroaches—often adapt to a change in environmental conditions in a short time. In contrast, populations of species that cannot produce large numbers of offspring rapidly—such as elephants, tigers, sharks, and humans—take a long time (typically thousands or even millions of years) to adapt through natural selection. You have to be able to throw the genetic dice fast.

Here is some *bad news* for most members of a population. Even when a favorable genetic trait is present in a population, most of the population would have to die or become sterile so individuals with the trait could predominate and pass the trait on. Thus, most players get kicked out of the genetic dice game before they can win. This means that most members of the human population would have to die for hundreds of thousands of generations for a new genetic trait to emerge. This is hardly a desirable solution to the environmental problems we face.

What Are Two Common Misconceptions about Evolution? Strong Does Not Cut It and There Is No Grand Design.

Evolution is about leaving the most descendants and there is no master plan leading to genetic perfection.

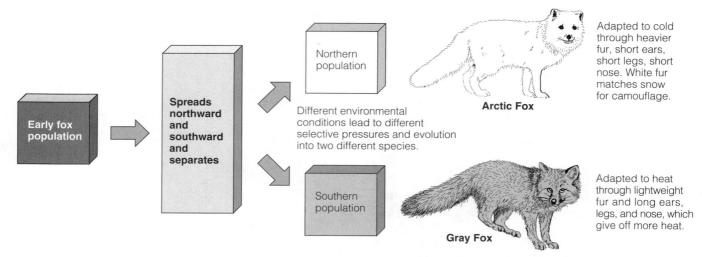

Figure 3-3 How *geographic isolation* can lead to reproductive isolation, divergence, and speciation.

There are two common misconceptions about evolution. One is that the "survival of the fittest" means "survival of the strongest." To biologists, *fitness* is a measure of reproductive success, not strength. Thus, the fittest individuals are those that leave the most descendants.

The other misconception is that evolution involves some grand plan of nature in which species become more perfectly adapted. From a scientific standpoint, no plan or goal of perfection has been identified in the evolutionary process.

3-3 SPECIATION, EXTINCTION, AND BIODIVERSITY

How Do New Species Evolve? Moving Out and Moving On

A new species arises when members of a population are isolated from other members so long that changes in their genetic makeup prevent them from producing fertile offspring if they get together again.

Under certain circumstances natural selection can lead to an entirely new species. In this process, called **speciation,** two species arise from one. For sexually reproducing species, a new species is formed when some members of a population can no longer breed with other members to produce fertile offspring.

The most common mechanism of speciation (especially among animals) takes place in two phases: geographic isolation and reproductive isolation. **Geographic isolation** occurs when different groups of the same population of a species become physically isolated from one another for long periods. For example, part of a population may migrate in search of food and

then begin living in another area with different environmental conditions (Figure 3-3). Populations also may become separated by a physical barrier (such as a mountain range, stream, lake, or road), by a change such as a volcanic eruption or earthquake, or when a few individuals are carried to a new area by wind or water.

The second phase of speciation is **reproductive isolation.** It occurs when mutation and natural selection change operate independently in the gene pools of geographically isolated populations. If this process continues long enough, members of the geographically and reproductively isolated populations of sexually reproducing species may become so different in genetic makeup that if they get together again they cannot produce live, fertile offspring. Then one species has become two, and speciation has occurred.

For some rapidly reproducing organisms, this type of speciation may occur within hundreds of years. For most species, such speciation takes from tens of thousands to millions of years.

What Is Extinction? Going, Going, Gone

A species becomes extinct when its populations cannot adapt to changing environmental conditions.

After speciation, the second process affecting the number and types of species on the earth is **extinction,** in which an entire species ceases to exist on the earth. When environmental conditions change drastically enough, a species must evolve (become better adapted), move to a more favorable area if possible, or become extinct.

Extinction is the ultimate fate of all species, just as death is for all individual organisms. Biologists estimate that 99.9% of all the species that have ever existed are now extinct and humans will not escape this ultimate fate.

As local environmental conditions change, a certain number of species disappear at a low rate, called **background extinction.** Based on the fossil record and analysis of ice cores, biologists estimate that the average annual background extinction rate is one to five species for each million species on the earth.

In contrast, **mass extinction** is a significant rise in extinction rates above the background level. It is a catastrophic, widespread (often global) event (such as climate change) in which large groups of existing species (perhaps 25–70%) are wiped out.

Scientists have also identified periods of **mass depletion** in which extinction rates were much higher than normal but not high enough to classify as a mass extinction. Fossil and geological evidence suggests that there have been five mass extinctions during the past 500 million years. But some recent evidence suggests that there may have been two mass extinctions and three mass depletions.

A mass extinction or mass depletion crisis for some species is an opportunity for other species. The existence of millions of species today means that speciation, on average, has kept ahead of extinction, especially during the last 250 million years (Figure 3-4). Study this figure carefully.

Evidence shows that the earth's mass extinctions and depletions have been followed by periods of recovery called **adaptive radiations** in which numerous new species evolve to fill new or vacated ecological roles or niches in changed environments. Fossil records suggest that it takes 1–10 million years for adaptive radiations to rebuild biological diversity after a mass extinction or depletion.

How Are Human Activities Affecting the Earth's Biodiversity? The Earth Giveth and We Taketh.

The scientific consensus is that human activities are decreasing the earth's biodiversity.

Speciation minus extinction equals *biodiversity,* the planet's genetic raw material for future evolution in response to changing environmental conditions. Extinction is a natural process. But much evidence indicates that humans have become a major force in the premature extinction of species.

According to biologists Stuart Primm and Edward O. Wilson, three independent measures estimate that during the 20th century, extinction rates increased by 100–1,000 times the natural background extinction rate. As human population and resource consumption increase over the next 50–100 years, we are expected to take over more and more of the earth's surface and net primary productivity (NPP) (Figure 2-20, p. 34). According to Wilson and Pimm, this may cause the premature extinction of at least one-fifth of the earth's current species by 2030 and half by the end of this century. This could constitute a new *mass depletion* and possibly a new *mass extinction.* Wilson says that if we make an "all-out effort to save the biologically richest parts of the world, the amount of loss can be cut at least by half."

On our short time scale, such major losses cannot be recouped by formation of new species; it took millions of years after each of the earth's past mass extinctions and depletions for life to recover to the previous level of biodiversity. We are also destroying or degrading ecosystems such as tropical forests, coral reefs, and wetlands that are centers for future speciation. See

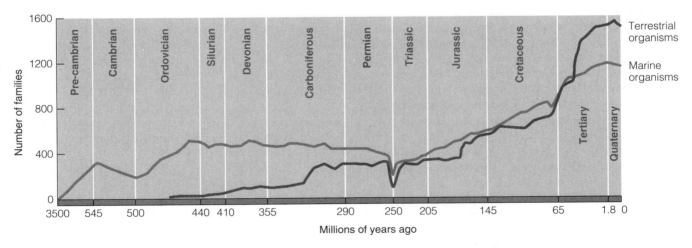

Figure 3-4 Natural capital: changes in the earth's biodiversity over geological time. The biological diversity of life on land and in the oceans has increased dramatically over the last 3.5 billion years, especially during the past 250 million years. Note that during the last 1.8 million years this increase has leveled off. During the next hundred years or so, will the human species be a major factor in decreasing the earth's precious biodiversity?

the Guest Essay on this topic by Norman Myers on the website for this chapter. Genetic engineering cannot stop this loss of biodiversity because genetic engineers rely on natural biodiversity for their genetic raw material.

We can summarize the 3.7-billion-year biological history of the earth in one sentence: *Organisms convert solar energy to food, chemicals cycle, and a variety of species with different biological roles (niches) has evolved in response to changing environmental conditions.*

Each species here today represents a long chain of evolution, and each of these species plays a unique ecological role in the earth's communities and ecosystems. These species, communities, and ecosystems also are essential for future evolution as populations of species continue to adapt to changes in environmental conditions by changing their genetic makeup.

3-4 WHAT IS THE FUTURE OF EVOLUTION?

How Have We Modified Genetic Traits? Getting the Types of Plants or Dogs We Want

We have learned how to selectively breed members of populations and to use genetic engineering to produce plants and animals with certain genetic traits.

We have used **artificial selection** to change the genetic characteristics of populations of a species. In this process, we select one or more desirable genetic traits in the population of a plant or animal, such as a type of wheat, fruit, or dog. Then we use *selective breeding* to end up with populations of the species containing large numbers of individuals with the desired traits.

Artificial selection has yielded food crops with higher yields, cows that give more milk, trees that grow faster, and a variety of different types of dogs and cats. But traditional crossbreeding is a slow process. And it can combine traits only from species that are close to one another genetically.

Now scientists are using genetic engineering to speed up our ability to manipulate genes. **Genetic engineering,** or **gene splicing,** is a set of techniques for isolating, modifying, multiplying, and recombining genes from different organisms. It enables scientists to transfer genes between different species that would never interbreed in nature. For example, genes from a fish species can be put into a tomato or strawberry.

The resulting organisms are called **genetically modified organisms (GMOs)** or **transgenic organisms.** Figure 3-5 outlines the steps involved in developing a genetically modified or transgenic plant. Study this figure carefully.

Compared to traditional crossbreeding, gene splicing takes about half as much time to develop a new crop or animal variety and costs less. Traditional crossbreeding involves mixing the genes of similar types of organisms through breeding. But genetic engineering allows us to transfer traits between different types of organisms.

Scientists have used gene splicing to develop modified crop plants, genetically engineered drugs, and pest-resistant plants. They have also created genetically engineered bacteria to clean up spills of oil and other toxic pollutants.

Genetic engineers have also learned how to produce a *clone,* or genetically identical version, of an individual in a population. Scientists have made clones of individuals in populations of domestic animals such as sheep and cows and may someday be able to clone humans—a possibility that excites some people and horrifies others.

Researchers envision using genetically engineered animals to act as biofactories for producing drugs, vaccines, antibodies, hormones, industrial chemicals such as plastics and detergents, and human body organs. This new field is called *biopharming.*

What Are Some Concerns about the Genetic Revolution? Genetic Wonderland or Genetic Wasteland

Genetic engineering has great promise for improving the human condition, but it is an unpredictable process and raises a number of privacy, ethical, legal, and environmental issues.

The hype about genetic engineering can lead us to believe that its results are controllable and predictable. In reality genetic engineering is messy and unpredictable. Genetic engineers can insert a gene into the nucleus of a cell, but with current technology they do not know whether the cell will incorporate the new gene into its DNA. They also do not know where the new gene will be located in the DNA molecule's structure and what effects this will have on the organism.

Thus, *genetic engineering is a trial and error process* with many failures and unexpected results. Indeed, the average success rate of genetic engineering experiments is only about 1%.

Applying our increasing genetic knowledge is filled with great promise, but it raises some serious ethical and privacy issues. For example, some people have genes that make them more likely to develop certain genetic diseases or disorders. We now have the power to detect these genetic deficiencies, even before birth.

This raises some important issues. If gene therapy is developed for correcting these deficiencies, who will

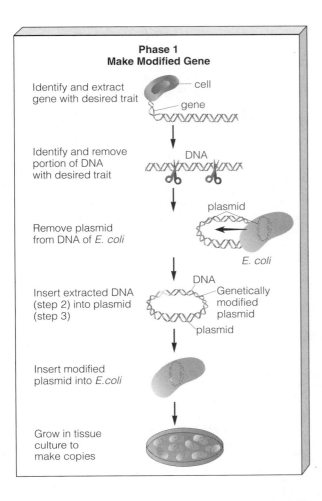

Phase 1
Make Modified Gene

Identify and extract gene with desired trait — cell — gene

Identify and remove portion of DNA with desired trait — DNA

Remove plasmid from DNA of *E. coli* — plasmid — *E. coli*

Insert extracted DNA (step 2) into plasmid (step 3) — DNA — Genetically modified plasmid — plasmid

Insert modified plasmid into *E.coli*

Grow in tissue culture to make copies

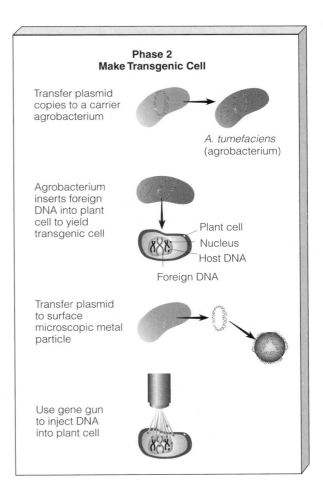

Phase 2
Make Transgenic Cell

Transfer plasmid copies to a carrier agrobacterium — *A. tumefaciens* (agrobacterium)

Agrobacterium inserts foreign DNA into plant cell to yield transgenic cell — Plant cell — Nucleus — Host DNA — Foreign DNA

Transfer plasmid to surface microscopic metal particle

Use gene gun to inject DNA into plant cell

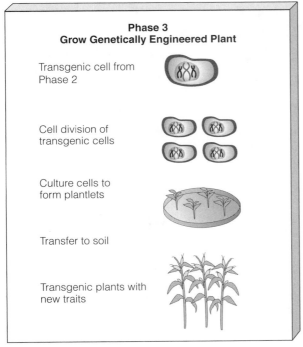

Phase 3
Grow Genetically Engineered Plant

Transgenic cell from Phase 2

Cell division of transgenic cells

Culture cells to form plantlets

Transfer to soil

Transgenic plants with new traits

Figure 3-5 *Genetic engineering.* Steps in genetically modifying a plant.

get it? Will it be mostly for the rich? Will this mean more abortions of genetically defective fetuses? Will health insurers refuse to insure people with certain genetic defects that could lead to health problems? Will employers refuse to hire them?

Some people dream of a day when our genetic prowess could eliminate death and aging altogether. As one's cells, organs, or other parts wear out or are damaged, they would be replaced with new ones. These replacement parts might be grown in genetic engineering laboratories or biopharms. Or people might choose to have a clone available for spare parts.

This raises a number of questions. Is it moral to do this? Who decides? Who regulates this? Will genetically designed humans and clones have the same legal rights as people?

X *HOW WOULD YOU VOTE?* Should we legalize the production of human clones if a reasonably safe technology for doing this becomes available? Cast your vote online at http://biology .brookscole.com/miller7.

What might be the environmental impacts of such genetic developments on resource use, pollution, and environmental degradation? If everyone could live with good health as long as they wanted for a price, sellers of body makeovers would encourage customers to line up. Each of these wealthy, long-lived people could have an enormous ecological footprint for perhaps centuries.

3-5 BIOMES: CLIMATE AND LIFE ON LAND

Why Do Different Organisms Live in Different Places? Think Climate

Different climates lead to different communities of organisms, especially vegetation.

Biologists have classified the terrestrial (land) portion of the biosphere into **biomes** ("BY-ohms"). They are large regions such as forests, deserts, and grasslands characterized by a distinct climate and specific forms of life, especially vegetation, adapted to it.

Why is one area of the earth's land surface a desert, another a grassland, and another a forest? Why do different types of deserts, grasslands, and forests exist?

The general answer to these questions is differences in **climate**: a region's long-term atmospheric conditions—typically over decades. *Average temperature* and *average precipitation* are the two main factors determining a region's climate. Figure 3-6 shows the

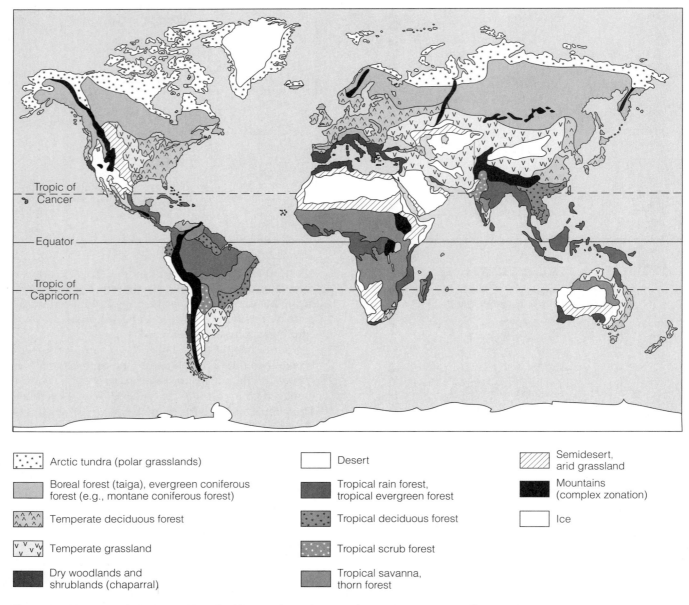

Arctic tundra (polar grasslands)

Boreal forest (taiga), evergreen coniferous forest (e.g., montane coniferous forest)

Temperate deciduous forest

Temperate grassland

Dry woodlands and shrublands (chaparral)

Desert

Tropical rain forest, tropical evergreen forest

Tropical deciduous forest

Tropical scrub forest

Tropical savanna, thorn forest

Semidesert, arid grassland

Mountains (complex zonation)

Ice

Figure 3-6 Natural capital: the earth's major *biomes*—the main types of natural vegetation in different undisturbed land areas—result primarily from differences in climate. Each biome contains many ecosystems whose communities have adapted to differences in climate, soil, and other environmental factors. In reality, people have removed or altered much of this natural vegetation in some areas for farming, livestock grazing, lumber and fuelwood, mining, and construction.

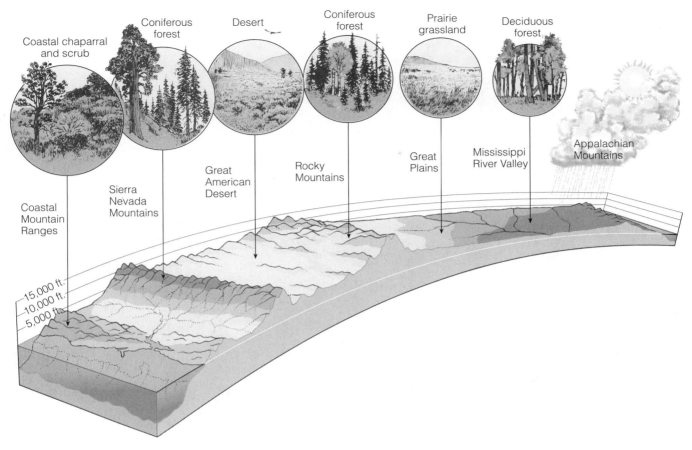

Figure 3-7 Natural capital: major biomes found along the 39th parallel across the United States. The differences reflect changes in climate, mainly differences in average annual precipitation and temperature (not shown).

global distribution of biomes. Study this figure carefully and identify the type of biome you live in. On maps such as the one in Figure 3-6, biomes are presented as having sharp boundaries and being covered with the same general type of vegetation. In reality, *biomes are not uniform.* They consist of a *mosaic of patches,* with somewhat different biological communities but with similarities unique to the biome. These patches occur mostly because the resources plants and animals need are not uniformly distributed. Go to a natural area in or near where you live or and see if you can find patches with different vegetation. Figure 3-7 shows major biomes in the United States as one moves through different climates along the 39th parallel.

For plants, precipitation generally is the limiting factor that determines whether a land area is *desert* (with low precipitation and sparse, widely spaced, mostly low vegetation), *grassland* (with enough precipitation to support grass but not large stands of trees), or *forest* (with enough precipitation to support stands of various tree species and smaller forms of vegetation).

Average annual precipitation and temperature are the most important factors in producing tropical, temperate, or polar deserts, grasslands, and forests (Figure 3-8, p. 56). Study this diagram carefully.

Figure 3-9 (p. 56) shows how climate and vegetation vary with **latitude** (distance from the equator) and **altitude** (elevation above sea level). If you climb a tall mountain from its base to its summit, you can observe changes in plant life similar to those you would encounter in traveling from the equator to the earth's poles.

What Impacts Do Humans Have on Deserts, Grasslands, and Forests?

Human activities are having major environmental impacts on most of the world's deserts, grasslands, and forest.

Figure 3-10 (p. 57) shows some major components and interactions in a temperate desert biome. Study this figure and note the types of species that live in this biome and how they are connected to one another.

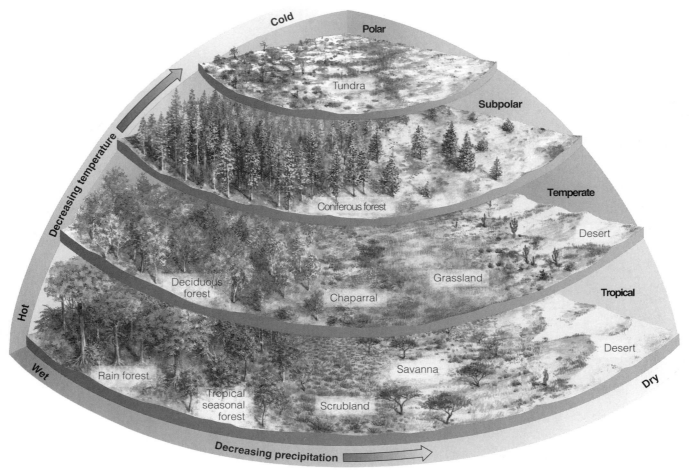

Figure 3-8 Natural capital: average precipitation and average temperature, acting together as limiting factors over a period of 30 or more years, determine the type of desert, grassland, or forest biome in a particular area. Although the actual situation is much more complex, this simplified diagram explains how climate determines the types and amounts of natural vegetation found in an area left undisturbed by human activities. (Used by permission of Macmillan Publishing Company, from Derek Elsom, *The Earth,* New York: Macmillan, 1992. Copyright © 1992 by Marshall Editions Developments Limited)

Figure 3-9 Generalized effects of altitude (left) and latitude (right) on climate and biomes. Parallel changes in vegetation type occur when we travel from the equator to the poles or from lowlands to mountaintops. This generalized diagram shows only one of many possible sequences.

| Producer to primary consumer | Primary to secondary consumer | Secondary to higher-level consumer | All producers and consumers to decomposers |

Figure 3-10 Natural capital: some components and interactions in a *temperate desert ecosystem*. When these organisms die, decomposers break down their organic matter into minerals that plants use. Arrows indicate transfers of matter and energy between producers, primary consumers (herbivores), secondary (or higher-level) consumers (carnivores), and decomposers. Organisms are not drawn to scale.

Figure 3-11 (p. 58) shows major human impacts on deserts. Look closely at this figure. What are the effects of your lifestyle on desert biomes? Deserts take a long time to recover from disturbances because of lack of water, their slow plant growth, low species diversity, and slow nutrient cycling (because of little bacterial activity in their soils). Desert vegetation destroyed by livestock overgrazing and off-road vehicles may take decades to grow back.

Figure 3-12 (p. 58) lists major human impacts on grasslands. What are the effects of your lifestyle on grassland biomes? Because of their thick and fertile soils, temperate grasslands are widely used to grow crops. But plowing disrupts their soils leaving them vulnerable to wind and water erosion.

Figure 3-13 (p. 58) lists major human impacts on the world's forests. Study this figure carefully. What impacts does your lifestyle have on forest biomes? Large areas of the world's temperate forests have been cleared to grow crops and build urban areas. Tropical forests are also being cleared rapidly for agriculture, timber, and mining.

Natural Capital Degradation
Deserts

Large desert cities

Soil destruction by off-road vehicles and urban development

Soil salinization from irrigation

Depletion of underground water supplies

Land disturbance and pollution from mineral extraction

Storage of toxic and radioactive wastes

Large arrays of solar cells and solar collectors used to produce electricity

Figure 3-11 Natural capital degradation: major human impacts on the world's deserts.

Natural Capital Degradation
Grasslands

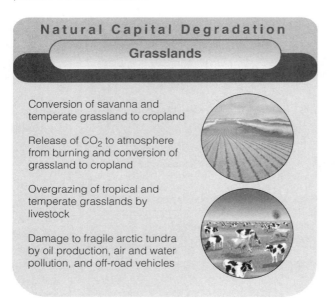

Conversion of savanna and temperate grassland to cropland

Release of CO_2 to atmosphere from burning and conversion of grassland to cropland

Overgrazing of tropical and temperate grasslands by livestock

Damage to fragile arctic tundra by oil production, air and water pollution, and off-road vehicles

Figure 3-12 Natural capital degradation: major human impacts on the world's grasslands.

3-6 LIFE IN WATER ENVIRONMENTS

What Are Aquatic Life Zones? Life in Water Environments

Life exists in freshwater and ocean aquatic life zones.

Scientists divide the watery parts of the biosphere into **aquatic life zones,** each containing numerous ecosys-

Natural Capital Degradation
Forests

Clearing and degradation of tropical forests for agriculture, livestock grazing, and timber harvesting

Clearing of temperate deciduous forests in Europe, Asia, and North America for timber, agriculture, and urban development

Clearing of evergreen coniferous forests in North America, Finland, Sweden, Canada, Siberia, and Russia

Conversion of diverse forests to less biodiverse tree plantations

Damage to soils from off-road vehicles

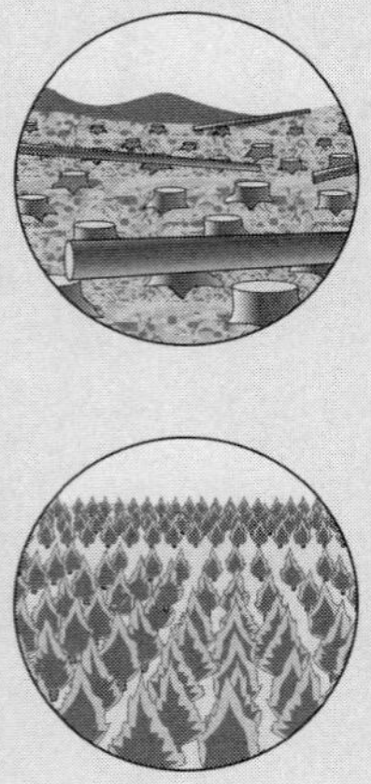

Figure 3-13 Natural capital degradation: major human impacts on the world's forests.

tems. Examples include *freshwater life zones* (such as lakes, streams, and inland wetlands) and *ocean* or *marine life zones* (such as coral reefs, coastal estuaries, and the deep ocean).

Most aquatic life zones can be divided into three layers: *surface, middle,* and *bottom.* A number of environmental factors determine the types and numbers of organisms found in these layers. Examples are *temperature, access to sunlight for photosynthesis, dissolved oxygen content,* and *availability of nutrients* such as carbon (as dissolved CO_2 gas), nitrogen (as NO_3^-), and phosphorus (mostly as PO_4^{3-}) for producers.

What Are the Major Saltwater Life Zones? The Ocean Planet

Oceans occupy almost three-fourths of the earth's surface and consist of a coastal zone and the open sea.

A more accurate name for Earth would be *Ocean* because saltwater oceans cover about 71% of the planet's surface (Figure 3-14). Figure 3-15 lists important ecological and economic services provided by these marine systems. We know more about the surface of the moon than about the oceans that cover most of the earth.

Oceans have two major life zones: the **coastal zone** and the **open sea** (Figure 3-16, p. 60). Although it makes up less than 10% of the world's ocean area, the coastal zone contains 90% of all marine species and is the site of most large commercial marine fisheries. This zone has numerous interactions with the land and thus human activities easily affect it.

Figure 3-14 Natural capital: the ocean planet. The salty oceans cover about 71% of the earth's surface. About 97% of the earth's water is in the interconnected oceans, which cover 90% of the planet's mostly ocean hemisphere (left) and 50% of its land-ocean hemisphere (right).

Most ecosystems found in the coastal zone have a very high net primary productivity per unit of area (Figure 2-20, bottom, p. 34). This occurs because of the zone's ample supplies of sunlight and plant nutrients (flowing from land and distributed by wind and ocean currents).

One highly productive area in the coastal zone is an *estuary*, a partially enclosed area of coastal water where sea water mixes with fresh water and nutrients from rivers, streams, and runoff from land. Another consists of *coastal wetlands:* land areas covered with water all or part of the year. They include *mangrove forest swamps* in tropical waters and *salt marshes* in temperate zones.

Coral reefs form in the shallow coastal zones of warm tropical and subtropical oceans. These beautiful natural wonders are among the world's most diverse and productive ecosystems and are homes for about one-fourth of all marine species.

The open sea is divided into three vertical *zones— euphotic, bathyal,* and *abyssal*—based primarily on the penetration of sunlight (Figure 3-16). This vast volume contains only about 10% of all marine species. Average net primary productivity per unit of area is quite low in the open sea (Figure 2-20, bottom, p. 34) except at an occasional equatorial upwelling, where currents bring up nutrients from the ocean bottom. However, because the open sea covers so much of the earth's surface, it makes the largest contribution to the earth's overall net primary productivity.

Figure 3-17 (p. 60) lists major human impacts on marine systems. Study this figure carefully. Does your lifestyle contribute to any of these impacts?

What Are the Major Freshwater Life Zones? Lakes, Wetlands, and Rivers

The freshwater ecosystems that cover less than 1% of the earth's surface provide important ecological and economic services.

Freshwater life zones consist of *standing* bodies of fresh water such as lakes, ponds, and inland wetlands and *flowing* systems such as streams and rivers. These freshwater systems provide a number of important ecological and economic services (Figure 3-18, p. 61). Study this important figure carefully.

Lakes are large natural bodies of standing fresh water formed when precipitation, runoff, or groundwater seepage fills depressions in the earth's surface. Lakes normally consist of distinct zones (Figure 3-19, p. 61), which provide habitats and niches for different species.

Ecologists classify lakes according to their nutrient content and primary productivity. A newly formed lake generally has a small supply of plant nutrients and is called an **oligotrophic** (poorly nourished) **lake.** Such nutrient-poor lakes are often deep and usually have crystal-clear blue or green water because their low net primary productivity supports few algae and other producers.

Over time, sediment washes into an oligotrophic lake, and plants grow and decompose to form bottom sediments. A lake with a large or excessive supply of nutrients (mostly nitrates and phosphates) needed by producers is called a **eutrophic** (well-nourished) **lake.**

Natural Capital

Marine Ecosystems

Ecological Services	Economic Services
Climate moderation	Food
CO₂ absorption	Animal and pet feed (fish meal)
Nutrient cycling	
Waste treatment and dilution	Pharmaceuticals
Reduced storm impact (mangrove, barrier islands, coastal wetlands)	Harbors and transportation routes
	Coastal habitats for humans
Habitats and nursery areas for marine and terrestrial species	Recreation
	Employment
Genetic resources and biodiversity	Offshore oil and natural gas
	Minerals
Scientific information	Building materials

Figure 3-15 Natural capital: major ecological and economic services provided by marine systems.

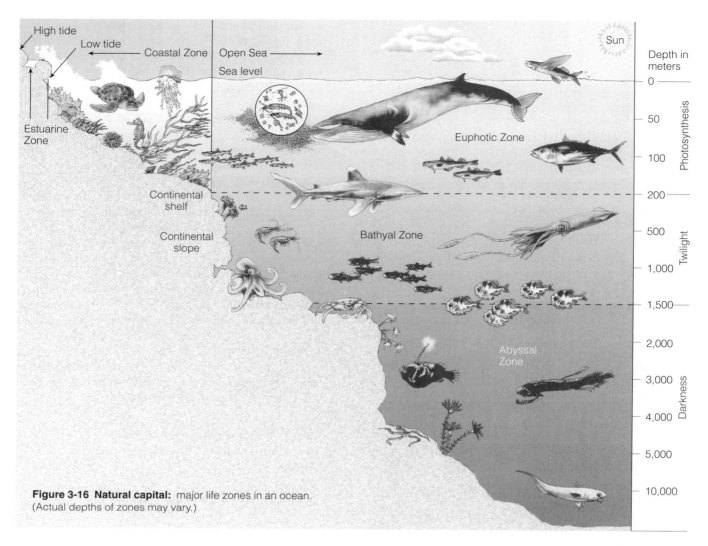

Figure 3-16 Natural capital: major life zones in an ocean. (Actual depths of zones may vary.)

Labels in figure:
High tide
Low tide
Coastal Zone
Open Sea
Sea level
Sun
Depth in meters
Estuarine Zone
Euphotic Zone
Photosynthesis
Continental shelf
Continental slope
Bathyal Zone
Twilight
Abyssal Zone
Darkness

Depth scale: 0, 50, 100, 200, 500, 1,000, 1,500, 2,000, 3,000, 4,000, 5,000, 10,000

Such nutrient-rich lakes typically are shallow, have a high net primary productivity, and have murky brown or green water with very poor visibility because of their high content of algae and other producers. Many lakes fall somewhere between the two extremes of nutrient enrichment and are called **mesotrophic lakes.**

Precipitation that does not sink into the ground or evaporate is **surface water.** It becomes **runoff** when it flows into streams. The land area that delivers runoff, sediment, and dissolved substances to a stream is called a **watershed,** or **drainage basin.** Small streams join to form rivers, and rivers flow downhill to the ocean (Figure 3-20, p. 62) as part of the hydrologic cycle.

In many areas, *streams* begin in mountainous or hilly areas that collect and release water falling to the earth's surface as rain or snow. The downward flow of water from mountain highlands to the sea takes place

Figure 3-17 Natural capital degradation: major human impacts on the world's marine systems.

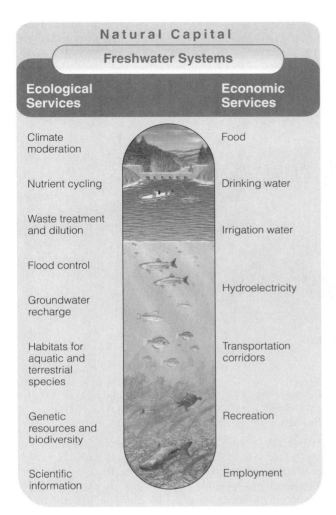

Natural Capital

Freshwater Systems

Ecological Services	Economic Services
Climate moderation	Food
Nutrient cycling	Drinking water
Waste treatment and dilution	Irrigation water
Flood control	Hydroelectricity
Groundwater recharge	
Habitats for aquatic and terrestrial species	Transportation corridors
Genetic resources and biodiversity	Recreation
Scientific information	Employment

Figure 3-18 Natural capital: major ecological and economic services provided by freshwater systems.

in three different aquatic life zones with different environmental conditions (Figure 3-20). Because of different environmental conditions in each zone, a river system is a series of different ecosystems with different average depths, flow rates, dissolved oxygen levels, temperatures, and aquatic species.

As streams flow downhill, they become powerful shapers of land. Over millions of years, the friction of moving water levels mountains and cuts deep canyons, and the rock and soil the water removes are deposited as sediment in low-lying areas.

We have constructed dams, power plants that need cooling water, sewage treatment plants, cities, recreation areas, shipping terminals, and we have established farmlands in the watersheds along the shores of rivers and streams, especially in their transition and floodplain zones. This greatly increases the flow of plant nutrients, sediment, and pollutants into these ecosystems. To protect a stream or river system from excessive inputs of nutrients and pollutants, we must protect its *watershed,* the land around it.

Inland wetlands are lands covered with fresh water all or part of the time (excluding lakes, reservoirs, and streams) and located away from coastal areas. They include *marshes* (dominated by grasses), *swamps* (dominated by trees and shrubs), *prairie potholes* (depressions carved out by glaciers), and *floodplains* (which receive excess water during heavy rains and floods).

Some wetlands are covered with water year-round. Others, called *seasonal wetlands,* usually are underwater or soggy for only a short time each year. They include prairie potholes, floodplain wetlands, and bottomland hardwood swamps. Some stay dry for years before being covered with water again.

Inland wetlands provide a number of important and free ecological and economic services such as filtering toxic wastes and pollutants, absorbing and storing excess water from storms, and providing habitats for a variety of species. But we continue to destroy or degrade many of these important systems. What factors in your life contribute to the destruction of inland wetlands?

In this chapter we have seen that the earth's life has

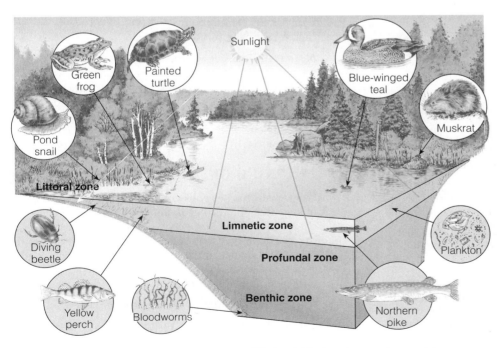

Figure 3-19 Natural capital: the distinct zones of life in a fairly deep temperate-zone lake.

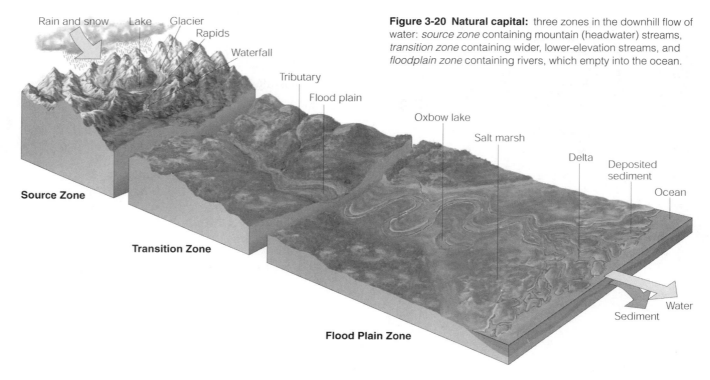

Figure 3-20 Natural capital: three zones in the downhill flow of water: *source zone* containing mountain (headwater) streams, *transition zone* containing wider, lower-elevation streams, and *floodplain zone* containing rivers, which empty into the ocean.

shown an extraordinary ability to diversify by adapting to new biological opportunities and, in the process, creating even more biological opportunities. In the earth's ballet of life that has been playing for about 3.7 billion years, life and death are interconnected. Some species appear and some disappear, but the show goes on.

We only recently became members of this evolutionary ballet. Will we temporarily interrupt the show, take out some of the dancers, and get kicked off the stage? Or will we learn the rules of the ballet and have a long run? We live in interesting and challenging times.

Nothing in biology makes sense except in the light of evolution.

THEODOSIUS DOBZHANSKY

CRITICAL THINKING

1. (a) How would you respond to someone who tells you that he or she does not believe in biological evolution because it is "just a theory"? **(b)** How would you respond to a statement that we should not worry about air pollution because through natural selection the human species will develop lungs that can detoxify pollutants?

2. How would you respond to someone who says that because extinction is a natural process, we should not worry about the loss of biodiversity?

3. An important adaptation of humans is a strong opposable thumb, which allows us to grip and manipulate things with our hands. As a demonstration of the importance of this trait, fold each of your thumbs into the palm of its hand and then tape them securely in that position

for an entire day. After the demonstration, make a list of the things you could not do without the use of your thumbs.

4. Explain why you are for or against each of the following: **(a)** requiring labels indicating the use of genetically modified components in any food item, **(b)** using genetic engineering to develop "superior" human beings, and **(c)** using genetic engineering to eliminate aging and death.

5. What type of biome do you live in or near? What have been the effects of human activities over the past 50 years on the characteristic vegetation and animal life normally found in the biome you live in? How is your own lifestyle affecting this biome?

6. You are a defense attorney arguing in court for sparing an undeveloped old-growth tropical rain forest and a coral reef from severe degradation or destruction by development. Write your closing statement for the defense of each of these ecosystems. If the judge decides you can save only one of the ecosystems, which one would you choose, and why?

LEARNING ONLINE

The website for this book contains helpful study aids and many ideas for further reading and research. They include a chapter summary, review questions for the entire chapter, flash cards for key terms and concepts, a multiple-choice practice quiz, interesting Internet sites, references, and a guide for accessing thousands of InfoTrac® College Edition articles. Log on to

http://biology.brookscole.com/miller7

Then click on the Chapter-by-Chapter area, choose Chapter 3, and select a learning resource.

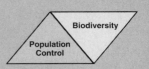

4 COMMUNITY ECOLOGY, POPULATION ECOLOGY, AND SUSTAINABILITY

What is this balance of nature that ecologists talk about?
STUART L. PIMM

4-1 TYPES OF SPECIES

What Different Roles Do Various Species Play in Communities? A Biological Play with Many Roles

Communities can contain native, nonnative, indicator, keystone, and foundation species that play different ecological roles.

Ecologists often use labels—such as *native, nonnative, indicator, keystone,* or *foundation*—to describe the major ecological roles or niches various species play in communities. Any given species may play more than one of these five roles in a particular community.

Native species are those that normally live and thrive in a particular community. Others that migrate into or are deliberately or accidentally introduced into an community are called **nonnative species, invasive species,** or **alien species.**

Many people tend to think of nonnative or invasive species as villains. But most introduced and domesticated species of crops and animals such as chickens, cattle, and fish from around the world and many wild game species are beneficial to us.

But some nonnative species can thrive and crowd out native species—an example of unintended consequences. Examples are fire ants and rapidly growing plant species such as kudzu.

What Are Indicator Species? Smoke Alarms

Some species can alert us to harmful changes that are taking place in biological communities.

Species that serve as early warnings of damage to a community are called **indicator species.** Birds are excellent biological indicators because they are found almost everywhere and are affected quickly by environmental change such as loss or fragmentation of their habitats and exposure to chemical pesticides. Research indicates that a major factor in the current decline of some species of migratory, insect-eating songbirds in North America is habitat loss or fragmentation. The tropical forests of Latin America and the Caribbean that are winter habitats for such birds are disappearing rapidly. Their summer habitats in North America also are disappearing or are being fragmented into patches that make the birds more vulnerable to attack by predators and parasites.

Some amphibians (frogs, toads, and salamanders), which live part of their lives in water and part on land, are also classified to be indicator species (Case Study, below). Butterflies are a good indicator species because their association with various plant species makes them vulnerable to habitat loss and fragmentation.

Case Study: Why Are Amphibians Vanishing? Warnings from Frogs

The disappearance of many of the world's amphibian species for a variety of reasons may indicate a decline in environmental quality in many parts of the world.

Amphibians (frogs, toads, and salamanders) live part of their lives in water and part on land, and some are classified as indicator species. Frogs, for example, are good indicator species because they are especially vulnerable to environmental disruption at various points in their life cycle, shown in Figure 4-1 (p. 64). As tadpoles they live in water and eat plants, and as adults they live mostly on land and eat insects that can expose them to pesticides. Their eggs have no protective shells to block ultraviolet (UV) radiation or pollution. As adults, they take in water and air through their thin, permeable skins that can readily absorb pollutants from water, air, or soil.

Since 1980, populations of hundreds of the world's estimated 5,280 amphibian species have been vanishing or declining in almost every part of the world, even in protected wildlife reserves and parks. According to the World Conservation Union, one-fourth of all known amphibian species are extinct, endangered, or vulnerable to extinction.

No single cause has been identified to explain the amphibian declines. However, scientists have identified a number of factors that can affect frogs and other amphibians at various points in their life cycles.

They include *habitat loss and fragmentation* (especially from draining and filling of inland wetlands, deforestation, and development), *prolonged drought* (which dries up breeding pools so few tadpoles survive), *pollution* (particularly from exposure to pesticides which can make them more vulnerable to bacterial, viral, and fungal diseases and cause an array of sexual abnormalities), *increases in ultraviolet radiation* caused by reductions in stratospheric ozone (which

Figure 4-1 Typical *life cycle of a frog.* Populations of various frog species can decline because of the effects of a variety of harmful factors at different points in their life cycle. Such factors include habitat loss, drought, pollution, increased ultraviolet radiation, parasitism, disease, overhunting for food (frog legs), and nonnative predators and competitors.

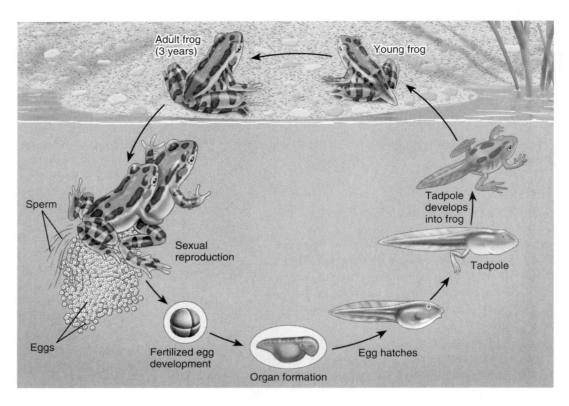

can harm young embryos of amphibians in shallow ponds), *parasites, overhunting* (especially in Asia and France, where frog legs are a delicacy), *viral and fungal diseases,* and *natural immigration or deliberate introduction of nonnative predators and competitors* (such as fish) and *disease organisms.* A combination of such factors probably is responsible for the decline or disappearance of most amphibian species.

So why should we care if various amphibian species become extinct? Scientists give three reasons. *First,* their extinction suggests that environmental health is deteriorating in parts of the world because amphibians are sensitive biological indicators of changes in environmental conditions such as habitat loss and degradation, pollution, UV exposure, and climate change.

Second, adult amphibians play important ecological roles in biological communities. For example, amphibians eat more insects (including mosquitoes) than do birds. In some habitats, extinction of certain amphibian species could lead to extinction of other species, such as reptiles, birds, aquatic insects, fish, mammals, and other amphibians that feed on them or their larvae.

Third, from a human perspective, amphibians represent a genetic storehouse of pharmaceutical products waiting to be discovered. Compounds in hundreds of secretions from amphibian skin have been isolated and some are used as painkillers and antibiotics and in treating burns and heart disease.

The plight of some amphibian indicator species is a warning signal. They do not need us, but we and other species need them.

What Are Keystone Species? Major Players Who Help Keep Communities Running Smoothly

Keystone species help determine the types and numbers of various other species in a community.

A keystone is the wedge-shaped stone placed at the top of a stone archway. Remove this stone and the arch collapses. In some communities, certain species called **keystone species** apparently serve a similar role. They have a much larger effect on the types and abundances of other species in a community than their numbers would suggest. According to this hypothesis, eliminating a keystone species can dramatically alter the structure and function of a community.

Keystone species play critical ecological roles. One is *pollination* of flowering plant species by bees, hummingbirds, bats, and other species. In addition, *top predator* keystone species feed on and help regulate the populations of other species. Examples are the wolf, leopard, lion, alligator, and great white shark (Case Study, p. 65).

The loss of a keystone species can lead to population crashes and extinctions of other species that depend on it for certain ecological services. According to biologist Edward O. Wilson, "The loss of a keystone species is like a drill accidentally striking a power line. It causes lights to go out all over."

Case Study: Why Are Sharks Important Species? Culling the Ocean and Helping Improve Human Health

Some shark species eat and remove sick and injured ocean animals and some can help us learn how to fight cancer and immune system disorders.

The world's 370 shark species vary widely in size. The smallest is the dwarf dog shark, about the size of a large goldfish. The largest is the whale shark, the world's largest fish. It can grow to 15 meters (50 feet) long and weigh as much as two full-grown African elephants.

Various shark species, feeding at the top of food webs, cull injured and sick animals from the ocean and thus play an important ecological role. Without such shark species, the oceans would be teeming with dead and dying fish.

Many people—influenced by movies (such as *Jaws*), popular novels, and widespread media coverage of a fairly small number of shark attacks per year—think of sharks as people-eating monsters. However, the three largest species—the whale shark, basking shark, and megamouth shark—are gentle giants. They swim through the water with their mouths open, filtering out and swallowing huge quantities of *plankton* (small free-floating sea creatures).

Every year, members of a few species of shark—mostly great white, bull, tiger, gray reef, lemon, hammerhead, shortfin mako, and blue—typically injure 60–100 people worldwide. Between 1990 and 2003, sharks killed eight people off U.S. coasts and 88 people worldwide—an average of seven people per year. Most attacks are by great white sharks, which feed on sea lions and other marine mammals and sometimes mistake divers and surfers for their usual prey. Whose fault is this?

Media coverage of shark attacks greatly distorts the danger from sharks. You are 30 times more likely to be killed by lightning than by a shark each year, and your chance of being killed by lightning is extremely small.

For every shark that injures a person, we kill at least 1 million sharks, a total of about 100 million sharks each year. Sharks are caught mostly for their fins and then thrown back into the water to die.

Shark fins are widely used in Asia as a soup ingredient and as a pharmaceutical cure-all. In high-end Hong Kong restaurants, a single bowl of shark fin soup can cost as much as $100.

According to a 2001 study by Wild Aid, shark fins sold in restaurants throughout Asia and in Chinese communities in cities such as New York, San Francisco, and London contain dangerously high levels of toxic mercury. Consumption of high levels of mercury is especially threatening for pregnant women, fetuses, and infants feeding on breast milk.

Sharks are also killed for their livers, meat (chiefly mako and thresher), hides (a source of exotic, high-quality leather), and jaws (especially great whites, whose jaws can sell for up to $10,000). They are also killed because we fear them. Some sharks (especially blue, mako, and oceanic whitetip) die when they are trapped in nets or lines deployed to catch swordfish, tuna, shrimp, and other commercially important species.

In addition to their important ecological roles, sharks save human lives. They are helping us learn how to fight cancer, which sharks almost never get. Scientists are also studying their highly effective immune system because it allows wounds to heal without becoming infected.

Sharks have three natural traits that make them prone to population declines from overfishing. They take a long time to reach sexual maturity (10–24 years), have only a few offspring (between 2 and 10) once every year or two, and have long gestation (pregnancy) periods (up to 24 months for some species).

Sharks are among the most vulnerable and least protected animals on the earth. Eight of the world's shark species are considered critically endangered, endangered, or vulnerable to extinction. In 2003, experts at the National Aquarium in Baltimore, Maryland, estimated that populations of a number of commercially valuable shark species have decreased by 90% since 1992.

In response to a public outcry over depletion of some shark species, the United States and several other countries have banned hunting sharks for their fins in their territorial waters. But such bans are difficult to enforce.

With more than 400 million years of evolution behind them, sharks have had a long time to get things right. Preserving their evolutionary genetic development begins with the knowledge that sharks do not need us, but we and other species need them.

What Are Foundation Species? Players Who Create New Habitats and Niches

Foundation species can create and enhance habitats that can benefit other species in a community.

Some ecologists think the keystone species concept should be expanded to include the roles of *foundation species*, which play a major role in shaping communities by creating and enhancing habitat that benefits other species.

For example, elephants push over, break, or uproot trees, creating forest openings in the savanna grasslands and woodlands of Africa. This promotes the growth of grasses and other forage plants that benefit

Figure 4-2 Sharing the wealth: *resource partitioning* of five species of common insect-eating warblers in the spruce forests of Maine. Each species minimizes competition with the others for food by spending at least half its feeding time in a distinct portion (shaded areas) of the spruce trees, and consuming somewhat different insect species. (After R. H. MacArthur, "Population Ecology of Some Warblers in Northeastern Coniferous Forests," *Ecology* 36 [1958]: 533–36)

smaller grazing species such as antelope. It also accelerates nutrient cycling rates. Some bats and bird foundation species can regenerate deforested areas and spread fruit plants by depositing plant seeds in their droppings.

4-2 SPECIES INTERACTIONS

How Do Species Interact? Ways to Get an Edge

Competition, predation, parasitism, mutualism, and commensalism are ways in which species can interact and increase their ability to survive.

When different species in a community have activities or resource needs in common, they may interact with one another. Members of these species may be harmed, helped, or unaffected by the interaction. Ecologists identify five basic types of interactions between species: *interspecific competition, predation, parasitism, mutualism,* and *commensalism.*

The most common interaction between species is *competition* for shared or scarce resources such as space and food. Ecologists call such competition between species **interspecific competition.** When intense competition for limited resources occurs, one of the competing species must migrate (if possible) to another area, shift its feeding habits or behavior through natural selection and evolution, suffer a sharp population decline, or become extinct in that area.

Humans are in competition with many other species for space, food, and other resources. As we convert more and more of the earth's land and aquatic resources and net primary productivity to our uses, we deprive many other species of resources they need to survive.

How Have Some Species Reduced or Avoided Competition? Share the Wealth by Becoming More Specialized

Some species evolve adaptations that allow them to reduce or avoid competition for resources with other species.

Over a time scale long enough for evolution to occur, some species competing for the same resources evolve adaptations that reduce or avoid competition. One way this happens is through **resource partitioning.** It occurs when species competing for similar scarce resources evolve more specialized traits that allow them to use shared resources at different times, in different ways, or in different places (Figure 4-2).

Here are some other examples of resource partitioning. When lions and leopards live in the same area, lions take mostly larger animals as prey and leopards take smaller ones. Hawks and owls feed on similar prey, but hawks hunt during the day and owls hunt at night. Some bird species feed on the ground, whereas others seek food in trees and shrubs.

How Do Predator and Prey Species Interact? Eating and Being Eaten

Species called predators feed on all or parts of other species called prey.

In **predation,** members of one species (the *predator*) feed directly on all or part of a living organism of another species (the *prey*). Together, the two kinds of organisms, such as lions (the predator or hunter) and zebras (the prey or hunted), are said to have a **predator–prey relationship.** Such relationships are depicted in Figures 2-16 (p. 32) and 2-17 (p. 33).

Most of the world's predation is not the kind we see in many nature documentaries such as lions killing zebras or bears plucking salmon from streams. Instead, most predation occurs unseen at the microscopic level in soils (Figure 2-22, p. 36) and sediments in aquatic systems.

Some people tend to view predators with contempt. When a hawk tries to capture and feed on a rabbit, some tend to root for the rabbit. Yet the hawk, like all predators, is merely trying to get enough food to feed itself and its young. And in doing this it is playing an important ecological role in controlling rabbit populations.

What Are Parasites and Why Are They Important? Living on or in Another Species

Although parasites can harm their host organisms, they can promote community biodiversity.

Parasitism occurs when one species (the *parasite*) feeds on part of another organism (the *host*) usually by living on or in the host. In this relationship, the parasite benefits and the host is harmed.

Parasitism can be viewed as a special form of predation. But unlike a conventional predator, a parasite usually is much smaller than its host (prey) and rarely kills its host. Also most parasites remain closely associated with, draw nourishment from, and may gradually weaken their host over time.

Tapeworms, disease-causing microorganisms, and other parasites live *inside* their hosts. Other parasites attach themselves to the *outside* of their hosts. Examples are ticks, fleas, mosquitoes, mistletoe plants, and fungi that cause diseases such as athlete's foot. Some parasites move from one host to another, as fleas and ticks do; others, such as tapeworms, spend their adult lives with a single host.

From the host's point of view, parasites are harmful, but parasites play important ecological roles. Collectively, the matrix of parasitic relationships in a community acts somewhat like glue that helps hold the species in a community together. Parasites also pro- mote biodiversity by helping keep some species from becoming so plentiful that they eliminate other species.

How Do Species Interact So Both Species Benefit? Win-Win Relationships

Pollination, bacteria in your gut that digest your food, and fungi that help plant roots take up nutrients are examples of species interactions that benefit both species.

In **mutualism,** two species interact in a way that benefits both. Such benefits include having pollen and seeds dispersed for reproduction, being supplied with food, or receiving protection.

Here are some examples of mutualistic relationships:

- Honeybees feed on a male flower's nectar, pick up pollen in the process, and then pollinate female flowers when they feed on them.

- Birds ride on the backs of large animals like African buffalo, elephants, and rhinoceroses (Figure 4-3a, p. 68). The birds remove and eat parasites from the animal's body and often make noises warning the animal when predators approach.

- Clownfish species live within sea anemones, whose tentacles sting and paralyze most fish that touch them (Figure 4-3b). The clownfish, which are not harmed by the tentacles, gain protection from predators and feed on the detritus left from the meals of the anemones. The sea anemones benefit because the clownfish protect them from some of their predators.

- Vast armies of bacteria in the digestive systems of animals break down (digest) their food. The bacteria receive a sheltered habitat and food from their host. In turn, they help break down (digest) their host's food. Such bacteria in your gut help digest the food you eat. Thank these little critters for helping keep you alive.

It is tempting to think of mutualism as an example of cooperation between species, but actually it involves each species benefiting by exploiting the other.

How Do Species Interact So One Benefits but the Other Is Not Harmed? Do No Harm

Some species interact in ways that help one species but neither help nor harm the other.

Commensalism is an interaction that benefits one species but neither helps nor harms the other species much, if at all. One example is a *redwood sorrel,* a small herb. It benefits from growing in the shade of tall redwood trees, with no known harmful effects on the redwood trees.

Another example is plants called *epiphytes* (such as some types of orchids and bromeliads) that attach themselves to the trunks or branches of large trees in tropical and subtropical forests. These so-called air plants benefit by having a solid base on which to grow.

Figure 4-3 Examples of *mutualism.* **(a)** Oxpeckers (or tickbirds) feed on the parasitic ticks that infest large thick-skinned animals such as a black rhinoceros. **(b)** A clownfish gains protection and food by living among deadly stinging sea anemones and helps protect the anemones from some of their predators.

(a) Oxpeckers and black rhinoceros

(b) Clown fish and sea anemone

They also live in an elevated spot that gives them better access to sunlight, water from the humid air and rain, and nutrients falling from the tree's upper leaves and limbs. This apparently does not harm the tree.

4-3 ECOLOGICAL SUCCESSION: COMMUNITIES IN TRANSITION

How Do Communities Respond to Change? Shifting Community Composition

Over time new environmental conditions can cause changes in community structure that lead to one group of species being replaced by other groups.

All communities change their structure and composition in response to changing environmental conditions. The gradual change in species composition of a given area is called **ecological succession.** During succession some species colonize an area and their populations become more numerous, whereas populations of other species decline and may even disappear.

Ecologists recognize two types of ecological succession depending on the conditions present at the beginning of the process. One is **primary succession,** which involves the gradual establishment of biotic communities on essentially lifeless ground where there is no soil in a terrestrial community (Figure 4-4)

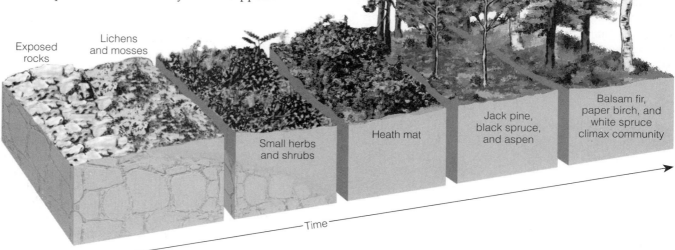

Exposed rocks

Lichens and mosses

Small herbs and shrubs

Heath mat

Jack pine, black spruce, and aspen

Balsam fir, paper birch, and white spruce climax community

Time

Figure 4-4 Starting from ground zero. *Primary ecological succession* over several hundred years of plant communities on bare rock exposed by a retreating glacier on Isle Royal in northern Lake Superior.

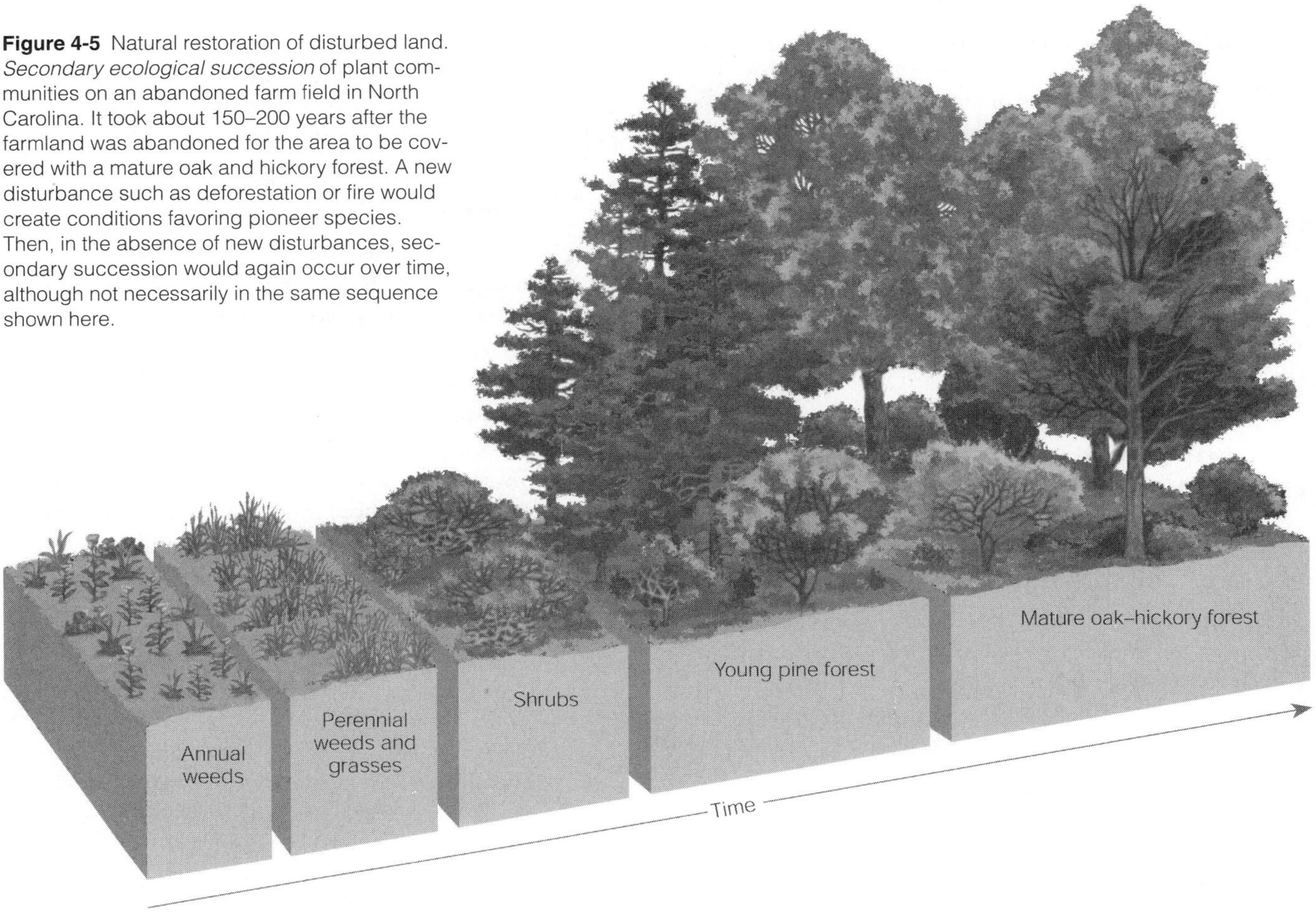

Figure 4-5 Natural restoration of disturbed land. *Secondary ecological succession* of plant communities on an abandoned farm field in North Carolina. It took about 150–200 years after the farmland was abandoned for the area to be covered with a mature oak and hickory forest. A new disturbance such as deforestation or fire would create conditions favoring pioneer species. Then, in the absence of new disturbances, secondary succession would again occur over time, although not necessarily in the same sequence shown here.

Annual weeds

Perennial weeds and grasses

Shrubs

Young pine forest

Mature oak–hickory forest

Time

or no bottom sediment in an aquatic community. Examples include bare rock exposed by a retreating glacier or severe soil erosion, newly cooled lava, an abandoned highway or parking lot, or a newly created shallow pond or reservoir.

Primary succession usually takes a long time. One reason is that before a community can become established on land, there must be soil. Depending mostly on the climate, it takes natural processes several hundred to several thousand years to produce fertile soil.

With the other, more common type, called **secondary succession,** a series of communities with different species can develop in places containing soil or bottom sediment. It begins in an area where the natural community of organisms has been disturbed, removed, or destroyed, but the soil or bottom sediment remains. Candidates for secondary succession include abandoned farmlands (Figure 4-5), burned or cut forests, heavily polluted streams, and land that has been dammed or flooded. Because some soil or sediment is present, new vegetation usually can begin to germinate within a few weeks. Seeds can be present in soils, or they can be carried from nearby plants by wind or by birds and other animals.

During primary or secondary succession, disturbances such as natural or human-caused fires or defor-

estation can convert a particular stage of succession to an earlier stage. Such disturbances create new conditions that encourage some species and discourage or eliminate others.

How Predictable Is Succession, and Is Nature in Balance? Things Are Always Changing.

Scientists cannot project the course of a given succession or view it as preordained progress toward a stable climax community that is in balance with its environment.

We may be tempted to conclude that ecological succession is an orderly sequence in which each stage leads automatically to the next, more stable stage. According to this classic view, succession proceeds along an expected path until a certain stable type of *climax community* occupies an area. Such a community is dominated by a few long-lived plant species and is in balance with its environment. This equilibrium model of succession is what ecologists meant when they talked about the *balance of nature.*

Over the last several decades, many ecologists have changed their views about balance and equilibrium in nature. Under the old *balance-of-nature* view, a large terrestrial community undergoing succession eventually became covered with an expected type of climax vegetation. But a close look at almost any community reveals that it consists of an ever-changing mosaic of vegetation patches at different stages of succession.

Such research indicates that we cannot predict the course of a given succession or view it as preordained progress toward an ideally adapted climax community. Rather, succession reflects the ongoing struggle by different species for enough light, nutrients, food, and space. This allows them to survive and gain reproductive advantages over other species in response to changes in environmental conditions.

4-4 POPULATION DYNAMICS AND CARRYING CAPACITY

What Limits Population Growth? Resources, Competitors, and Predators

No population can grow indefinitely because resources such as light, water, and nutrients are limited and because of the presence of competitors or predators.

Four variables—*births, deaths, immigration,* and *emigration*—govern changes in population size. A population gains individuals by birth and immigration and loses them by death and emigration:

$$\text{Population change} = (\text{Births} + \text{Immigration}) - (\text{Deaths} + \text{Emigration})$$

Populations vary in their capacity for growth, also known as the **biotic potential** of a population. The **intrinsic rate of increase (r)** is the rate at which a population would grow if it had unlimited resources. Most populations grow at a rate slower than this maximum.

Individuals in populations with a high rate of growth typically *reproduce early in life, have short generation times* (the time between successive generations), *can reproduce many times* (have a long reproductive life), and *have many offspring each time they reproduce.*

Some species have an astounding biotic potential. Without any controls on its population growth, the descendants of a single female housefly could total about 5.6 trillion flies within about 13 months. If this rapid exponential growth kept up, within a few years there would be enough flies to cover the earth's entire surface!

Fortunately, this is not a realistic scenario because *no population can grow indefinitely.* In the real world, a rapidly growing population reaches some size limit imposed by one or more limiting factors, such as light, water, space, or nutrients, or by too many competitors or predators. *There are always limits to population growth in nature.* This important lesson from nature is the main message of this section.

Environmental resistance consists of all factors that act to limit the growth of a population. Together, biotic potential and environmental resistance determine the **carrying capacity (K).** It is the number of individuals of a given species that can be sustained indefinitely in a given space (area or volume). The growth rate of a population decreases as its size nears the carrying capacity of its environment because resources such as food and water begin to dwindle.

What Is the Difference between Exponential and Logistic Population Growth? J-Curves and S-Curves

With ample resources a population can grow rapidly, but as resources become limited its growth rate slows and levels off.

A population with few if any resource limitations grows exponentially. **Exponential growth** starts slowly and grows faster as the population increases because the base size of the population is increasing. Plotting the number of individuals against time yields a J-shaped growth curve (Figure 4-6, lower part of curve). Whether the curve looks steep or "fast" depends on the period of observation.

Logistic growth involves exponential population growth followed by a steady decrease in population growth with time until the population size levels off (Figure 4-6, top half of curve). This occurs as the population encounters environmental resistance and approaches the carrying capacity of its environment. After leveling off, a population with this type of growth typically fluctuates slightly above and below the carrying capacity.

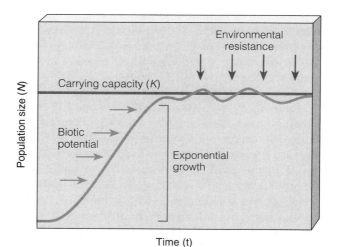

Figure 4-6 No population can grow forever. *Exponential growth* (lower part of the curve) occurs when resources are not limiting and a population can grow at its *intrinsic rate of increase* (r) or biotic potential. Such exponential growth is converted to *logistic growth,* in which the growth rate decreases as the population gets larger and faces environmental resistance. With time, the population size stabilizes at or near the *carrying capacity* (K) of its environment and results in the sigmoid (S-shaped) population growth curve shown in this figure. Depending on resource availability, the size of a population often fluctuates around its carrying capacity.

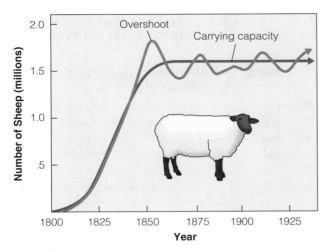

Figure 4-7 *Logistic growth* of a sheep population on the island of Tasmania between 1800 and 1925. After sheep were introduced in 1800, their population grew exponentially because of ample food. By 1855, they overshot the land's carrying capacity. Their numbers then stabilized and fluctuated around a carrying capacity of about 1.6 million sheep.

A plot of the number of individuals against time yields a sigmoid, or *S*-shaped, logistic growth curve. Figure 4-7 shows such a case involving sheep on the island of Tasmania, south of Australia, in the early 19th century.

What Happens If the Population Size Exceeds the Carrying Capacity? Diebacks

When a population exceeds its resource supplies many of its members die unless they can switch to new resources or move to an area with more resources.

The populations of some species do not make a smooth transition from exponential growth to logistic growth. Such populations use up their resource supplies and temporarily *overshoot*, or exceed, the carrying capacity of their environment. This occurs because of a *reproductive time lag:* the period needed for the birth rate to fall and the death rate to rise in response to resource overconsumption. Sometimes it takes a while to get the message out.

In such cases the population suffers a *dieback*, or *crash*, unless the excess individuals can switch to new resources or move to an area with more resources. Such a crash occurred when reindeer were introduced onto a small island off the southwest coast of Alaska (Figure 4-8).

Sometimes when a population exceeds the carrying capacity of an area, it can cause damage that reduces the area's carrying capacity. For example, overgrazing by cattle on dry western lands in the United States has reduced grass cover in some areas. This has allowed sagebrush—which cattle cannot eat—to move in, thrive, and replace grasses. This reduces the land's carrying capacity for cattle.

Humans are not exempt from population overshoot and dieback. Ireland experienced a population crash after a fungus destroyed the potato crop in 1845. About 1 million people died, and 3 million people migrated to other countries.

Technological, social, and other cultural changes have extended the earth's carrying capacity for the human species. We have increased food production and used large amounts of energy and matter resources to make normally uninhabitable areas habitable. A critical question is how long will we be able to keep doing this on a planet with a finite size, finite resources, and a human population whose size and per capita resource use is growing exponentially?

✗ *How Would You Vote?* Can we continue expanding the earth's carrying capacity for humans? Cast your vote online at http://biology.brookscole.com/miller7.

What Types of Reproductive Patterns Do Species Have? Opportunists and Competitors

While some species have a large number of small offspring and give them little parental care, other species have a few larger offspring and take care of them until they can reproduce.

Reproductive individuals in populations of have an inherent evolutionary drive to help ensure that as many members of the next generation as possible will carry their genes. This increases the chances that their population will undergo evolution through natural selection.

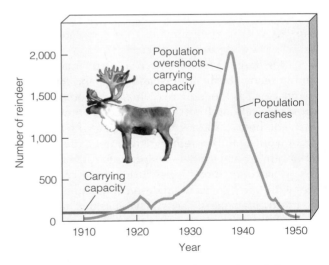

Figure 4-8 Exponential growth, overshoot, and population crash of reindeer introduced to a small island off the southwest coast of Alaska. When 26 reindeer (24 of them female) were introduced in 1910, lichens, mosses, and other food sources were plentiful. By 1935, the herd's population had soared to 2,000, overshooting the island's carrying capacity. This led to a population crash, with the herd plummeting to only eight reindeer by 1950.

Species use different reproductive patterns to help ensure their survival. At one extreme are species that reproduce early and put most of their energy into reproduction. Examples are algae, bacteria, rodents, annual plants (such as dandelions), and most insects.

These species have many, usually small, offspring and give them little or no parental care or protection. They overcome the massive loss of their offspring by producing so many young that a few will survive to reproduce many offspring to begin the cycle again.

Such species tend to be *opportunists.* They reproduce and disperse rapidly when conditions are favorable or when a disturbance opens up a new habitat or niche for invasion, as in the early stages of ecological succession.

Environmental changes caused by disturbances can allow opportunist species to gain a foothold. However, once established, their populations may crash because of further changing or unfavorable environmental conditions or invasion by more competitive species. Therefore, most opportunist species go through irregular and unstable boom-and-bust cycles in their population size.

At the other extreme are *competitor* species that tend to reproduce late in life and have a small number of offspring with fairly long life spans. Typically, the offspring of such species develop inside their mothers (where they are safe), are born fairly large, mature slowly, and are cared for and protected by one or both parents until they reach reproductive age. This reproductive pattern results in a few big and strong individuals that can compete for resources and reproduce a few young to begin the cycle again.

Such species tend to do well in competitive conditions when their population size is near the carrying capacity of their environment. Their populations typically follow a logistic growth curve.

Most large mammals (such as elephants, whales, and humans), birds of prey, and large and long-lived plants (such as the saguaro cactus, redwood trees, and most tropical rain forest trees) are competitor species. Many of these species—especially those with long generation times and low reproductive rates like elephants, rhinoceroses, and sharks—are prone to extinction.

Most competitor species thrive best in communities with fairly constant environmental conditions. In contrast, opportunists thrive in habitats that have experienced disturbances such as a tree falling, a forest fire, or the clearing of a forest or grassland for raising crops.

The reproductive pattern of a species may give it a temporary advantage. But *the availability of suitable habitat for individuals of a population in a particular area is what determines its ultimate population size.* Regardless of how fast a species can reproduce, there can be no more dandelions than there is dandelion habitat and no more zebras than there is zebra habitat in a particular area.

4-5 HUMAN IMPACTS ON NATURAL SYSTEMS: LEARNING FROM NATURE

How Have Humans Modified Natural Ecosystems? Our Big Footprints

We have used technology to alter much of the rest of nature to meet our growing needs and wants in nine major ways that threaten the survival of many other species and could reduce the quality of life for our own species.

To survive and provide resources for growing numbers of people, we have modified, cultivated, built on, or degraded a greatly increasing number and area of the earth's natural systems. Excluding Antarctica, our activities have, to some degree, directly affected about 83% of the earth's land surface. Figure 4-9 compares some of the characteristics of natural and human-dominated systems.

We have used technology to alter much of the rest of nature to meet our growing needs and wants in nine major ways. One is *reducing biodiversity by destroying, fragmenting, and degrading wildlife habitats.* This happens when we clear forests, dig up grasslands, and fill in wetlands to grow food or to construct buildings, highways, and parking lots.

A second is *reducing biodiversity by simplifying and homogenizing natural communities.* Communities dominated by humans tend to have fewer species and fewer community interactions than do undisturbed communities. When we plow grasslands and clear forests, we often replace their thousands of interrelated plant and animal species with one crop or one kind of tree—called a *monoculture.* Then we spend a lot of time, energy, and money trying to protect such monocultures against threats such as invasions by *opportunist species* of plants (weeds) and *pests*—mostly insects to which a monoculture crop is like an all-you-can-eat restaurant. Another threat is invasions by *pathogens*—fungi, viruses, or bacteria—that harm the plants and animals we want to raise.

A third type of alteration is *using, wasting, or destroying an increasing percentage of the earth's net primary productivity that supports all consumer species (including humans).* This factor is the main reason we are crowding out or eliminating the habitats and food supplies of a growing number of species.

A fourth type of intervention has unintentionally *strengthened some populations of pest species and disease-causing bacteria.* This has occurred through overuse of pesticides and antibiotics that has speeded up natural selection and caused genetic resistance to these chemicals.

A fifth effect has been to *eliminate some predators.* Some ranchers want to eradicate bison or prairie dogs that compete with their sheep or cattle for grass. They also want to eliminate wolves, coyotes, eagles, and

Property	Natural Systems	Human-Dominated Systems
Complexity	Biologically diverse	Biologically simplified
Energy source	Renewable solar energy	Mostly nonrenewable fossil fuel energy
Waste production	Little, if any	High
Nutrients	Recycled	Often lost or wasted
Net primary productivity	Shared among many species	Used, destroyed, or degraded to support human activities

Figure 4-9 Some typical characteristics of natural and human dominated systems.

other predators that occasionally kill sheep. A few big-game hunters push for elimination of predators that prey on game species.

Sixth, *we have deliberately or accidentally introduced new or nonnative species into communities.* Most of these species, such as food crops and domesticated live-stock, are beneficial to us but a few are harmful to us and other species.

Seventh, we have *overharvested some renewable resources.* Ranchers and nomadic herders sometimes allow livestock to overgraze grasslands until erosion converts these communities to less productive semi-deserts or deserts. Farmers sometimes deplete soil nutrients by excessive crop growing. Some fish species are overharvested. Illegal hunting or poaching endangers wildlife species with economically valuable parts such as elephant tusks, rhinoceros horns, and tiger skins. In some areas, fresh water is being pumped out of underground aquifers faster than it is replenished.

Eighth, some human activities *interfere with the normal chemical cycling and energy flows in ecosystems.* Soil nutrients can erode from monoculture crop fields, tree plantations, construction sites, and other simplified communities, and overload and disrupt other communities such as lakes and coastal communities. Chemicals such as chlorofluorocarbons (CFCs) released into the atmosphere can increase the amount of harmful ultraviolet energy reaching the earth by reducing ozone levels in the stratosphere. Emissions of carbon dioxide and other greenhouse gases—from burning fossil fuels and from clearing and burning forests and grasslands—can trigger global climate change by altering energy flow through the troposphere.

Ninth, while most natural systems run on sunlight, *human-dominated ecosystems have become increasingly dependent on nonrenewable energy from fossil fuels.* Fossil fuel systems typically produce pollution, add the greenhouse gas carbon dioxide to the atmosphere, and waste much more energy than they need to.

To survive we must exploit and modify parts of nature. However, we are beginning to understand that any human intrusion into nature has multiple effects, most of them unintended and unpredictable .

We face two major challenges. *First,* we need to maintain a balance between simplified, human-altered communities and the more complex natural communities on which we and other species depend. *Second,* we need to slow down the rates at which we are altering nature for our purposes. If we simplify, homogenize, and degrade too much of the planet to meet our needs and wants, what is at risk is not the resilient earth but the quality of life for our own species and the existence of other species we drive to premature extinction.

What Can We Learn from Ecology about Living More Sustainably? Copy Nature

We can develop more sustainable economies and societies by mimicking the four major ways that nature has adapted and sustained itself for several billion years.

How can we live more sustainably? Ecologists say, find out how nature has survived and adapted for several billion years and copy this strategy. Figure 4-10 (also found in color on the bottom half of the back cover) summarizes the four major ways in which life on earth has survived and adapted for several billions of years. Figure 4-11 (left) gives an expanded description of these principles and Figure 4-11 (right) summarizes how we can live more sustainably by mimicking these fundamental but amazingly simple lessons from nature in designing our societies, products, and economies *Figures 4-10 (p. 74) and 4-11 (p. 74) summarize the major message of this book. Study them carefully.*

Biologists have used these lessons from their ecological study of nature to formulate four guidelines for developing more sustainable societies and lifestyles:

■ *Our lives, lifestyles, and economies are totally dependent on the sun and the earth.* We need the earth, but the earth does not need us. As a species, we are expendable.

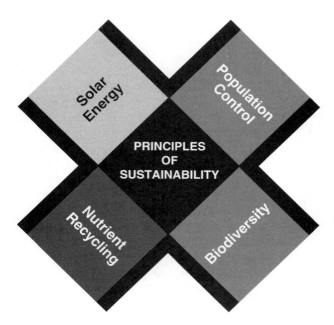

Figure 4-10 Sustaining natural capital: four interconnected principles of sustainability derived from learning how nature sustains itself. This diagram also appears in color on the bottom half of the back cover of this book.

■ *Everything is connected to, and interdependent with, everything else.* The primary goal of ecology is to discover what connections in nature are the strongest, most important, and most vulnerable to disruption for us and other species.

■ *We can never do merely one thing.* Any human intrusion into nature has unexpected and mostly unintended side effects (Connections, right). When we alter nature, we need to ask, "Now what will happen?"

■ *We cannot indefinitely sustain a civilization that depletes and degrades the earth's natural capital, but we can sustain one that lives off the biological income provided by that capital.*

Increasingly, environmental scientists and ecologists are urging that we base our efforts to prevent damage to the earth's life-support system on the **precautionary principle:** When evidence indicates that an activity can seriously harm human health or the environment, we should take precautionary measures to prevent harm even if some of the cause-and-effect relationships have not been fully established scientifically. It is based on the commonsense idea behind many adages such as "Better safe than sorry," "Look before you leap," "First, do no harm," and "Slow down for speed bumps."

As an analogy, we know that eating too much of certain types of foods and not getting enough exercise can greatly increase our chances of a heart attack, diabetes, and other disorders. But the exact connections between these health problems, chemicals in various foods, exercise, and genetics are still under study and often debated. We could use this uncertainty and unpredictability as an excuse to continue overeating and not exercising. However, the wise course is to eat better and exercise more to help *prevent* potentially serious health problems.

Recently, the precautionary principle has formed the basis of several international environmental treaties. One example is the global treaty developed by 122 countries in 2000 to ban or phase out 12 *persistent organic pollutants (POPs).*

However, some analysts point out that we should be selective in applying the precautionary principle. We need to project possible unintended effects as carefully as possible. But we can never know all of the unintended effects of our actions and technologies. Thus, we must be willing to take some risks. Otherwise, we would stifle creativity and innovation and severely limit the development of new technologies and products.

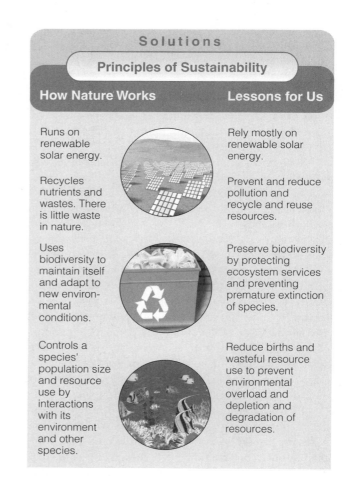

Figure 4-11 Solutions: implications of the four principles of sustainability (left), derived from observing nature, for the long-term sustainability of human societies (right). These four operating principles of nature are connected to one another. Failure of any single principle can lead to temporary or long-term unsustainability, and disruption of ecosystems and human economies and societies.

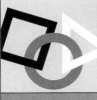

Ecological Surprises

Malaria once infected nine out of ten people in North Borneo, now known as Sabah. In 1955, the World Health Organization (WHO) began spraying the island with dieldrin (a DDT relative) to kill malaria-carrying mosquitoes. The program was so successful that the dreaded disease was nearly eliminated.

But unexpected things began to happen. The dieldrin also killed other insects, including flies and cockroaches living in houses. The islanders applauded. But then small insect-eating lizards that also lived in the houses died after gorging themselves on dieldrin-contaminated insects.

Next, cats began dying after feeding on the lizards. Then, in the absence of cats, rats flourished and overran the villages. When the people became threatened by sylvatic plague carried by rat fleas, the WHO parachuted healthy cats onto the island to help control the rats. Operation Cat Drop worked.

But then the villagers' roofs began to fall in. The dieldrin had killed wasps and other insects that fed on a type of caterpillar that either avoided or was not affected by the insecticide. With most of its predators eliminated, the caterpillar population exploded, munching its way through its favorite food: the leaves used in thatched roofs.

Ultimately, this episode ended happily: Both malaria and the unexpected effects of the spraying program were brought under control. Nevertheless, the chain of unintended and unforeseen events emphasizes the unpredictability of interfering with a community. It reminds us that when we intervene in nature, we need to ask, "Now what will happen?"

Critical Thinking

Do you believe the beneficial effects of spraying pesticides on Sabah outweighed the resulting unexpected and harmful effects? Explain.

Solutions: How Can We Develop More Sustainable Economies?

We can live more sustainably by shifting from high-throughput to low-throughput economies.

Most of today's advanced industrialized countries have **high-throughput economies** that attempt to sustain ever-increasing economic growth by increasing the flow of matter and energy resources through their economic systems (Figure 4-12). These resources flow through the economies of such societies to planetary *sinks* (air, water, soil, organisms), where pollutants and wastes end up and can accumulate to harmful levels.

What happens if more and more people continue to use and waste more and more energy and matter resources at an increasing rate? In other words, what happens if most of the world's people get infected with the affluenza virus?

The law of conservation of matter and the two laws of thermo-dynamics tell us that eventually this consumption will exceed the capacity of the environment to dilute and degrade waste matter and absorb waste heat. However, they do not tell us how close we are to reaching such limits.

A temporary solution to this problem is to convert a linear high-throughput economy into a circular **matter-recycling-and reuse economy** that mimics nature (Figure 4-10) by recycling and reusing most of our matter outputs back through the economy instead of dumping them into the environment.

Changing to a matter-recycling-and-reuse economy is an important way to buy some time. But this does not allow more and more people to use more and more resources indefinitely, even if all of them were somehow perfectly recycled and reused. The reason is that the two laws of thermodynamics tell us that recycling and reusing matter resources always requires using high-quality energy (which cannot be recycled) and adds waste heat to the environment.

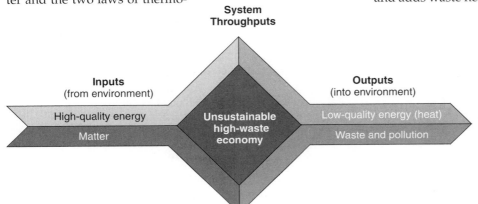

Figure 4-12 The *high-throughput economies* of most developed countries are based on continually increasing the rates of energy and matter flow. This produces valuable goods and services but also converts high-quality matter and energy resources into waste, pollution, and low-quality heat.

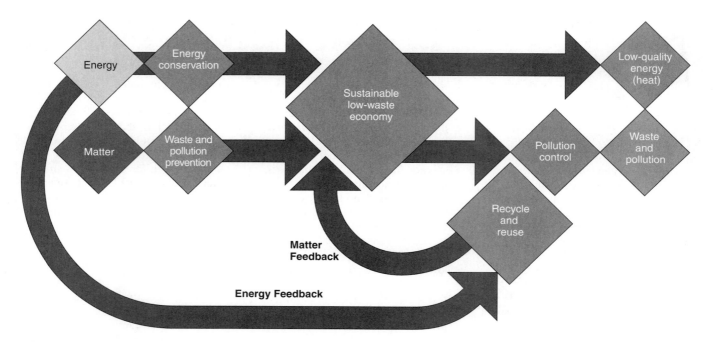

Figure 4-13 Solutions: lessons from nature. A *low-throughput economy*, based on energy flow and matter recycling, works with nature to reduce the throughput of matter and energy resources. This is done by (1) reusing and recycling most nonrenewable matter resources, (2) using renewable resources no faster than they are replenished, (3) using matter and energy resources efficiently, (4) reducing unnecessary consumption, (5) emphasizing pollution prevention and waste reduction, and (6) controlling population growth.

The three scientific laws governing matter and energy changes and the four principles of sustainability derived from nature (Figure 4-11) suggest that the best long-term solution to our environmental and resource problems is to shift from an economy based on maximizing matter and energy flow (throughput) to a more sustainable **low-throughput (low-waste) economy,** as summarized in Figure 4-13.

In the next chapter we will apply the principles of population dynamics and sustainability discussed in this chapter to the growth of the human population. In the two chapters after that we apply them to understanding and sustaining the earth's terrestrial and aquatic biodiversity.

We cannot command nature except by obeying her.

Sir Francis Bacon

CRITICAL THINKING

1. How would you reply to someone who argues we should not worry about our effects on natural systems because succession will heal the wounds of human activities and restore the balance of nature?

2. How would you determine whether a particular species found in a given area is a keystone species?

3. Explain why a simplified community such as a cornfield usually is much more vulnerable to harm from in-

sects and plant diseases than a more complex, natural community such as a grassland. Does this mean that we should never convert a grassland to a cornfield? Explain. What restrictions, if any, would you put on such conversions?

4. A bumper sticker reads "Nature always bats last and owns the stadium." What does this mean in ecological terms? What is its lesson for the human species?

5. Explain why you agree or disagree with the four principles of sustainability listed in Figure 4-11 (left) and their lessons for human societies listed in Figure 4-11 (right). Identify aspects of your lifestyle that follow or violate each of these four sustainability principles. Would you be willing to change the aspects of your lifestyle that violate these sustainability principles? Explain.

LEARNING ONLINE

The website for this book contains helpful study aids and many ideas for further reading and research. They include a chapter summary, review questions for the entire chapter, flash cards for key terms and concepts, a multiple-choice practice quiz, interesting Internet sites, references, and a guide for accessing thousands of InfoTrac® College Edition articles. Log on to

http://biology.brookscole.com/miller7

Then click on the Chapter-by-Chapter area, choose Chapter 4, and select a learning resource.

5 APPLYING POPULATION ECOLOGY: THE HUMAN POPULATION

The problems to be faced are vast and complex, but come down to this: 6.4 billion people are breeding exponentially. The process of fulfilling their wants and needs is stripping earth of its biotic capacity to produce life; a climactic burst of consumption by a single species is overwhelming the skies, earth, waters, and fauna.

PAUL HAWKEN

5-1 FACTORS AFFECTING HUMAN POPULATION SIZE

How Is Population Size Affected by Birth Rates and Death Rates? Entrances and Exits

Population increases because of births and immigration and decreases through deaths and emigration.

Human populations grow or decline through the interplay of three factors: *births*, *deaths*, and *migration*. **Population change** is calculated by subtracting the number of people leaving a population (through death and emigration) from the number entering it (through birth and immigration) during a specified period of time (usually a year):

$$\frac{\text{Population}}{\text{change}} = (\text{Births} + \text{Immigration}) - (\text{Deaths} + \text{Emigration})$$

When births plus immigration exceed deaths plus emigration, population increases; when the reverse is true, population declines.

Instead of using the total numbers of births and deaths per year, demographers use the **birth rate,** or **crude birth rate** (the number of live births per 1,000 people in a population in a given year), and the **death rate,** or **crude death rate** (the number of deaths per 1,000 people in a population in a given year). Figure 5-1 shows the crude birth and death rates for various groupings of countries in 2004.

How Fast Is the World's Population Growing? Good and Bad News

The rate at which the world's population increases has slowed, but the population is still growing fairly rapidly.

Birth rates and death rates are coming down worldwide, but death rates have fallen more sharply than birth rates. As a result, more births are occurring than deaths; every time your heart beats, 2.3 more babies are added to the world's population. At this rate, we share

the earth and its resources with about 219,000 more people per day—97% of them in developing countries.

The rate of the world's annual population change usually is expressed as a percentage:

$$\begin{aligned}\text{Annual rate of}\\ \text{natural population}\\ \text{change (\%)}\end{aligned} = \frac{\text{Birth rate} - \text{Death rate}}{1{,}000 \text{ persons}} \times 100$$

$$= \frac{\text{Birth rate} - \text{Death rate}}{10}$$

Exponential population growth has not disappeared but is occurring at a slower rate. The rate of the world's annual population growth (natural increase) dropped by almost half between 1963 and 2004, from 2.2% to 1.25%. This is *good news,* but during the same period the population base doubled from 3.2 billion to 6.4 billion. This drop in the rate of population increase is somewhat like learning that the truck heading straight

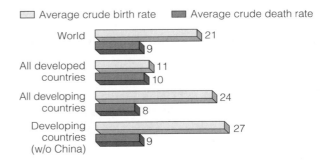

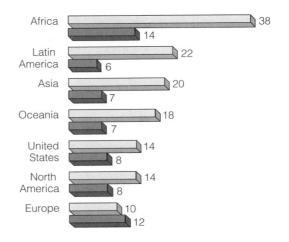

Figure 5-1 Average crude birth and death rates for various groupings of countries in 2004. (Data from Population Reference Bureau)

at you has slowed from 100 kilometers per hour (kph) to 43 kph while its weight has doubled.

An exponential growth rate of 1.25% may seem small. But in 2004, it added about 80 million people per year to the world's population, compared to 69 million in 1963 when the world's population growth reached its peak. An increase of 80 million people per year is roughly equal to adding another New York City every month, a Germany every year, and a United States every 3.7 years.

Also, there is a big difference between exponential population growth rates in developed and developing countries. In 2004, the population of developed countries was growing at a rate of 0.1%. That of the developing countries was 1.5%—fifteen times faster.

As a result of these trends, the population of the developed countries, currently at 1.2 billion, is expected to change little in the next 50 years. In contrast, the population of the developing countries is projected to rise steadily from 5.2 billion in 2004 to 7.9 billion in 2050. The six nations expected to experience most of this growth are, in order: India, China, Pakistan, Nigeria, Bangladesh, and Indonesia.

In numbers of people, China with 1.3 billion in 2004—about one of every five people in the world—and India with 1.1 billion dwarf all other countries. Together they have 37% of the world's population. The United States, with 294 million people in 2004, had the world's third largest population but only 4.6% of the world's people.

How Have Global Fertility Rates Changed? Having Fewer Babies per Woman

The average number of children that a woman bears has dropped sharply since 1950, but the number is not low enough to stabilize the world's population in the near future.

Fertility is the number of births that occur to an individual woman or in a population. Two types of fertility rates affect a country's population size and growth rate. The first type, **replacement-level fertility,** is the number of children a couple must bear to replace themselves. It is slightly higher than two children per couple (2.1 in developed countries and as high as 2.5 in some developing countries), mostly because some female children die before reaching their reproductive years.

Does reaching replacement-level fertility mean an immediate halt in population growth? No, because so many future parents are alive. If each of today's couples had an average of 2.1 children and their children also had 2.1 children, the world's population would continue to grow for 50 years or more (assuming death rates do not rise).

The second type of fertility rate is the **total fertility rate (TFR):** the average number of children a woman typically has during her reproductive years. In 2004,

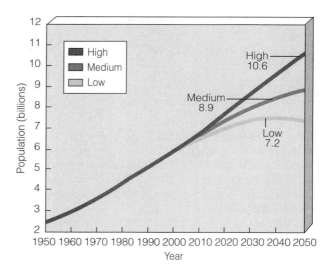

Figure 5-2 UN world population projections to 2050, assuming the world's total fertility rate is 2.6 (high), 2.1 (medium), or 1.5 (low) children per woman. The most likely projection is the medium one—8.9 billion by 2050. (Data from United Nations, *World Population Prospects: The 2000 Revision,* 2001)

the average global TFR was 2.8 children per woman. It was 1.6 in developed countries (down from 2.5 in 1950) and 3.1 in developing countries (down from 6.5 in 1950). This drop in the average number of children born to women in developing countries is impressive, but this level of fertility is still far above the replacement level of 2.1.

So how many of us are likely to be here in 2050? Answer: From 7.2 to 10.6 billion, depending on the world's projected average TFR (Figure 5-2). The medium projection is 8.9 billion people. About 97% of this growth is projected to take place in developing countries, where acute poverty (living on less than $1 per day) is a way of life for about 1.4 billion people.

How Have Fertility Rates Changed in the United States? Ups and Downs

Population growth in the United States has slowed but is not close to leveling off.

The population of the United States has grown from 76 million in 1900 to 294 million in 2004, despite oscillations in the country's TFR (Figure 5-3). In 1957, the peak of the baby boom after World War II, the TFR reached 3.7 children per woman. Since then it has generally declined, remaining at or below replacement level since 1972.

The drop in the TFR has led to a decline in the rate of population growth in the United States. But the country's population is still growing faster than that of any other developed country and is not close to leveling off.

About 2.9 million people were added to the U.S. population in 2004. About 59% of this growth was the

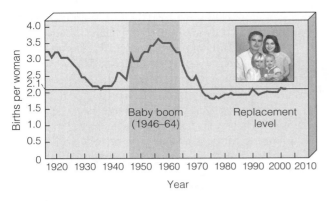

Figure 5-3 Total fertility rates for the United States between 1917 and 2004. (Data from Population Reference Bureau and U.S. Census Bureau)

result of more births than deaths and the rest came from legal and illegal immigration.

According to U.S. Bureau of Census medium projections, the U.S. population is likely to increase from 294 million in 2004 to 420 million by 2050 and reach 571 million by 2100. In contrast, population growth has slowed in other major developed countries since 1950 and most are expected to have declining populations after 2010. Because of a high per capita rate of resource use, each addition to the U.S. population has an enormous environmental impact (Figure 1-10, p. 13).

How do the population dynamics of the United States compare with those of its neighbors Canada and Mexico? Find out by looking at Figure 5.4.

What Factors Affect Birth Rates and Fertility Rates? Reducing Births

The number of children women have is affected by the cost of raising and educating children, educational and employment opportunities for women, infant deaths, marriage age, and availability of contraceptives and abortions.

Many factors affect a country's average birth rate and TFR. One is the *importance of children as a part of the labor force.* Proportions of children working tend to be higher in developing countries—especially in rural areas, where children begin working to help raise crops at an early age.

Another economic factor is the *cost of raising and educating children.* Birth and fertility rates tend to be lower in developed countries, where raising children

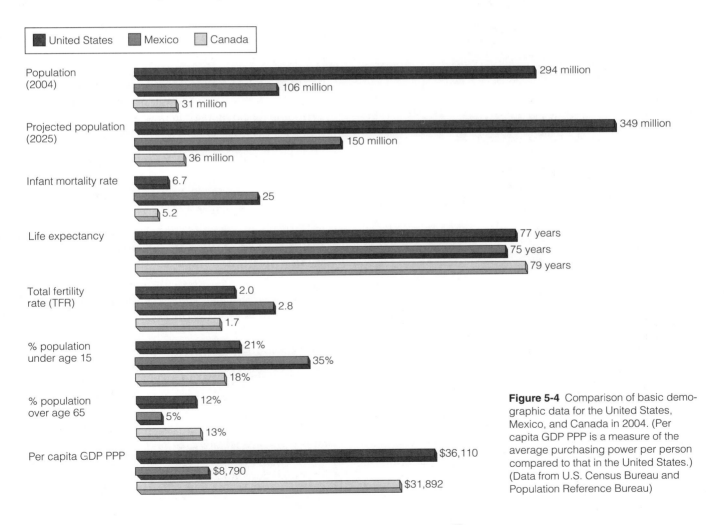

Figure 5-4 Comparison of basic demographic data for the United States, Mexico, and Canada in 2004. (Per capita GDP PPP is a measure of the average purchasing power per person compared to that in the United States.) (Data from U.S. Census Bureau and Population Reference Bureau)

Extremely Effective

Method	Effectiveness
Total abstinence	100%
Sterilization	99.6%
Vaginal ring	98–99%

Highly Effective

Method	Effectiveness
IUD with slow-release hormones	98%
IUD plus spermicide	98%
Vaginal pouch ("female condom")	97%
IUD	95%
Condom (good brand) plus spermicide	95%
Oral contraceptive	93%

Effective

Method	Effectiveness
Cervical cap	89%
Condom (good brand)	86%
Diaphragm plus spermicide	84%
Rhythm method (Billings, Sympto-Thermal)	84%
Vaginal sponge impregnated with spermicide	83%
Spermicide (foam)	82%

Moderately Effective

Method	Effectiveness
Spermicide (creams, jellies, suppositories)	75%
Rhythm method (daily temperature readings)	74%
Withdrawal	74%
Condom (cheap brand)	70%

Unreliable

Method	Effectiveness
Douche	40%
Chance (no method)	10%

Figure 5-5 Typical effectiveness rates of birth control methods in the United States. Percentages are based on the number of undesired pregnancies per 100 couples using a specific method as their sole form of birth control for a year. For example, an effectiveness rating of 93% for oral contraceptives means that for every 100 women using the pill regularly for 1 year, 7 will get pregnant. Effectiveness rates tend to be lower in developing countries, primarily because of lack of education. Globally, about 39% of the world's people using contraception rely on sterilization (32% of females and 7% of males), followed by IUDs (22%), the pill (14%), and male condoms (7%). Preferences in the United States are female sterilization (26%), the pill (25%), male condoms (19%), and male sterilization (10%). (Data from Alan Guttmacher Institute, Henry J. Kaiser Family Foundation, and the United Nations Population Division)

is much more costly because they do not enter the labor force until they are in their late teens or 20s.

The *availability of private and public pension systems* affects how many children couples have. Pensions eliminate parents' need to have many children to help support them in old age.

Urbanization plays a role. Why? Because people living in urban areas usually have better access to family planning services and tend to have fewer children than those living in rural areas where children are needed to perform essential tasks.

Another important factor is *the educational and employment opportunities available for women.* TFRs tend to be low when women have access to education and paid employment outside the home. In developing countries, women with no education generally have two more children than women with a secondary school education.

Another factor is the *infant mortality rate.* In areas with low infant mortality rates, people tend to have a smaller number of children because fewer children die at an early age.

Average age at marriage (or, more precisely, the average age at which women have their first child) also plays a role. Women normally have fewer children when their average age at marriage is 25 or older.

Birth rates and TFRs are also affected by the *availability of legal abortions.* Each year about 190 million women become pregnant. The United Nations and the World Bank estimate that about 46 million of these women get abortions: 26 million of them legal and 20 million illegal (and often unsafe).

The *availability of reliable birth control methods* (Figure 5-5) allows women to control the number and spacing of the children they have. *Religious beliefs, traditions, and cultural norms* also play a role. In some countries, these factors favor large families and strongly oppose abortion and some forms of birth control.

What Factors Affect Death Rates? Reducing Deaths

Death rates have declined because of increased food supplies, better nutrition, advances in medicine, improved sanitation and personal hygiene, and safer water supplies.

The rapid growth of the world's population over the past 100 years is not the result of a rise in the crude birth rate. Instead, it has been caused largely by a decline in crude death rates, especially in developing countries.

More people started living longer and fewer infants died because of increased food supplies and distribution, better nutrition, medical advances such as immunizations and antibiotics, improved sanitation, and safer water supplies (which curtailed the spread of many infectious diseases).

Two useful indicators of the overall health of people in a country or region are **life expectancy** (the average number of years a newborn infant can expect to live) and the **infant mortality rate** (the number of babies out of every 1,000 born who die before their first birthday).

Great news. The global life expectancy at birth increased from 48 years to 67 years (76 years in developed countries and 65 years in developing countries) between 1955 and 2004. It is projected to reach 74 by 2050. Between 1900 and 2004, life expectancy in the United States increased from 47 to 77 years and is projected to reach 82 years by 2050.

Bad news. In the world's poorest and least developed countries, mainly in Africa, life expectancy is 49 years or less. In many African countries, life expectancy is expected to fall further because of more deaths from AIDS.

Infant mortality is viewed as the best single measure of a society's quality of life because it reflects a country's general level of nutrition and health care. A high infant mortality rate usually indicates insufficient food (undernutrition), poor nutrition (malnutrition), and a high incidence of infectious disease (usually from contaminated drinking water and weakened disease resistance from undernutrition and malnutrition).

Good news. Between 1965 and 2004, the world's infant mortality rate dropped from 20 per 1,000 live births to 7 in developed countries and from 118 to 62 in developing countries. *Bad news.* At least 8 million infants (most in developing countries) die of preventable causes during their first year of life—an average of 22,000 mostly unnecessary infant deaths per day. This is equivalent to 55 jumbo jets, each loaded with 400 infants under age 1, crashing each day with no survivors!

The U.S. infant mortality rate declined from 165 in 1900 to 6.7 in 2004. This sharp decline was a major factor in the marked increase in U.S. average life expectancy during this period.

Still, some 40 countries had lower infant mortality rates than the United States in 2004. Three factors that keep the U.S. infant mortality rate higher than it could be: *inadequate health care for poor women during pregnancy and for their babies after birth, drug addiction among pregnant women,* and *a high birth rate among teenagers.*

Case Study: Should the United States Encourage or Discourage Immigration? An Important Issue

Immigration has played, and continues to play, a major role in the growth and cultural diversity of the U.S. population.

Since 1820, the United States admitted almost twice as many immigrants and refugees as all other countries combined! However, the number of legal immigrants (including refugees) has varied during different periods because of changes in immigration laws and rates of economic growth (Figure 5-6). Currently, legal and illegal immigration account for about 41% of the country's annual population growth.

Between 1820 and 1960, most legal immigrants to the United States came from Europe. Since 1960, most have come from Latin America (51%) and Asia (30%), followed by Europe (13%). Latinos (67% of them from Mexico) made up 14% of the U.S. population in 2003. By 2050, Latinos are projected to make up one of every four people in the United States.

In 1995, the U.S. Commission on Immigration Reform recommended reducing the number of legal immigrants from about 900,000 to 700,000 per year for a transition period and then to 550,000 a year. Some analysts want to limit legal immigration to about 20% of the country's annual population growth. They would accept immigrants only if they can support themselves, arguing that providing immigrants with public services makes the United States a magnet for the world's poor.

There is also support for efforts to sharply reduce illegal immigration. But some are concerned that a crackdown on the country's 8–10 million illegal immigrants can lead to discrimination against legal immigrants.

Proponents of reducing immigration argue that it would allow the United States to stabilize its population sooner and help reduce the country's enormous environmental impact. The public strongly supports reducing U.S. immigration levels. A January 2002 Gallup poll found that 58% of the people polled believed that immigration rates should be reduced (up from 45% in

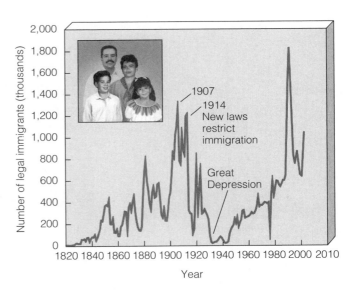

Figure 5-6 Legal immigration to the United States, 1820–2002. The large increase in immigration since 1989 resulted mostly from the Immigration Reform and Control Act of 1986, which granted legal status to illegal immigrants who could show they had been living in the country for several years. (Data from U.S. Immigration and Naturalization Service)

January 2001). A 1993 Hispanic Research Group survey found that 89% of Hispanic Americans supported an immediate moratorium on immigration.

Others oppose reducing current levels of legal immigration. They argue that this would diminish the historical role of the United States as a place of opportunity for the world's poor and oppressed. In addition, immigrants pay taxes, take many menial and low-paying jobs that other Americans shun, open businesses, and create jobs. Moreover, according to the U.S. Census Bureau, after 2020 higher immigration levels will be needed to supply enough workers as baby boomers retire.

✗ _How Would You Vote?_ Should immigration into the United States (or the country where you live) be reduced? Cast your vote online at http://biology.brookscole.com/miller7.

5-2 POPULATION AGE STRUCTURE

What Are Age-Structure Diagrams? Sorting People by Age Groups

The number of people in young, middle, and older age groups determines how fast populations grow or decline.

As mentioned earlier, even if the replacement-level fertility rate of 2.1 were magically achieved globally tomorrow, the world's population would keep growing for at least another 50 years (assuming no large increase in the death rate). The reason for this is a population's **age structure:** the distribution of males and females in each age group.

Population experts (demographers) construct a population age-structure diagram by plotting the percentages or numbers of males and females in the total population in each of three age categories: _prereproductive_ (ages 0–14), _reproductive_ (ages 15–44), and _postreproductive_ (ages 45 and up). Figure 5-7 presents generalized age-structure diagrams for countries with rapid, slow, zero, and negative population growth rates. Which of these figures best represents the country where you live?

How Does Age Structure Affect Population Growth? Teenagers Are the Population Wave of the Future.

The number of people under age 15 is the major factor determining a country's future population growth.

Any country with many people below age 15 (represented by a wide base in Figure 5-7, left) has a powerful built-in momentum to increase its population size unless death rates rise sharply. The number of births rises even if women have only one or two children because a large number of girls will soon be moving into their reproductive years.

What is perhaps the world's most important population statistic? Answer: _30% of the people on the planet were under 15 years old in 2004._ These 1.9 billion young people are poised to move into their prime reproductive years. In developing countries the number is even higher: 33%, compared with 17% in developed countries. We live in a _demographically divided world_, as shown by population data for the United States, Brazil, and Nigeria (Figure 5-8).

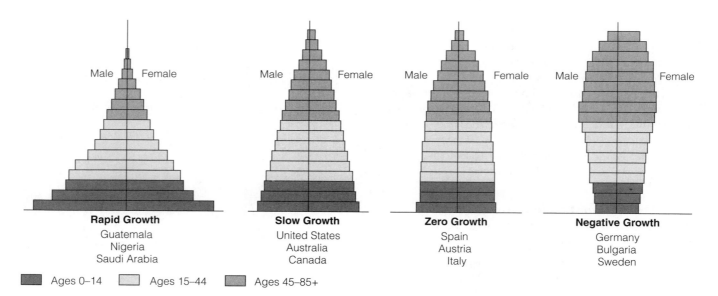

Rapid Growth	Slow Growth	Zero Growth	Negative Growth
Guatemala	United States	Spain	Germany
Nigeria	Australia	Austria	Bulgaria
Saudi Arabia	Canada	Italy	Sweden

■ Ages 0–14   □ Ages 15–44   ▨ Ages 45–85+

Figure 5-7 Generalized population age-structure diagrams for countries with rapid, (1.5–3%), slow (0.3–1.4%), zero (0–0.2%), and negative population growth rates (a declining population). (Data from Population Reference Bureau)

Legend:
- United States (highly developed)
- Brazil (moderately developed)
- Nigeria (less developed)

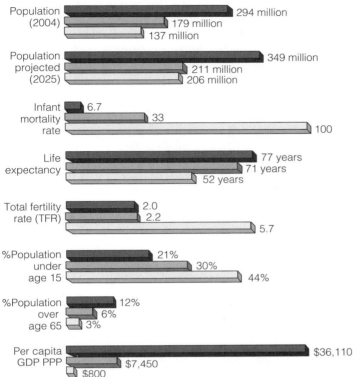

Population (2004): 294 million, 179 million, 137 million

Population projected (2025): 349 million, 211 million, 206 million

Infant mortality rate: 6.7, 33, 100

Life expectancy: 77 years, 71 years, 52 years

Total fertility rate (TFR): 2.0, 2.2, 5.7

%Population under age 15: 21%, 30%, 44%

%Population over age 65: 12%, 6%, 3%

Per capita GDP PPP: $36,110, $7,450, $800

Figure 5-8 Comparison of key demographic indicators in highly developed (United States), moderately developed (Brazil), and less developed (Nigeria) countries in 2004. (Data from Population Reference Bureau)

How Can Age-Structure Diagrams Be Used to Make Population and Economic Projections? Looking into a Crystal Ball

Changes in the distribution of a country's age groups have long-lasting economic and social impacts.

Between 1946 and 1964, the United States had a *baby boom* that added 79 million people to its population. Over time, this group looks like a bulge moving up through the country's age structure (Figure 5-9).

Baby boomers now make up nearly half of all adult Americans. As a result, they dominate the population's demand for goods and services. They also play an increasingly important role in deciding who gets elected and what laws are passed. Baby boomers who created the youth market in their teens and 20s are now creating the 50-something market and will soon move on to create a 60-something market. In 2011, the first baby boomers will turn 65 and the number of Americans over age 65 will grow sharply through 2029.

According to some analysts, the retirement of baby boomers is likely to create a shortage of workers in the United States unless immigrant workers replace some of these retirees. Retired baby boomers are likely to use their political clout to force the smaller number of people in the baby-bust generation that followed them (Figure 5-3) to pay higher income, health-care, and Social Security taxes.

In other respects, the baby-bust generation should have an easier time than the baby-boom generation. Fewer people will be competing for educational opportunities, jobs, and services. Also, labor shortages

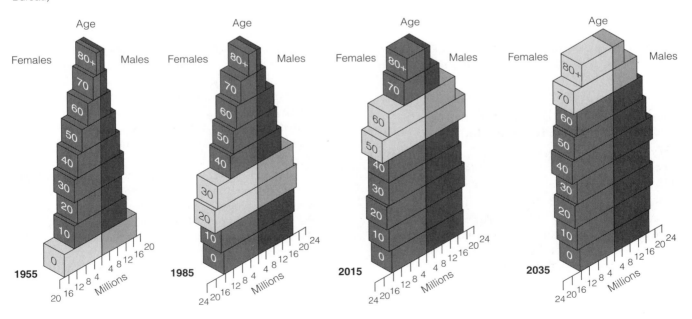

Figure 5-9 Tracking the baby-boom generation in the United States. (Data from Population Reference Bureau and U.S. Census Bureau

may drive up their wages, at least for jobs requiring education or technical training beyond high school.

From these few projections, we can see that any booms or busts in the age structure of a population create social and economic changes that ripple through a society for decades.

What Are Some Effects of Population Decline from Reduced Fertility? Sliding Down a Hill Too Fast Can Hurt.

Rapid population decline can lead to long-lasting economic and social problems.

The populations of most of the world's countries are projected to grow throughout most of this century. By 2004, however, 40 countries had populations that were either stable (annual growth rates at or below 0.3%) or declining. All, except Japan, are in Europe. This means that about 14% of humanity (896 million people) lives in countries with stable or declining populations. By 2050, the UN projects that the population size of most developed countries (but not the United States) will have stabilized.

As the age structure of the world's population changes and the percentage of people age 60 or older increases, more countries will begin experiencing population declines. If population decline is gradual, its harmful effects usually can be managed.

But rapid population decline, like rapid population growth, can lead to severe economic and social problems. A country undergoing rapid population decline because of a "baby bust" or a "birth dearth" has a sharp rise in the proportion of older people. They consume an increasingly larger share of medical care, social security funds, and other costly public services funded by a decreasing number of working taxpayers. Such countries can also face labor shortages unless they rely more on greatly increased automation or immigration of foreign workers.

What Are Some Effects of Population Decline from a Rise in Death Rates? The AIDS Tragedy

Large numbers of deaths from AIDS disrupts a country's social and economic structure by removing large numbers of young adults from its age structure.

Globally, between 2000 and 2050, AIDS is projected to cause the premature deaths of 278 million people in 53 countries—38 of them in Africa. These premature deaths are almost equal to the entire current population of the United States. Read this paragraph again and think about the enormity of this tragedy.

Hunger and malnutrition kill mostly infants and children, but AIDS kills many young adults. This change in the young adult age structure of a country has a number of harmful effects. One is a sharp drop in average life expectancy. In 16 African counties, where up to a third of the adult population is infected with HIV, life expectancy could drop to 35–40 years of age.

Another effect is a loss of a country's most productive young adult workers and trained personnel such as scientists, farmers, engineers, teachers, and government, business, and health-care workers. This causes a sharp drop in the number of productive adults available to support the young and the elderly and to grow food.

Analysts call for the international community—especially developed countries—to develop and fund a massive program to help countries ravaged by AIDS in Africa and elsewhere. The program would have two major goals. One is to reduce the spread of HIV through a combination of improved education and health care. The other is to provide financial assistance for education and health care as well as volunteer teachers and health-care and social workers to help compensate for the missing young adult generation.

5-3 SOLUTIONS: INFLUENCING POPULATION SIZE

What Are the Advantages and Disadvantages of Reducing Births? An Important Controversy

There is disagreement over whether the world should encourage or discourage population growth.

The projected increase of the human population from 6.4 to 8.9 billion or more between 2004 and 2050 raises an important question: *Can the world provide an adequate standard of living for 2.5 billion more people without causing widespread environmental damage?*

Controversy surrounds this and two related questions: whether the earth is overpopulated, and what measures, if any, should be taken to slow population growth? To some the planet is already overpopulated. To others we should encourage population growth to help stimulate economic growth by having more consumers.

Those who do not believe the earth is overpopulated point out that the average life span of the world's 6.4 billion people is longer today than at any time in the past and is projected to get longer. They say that the world can support billions more people. They also see more people as the world's most valuable resource for solving the problems we face and stimulating economic growth by becoming consumers.

Some believe all people should be free to have as many children as they want. Some view any form of population regulation as a violation of their religious beliefs. Others see it as an intrusion into their privacy and personal freedom. Some developing countries and some members of minorities in developed countries regard population control as a form of genocide to keep their numbers and power from rising.

Proponents of slowing and eventually stopping population growth have a different view. They point out that we fail to provide the basic necessities for one out of six people on the earth today. If we cannot or will not do this now, they ask, how will we be able to do this for the projected 2.5 billion more people by 2050? See the Guest Essay by Garrett Hardin on this topic on the website for this chapter.

Proponents of slowing population growth warn of two serious consequences if we do not sharply lower birth rates. One possibility is a higher death rate because of declining health and environmental conditions in some areas—something that is already happening in parts of Africa. Another is increased resource use and environmental harm as more consumers increase their already large ecological footprint in developed countries and in some developing countries, such as China and India, which are undergoing rapid economic growth.

Population increase and the consumption that goes with it can increase environmental stresses such as *infectious disease, biodiversity losses, loss of tropical forests, fisheries depletion, increasing water scarcity, pollution of the seas,* and *climate change.*

Proponents of this view recognize that population growth is not the only cause of these problems. But they argue that adding several hundred million more people in developed countries and several billion more in developing countries can only intensify existing environmental and social problems.

These analysts believe people should have the freedom to produce as many children as they want, but only if it does not reduce the quality of other people's lives now and in the future, either by impairing the earth's ability to sustain life, or by causing social disruption. They point out that limiting the freedom of individuals to do anything they want, in order to protect the freedom of other individuals, is the basis of most laws in modern societies.

X *HOW WOULD YOU VOTE?* Should the population of the country where you live be stabilized as soon as possible? Cast your vote online at http://biology.brookscole.com/miller7.

How Can Economic Development Help Reduce Birth Rates? Economics Worked Once, but Will It Work Again?

History indicates that as countries become economically developed their birth and death rates decline.

Demographers have examined the birth and death rates of western European countries that industrial-ized during the 19th century. From these data they developed a hypothesis of population change known as the **demographic transition:** as countries become industrialized, first their death rates and then their birth rates decline.

According to this hypothesis, the transition takes place in four distinct stages (Figure 5-10, p. 86). First is the *preindustrial stage,* when there is little population growth because harsh living conditions lead to both a high birth rate (to compensate for high infant mortality) and a high death rate.

Next is the *transitional stage,* when industrialization begins, food production rises, and health care improves. Death rates drop and birth rates remain high, so the population grows rapidly (typically 2.5–3% a year).

During the third phase, called the *industrial stage,* the birth rate drops and eventually approaches the death rate as industrialization, medical advances, and modernization become widespread. Population growth continues, but at a slower and perhaps fluctuating rate, depending on economic conditions. Most developed countries and a few developing countries are now in this third stage.

The last phase is the *postindustrial stage,* when the birth rate declines further, equaling the death rate and reaching zero population growth. Then the birth rate falls below the death rate and population size decreases slowly. Forty countries containing about 14% of the world's population have entered this stage and more of the world's developed countries are expected to enter this phase by 2050.

In most developing countries today, death rates have fallen much more than birth rates. In other words, these developing countries are still in the transitional stage, halfway up the economic ladder, with high population growth rates.

Some economists believe developing countries will make the demographic transition over the next few decades. But some population analysts fear the still-rapid population growth in many developing countries will outstrip economic growth and overwhelm some local life-support systems. This could cause many of these countries to be caught in a *demographic trap* at stage 2, the transition stage. This is now happening as death rates rise in a number of developing countries, especially in Africa. Indeed, countries in Africa being ravaged by the HIV/AIDS epidemic are falling back to stage 1.

Analysts also point out that some of the conditions that allowed developed countries to develop are not available to many of today's developing countries. One problem is a shortage of skilled workers needed to produce the high-tech products necessary to compete in today's global economy. Another is a lack of the capital and other resources that allow rapid economic development.

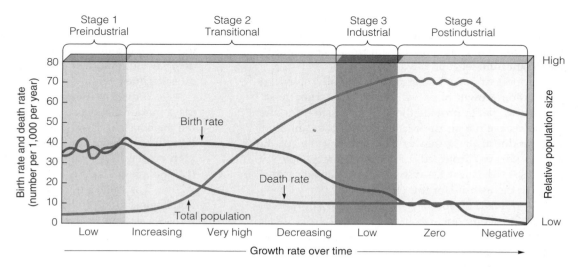

Figure 5-10 Generalized model of the *demographic transition.*

Two other problems hinder economic development in many developing countries. One is a sharp rise in their debt to developed countries. Much of the income of such countries has to be used to pay the interest on their debts. This leaves too little money for improving social, health, and environmental conditions.

Another problem is that since 1980, developing countries have experienced a drop in economic assistance from developed countries. Indeed, since the mid-1980s, developing countries have paid developed countries $40–50 billion a year (mostly in debt interest) more than they have received from these countries.

How Can Family Planning Help Reduce Birth and Abortion Rates and Save Lives? Planning for Babies Works

Family planning has been a major factor in reducing the number of births and abortions throughout most of the world.

Family planning provides educational and clinical services that help couples choose how many children to have and when to have them. Such programs vary from culture to culture, but most provide information on birth spacing, birth control, and health care for pregnant women and infants.

Family planning has helped increase the proportion of married women in developing countries who use modern forms of contraception from 10% of married women of reproductive age in the 1960s to 51% of these women in 2004. Studies also show that family planning is responsible for at least 55% of the drop in TFRs in developing countries, from 6 in 1960 to 3.1 in 2004. For example, a national family planning program was the major factor in a sharp drop in population growth and the TFR in Thailand between 1971 and 1986. Family planning has also reduced the num-

ber of legal and illegal abortions per year and the risk of maternal and fetus death from pregnancy.

Despite such successes, there is also some *bad news. First,* according to John Bongaarts of the Population Council and the UNFPA, 42% of all pregnancies in the developing countries are unplanned and 26% end with abortion. *Second,* an estimated 150 million women in developing countries want to limit the number and determine the spacing of their children, but they lack access to contraceptive services. According to the United Nations, extending family planning services to these women and to those who will soon be entering their reproductive years could prevent an estimated 5.8 million births a year and more than 5 million abortions a year!

Some analysts call for expanding family planning programs to include teenagers and sexually active unmarried women, who are excluded in many existing programs. For teenagers, many advocate much greater emphasis on abstinence.

Another suggestion is to develop programs that educate men about the importance of having fewer children and taking more responsibility for raising them. Proponents also call for greatly increased research on developing new, more effective, and more acceptable birth control methods for men.

Finally, a number of analysts urge pro-choice and pro-life groups to join forces in greatly reducing unplanned births and abortions, especially among teenagers.

How Can Empowering Women Help Reduce Birth Rates? Ensuring Education, Jobs, and Rights

Women tend to have fewer children if they are educated, have a paying job outside the home, and do not have their human rights suppressed.

What key factors lead women to have fewer and healthier children? Three things: education, paying jobs outside the home, and living in societies where their rights are not suppressed.

Women make up roughly half of the world's population. They do almost all of the world's domestic work and childcare with little or no pay. Women also provide more unpaid health care than all the world's organized health services combined.

They also do 60–80% of the work associated with growing food, gathering fuelwood, and hauling water in rural areas of Africa, Latin America, and Asia. As one Brazilian woman put it, "For poor women the only holiday is when you are asleep."

Globally, women account for two-thirds of all hours worked but receive only 10% of the world's income, and they own less than 2% of the world's land. In most developing countries, women do not have the legal right to own land or to borrow money. Women also make up 70% of the world's poor and 60% of the 875 million illiterate adults worldwide who can neither read nor write.

According to United Nations Population Agency's executive director Thorya Obaid, "Many women in the developing world are trapped in poverty by illiteracy, poor health, and unwanted high fertility. All of these contribute to environmental degradation and tighten the grip of poverty. If we are serious about sustainable development, we must break this vicious cycle."

Breaking out of this trap means giving women everywhere full legal rights and the opportunity to become educated and earn income outside the home. Achieving this would slow population growth, promote human rights and freedom, reduce poverty, and slow environmental degradation—a win-win result.

Empowering women by seeking gender equality will take some major social changes. This will be difficult to achieve in male-dominated societies but it can be done.

Good news. An increasing number of women in developing countries are taking charge of their lives and reproductive behavior. They are not waiting around for the slow processes of education and cultural change. As it expands, such bottom-up change by individual women will play an important role in stabilizing population and providing women with equal rights.

How Can Population Growth Be Reduced? A New Vision

Experience suggests that the best way to slow population growth is a combination of investing in family planning, reducing poverty, and elevating the status of women.

In 1994, the United Nations held its third once-in-a-decade Conference on Population and Development in Cairo, Egypt. One of the conference's goals was to en-courage action to stabilize the world's population at 7.8 billion by 2050 instead of the projected 8.9 billion.

The major goals of the resulting population plan, endorsed by 180 governments, are to do the following by 2015:

- Provide universal access to family planning services and reproductive health care
- Improve health care for infants, children, and pregnant women
- Develop and implement national population polices
- Improve the status of women and expand education and job opportunities for young women
- Provide more education, especially for girls and women
- Increase the involvement of men in child-rearing responsibilities and family planning
- Sharply reduce poverty
- Sharply reduce unsustainable patterns of production and consumption

This is a tall order. But it can be done if developed and developing nations work together to implement such reforms. Some *good news* is that the experience of Japan, Thailand, South Korea, Taiwan, Iran, and China indicates that a country can achieve or come close to replacement-level fertility within a decade or two. Such experience also suggests that the best way to slow population growth is a combination of investing in family planning, reducing poverty, and elevating the status of women.

5-4   CASE STUDIES: INDIA AND CHINA

What Success Has India Had in Controlling Its Population Growth? Some Progress but Not Enough

For over five decades India has tried to control its population growth with only modest success.

The world's first national family planning program began in India in 1952, when its population was nearly 400 million. In 2004, after 52 years of population control efforts, India was the world's second most populous country, with a population of 1.1 billion.

In 1952, India added 5 million people to its population. In 2004, it added 18 million. Figure 5-11 (p. 88) compares demographic data for India and China.

India faces a number of already serious poverty, malnutrition, and environmental problems that could worsen as its population continues to grow rapidly. By global standards, about one of every four people in

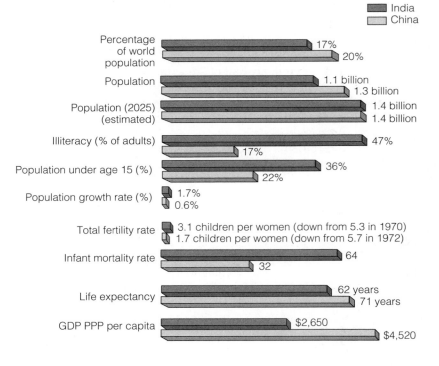

India is poor. And nearly half of India's labor force is unemployed or can find only occasional work.

India currently is self-sufficient in food grain production. Still, about 40% of its population and 53% of its children suffer from malnutrition, mostly because of poverty.

Furthermore, India faces serious resource and environmental problems. With 17% of the world's people, it has just 2.3% of the world's land resources and 2% of the world's forests. About half of the country's cropland is degraded as a result of soil erosion, waterlogging, salinization, overgrazing, and deforestation. In addition, over two-thirds of India's water is seriously polluted and sanitation services often are inadequate.

Without its long-standing family planning program, India's population and environmental problems would be growing even faster. Still, to its supporters the results of the program have been disappointing for several reasons: poor planning, bureaucratic inefficiency, the low status of women (despite constitutional guarantees of equality), extreme poverty, and lack of administrative and financial support.

The government has provided information about the advantages of small families for years. Yet Indian women still have an average of 3.1 children. One reason is that most poor couples believe they need many children to do work and care for them in old age. Another is the strong cultural preference for male children, which means some couples keep having children until they produce one or more boys. These factors in part explain why even though 90% of Indian couples know of at least one modern birth control method, only 43% actually use one.

What Success Has China Had in Controlling Its Population Growth? Good Progress, Enforced with an Iron Hand

Since 1970 China has used a government-enforced program to cut its birth rate in half and sharply reduce its fertility rate.

Since 1970, China has made impressive efforts to feed its people and bring its population growth under con-

trol. Between 1972 and 2004, China cut its crude birth rate in half and cut its TFR from 5.7 to 1.7 children per woman (Figure 5-11).

To achieve its sharp drop in fertility, China has established the world's most extensive, intrusive, and strict population control program. Couples are strongly urged to postpone marriage and to have no more than one child. Married couples who pledge to have no more than one child receive extra food, larger pensions, better housing, free medical care, salary bonuses, free school tuition for their one child, and preferential treatment in employment when their child enters the job market. Couples who break their pledge lose such benefits.

The government also provides married couples with ready access to free sterilization, contraceptives, and abortion. This helps explain why about 83% of married women in China use modern contraception.

Government officials realized in the 1960s that the only alternative to strict population control was mass starvation. China is a dictatorship. Thus, unlike India, it has been able to impose a fairly consistent population policy throughout its society.

However, there is a strong preference in China (and in India) for male children because there is no real social security system. A folk saying goes, "Rear a son, protect yourself in old age." As a result, many pregnant women use ultrasound to determine the gender and often get an abortion if the fetus is female. This has lead to a growing *gender imbalance* in China's population, with a projected 30–40 million surplus of men by 2020.

China has 20% of the world's population. But it has only 7% of the world's fresh water and cropland, 4% of its forests, and 2% of its oil. Soil erosion in China is serious and apparently getting worse.

Population experts expect China's population to peak around 2040 and then begin a slow decline. This has led some members of China's parliament to call for amending the country's one-child policy so that some urban couples can have a second child. The goal would be to provide more workers to help support China's aging population.

What lesson can other countries learn from China? One possibility is to try to curb population growth before they must choose between mass starvation and coercive measures that severely restrict human freedom.

5-5 POPULATION DISTIRIBUTION: URBAN GROWTH AND PROBLEMS

How Fast Are Urban Areas Growing? More People and More Poverty

Urban populations are growing rapidly throughout the world and many cities in developing countries have become centers of poverty.

Today, almost half of the world's people live in densely populated urban areas as rural people have migrated to cities with the hope of finding jobs and a better life. Rural people are *pulled* to urban areas in search of jobs, food, housing, a better life, entertainment, and freedom from religious, racial, and political conflicts. Some are also *pushed* from rural areas into urban areas by factors such as poverty, lack of land to grow food, declining agricultural jobs, famine, and war.

Five major trends are important in understanding the problems and challenges of urban growth: First, *the proportion of the global population living in urban areas is increasing.* Between 1850 and 2004, the percentage of people living in urban areas increased from 2% to 48%. If UN projections are correct, by 2030 about 60% of the world's people will be living in urban areas. This means that between 2004 and 2030 the world's urban population is projected to increase from 3.1 billion to 5 billion. Almost all of this growth will occur in already overcrowded cities in developing countries (Figure 5-12). Use Figure 5-12 to list in order the world's five most populous cities in 2004 and the five most populous cities projected for 2015.

Second, *the number of large cities is mushrooming.* In 2004, more than 400 cities had a million or more people and this is projected to increase to 564 by 2015. Today there are 18 *megacities* or *meagalopolises* (up from 8 in 1985) with 10 million or more people—most of them in developing countries (Figure 5-12). As they grow and sprawl outward, separate urban areas may

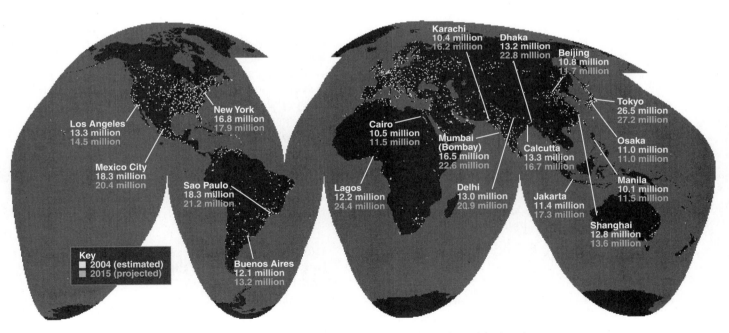

Figure 5-12 Major urban areas throughout the world based on satellite images of the earth at night that show city lights. Currently, the 48% of the world's people living in urban areas occupy about 2% of the earth's land area. Note that most of the world's urban areas are found along the coasts of continents, and most of Africa and much of the interior of South America, Asia, and Australia are dark at night. This figure also shows the populations of the world's 18 *megacities* with 10 or more million people in 2004, and their projected populations in 2015. Note that all but four are located in developing countries. (Data from National Geophysics Data Center, National Oceanic and Atmospheric Administration, and United Nations)

merge to form a *megalopolis*. For example, the remaining open space between Boston, Massachusetts, and Washington, D.C., is rapidly urbanizing and coalescing. The result is an almost 800-kilometer-long (500-mile-long) urban area that is sometimes called *Bowash* (Figure 5-13).

A third trend is that *the urban population is increasing rapidly in developing countries*. Between 2004 and 2030, the percentage of people living in urban areas in developing countries is expected to increase from 41% to 56%. However, about 75% of the people in South America live in cities, mostly along the coasts.

Fourth, *urban growth is much slower in developed countries* (with 76% urbanization) *than in developing countries*. Still, developed countries are projected to reach 84% urbanization by 2030.

Fifth, *poverty is becoming increasingly urbanized as more poor people migrate from rural to urban areas, mostly in developing countries*. The United Nations estimates that at least 1 billion people live in the crowded urban areas of developing countries. If you visit poor areas of such a city your senses may be overwhelmed with a chaotic but vibrant crush of people, vehicles of all sorts, street vendors, traffic jams, noise, smells, smoke from wood and coal fires, and people sleeping on streets or living in crowed, unsanitary, rickety, and unsafe slums and shantytowns.

How Urbanized Is the United States? City-Dwellers Dominate

About eight of every ten Americans live in urban areas with about half of them living in increasingly sprawling suburbs.

Between 1800 and 2004, the percentage of the U.S. population living in urban areas increased from 5% to 79%.

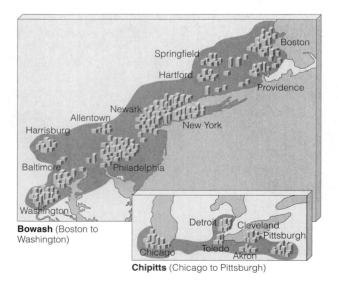

Bowash (Boston to Washington)

Chipitts (Chicago to Pittsburgh)

Figure 5-13 Two megalopolises: *Bowash*, consisting of urban sprawl and coalescence between Boston and Washington, D.C., and *Chipitts*, extending from Chicago to Pittsburgh.

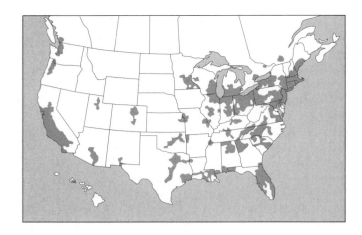

Figure 5-14 Major urban regions in the United States. About 79% of Americans live in urban areas occupying about 1.7% of the land area of the lower 48 states. Nearly half (48%) of Americans live in *consolidated metropolitan areas* with 1 million or more people. These areas are projected to merge into the megalopolises shown as shaded areas in this figure. (Data from U.S. Census Bureau)

The population has shifted in four phases. First, *people migrated from rural areas to large central cities*. Currently, three-fourths of Americans live in 271 *metropolitan areas* (cities and towns with at least 50,000 people), and nearly half of the country's population lives in consolidated metropolitan areas containing 1 million or more residents (Figure 5-14).

Second, many people *migrated from large central cities to suburbs and smaller cities*. Currently, about 51% of the U.S. population lives in the suburbs and 30% live in central cities.

Third, many people *migrated from the North and East to the South and West*. Since 1980, about 80% of the U.S. population increase has occurred in the South and West, particularly near the coasts. California in the West, with 34.5 million people, is the most populous state, followed by Texas in the Southwest with 21.3 million people. This shift is expected to continue.

Fourth, *some people have migrated from urban areas back to rural areas* since the 1970s, and especially since 1990.

How Has the Quality of Urban Life in the United States Changed? Progress and Challenges

The quality of urban life improved significantly for most Americans during the last century, but there is still a long way to go.

Since 1920, many of the worst urban environmental problems in the United States have been reduced significantly. Most people have better working and housing conditions, and air and water quality have improved.

Better sanitation, public water supplies, and medical care have slashed death rates and the prevalence of sickness from malnutrition and transmittable diseases. And concentrating most of the population in urban areas has helped protect the country's biodiversity by reducing the destruction and degradation of wildlife habitat.

However, a number of U.S. cities, especially older ones, have *deteriorating services* and *aging infrastructures* (streets, schools, bridges, housing, and sewers). Many also face *budget crunches* from rising costs as some businesses and people move to the suburbs or rural areas and reduce revenues from property taxes. And there is *rising poverty* in the centers of many older cities, where unemployment typically is 50% or higher.

What Is Urban Sprawl and What Are Its Effects? Paving Paradise and Driving to Get Anywhere

Where there is ample and affordable land, urban areas tend to sprawl outward, swallowing up surrounding countryside.

Another major problem in the United States and some other countries with lots of room for expansion is **urban sprawl.** Growth of low-density development on the edges of cities and towns gobbles up surrounding countryside—frequently prime farmland or forests—and increases dependence on cars. The result is a far-flung hodgepodge of housing developments, shopping malls, parking lots, and office complexes—loosely connected by multilane highways and freeways. Urban sprawl is the product of ample and affordable land, automobiles, cheap gasoline, and poor urban planning.

Figure 5-15 shows some of the undesirable consequences of urban sprawl. Study this figure carefully. Sprawl has increased travel time in automobiles, decreased energy efficiency, increased urban flooding problems, and destroyed prime cropland, forests, open space, and wetlands. It has also led to the economic death of many central cities.

To pay for heavily mortgaged houses and cars, many adults in a typical suburban family have to

Land and Biodiversity

Loss of cropland

Loss of forests and grasslands

Loss of wetlands

Loss and fragmentation of wildlife habitats

Increased wildlife roadkill

Increased soil erosion

Human Health and Aesthetics

Contaminated drinking water and air

Weight gain

Noise pollution

Sky illumination at night

Traffic congestion

Water

Increased runoff

Increased surface water and groundwater pollution

Increased use of surface water and groundwater

Decreased storage of surface water and groundwater

Increased flooding

Decreased natural sewage treatment

Energy, Air, and Climate

Increased energy use and waste

Increased air pollution

Increased greenhouse gas emissions

Enhanced global warming

Warmer microclimate (heat island effect)

Economic Effects

Higher taxes

Decline of downtown business districts

Increased unemployment in central city

Loss of tax base in central city

Figure 5-15 Some of undesirable impacts of urban sprawl or car-dependent development. Do you live in an area suffering from urban sprawl?

spend most of their non-working hours driving to and from work or running errands over a vast suburban landscape. This leaves many of them with little energy and time for their children or themselves, or getting to know their neighbors.

5-6 URBAN ENVIRONMENTAL AND RESOURCE PROBLEMS

What Are the Advantages of Urbanization? Concentrating People Helps

Urban areas can offer more job opportunities and better education and health, and can help protect biodiversity by concentrating people.

Urbanization has many benefits. From an *economic standpoint,* cities are centers of economic development, education, technological developments, and jobs, and have served as centers of industry, commerce, and transportation.

In terms of *health,* urban residents in many parts of the world live longer than do rural residents and have lower infant mortality rates and fertility rates than do rural populations. In addition, urban dwellers generally have better access to medical care, family planning, education, and social services than do people in rural areas.

Urban areas also have some environmental advantages. For example, recycling is more economically feasible because large concentrations of recyclable materials and per capita expenditures on environmental protection are higher in urban areas. Also, concentrating people in urban areas helps preserve biodiversity by reducing the stress on wildlife habitats.

What Are the Disadvantages of Urbanization? Concentrating People Has Some Harmful Effects.

Cities are rarely self-sustaining, and they threaten biodiversity, lack trees, grow little of their food, concentrate pollutants and noise, spread infectious diseases, and are centers of poverty, crime, and terrorism.

Although urban dwellers occupy only about 2% of the earth's land area, they consume about three-fourths of the earth's resources. Because of this and their high waste output (Figure 5-16), most of the world's cities are not self-sustaining systems.

Inputs

Energy
Food
Water
Raw materials
Manufactured goods
Money
Information

Outputs

Solid wastes
Waste heat
Air pollutants
Water pollutants
Greenhouse gases
Manufactured goods
Noise
Wealth
Ideas

Figure 5-16 National capital degradation: urban areas rarely are sustainable systems. The typical city depends on large nonurban areas of land and water for huge inputs of matter and energy resources and for large outputs of waste matter and heat. For example, according to an analysis by Mathis Wackernagel and William Rees, an area 58 times as large as that of London is needed to supply its residents with resources. They estimate that meeting the needs of all the world's people at the same rate of resource use as that of London would take at least three more earths.

Large areas of land must be disturbed and degraded to provide urban dwellers with food, water, energy, minerals, and other resources. This decreases and degrades the earth's biodiversity. Also as cities expand, they destroy rural cropland, fertile soil, forests, wetlands, and wildlife habitats. At the same time, they provide little of the food they use. From an environmental standpoint, urban areas are somewhat like gigantic vacuum cleaners, sucking up much of the world's matter, energy, and living resources and spewing out pollution, wastes, and heat.

Thus, urban areas have large ecological footprints that extend far beyond their boundaries. If you live in a city, you can calculate its ecological footprint by going to the website www.redefining progress.org/. Also, see the Guest Essay on this topic by Michael Cain on this chapter's website.

In urban areas, most trees, shrubs, or other plants are destroyed to make way for buildings, roads, and parking lots. Thus, most cities largely lose the benefits provided by vegetation that would absorb air pollutants, give off oxygen, help to cool the air through transpiration, provide shade, reduce soil erosion, muffle noise, provide wildlife habitats, and give aesthetic pleasure. As one observer remarked, "Most cities are places where they cut down most of the trees and then name the streets after them."

As cities grow and their water demands increase, expensive reservoirs and canals must be built and deeper wells drilled. This can deprive rural and wild areas of surface water and deplete groundwater faster than it is replenished.

Flooding also tends to be greater in cities, in some cases because they are built on floodplain areas or along low-lying coastal areas subject to natural flooding. Another reason is that covering land with buildings, asphalt, and concrete causes precipitation to run off quickly and overload storm drains. In addition, urban development often destroys or degrades wetlands that act as natural sponges to help absorb excess water. Another threat is that many of the world's largest cities are in coastal areas (Figure 5-12) that could be flooded sometime in this century if sea levels rise due to global warming.

Because of their high population densities and high resource consumption, urban dwellers produce most of the world's air pollution, water pollution, and solid and hazardous wastes. Also pollutant levels in urban areas are generally higher than in rural areas because they are produced in a smaller area and cannot be dispersed and diluted as readily as are those produced in rural areas. In addition, high population densities in urban areas can increase the spread of *infectious diseases* (especially if adequate drinking water and sewage systems are not available), physical injuries (mostly from industrial and traffic accidents), and excessive noise (Figure 5-17).

Cities generally are warmer, rainier, foggier, and cloudier than suburbs and nearby rural areas. The enormous amounts of heat generated by cars, factories, furnaces, lights, air conditioners, and heat-absorbing dark roofs and roads in cities create an *urban heat island* surrounded by cooler suburban and rural areas. As cities grow and merge, the heat islands merge and can keep polluted air from being diluted and cleansed.

Also, the artificial light created by cities hinders astronomers from conducting their research and affects various plant and animal species. Species affected by such *light pollution* include endangered sea turtles who lay their eggs on beaches at night and migrating

Permanent damage begins after 8-hour exposure

Noise Levels (in dbA)

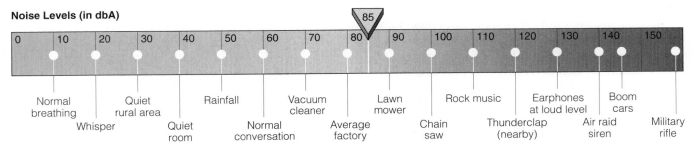

Normal breathing — Whisper — Quiet rural area — Quiet room — Rainfall — Normal conversation — Vacuum cleaner — Average factory — Lawn mower — Chain saw — Rock music — Thunderclap (nearby) — Earphones at loud level — Air raid siren — Boom cars — Military rifle

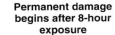

Figure 5-17 *Noise levels* (in decibel-A [dbA] sound pressure units) of some common sounds. You are being exposed to a sound level high enough to cause permanent hearing damage if you need to raise your voice to be heard above the racket, if a noise causes your ears to ring, or if nearby speech seems muffled. Prolonged exposure to lower noise levels and occasional loud sounds may not damage your hearing but can greatly increase internal stress. Noise pollution can be reduced by modifying noisy activities and devices, shielding noisy devices or processes, shielding workers or other receivers from the noise, moving noisy operations or things away from people, and using antinoise (a technology that cancels out one noise with another).

birds that are lured off course by the lights of high-rise buildings and fatally collide with them.

Urban areas can intensify poverty and social problems. Crime rates also tend to be higher in urban areas than in rural areas. And urban areas are more likely and desirable targets for terrorist acts.

Case Study: How Do the Urban Poor in Developing Countries Live? Life on the Edge with Ingenuity and Hope

Most of the urban poor in developing countries live in crowded, unhealthy, and dangerous conditions, but many are better off than the rural poor.

Many of the world's poor live in *squatter settlements* and *shantytowns* on the outskirts of most cities in developing countries, some perched precariously on steep hillsides subject to landslides. In these illegal settlements, people take over unoccupied land and build shacks from corrugated metal, plastic sheets, scrap wood, discarded packing crates, and other scavenged building materials. Still others live or sleep on the streets, having nowhere else to go.

Squatters living near the edge of survival in these areas usually lack clean water supplies, sewers, electricity, and roads, and often are subject to severe air and water pollution and hazardous wastes from nearby factories. Their locations may also be especially prone to landslides, flooding, earthquakes, or volcanic eruptions.

Most cities cannot afford to provide squatter settlements and shantytowns with basic services and protections, and their officials fear that improving services will attract even more of the rural poor. Many city governments regularly bulldoze squatter shacks and send police to drive the illegal settlers out. The people then move back in or develop another shantytown somewhere else.

Despite joblessness, squalor, overcrowding, and environmental and health hazards, most squatter and slum residents are better off than the rural poor. With better access to family planning programs, they tend to have fewer children and better access to schools. Many squatter settlements provide a sense of community and a vital safety net of neighbors, friends, and relatives for the poor.

Mexico City is an example of an urban area in crisis. About 18.3 million people—roughly one of every six Mexicans—live there (Figure 5-12). It is the world's second most populous city, and each year about 210,000 new residents arrive.

Mexico City suffers from severe air pollution, close to 50% unemployment, deafening noise, overcrowding, traffic congestion, inadequate public transporta-

tion, and a soaring crime rate. More than one-third of its residents live in slums called *barrios* or in squatter settlements without running water or electricity.

At least 3 million people have no sewer facilities. This means huge amounts of human waste are deposited in gutters, vacant lots, and open sewers every day, attracting armies of rats and swarms of flies. When the winds pick up dried excrement, a *fecal snow* often falls on parts of the city. Open garbage dumps also contribute dust and bacteria to the atmosphere. This bacteria-laden fallout leads to widespread salmonella and hepatitis infections, especially among children.

Mexico City has one of the world's worst photochemical smog problems because of a combination of too many cars and polluting industries, a sunny climate, and topographical bad luck. The city lies in a high-elevation bowl-shaped valley surrounded on three sides by mountains—conditions that trap air pollutants at ground level. Since 1982, the amount of contamination in the city's air has more than tripled, and breathing that air is said to be roughly equivalent to smoking three packs of cigarettes a day.

The city's air and water pollution cause an estimated 100,000 premature deaths per year. Writer Carlos Fuentes has nicknamed this megacity "Makesicko City."

Some progress has been made. The percentage of days each year in which air pollution standards are violated has fallen from 50% to 20%. But the city has an inadequate mass transportation system and still has weak, poorly enforced air pollution standards for industries and motor vehicles. If you were in charge of Mexico City, what are the three most important things you would do?

X *HOW WOULD YOU VOTE?* Should squatters around cities of developing countries be given title to land they do not own? Cast your vote online at http://biology.brookscole.com/miller7.

5-7 TRANSPORTATION AND URBAN DEVELOPMENT

How Do Land Availability and Transportation Systems Affect Urban Development? Stack or Sprawl

Land availability determines whether a city must grow vertically or spread out horizontally and whether it relies mostly on mass transportation or the automobile.

If a city cannot spread outward, it must grow vertically—upward and downward (below ground)—so it occupies a small land area with a high population density. Most people living in such *compact cities* like Hong

Kong and Tokyo walk, ride bicycles, or use energy-efficient mass transit.

A combination of cheap gasoline, plentiful land, and a network of highways produces *dispersed cities.* They are found in countries such the United States, Canada, and Australia where ample land often is available for outward expansion. Sprawling cities depend on the automobile for most travel, and have a number of undesirable effects (Figure 5-15). But motor vehicles are increasing in both compact and dispersed cities.

What Is the Role of Motor Vehicles in the United States? Cars Rule

Passenger vehicles account for almost all urban transportation in the United States, and each year Americans drive as far as everyone else in the world combined.

America showcases the advantages and disadvantages of living in a society dominated by motor vehicles. With 4.6% of the world's people, the United States has almost a third of the world's motor vehicles, a third of them fuel-inefficient sport utility vehicles (SUVs), pickup trucks, and vans.

Mostly because of urban sprawl and convenience, passenger vehicles are used for 98% of all urban transportation and 91% of travel to work in the United States. About 75% of Americans drive to work alone, 5% commute to work on public transit, and 0.5% bicycle to work. Mostly because of urban sprawl and a network of highways, Americans drive about 4 trillion kilometers (2.5 trillion miles) each year, about the same distance driven by all other drivers in the world! Each year American vehicles consume about 43% of the world's gasoline. According to the American Public Transit System, if Americans increased their use of mass transit from the current 5% to 10%, it would reduce U.S. dependence on oil by 40%.

Many governments in rapidly industrializing countries such as China want to develop an automobile-centered transportation system. Suppose China succeeds in having one or two cars in every garage and consumes oil at the U.S. rate. According to environmental leader Lester R. Brown, China would then need slightly more oil each year than the world now produces and would have to pave an area equal to half of the land it now uses to produce food.

What Are the Advantages and Disadvantages of Motor Vehicles? A Troubled Love Affair

Motor vehicles provide personal benefits and help run economies, but they also kill lots of people, pollute the air, promote urban sprawl, and lead to time- and gas-wasting traffic jams.

On a personal level, motor vehicles provide mobility and are a convenient and comfortable way to get from one place to another. They also are symbols of power, sex, social status, and success for many people. For some, they also provide escape from an increasingly hectic world.

From an economic standpoint, much of the world's economy is built on producing motor vehicles and supplying roads, services, and repairs for them. In the United States, for example, $1 of every $4 spent and one of every six nonfarm jobs is connected to the automobile.

Despite their important benefits, motor vehicles have many harmful effects on people and the environment. They have killed almost 18 million people since 1885, when Karl Benz built the first automobile. Throughout the world, they kill an estimated 1.2 million people each year—an average of 3,300 deaths per day—and injure another 15 million people.

In the United States, motor vehicle accidents kill more than 43,000 people a year and injure another 5 million, at least 300,000 of them severely. *Car accidents have killed more Americans than all wars in the country's history.*

Motor vehicles are the world's largest source of air pollutants. According to the Environmental Protection Agency, motor vehicles are also the fastest growing source of climate-changing carbon dioxide emissions—now producing almost one-fourth of them. In addition, they account for two-thirds of the oil used in the United States and one-third of the world's oil consumption.

Motor vehicles have helped create urban sprawl. At least a third of urban land worldwide and half in the United States is devoted to roads, parking lots, gasoline stations, and other automobile-related uses. This prompted urban expert Lewis Mumford to suggest that the U.S. national flower should be the concrete cloverleaf.

Another problem is congestion. If current trends continue, U.S. motorists will spend an average of 2 years of their lives in traffic jams. Building more roads may not be the answer. Many analysts agree with the idea, as stated by economist Robert Samuelson, that "cars expand to fill available concrete."

How Can We Reduce Automobile Use? Honest Accounting

We can reduce automobile use by having users pay for its harmful effects, but this is politically unpopular.

Environmentalists and a number of economists suggest that one way to reduce the harmful effects of automobile use is to make drivers pay directly for most of the harmful costs of automobile use—a *user-pays* approach based on honest environmental accounting. One way to phase in such full-cost pricing is to include the estimated harmful costs of driving as a tax on gasoline. Such taxes would amount to about $1.30–2.10 per liter ($5–8 per gallon) of gasoline in the United States and would spur the use of more energy-efficient motor vehicles.

Proponents urge governments to use gasoline tax revenues to help finance mass transit systems, bike paths, and sidewalks. The government could reduce taxes on income and wages to offset the increased taxes on gasoline and thus make such a *tax shift* more politically acceptable. Another way to reduce automobile use and congestion is to raise parking fees and charge tolls on roads, tunnels, and bridges—especially during peak traffic times.

Most analysts doubt that these approaches are feasible in the United States, for three reasons. *First,* they face strong political opposition from two groups, one being the public, largely unaware of the huge hidden costs they are already paying. The other group is the powerful transportation-related industries such as oil and tire companies, road builders, carmakers, and many real estate developers. However, taxpayers might accept sharp increases in gasoline taxes if the extra costs were offset by decreases in taxes on wages and income.

Second, fast, efficient, reliable, and affordable mass transit options and bike paths are not widely available in most of the United States. In addition, the dispersed nature of most U.S. urban areas makes most people dependent on cars.

Third, most people who can afford cars are virtually addicted to them, and many people in the U.S. and elsewhere hope to buy one someday.

What Are Alternatives to the Car? Use Your Muscles and Travel with Others

Alternatives include walking, bicycling, driving scooters, and taking subways, trolleys, trains, and buses.

There are a number of alternatives, each with advantages and disadvantages. Examples are *bicycles* (Figure 5-18), *motor scooters* (Figure 5-19), *mass transit rail systems in urban areas* (Figure 5-20), *bus systems in urban areas* (Figure 5-21), and *rapid rail systems between urban areas* (Figure 5-22). For each of these options do you believe that the advantages outweigh the disadvantages? Explain.

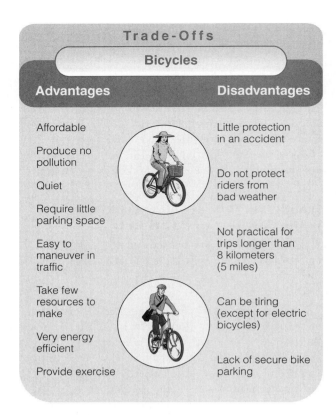

Figure 5-18 Trade-offs: advantages and disadvantages of *bicycles*. Pick the single advantage and disadvantage that you think are the most important.

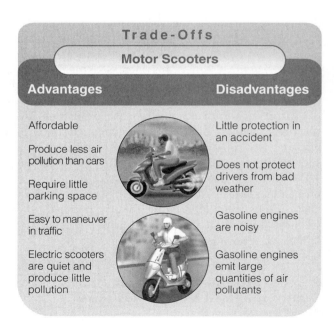

Figure 5-19 Trade-offs: advantages and disadvantages of *motor scooters*. Pick the single advantage and disadvantage that you think are the most important.

Trade-Offs

Mass Transit Rail

Advantages	Disadvantages
More energy efficient than cars	Expensive to build and maintain
Produces less air pollution than cars	Cost effective only along a densely populated narrow corridor
Requires less land than roads and parking areas for cars	Commits riders to transportation schedules
Causes fewer injuries and deaths than cars	Can cause noise and vibration for nearby residents
Reduces car congestion in cities	

Figure 5-20 Trade-offs: advantages and disadvantages of *mass transit rail systems in urban areas*. Pick the single advantage and disadvantage that you think are the most important.

Trade-Offs

Buses

Advantages	Disadvantages
More flexible than rail system	Can lose money because they need low fares to attract riders
Can be rerouted as needed	Often get caught in traffic unless operating in express lanes
Cost less to develop and maintain than heavy-rail system	Commits riders to transportation schedules
Can greatly reduce car use and pollution	Noisy

Figure 5-21 Trade-offs: advantages and disadvantages of *bus systems in urban areas*. Pick the single advantage and disadvantage that you think are the most important.

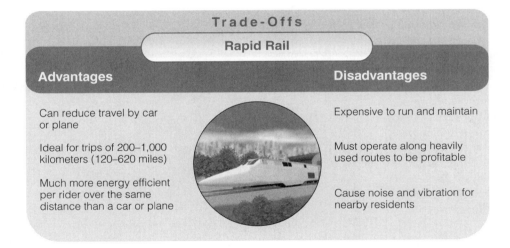

Trade-Offs

Rapid Rail

Advantages	Disadvantages
Can reduce travel by car or plane	Expensive to run and maintain
Ideal for trips of 200–1,000 kilometers (120–620 miles)	Must operate along heavily used routes to be profitable
Much more energy efficient per rider over the same distance than a car or plane	Cause noise and vibration for nearby residents

Figure 5-22 Trade-offs: advantages and disadvantages of *rapid rail systems between urban areas*. Pick the single advantage and disadvantage that you think are the most important.

5-8 MAKING URBAN AREAS MORE LIVABLE AND SUSTAINABLE

How Is Smart Growth Being Used to Control Growth and Sprawl? Channeling Growth and Reining In the Car

Smart growth can control growth patterns, discourage urban sprawl, reduce car dependence, and protect ecologically sensitive areas.

There is growing use of the concept of **smart growth,** or **new urbanism,** to encourage more environmentally sustainable development that requires less dependence on cars, controls and directs sprawl, and reduces wasteful resource use. It recognizes that urban growth will occur. But it uses zoning laws and an array of other tools to channel growth to areas where it can cause less harm, discourage sprawl, protect ecologically sensitive and important lands and waterways, and develop more environmentally sustainable urban

Figure 5-23 Solutions: *smart growth* or *new urbanism tools* used to prevent and control urban growth and sprawl.

Solutions

Smart Growth Tools

Limits and Regulations

Limit building permits

Urban growth boundaries

Greenbelts around cities

Public review of new development

Zoning

Encourage mixed use

Concentrate development along mass transportation routes

Promote high-density cluster housing developments

Planning

Ecological land-use planning

Environmental impact analysis

Integrated regional planning

State and national planning

Protection

Preserve existing open space

Buy new open space

Buy development rights that prohibit certain types of development on land parcels

Taxes

Tax land, not buildings

Tax land on value of actual use (such as forest and agriculture) instead of highest value as developed land

Tax Breaks

For owners agreeing legally to not allow certain types of development (conservation easements)

For cleaning up and developing abandoned urban sites (brownfields)

Revitalization and New Growth

Revitalize existing towns and cities

Build well-planned new towns and villages within cities

areas and neighborhoods that are more enjoyable places to live. Figure 5-23 lists smart growth tools used to prevent and control urban growth and sprawl. Study this figure carefully. Which, if any, of these tools are being used in your community?

Some communities are using principles of new urbanism to develop entire villages and recreate mixed neighborhoods within existing cities. The principles include *walkability* with most things within a 10-minute walk of home and work; *mixed-use and diversity*, where there is a mix of pedestrian-friendly shops, offices, apartments, and homes and people of different ages, classes, cultures, and races; *quality urban design* emphasizing beauty, aesthetics, and architectural diversity; *environmental sustainability* based on development with minimal environmental impact; and *smart transportation* with high-quality trains connecting neighborhoods, towns, and cities. The goal is to create places that uplift, enrich, and inspire the human spirit.

How Can We Make Cities More Sustainable, Desirable Places to Live? The Ecocity Concept

An ecocity allows people to walk, bike, or take mass transit for most of their travel, and it recycles and reuses most of its wastes, grows much of its own food, and protects biodiversity by preserving surrounding land.

According to most environmentalists and urban planners, the primary problem is not urbanization but our failure to make cities more sustainable and livable. They call for us to make new and existing urban areas more self-reliant, sustainable, and enjoyable places to live through good ecological design. See the Guest Essay on this topic by David Orr on the website for this chapter.

A more environmentally sustainable city, called an *ecocity* or *green city*, emphasizes:

- preventing pollution and reducing waste;

- using energy and matter resources efficiently;

- recycling, reusing, and composting at least 60% of all municipal solid waste;

- using solar and other locally available, renewable energy resources; and

- protecting and encouraging biodiversity by preserving surrounding land.

An ecocity is a people-oriented city, not a car-oriented city. Its residents are able to walk, bike, or use low-polluting mass transit for most of their travel. An ecocity requires that all buildings, vehicles, and appliances meet high energy-efficiency standards. Trees and plants adapted to the local climate and soils are planted throughout to provide shade and beauty, supply wildlife habitats, and reduce pollution, noise, and soil erosion. Small organic gardens and a variety of plants adapted to local climate conditions often replace monoculture grass lawns.

Abandoned lots, industrial sites, and polluted creeks and rivers are cleaned up and restored. Nearby forests, grasslands, wetlands, and farms are preserved. Much of an ecocity's food comes from nearby organic farms, solar greenhouses, community gardens, and small gardens on rooftops, in yards, and in window boxes. People designing and living in ecocities take seriously the advice Lewis Mumford gave more than three decades ago: "Forget the damned motor car and build cities for lovers and friends."

The ecocity is not a futuristic dream. Examples of cities that have attempted to become more environmentally sustainable and livable include Curitiba, Brazil (Solutions, below), Waitakere City, New Zealand; Leichester, England; Portland, Oregon; Davis, California; Olympia, Washington; and Chattanooga, Tennessee.

Case Study: How Has Curitiba, Brazil, Become One of the World's Most Sustainable Major Cities? Innovative Solutions to Urban Problems

One of the world's most livable and sustainable major cities is Curitiba, Brazil, with more than 2.5 million people.

Curitiba, known as the ecological capital of Brazil, decided in 1969 to focus on mass transit. The city probably has the world's best bus system, which each day carries more than three-fourths of its people throughout the city along express lanes dedicated to buses. Only high-rise apartment buildings are allowed near major bus routes, and each building must devote the bottom two floors to stores, which reduces the need for residents to travel.

Bike paths run throughout most of the city. Cars are banned from 49 blocks of the city's downtown area, which has a network of pedestrian walkways connected to bus stations, parks, and bike paths. Because Curitiba relies less on automobiles, it uses less energy per person and has less air pollution, greenhouse gas emissions, and traffic congestion than most comparable cities.

Trees have been planted everywhere. No tree in the city can be cut down without a permit, and two trees must be planted for each one cut down.

The city recycles roughly 70% of its paper and 60% of its metal, glass, and plastic, which is sorted by households for collection three times a week. Recov-

ered materials are sold mostly to the city's more than 500 major industries that must meet strict pollution standards. Most of these industries are in an industrial park outside the city limits. A major bus line runs to the park, but many of the workers live nearby and can walk or bike to work.

The city uses old buses as roving classrooms to give the poor basic skills needed for jobs. Other retired buses have become classrooms, health clinics, soup kitchens, and some of the city's 200 day-care centers, which are open 11 hours a day and are free for low-income parents.

The poor receive free medical, dental, and childcare, and 40 feeding centers are available for street children. The city has a *build-it-yourself* system that gives each poor family a plot of land, building materials, two trees, and an hour's consultation with an architect.

In Curitiba, virtually all households have electricity, drinking water, and trash collection. About 95% of its citizens can read and write, and 83% of the adults have at least a high school education. All schoolchildren study ecology. Polls show that 99% of the city's inhabitants would not want to live anywhere else.

This ecocity is the brainchild of architect and former college teacher Jaime Lerner, who has served as the city's mayor three times since 1969. Under his leadership, the municipal government dedicated itself to two goals. *First,* find solutions to problems that are simple, innovative, fast, cheap, and fun. *Second,* establish a government that is honest, accountable, and open to public scrutiny.

An exciting challenge during this century will be to reshape existing cities and design new ones like Curitiba that are more livable and sustainable and have a lower environmental impact.

The city is not an ecological monstrosity. It is rather the place where both the problems and the opportunities of modern technological civilization are most potent and visible.

PETER SELF

CRITICAL THINKING

1. Why is it rational for a poor couple in a developing country such as India to have four or five children? What changes might induce such a couple to consider their behavior irrational?

2. Choose what you consider to be a major local, national, and global environmental problem, and describe the role of population growth in this problem. Compare your answer with those of your classmates.

3. Suppose that all women in the world today began bearing children at replacement-level fertility rates of 2.1 children per woman. Explain why this would not immediately stop global population growth. About how long would it take for population growth to stabilize (assuming death rates do not rise)?

4. Do you believe the population of **(a)** your own country and **(b)** the area where you live is too high? Explain.

5. Should everyone have the right to have as many children as they want? Explain.

6. Some people believe the most important goal is to sharply reduce the rate of population growth in developing countries, where 97% of the world's population growth is expected to take place. Some people in developing countries agree that population growth in these countries can cause local environmental problems. But they contend that the most serious environmental problem the world faces is disruption of the global life-support system for the human species by high levels of resource consumption per person in developed countries, which use about 80% of the world's resources. What is your view on this issue? Explain.

7. How environmentally sustainable is the area where you live? List five ways to make it more environmentally sustainable.

8. Do you believe the United States or the country where you live should develop a comprehensive and integrated mass transit system over the next 20 years, including building an efficient rapid-rail network for travel within and between its major cities? How would you pay for such a system?

9. If you own a car or hope to own one, what conditions, if any, would encourage you to rely less on the automobile and to travel to school or work by bicycle, on foot, by mass transit, or by a carpool or vanpool?

10. Congratulations! You are in charge of the world. List the three most important features of your **(a)** population policy and **(b)** urban policy.

LEARNING ONLINE

The website for this book contains helpful study aids and many ideas for further reading and research. They include a chapter summary, review questions for the entire chapter, flash cards for key terms and concepts, a multiple-choice practice quiz, interesting Internet sites, references, and a guide for accessing thousands of InfoTrac® College Edition articles. Log on to

http://biology.brookscole.com/miller7

Then click on the Chapter-by-Chapter area, choose Chapter 5, and select a learning resource.

6 SUSTAINING BIODIVERSITY: THE ECOSYSTEM APPROACH

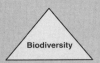

Forests precede civilizations, deserts follow them.

FRANÇOIS-AUGUSTE-RENÉ DE CHATEAUBRIAND

6-1 HUMAN IMPACTS ON BIODIVERSITY

How Have Human Activities Affected Global Biodiversity? Increasing Our Ecological Footprint

We have depleted and degraded some of the earth's biodiversity and these threats are expected to increase.

Figure 6-1 lists factors that tend to increase or decrease biodiversity. Many of our activities decrease biodiversity, as summarized in Figure 6-2. According to biodiversity expert Edward O. Wilson, "The natural world is everywhere disappearing before our eyes—cut to pieces, mowed down, plowed under, gobbled up, replaced by human artifacts."

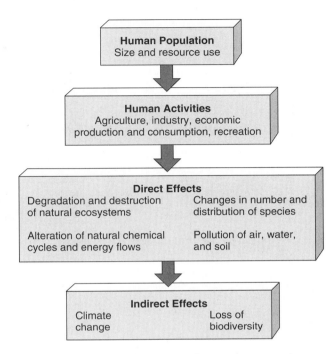

Figure 6-2 Natural capital degradation: major connections between human activities and the earth's biodiversity.

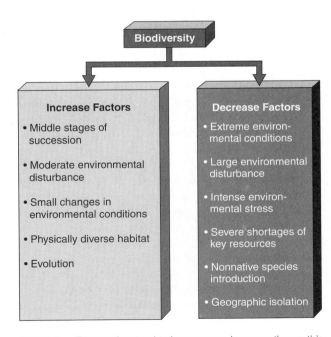

Figure 6-1 Factors that tend to increase or decrease the earth's biodiversity.

Consider a few examples of how human activities have decreased and degraded the earth's terrestrial biodiversity. According to a 2000 study by the Wildlife Conservation Society and the Center for International Earth Science Information Network at Columbia University, we have taken over, disturbed, or degraded to some extent at least half, and probably 83%, of the earth's land surface (excluding Antarctica and Greenland). We have done most of this by filling in wetlands and converting grasslands and forests to crop fields and urban areas.

In the United States, at least 95% of the virgin forests in the lower 48 states have been logged for lumber and to make room for agriculture, housing, and industry. In addition, 98% of tallgrass prairie in the Midwest and Great Plains has disappeared, and 99% of California's native grassland and 85% of its original redwood forests are gone. More than half of the country's wetlands have been destroyed.

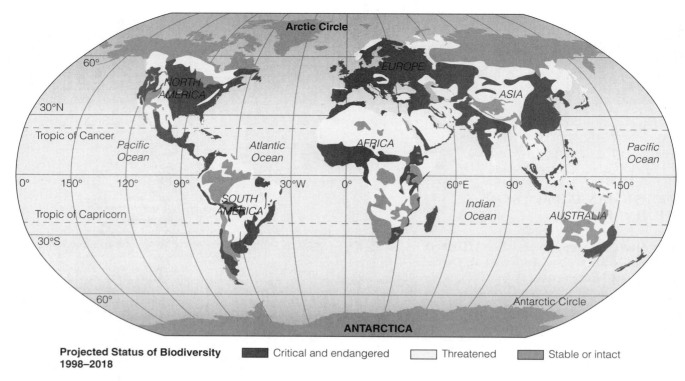

Projected Status of Biodiversity 1998–2018

■ Critical and endangered ☐ Threatened ▨ Stable or intact

Figure 6-3 Natural capital degradation: projected status of the earth's biodiversity, between 1998 and 2018. (Data from World Resources Institute, World Conservation Monitoring Center, and Conservation International)

By some estimates, humans use, waste, or destroy about 10–55% of the net primary productivity of the planet's terrestrial ecosystems.

Human activities are also degrading the world's *aquatic biodiversity*. About half of the world's wetlands and half of the wetlands in the United States were lost during the last century. An estimated 27% of the world's diverse coral reefs have been severely damaged. By 2050, another 70% may be severely damaged or eliminated.

About three-fourths of the world's 200 commercially valuable marine fish species are either overfished or fished to their estimated sustainable yield. According to a 2002 preliminary report by the U.S. Commission on Ocean Policy, about 40% of U.S. commercial fish stocks are depleted or overfished.

Human activities also contribute to the *premature extinction of species*. Biologists estimate that the current global extinction rate of species is at least 100 times and probably 1,000 to 10,000 times what it was before humans existed. These threats to the world's biodiversity are projected to increase sharply by 2018 (Figure 6-3).

Figure 6-4 outlines the goals, strategies, and tactics for preserving and restoring the terrestrial ecosystems and aquatic systems that provide habitats and resources for the world's species (as discussed in this chapter), and preventing the premature extinction of species (as discussed in Chapter 7).

Why Should We Care about Biodiversity? Sustaining a Vital Part of the World's Life-Support System

Biodiversity should be protected from degradation by human activities because it exists and because of its usefulness to us and other species.

Biodiversity researchers contend that we should act to preserve the earth's overall diversity because its genes, species, ecosystems, and ecological processes have two types of value. One is **intrinsic value** because these components of biodiversity exist, regardless of their use to us.

The other is **instrumental value** because of their usefulness to us. There are two major types of instrumental values. One consists of *use values* that benefit us in the form of economic goods and services, ecological services, recreation, scientific information, and preserving options for such uses in the future.

Another type consists of *nonuse values*. One is the *existence value* in knowing that a redwood forest, wilderness, or endangered species exists, even if we will never see it or get direct use from it. *Aesthetic value* is another nonuse value because many people appreciate a tree, a forest, a wild species, or a vista because of its beauty. *Bequest value* is a third type of nonuse value. It is based on a willingness of some people to pay to protect some forms of natural capital for use by future generations.

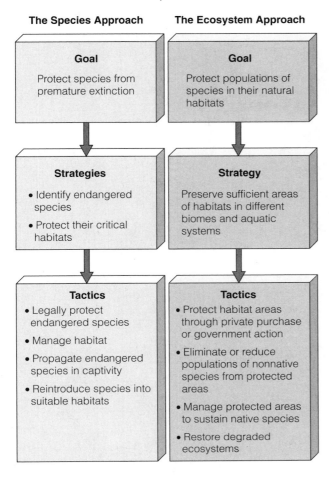

The Species Approach

Goal

Protect species from premature extinction

↓

Strategies

- Identify endangered species
- Protect their critical habitats

↓

Tactics

- Legally protect endangered species
- Manage habitat
- Propagate endangered species in captivity
- Reintroduce species into suitable habitats

The Ecosystem Approach

Goal

Protect populations of species in their natural habitats

↓

Strategy

Preserve sufficient areas of habitats in different biomes and aquatic systems

↓

Tactics

- Protect habitat areas through private purchase or government action
- Eliminate or reduce populations of nonnative species from protected areas
- Manage protected areas to sustain native species
- Restore degraded ecosystems

Figure 6-4 Goals, strategies, and tactics for protecting biodiversity.

6-2 PUBLIC LANDS IN THE UNITED STATES

What Are the Major Types of U.S. Public Lands? Land for Current and Future Generations

More than a third of the land in the United States consists of publicly owned national forests, resource lands, parks, wildlife refuges, and protected wilderness areas.

No nation has set aside as much of its land for public use, resource extraction, enjoyment, and wildlife as has the United States. The federal government manages roughly 35% of the country's land that belongs to every American. About 73% of this federal public land is in Alaska, and another 22% is in the western states (Figure 6-5, p. 104).

Some federal public lands are used for many purposes. One example is the *National Forest System*, which consists of 155 forests and 22 grasslands. These forests, managed by the U.S. Forest Service (USFS), are used for logging, mining, livestock grazing, farm-

ing, oil and gas extraction, recreation, hunting, fishing, and conservation of watershed, soil, and wildlife resources.

A second example is the *National Resource Lands*, managed by the Bureau of Land Management (BLM). These lands are used primarily for mining, oil and gas extraction, and livestock grazing.

A third system consists of 542 *National Wildlife Refuges* that are managed by the U.S. Fish and Wildlife Service (USFWS). Most refuges protect habitats and breeding areas for waterfowl and big game to provide a harvestable supply for hunters; a few protect endangered species from extinction. Permitted activities in most refuges include hunting, trapping, fishing, oil and gas development, mining, logging, grazing, some military activities, and farming.

Uses of other public lands are more restricted. One example is the *National Park System* managed by the National Park Service (NPS). It includes 56 major parks (mostly in the West) and 331 national recreation areas, monuments, memorials, battlefields, historic sites, parkways, trails, rivers, seashores, and lakeshores. Only camping, hiking, sport fishing, and boating can take place in the national parks, but sport hunting, mining, and oil and gas drilling is allowed in National Recreation Areas.

The most restricted public lands are 630 roadless areas that make up the *National Wilderness Preservation System*. These areas lie within the national parks, national wildlife refuges, national forests, and national resource lands and are managed by agencies in charge of those lands. Most of these areas are open only for recreational activities such as hiking, sport fishing, camping, and nonmotorized boating.

Case Study: How Should U.S. Public Lands Be Managed? An Ongoing Controversy

Since the 1800s there has been controversy over how U.S. public lands should be used because of the valuable resources they contain.

Federal public lands contain valuable oil, natural gas, coal, timber, and mineral resources.. Since the 1800s there has been controversy over how the resources on these lands should be used and managed.

Most conservation biologists, environmental economists, and many free-market economists believe the following four principles should govern use of public land:

- Protecting biodiversity, wildlife habitats, and the ecological functioning of public land ecosystems should be the primary goal.

- No one should receive subsidies or tax breaks for using or extracting resources on public lands—a user-pays approach.

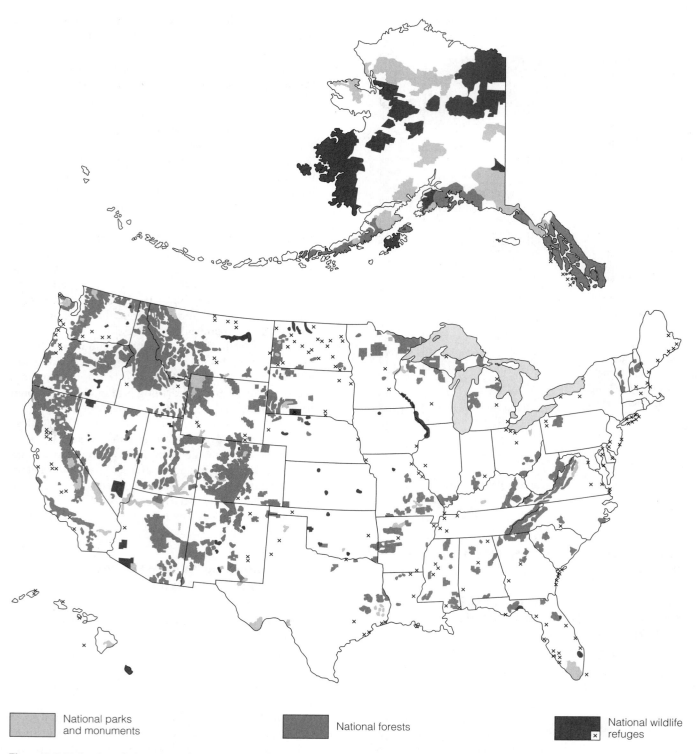

Figure 6-5 Natural capital: national forests, national parks, and wildlife refuges managed by the U.S. federal government. U.S. citizens jointly own these and other public lands. (Data from U.S. Geological Survey)

National parks and monuments

National forests

National wildlife refuges

- The American people deserve fair compensation for the use of their property.

- All users or extractors of resources on public lands should be fully responsible for any environmental damage they cause.

There is strong and effective opposition to these ideas. Economists, developers, and resource extractors tend to view public lands in terms of their usefulness in providing mineral, timber, and other resources and their ability to increase short-term economic growth.

They have succeeded in blocking implementation of the four principles just listed. For example, in recent years, the government has given more than $1 billion a year in subsidies to privately owned mining, fossil fuel extraction, logging, and grazing interests using U.S. public lands.

Some developers and resource extractors go further and have mounted a campaign to get the U.S. Congress to pass laws that would:

- Sell public lands or their resources to corporations or individuals, usually at less than market value.

- Slash federal funding for regulatory administration of public lands.

- Cut all old-growth forests in the national forests and replace them with tree plantations.

- Open all national parks, national wildlife refuges, and wilderness areas to oil drilling, mining, off-road vehicles, and commercial development.

- Do away with the National Park Service and launch a 20-year construction program of new concessions and theme parks run by private firms in the national parks.

- Continue mining on public lands under the provisions of the 1872 Mining Law, which allows mining interests to pay no royalties to taxpayers for hard-rock minerals they remove.

- Repeal the Endangered Species Act or modify it to allow economic factors to override protection of endangered and threatened species.

- Redefine government-protected wetlands so about half of them would no longer be protected.

- Prevent individuals or groups from legally challenging these uses of public land for private financial gain.

X *How WOULD YOU VOTE?* Should much more of U.S. public lands (or government-owned lands in the country where you live) be opened up to the extraction of timber, mineral, and energy resources? Cast your vote online at http://biology .brookscole.com/miller7.

6-3 MANAGING AND SUSTAINING FORESTS

What Are the Major Types of Forests? Old-Growth, Second-Growth, and Tree Plantations

Some forests have not been disturbed by human activities, others have grown back after being cut, and some consist of planted stands of a particular tree species.

Forests with at least 10% tree cover occupy about 30% of the earth's land surface (excluding Greenland and Antarctica). Figure 3-6 (p. 54) shows the distribution of the world's boreal, temperate, and tropical forests. These forests provide many important ecological and economic services (Figure 6-6).

Forest managers and ecologists classify forests into three major types based on age and structure. One type consists of **old-growth forests:** uncut forests or regenerated forests that have not been seriously disturbed by human activities or natural disasters for at

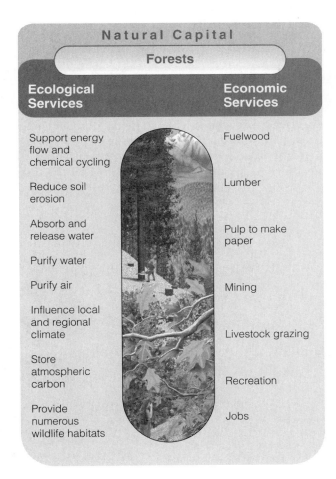

Figure 6-6 Natural capital: major ecological and economic services provided by forests.

least several hundred years. Old-growth forests are storehouses of biodiversity because they provide ecological niches for a multitude of wildlife species.

A second type is a **second-growth forest:** a stand of trees resulting from secondary ecological succession (Figure 4-5, p. 69). They develop after the trees in an area have been removed by *human activities* (such as clear-cutting for timber or conversion to cropland) or by *natural forces* (such as fire, hurricanes, or volcanic eruption).

A **tree plantation,** also called a **tree farm,** is a third type. It is a managed tract with uniformly aged trees of one species that are harvested by clear-cutting as soon as they become commercially valuable. It is then replanted and clear-cut again in a regular cycle (Figure 6-7). Currently, about 63% of the world's forests are secondary-growth forests, 22% are old-growth forests, and 5% are tree plantations (that produce about one-fifth of the world's commercial wood).

What Are the Major Types of Forest Management? Simple Tree Plantations and Diverse Forests

Some forests consist of one or two species of commercially important tree species that are cut down and replanted, and others contain diverse tree species harvested individually or in small groups.

There are two forest management systems. One is **even-aged management,** which involves maintaining trees in a given stand at about the same age and size. In this approach, sometimes called *industrial forestry,* a simplified *tree plantation* replaces a biologically diverse old-growth or second-growth forest. The plantation consists of one or two fast-growing and economically desirable species that can be harvested every 6–10 years, depending on the species (Figure 6-7).

A second type is **uneven-aged management,** which involves maintaining a variety of tree species in a stand with many ages and sizes, to foster natural regeneration. Here the goals are biological diversity, long-term sustainable production of high-quality timber, selective cutting of individual mature or intermediate-aged trees, and multiple use of the forest for timber, wildlife, watershed protection, and recreation.

According to a 2001 study by the World Wildlife Fund, intensive but sustainable management of as little as one-fifth of the world's forests—an area twice the size of India—could meet the world's current and future demand for commercial wood and fiber. This intensive use of the world's tree plantations and some of its secondary forests would leave the world's remaining old-growth forest untouched.

How Are Trees Harvested? Be Selective or Chop Them All Down

Trees can be harvested individually from diverse forests, or an entire forest stand can be cut down in one or several phases.

The first step in forest management is to build roads for access and timber removal. Even carefully designed logging roads have a number of harmful effects (Figure 6-8). They include increased erosion and sediment runoff into waterways, habitat fragmentation, and biodiversity loss. Logging roads also expose forests to invasion by nonnative pests, diseases, and

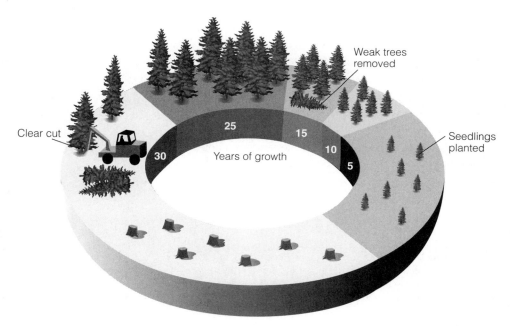

Figure 6-7 Short (25- to 30-year) rotation cycle of cutting and regrowth of a monoculture tree plantation in modern industrial forestry. In tropical countries, where trees can grow more rapidly year round, the rotation cycle can be 6–10 years.

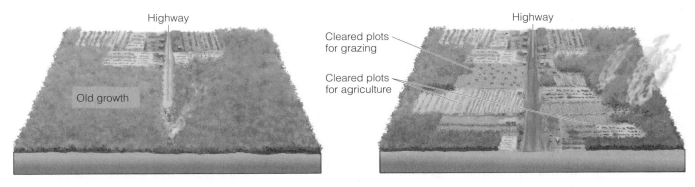

Figure 6-8 Natural capital degradation: building roads into previously inaccessible forests paves the way to fragmentation, destruction, and degradation.

wildlife species. They also open once-inaccessible forests to farmers, miners, ranchers, hunters, and off-road vehicle users. In addition, logging roads on public lands in the United States disqualify the land for protection as wilderness.

Once loggers reach a forest area, they use various methods to harvest the trees (Figure 6-9, p. 108). With *selective cutting*, intermediate-aged or mature trees in an uneven-aged forest are cut singly or in small groups (Figure 6-9a).

Some tree species that grow best in full or moderate sunlight are cleared completely in one or several cuttings by *shelterwood cutting* (Figure 6-9b), *seed-tree cutting* (Figure 6-9c), or *clear-cutting* (in which all trees on a site are removed in a single cut; Figure 6-9d). Figure 6-10 (p. 109) lists the advantages and disadvantages of clear-cutting.

A clear-cutting variation that can allow a more sustainable timber yield without widespread destruction is *strip cutting* (Figure 6-9e). It involves clear-cutting a strip of trees along the contour of the land, with the corridor narrow enough to allow natural regeneration within a few years. After regeneration, loggers cut another strip above the first, and so on. This allows clear-cutting of a forest in narrow strips over several decades with minimal damage.

What Are the Harmful Environmental Effects of Deforestation? Biodiversity Loss and Climate Change

Cutting down large areas of forests reduces biodiversity and the ecological services forests provide, and can contribute to regional and global climate change.

Deforestation is the temporary or permanent removal of large expanses of forest for agriculture or other uses. Harvesting timber and fuelwood from forests provides many economic benefits (Figure 6-6, right). However, deforestation can have many harmful environmental

effects (Figure 6-11, p. 109) that can reduce the ecological services provided by forests (Figure 6-6, left).

If deforestation occurs over a large enough area, it can cause a region's climate to become hotter and drier and prevent the return of a forest. Deforestation can also contribute to projected global warming if trees are removed faster than they grow back. When forests are cleared for agriculture or other purposes and burned, the carbon stored in the trees' biomass is released into the atmosphere as the greenhouse gas carbon dioxide (CO_2).

What Is Happening to the World's Forests? Mixed News

Human activities have reduced the earth's forest cover by 20–50%, and deforestation is continuing at a fairly rapid rate, except in most temperate forests in North America and Europe.

Forests are renewable resources as long as the rate of cutting and degradation does not exceed the rate of regrowth. Here are two pieces of *bad news:*

First, surveys by the World Resources Institute (WRI) indicate that over the past 8,000 years human activities have reduced the earth's original forest cover by 20–50%. *Second,* surveys by the UN Food and Agricultural Organization (FAO) and the World Resources Institute indicate that the global rate of forest cover loss during the 1990s was between 0.2% and 0.5% a year, and at least another 0.1–0.3% of the world's forests were degraded. If correct, the world's forests are being cleared and degraded at a rate of 0.3–0.8% a year, with much higher rates in some areas. Over four-fifths of these losses took place in the tropics. The World Resources Institute estimates that if current deforestation rates continue, about 40% of the world's remaining intact forests will have been logged or converted to other uses within 10–20 years, if not sooner.

Here are two pieces of *good news: First,* the total area of many temperate forests in North America and

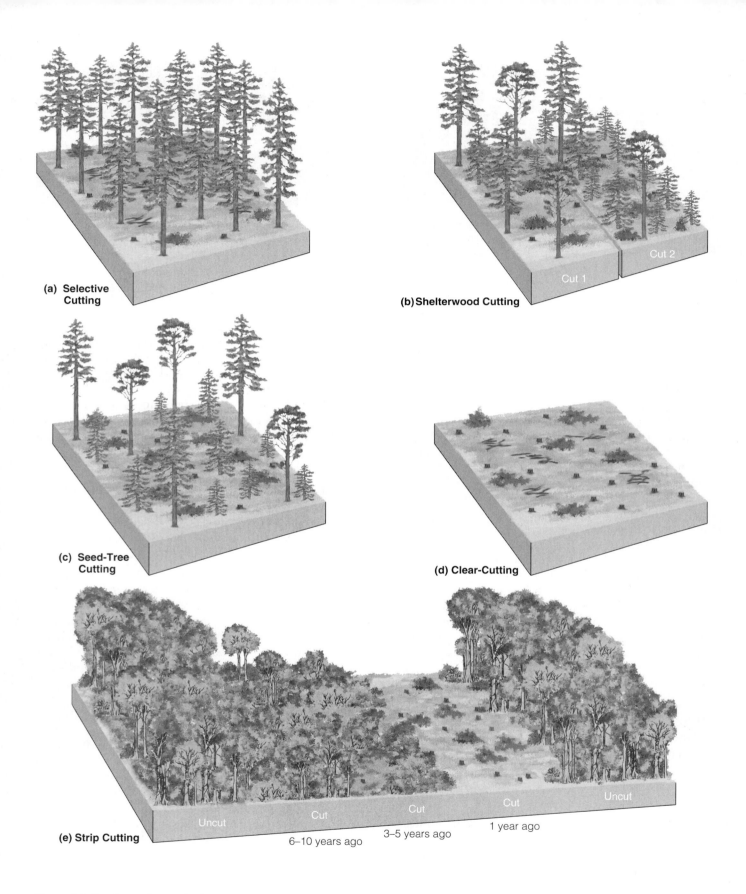

(a) Selective Cutting

(b) Shelterwood Cutting

Cut 1

Cut 2

(c) Seed-Tree Cutting

(d) Clear-Cutting

(e) Strip Cutting

Uncut

Cut

6–10 years ago

Cut

3–5 years ago

Cut

1 year ago

Uncut

Figure 6-9 Tree-harvesting methods.

Trade-Offs

Clear-Cutting Forests

Advantages	Disadvantages

Advantages	Disadvantages
Higher timber yields	Reduces biodiversity
Maximum economic return in shortest time	Disrupts ecosystem processes
Can reforest with genetically improved fast-growing trees	Destroys and fragments some wildlife habitats
Short time to establish new stand of trees	Leaves moderate to large openings
Needs less skill and planning	Increases soil erosion
Best way to harvest tree plantations	Increases sediment water pollution and flooding when done on steep slopes
Good for tree species needing full or moderate sunlight for growth	Eliminates most recreational value for several decades

Figure 6-10 Trade-offs: advantages and disadvantages of clear-cutting forests. Pick the single advantage and disadvantage that you think are the most important.

Environmental Degradation

Deforestation

- Decreased soil fertility from erosion
- Runoff of eroded soil into aquatic systems
- Premature extinction of species with specialized niches
- Loss of habitat for migratory species such as birds and butterflies
- Regional climate change from extensive clearing
- Releases CO_2 into atmosphere from burning and tree decay
- Accelerates flooding

Figure 6-11 Natural capital degradation: harmful environmental effects of deforestation that can reduce the ecological services provided by forests.

Europe has increased slightly because of reforestation from secondary ecological succession on cleared forest areas and abandoned croplands.

Second, some of the cut areas of tropical forest have increased tree cover from regrowth and planting of tree plantations. But ecologists do not believe that tree plantations with their much lower biodiversity should be counted as forest any more than croplands should be counted as grassland. According to ecologist Michael L. Rosenzweig, "Forest plantations are just cornfields whose stalks have gotten very tall and turned to wood. They display nothing of the majesty of natural forests."

How Can We Manage Forests More Sustainably? Making Sustaining Forests Profitable

We can use forests more sustainably by including the economic value of their ecological services, harvesting trees no faster than they are replenished, and protecting old-growth and vulnerable areas.

Biodiversity researchers and a growing number of foresters call for more sustainable forest management. Figure 6-12 lists ways to do this. Which two of these solutions do you believe are most important? Explain.

Currently, forests are valued mostly for their economic services (Figure 6-6, right). But suppose we took into account the estimated monetary value of the ecological services provided by forests (Figure 6-6, left). According to a 1997 appraisal by a team of ecologists,

Solutions

Sustainable Forestry

- Grow more timber on long rotations
- Rely more on selective cutting and strip cutting
- No clear-cutting, seed-tree, or shelterwood cutting on steeply sloped land
- No fragmentation of remaining large blocks of forest
- Sharply reduce road building into uncut forest areas
- Leave most standing dead trees and fallen timber for wildlife habitat and nutrient recycling
- Certify timber grown by sustainable methods
- Include ecological services of trees and forests in estimating economic value

Figure 6-12 Solutions: ways to manage forests more sustainably.

economists, and geographers, the world's forests provide us with ecological services worth about $4.7 trillion per year—hundreds of times more than the economic value of forests.

Based on this more balanced accounting system, most of the world's old-growth and second-growth forests would not be clear-cut. Instead, their ecological services could be sustained indefinitely by selectively harvesting trees no faster than they are replenished.

Currently, the value of these ecological services are left out of our economic decisions because of overemphasis on short-term wants and needs and government subsidies that encourage destruction and degradation of forests for short-term economic gain. Also, most people are unaware of the value of the ecological services and income provided by nature.

Another way to encourage sustainable use of forests is to establish and use methods to evaluate timber that has been grown sustainably (Solutions, below).

Solutions: Certifying Sustainably Grown Timber

Organizations have developed standards for certifying that timber has been harvested sustainably and that wood products have been produced from sustainably harvested timber.

Collins Pine owns and manages a large area of productive timberland in northeastern California. Since 1940, the company has used selective cutting to help maintain ecological, economic, and social sustainability of its timberland.

Since 1993, Scientific Certification Systems (SCS) has evaluated the company's timber production. SCS is part of the nonprofit Forest Stewardship Council (FSC). It was formed in 1993 to develop a list of environmentally sound practices for use in certifying timber and products made from such timber.

Each year, SCS evaluates Collins's landholdings to ensure that cutting has not exceeded long-term forest regeneration, roads and harvesting systems have not caused unreasonable ecological damage, soils are not damaged, downed wood (boles) and standing dead trees (snags) are left to provide wildlife habitat, and the company is a good employer and a good steward of its land and water resources.

In 2002, Mitsubishi, one of the world's largest forestry companies, announced that it would have third parties certify its forestry operations using standards developed by the Forest Stewardship Council. And Home Depot, Lowes, Andersen, and other major sellers of wood products in the United States have agreed to sell only wood certified as being sustainably grown by independent groups such as the Forest Stewardship Council (to the degree that certified wood is available).

6-4 FOREST RESOURCES AND MANAGEMENT IN THE UNITED STATES

What Is the Status of Forests in the United States? Encouraging News

U.S. forests cover more area than they did in 1920, more wood is grown than is cut, and the country has set aside large areas of protected forests.

Forests cover about 30% of the U.S. land area, providing habitats for more than 80% of the country's wildlife species and supplying about two-thirds of the nation's surface water.

Forests (including tree plantations) in the United States cover more area than they did in 1920. Many of the old-growth forests that were cleared or partially cleared between 1620 and 1960 have grown back naturally through secondary ecological succession as fairly diverse second-growth (and in some cases third-growth) forest in every region of the United States, except much of the West. In 1995, environmental writer Bill McKibben cited forest regrowth in the United States—especially in the East—as "the great environmental story of the United States, and in some ways the whole world."

Also, every year more wood is grown in the United States than is cut and each year the total area planted with trees increases. In addition, the United States was the world's first country to set aside large areas of forest in protected areas. By 2000, protected forests made up about 40% of the country's total forest area, mostly in the national forests.

However, since the mid-1960s, an increasing area of the nation's remaining old-growth and fairly diverse second-growth forests has been clear-cut and replaced with biologically simplified tree plantations. According to biodiversity researchers, this reduces overall forest biodiversity and disrupts ecosystem processes such as energy flow and chemical cycling. Some environmentally concerned citizens have protested the cutting down of ancient trees and forests (Individuals Matter, right).

Case Study: How Should U.S. National Forests Be Managed? Another Controversy

There is controversy over whether U.S. national forests should be managed primarily for timber, their ecological services, recreation, or a mix of these uses.

For decades, there has been intense controversy over the use of resources in the national forests. Timber companies push to cut as much of this timber in these forests as possible at low prices. Biodiversity experts and environmentalists call for sharply reducing or eliminating tree harvesting in national forests and using more sustainable forest management practices (Figure 6-12) for timber cutting in these forests. They believe that national forests should be managed primarily to provide recreation and to sustain biodiversity, water resources, and other ecological services.

The Forest Service's timber-cutting program loses money because revenue from timber sales does not cover the costs of road building, timber sale preparation, administration, and other overhead costs. Because of such government subsidies, timber sales from U.S. federal lands have lost money for taxpayers in 97 of the last 100 years!

Figure 6-13 lists advantages and disadvantages of logging in national forests. According to a 2000 study by the accounting firm Econorthwest, recreation, hunting, and fishing in national forests add ten times more money to the national economy and provide seven times more jobs than does extraction of timber and other resources.

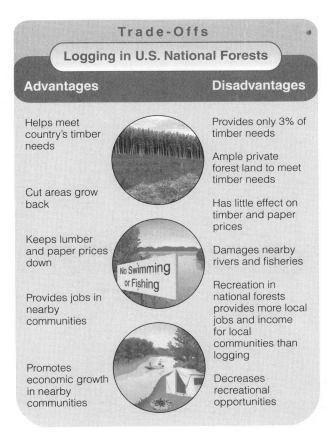

Figure 6-13 Trade-offs: advantages and disadvantages of allowing logging in U.S. national forests. Pick the single advantage and disadvantage that you think are the most important.

Butterfly in a Redwood Tree

Butterfly is the nickname given to Julia Hill. This young woman spent two years of her life on a small platform near the top of a giant redwood tree in California to protest the clear-cutting of a forest of these ancient trees, some of them more than 1,000 years old.

She and other protesters were illegally occupying these trees as a form of *nonviolent civil disobedience* used decades ago by Mahatma Gandhi in his efforts to end the British occupation of India. Butterfly had never participated in any environmental protest or act of civil disobedience.

She went to the site to express her belief that it was wrong to cut down these ancient giants for short-term economic gain, even if you own them. She planned to stay only for a few days.

But after seeing the destruction and climbing one of these magnificent trees, she ended up staying in the tree for two years to bring publicity to what was happening and help save the surrounding trees. She became a media symbol of the protest and during her stay used a cell phone to communicate with members of the mass media throughout the world to help develop public support for saving the trees.

Can you imagine spending two years of your life in a tree on a platform not much bigger than a king-sized bed 55 meters (180 feet) above the ground and enduring high winds, intense rainstorms, snow, and ice? She was not living in a quiet pristine forest. All round her was the noise of trucks, chainsaws, and helicopters trying to scare her into returning to the ground.

She lost her courageous battle to save the surrounding forest but persuaded Pacific Lumber MAXXAM to save her tree (called Luna) and a 60-meter (200-foot) buffer zone around it. Not too long after she descended from the tree someone used a chainsaw to seriously damage it, and cables and steel plates have been used to preserve it.

But maybe she and the earth did not lose. A book she wrote about her stand, and her subsequent travels to campuses all over the world, have inspired a number of young people to stand up for protecting biodiversity and other environmental causes.

She was leading by following in the tradition of Gandhi, who said, "My life is my message." Would you spend a day or a week of your life protesting something that you believed to be wrong?

6-5 TROPICAL DEFORESTATION

How Fast Are Tropical Forests Being Cleared and Degraded? Protecting the Priceless

Large areas of ecologically and economically important forests are being cleared and degraded at a fast rate.

Tropical forests cover about 6% of the earth's land area—roughly the area of the lower 48 states. Climatic and biological data suggest that mature tropical forests once covered at least twice as much area as they do today, with most of the destruction occurring since 1950. Satellite scans and ground-level surveys used to estimate forest destruction indicate that large areas of tropical forests are being cut rapidly in parts of South America (especially Brazil), Africa, and Asia.

Studies indicate that more than half of the world's species of terrestrial plants and animals live in tropical rain forests. Brazil has about 40% of the world's remaining tropical rain forest in the vast Amazon basin, which is about two-thirds the size of the continental United States. In 1970, deforestation affected only 1% of the area of the Amazon basin. By 2003, almost 20% had been deforested or degraded. In 2003, conservation biologists and Brazilian environmental officials were shocked to learn that the deforestation rate in the Amazon rose by 40% between 2001 and 2002.

According to a 2001 study by Penn State researcher James Alcock, without immediate and aggressive action to reduce current forest destruction and degradation practices, Brazil's original Amazon rain forests may largely disappear within 40–50 years.

There are disagreements about how rapidly tropical forests are being deforested and degraded because of three factors. *First*, it is difficult to interpret satellite images. *Second*, some countries hide or exaggerate deforestation rates for political and economic reasons. *Third*, governments and international agencies define forest, deforestation, and forest degradation in different ways.

For these reasons, estimates of global tropical forest loss vary from 50,000 square kilometers (19,300 square miles) to 170,000 square kilometers (65,600 square miles) per year. This is high enough to lose or degrade half of the world's remaining tropical forests in 35–117 years.

What Causes Tropical Deforestation and Degradation? The Big Five

The primary causes of tropical deforestation and degradation are population growth, poverty, environmentally harmful government subsidies, debts owed to developed countries, and failure to value their ecological services.

Tropical deforestation results from a number of interconnected primary and secondary causes (Figure 6-14). Population growth and poverty combine to drive subsistence farmers and the landless poor to tropical forests, where they try to grow enough food to survive. Government subsidies can accelerate deforestation by making timber or other tropical forest resources cheap, relative to the economic value of the ecological services they provide.

Governments in Indonesia, Mexico, and Brazil also encourage the poor to colonize tropical forests by giving them title to land that they clear. This can help reduce poverty but can lead to environmental degradation unless the new settlers are taught how to use such forests more sustainably, which is rarely done. In addition, international lending agencies encourage developing countries to borrow huge sums of money from developed countries to finance projects such as roads, mines, logging operations, oil drilling, and dams in

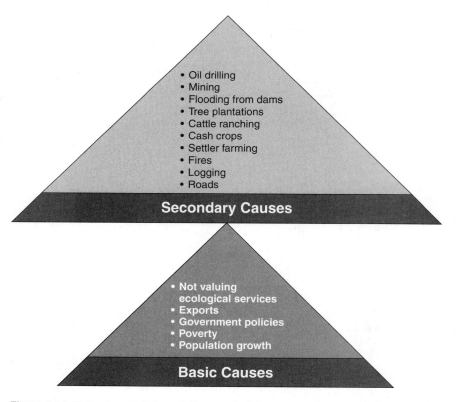

Figure 6-14 Natural capital degradation: major interconnected causes of the destruction and degradation of tropical forests. The importance of specific secondary causes varies in different parts of the world.

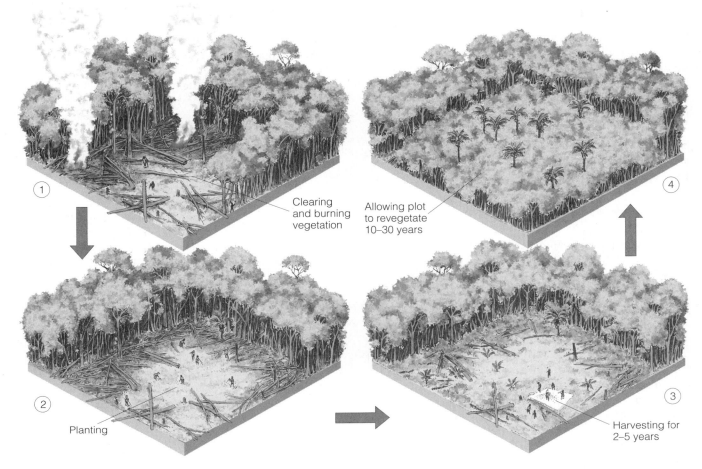

Figure 6-15 The first crop-growing technique may have been a combination of slash-and-burn and shifting cultivation in tropical forests. This method is sustainable only if small plots of the forest are cleared, cultivated for no more than 5 years, and then allowed to regenerate for 10–30 years to renew soil fertility. Indigenous cultures have developed many variations of this technique and have found ways to use some former plots nondestructively while they are being regenerated.

tropical forests. Another cause is failure to value the ecological services of forests (Figure 6-6, left).

The depletion and degradation of a tropical forest begins when a road is cut deep into the forest interior for logging and settlement. Loggers then use selective cutting to remove the best timber. This topples many other trees because of their shallow roots and the network of vines connecting trees in the forest's canopy. Timber exports to developed countries contribute significantly to tropical forest depletion and degradation. But domestic use accounts for more than 80% of the trees cut in developing countries.

After the best timber has been removed, timber companies often sell the land to ranchers. Within a few years, they typically overgraze it and sell it to settlers who have migrated to the forest hoping to grow enough food to survive. Then they move their land-degrading ranching operations to another forest area. According

to a 2004 report by the Center for International Forestry Research, the rapid spread of cattle ranching is the biggest threat to the Amazon's tropical forests.

The settlers cut most of the remaining trees, burn the debris after it has dried for about a year, and plant crops using slash-and-burn agriculture (Figure 6-15). They can also endanger some wild species by hunting them for meat. After a few years of crop growing and rain erosion, the nutrient-poor tropical soil is depleted of nutrients. Then the settlers move on to newly cleared land.

In some areas—especially Africa and Latin America—large sections of tropical forest are cleared for raising cash crops such as sugarcane, bananas, pineapples, strawberries, and coffee—mostly for export to developed countries. Tropical forests are also cleared for mining and oil drilling and to build dams on rivers that flood large areas of the forest.

Healthy rain forests do not burn. But increased logging, settlements, grazing, and farming along roads built in these forests results in fragments of forest (Figure 6-8, right) that dry out. This makes such areas easier to ignite by lightning and for farmers and ranchers to burn. In addition to destroying and degrading biodiversity, this releases large amounts of carbon dioxide into the atmosphere.

Solutions: How Can We Reduce Deforestation and Degradation of Tropical Forests? Prevention Is Best.

There are a number of ways to slow and reduce the deforestation and degradation of tropical forests.

Analysts have suggested various ways to protect tropical forests and use them more sustainably (Figure 6-16). Which two of the solutions in this figure do you believe are the most important?

One method is to help new settlers in tropical forests learn how to practice small-scale sustainable agriculture and forestry. Another method is to sustainably harvest some of the renewable resources such as fruits and nuts in tropical rain forests.

We can also use *debt-for-nature swaps* to make it financially profitable for countries to protect tropical forests. In such a swap, participating countries act as custodians of protected forest reserves in return for foreign aid or debt relief. Another important tool is using an international system for evaluating and certifying tropical timber produced by sustainable methods.

Loggers can also use gentler methods for harvesting trees. For example, cutting canopy vines (lianas) before felling a tree can

reduce damage to neighboring trees by 20–40%, and using the least obstructed paths to remove the logs can halve the damage to other trees. In addition, governments and individuals can mount efforts to reforest and rehabilitate degraded tropical forests and watersheds (Individuals Matter, right).

Another suggestion is to clamp down on illegal logging.

Solutions: The Incredible Neem Tree

The neem tree could eventually be of benefit almost everyone on the earth.

Suppose a single plant existed that could quickly reforest degraded land, supply fuelwood and lumber in dry areas, provide natural alternatives to toxic pesticides, be used to treat numerous diseases, and help control human population growth? There is: the *neem tree*, a broadleaf evergreen member of the mahogany family.

This remarkable tropical species, native to India and Burma, is ideal for reforestation because it can grow to maturity in only 5–7 years. It grows well in poor soil in semiarid lands such as those in Africa, providing abundant fuelwood, lumber, and lamp oil.

It also contains various natural pesticides. Chemicals from its leaves and seeds can repel or kill more than 200 insect pest species. Extracts from neem seeds and leaves can fight bacterial, viral, and fungal infections. Villagers call the tree a "village pharmacy" be-

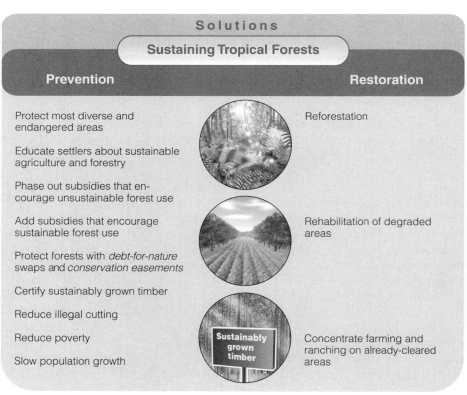

Figure 6-16 Solutions: ways to protect tropical forests and use them more sustainably.

Solutions
Sustaining Tropical Forests

Prevention

- Protect most diverse and endangered areas
- Educate settlers about sustainable agriculture and forestry
- Phase out subsidies that encourage unsustainable forest use
- Add subsidies that encourage sustainable forest use
- Protect forests with *debt-for-nature* swaps and *conservation easements*
- Certify sustainably grown timber
- Reduce illegal cutting
- Reduce poverty
- Slow population growth

Restoration

- Reforestation
- Rehabilitation of degraded areas
- Concentrate farming and ranching on already-cleared areas

Kenya's Green Belt Movement

In Kenya, Wangari Maathai (Figure 6-A) founded the Green Belt Movement in 1977. The goals of this highly regarded women's self-help group are to establish tree nurseries, raise seedlings, and plant and protect a tree for each of Kenya's 32 million people. By 2002, the 50,000 members of this grassroots group had established 6,000 village nurseries and planted and protected more than 20 million trees.

The success of this project has sparked the creation of similar programs in more than 30 other African countries. This inspiring leader has said,

I don't really know why I care so much. I just have something inside me that tells me that there is a problem and I have got to do something about it. And I'm sure it's

the same voice that is speaking to everyone on this planet, at least everybody who seems to be concerned about the fate of the world, the fate of this planet.

Figure 6-A Wangari Maathai, the first Kenyan woman to earn a Ph.D. (in anatomy) and to head an academic department (veterinary medicine) at the University of Nairobi, organized the internationally acclaimed Green Belt Movement in 1977. For her work in protecting the environment, she has received many honors, including the Goldman Prize, the Right Livelihood Award, the UN Africa Prize for Leadership, and the Golden Ark Award. After years of being harassed, beaten, and jailed for opposing government policies, she was elected to Kenya's parliament as a member of the Green Party in 2002. In 2003, she was appointed Assistant Minister for Environment, Natural Resources, and Wildlife.

---

cause its chemicals can relieve so many different health problems. People also use the tree's twigs as an antiseptic toothbrush and the oil from its seeds to make toothpaste and soap.

That is not all. Neem-seed oil evidently acts as a strong spermicide and may help in the development of a much-needed male birth control pill. According to a study by the U.S. National Academy of Sciences, the neem tree "may eventually benefit every person on the planet."

Despite its numerous advantages, ecologists caution against widespread planting of the neem tree outside its native range. As a nonnative species, it could take over and displace native species because of its rapid growth and resistance to pests. But proponents say the benefits of neem trees far outweigh such risks. What do you think?

6-6 NATIONAL PARKS

What Are National Parks and How Are They Threatened? Under Assault

Countries have established over 1,100 national parks, but most are threatened by human activities.

Today, more than 1,100 national parks larger than 10 square kilometers (4 square miles) each are located in more than 120 countries.

According to a 1999 study by the World Bank and the World Wildlife Fund, only 1% of the parks in developing countries receive protection.

Local people invade most of these parks in search of wood, cropland, game animals, and other natural products for their daily survival. Loggers, miners, and wildlife poachers (who kill animals to obtain and sell items such as rhino horns, elephant tusks, and furs) also invade many of these parks. Park services in developing countries typically have too little money and too few personnel to fight these invasions, either by force or by education.

Another problem is that most national parks are too small to sustain many large-animal species. Also, many parks suffer from invasions by nonnative species that can reduce the populations of native species and cause ecological disruption.

Case Study: National Parks in the United States

National parks in the United States face many threats.

The U.S. national park system, established in 1912, has 56 national parks (sometimes called the country's *crown jewels*), most of them in the West (Figure 6-5). State, county, and city parks supplement these national parks. Most state parks are located near urban areas and have about twice as many visitors per year as the national parks.

Popularity is one of the biggest problems of national and state parks in the United States. During the summer, users entering the most popular U.S. national and state parks often face hour-long backups and experience noise, congestion, eroded trails, and stress instead of peaceful solitude.

Many visitors expect parks to have grocery stores, laundries, bars, golf courses, video arcades, and other facilities found in urban areas. U.S. Park Service rangers spend an increasing amount of their time on law enforcement and crowd control instead of conservation management and education. Many overworked and underpaid rangers are leaving for better-paying jobs. In some parks, noisy dirt bikes, dune buggies, snowmobiles, and other off-road vehicles (ORVs) degrade the aesthetic experience for many visitors, destroy or damage fragile vegetation, and disturb wildlife.

Predators such as wolves all but vanished from various parks mostly because of excessive hunting and poisoning by ranchers and federal officials. In some parks, populations of the species the wolves once controlled have exploded, destroying vegetation and

crowding out other species. To help correct this situation, the U.S. Fish and Wildlife Service established a controversial program to reintroduce the gray wolf into the Yellowstone National Park area (Case Study, below).

Many parks suffer damage from the migration or deliberate introduction of nonnative species. European wild boars (imported to North Carolina in 1912 for hunting) threaten vegetation in part of the Great Smoky Mountains National Park. Nonnative mountain goats in Washington's Olympic National Park trample native vegetation and accelerate soil erosion. While some nonnative species have moved into parks, some economically valuable native species of animals and plants (including many threatened or endangered species) are killed or removed illegally by poachers in almost half of U.S. national parks.

Reintroducing Wolves to Yellowstone

CASE STUDY

At one time, the gray wolf ranged over most of North America. But between 1850 and 1900, an estimated 2 million wolves were shot, trapped, and poisoned by ranchers, hunters, and government employees. The idea was to make the West and the Great Plains safe for livestock and for big-game animals prized by hunters.

It worked. When Congress passed the U.S. Endangered Species Act in 1973, only about 400–500 gray wolves remained in the lower 48 states, primarily in Minnesota and Michigan.

Ecologists recognize the important role this keystone predator species once played in parts of the West and the Great Plains. These wolves culled herds of bison, elk, caribou, and mule deer, and kept down coyote populations. They also provided uneaten meat for scavengers such as ravens, bald eagles, ermines, and foxes.

In recent years, herds of elk, moose, and antelope have expanded. Their larger numbers have devastated some vegetation, increased erosion, and threatened the

niches of other wildlife species. Reintroducing a keystone species such as the gray wolf into a terrestrial ecosystem is one way to help sustain the biodiversity of the ecosystem and prevent further environmental degradation.

In 1987, the USFWS proposed reintroducing gray wolves into the Yellowstone ecosystem. This brought angry protests. Some objections came from ranchers who feared the wolves would attack their cattle and sheep; one enraged rancher said that the idea was "like reintroducing smallpox." Other protests came from hunters who feared the wolves would kill too many big-game animals, and from mining and logging companies who worried the government would halt their operations on wolf-populated federal lands.

Since 1995, federal wildlife officials have caught gray wolves in Canada and relocated them in Yellowstone National Park and northern Idaho. By 2004, there were about 760 gray wolves in these two areas.

Their presence is causing a cascade of ecological changes in Yellowstone. With wolves around, elk are gathering less near streams and

rivers. This has spurred the growth of aspen and willow trees that attract beavers, and elk killed by wolves are an important food source for grizzlies.

The wolves have cut coyote populations in half. This has increased populations of smaller animals such as ground squirrels and foxes hunted by coyotes. This provides more food for eagles and hawks. Between 1995 and 2002, the wolves also killed 792 sheep, 278 cattle, and 62 dogs in the Northern Rockies.

In 2004, the U.S. Fish and Wildlife Service proposed removing wolves from protection under the Endangered Species Act in Idaho and Montana. This would allow private citizens in these states to kill wolves that are attacking livestock or pets on private lands. Conservationists say this action is premature, warning that it could undermine one of the nation's most successful conservation efforts.

Critical Thinking

Do you agree or disagree with the program to reestablish populations of the gray wolf in the Yellowstone ecosystem? Explain. What are the major drawbacks of this program?

Solutions

National Parks

- Integrate plans for managing parks and nearby federal lands

- Add new parkland near threatened parks

- Buy private land inside parks

- Locate visitor parking outside parks and use shuttle buses for entering and touring heavily used parks

- Increase funds for park maintenance and repairs

- Survey wildlife in parks

- Raise entry fees for visitors and use funds for park management and maintenance

- Limit number of visitors to crowded park areas

- Increase number and pay of park rangers

- Encourage volunteers to give visitor lectures and tours

- Seek private donations for park maintenance and repairs

Figure 6-17 Solutions: suggestions for sustaining and expanding the national park system in the United States. (Data from Wilderness Society and National Parks and Conservation Association)

Nearby human activities that threaten wildlife and recreational values in many national parks include mining, logging, livestock grazing, coal-burning power plants, water diversion, and urban development. Polluted air, drifting hundreds of kilometers, kills ancient trees in California's Sequoia National Park and often blots out the awesome views at Arizona's Grand Canyon. According to the National Park Service, air pollution affects scenic views in national parks more than 90% of the time.

Figure 6-17 lists suggestions that various analysts have made for sustaining and expanding the national park system in the United States. Which two of these solutions do you believe are the most important?

6-7 NATURE RESERVES

How Much Land Should We Protect from Damaging Forms of Human Exploitation? The Answer Is More.

Ecologists believe that we should protect more land to help sustain the earth's biodiversity.

Most ecologists and conservation biologists believe the best way to preserve biodiversity is through a worldwide network of protected areas. Currently about 12% of the earth's land area has been protected strictly or partially in nature reserves, parks, wildlife refuges, wilderness, and other areas. In other words, we have reserved 88% of the earth's land for us, and most of the remaining 12% we have protected is ice, tundra, or desert where we do not want to live because it's too cold or too hot.

And this 12% figure is misleading because no more than 5% of these areas are actually protected. Thus we have strictly protected only about 7% of the earth's terrestrial areas from potentially harmful human activities.

Conservation biologists call for protection of at least 20% of the earth's land area in a global system of biodiversity reserves that includes multiple examples of all the earth's biomes. Doing this will take action and funding by national governments (Case Study, p. 118), private groups, and cooperative ventures involving governments, businesses, and private conservation groups.

Private groups play an important role in establishing wildlife refuges and other reserves to protect biological diversity. For example, since its founding by a group of professional ecologists in 1951, the *Nature Conservancy*—with more than 1 million members worldwide—has created the world's largest system of private natural areas and wildlife sanctuaries in 30 countries.

Most developers and resource extractors oppose protecting even the current 12% of the earth's remaining undisturbed ecosystems. They contend that most of these areas contain valuable resources that would add to economic growth.

Ecologists and conservation biologists disagree. They view protected areas as islands of biodiversity that help sustain all life and economies, and serve as centers of future evolution. See Norman Myer's Guest Essay on this topic on the website for this chapter.

X *HOW WOULD YOU VOTE?* Should at least 20% of the earth's land area be strictly protected from economic development? Cast your vote online at http://biology.brookscole.com/miller7.

Whenever possible, conservation biologists call for using the *buffer zone concept* to design and manage nature reserves. This means protecting an inner core of a reserve by two buffer zones in which local people can extract resources in ways that are sustainable and that do not harm the inner core. Doing this can enlist local people as partners in protecting a reserve from unsustainable uses. The United Nations uses this principle in

Biosphere Reserve

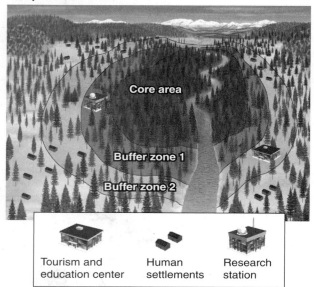

Figure 6-18 Solutions: a model *biosphere reserve*. Each reserve contains a protected inner core surrounded by two buffer zones that local and indigenous people can use for sustainable logging, food growing, cattle grazing, hunting, fishing, and eco-tourism.

establishing its global network of 425 biosphere reserves in 95 countries (Figure 6-18).

Case Study: What Has Costa Rica Done to Protect Some of Its Land from Degradation? A Global Conservation Leader

Costa Rica has devoted a larger proportion of land than any other country to conserving its significant biodiversity.

Tropical forests once completely covered Central America's Costa Rica, which is smaller in area than West Virginia and about one-tenth the size of France. Between 1963 and 1983, politically powerful ranching families cleared much of the country's forests to graze cattle. They exported most of the beef produced to the United States and western Europe.

Despite such widespread forest loss, tiny Costa Rica is a superpower of biodiversity, with an estimated 500,000 plant and animal species. A single park in Costa Rica is home to more bird species than all of North America.

In the mid-1970s, Costa Rica established a system of reserves and national parks that by 2003 included about a quarter of its land—6% of it in reserves for indigenous peoples. Costa Rica now devotes a larger proportion of land to biodiversity conservation than has any other country!

The country's parks and reserves are consolidated into eight *megareserves* designed to sustain about 80%

of Costa Rica's biodiversity (Figure 6-19). Each reserve contains a protected inner core surrounded by two buffer zones that local and indigenous people can use for sustainable logging, food growing, cattle grazing, hunting, fishing, and eco-tourism.

Costa Rica's biodiversity conservation strategy has paid off. Today, the $1-billion-a-year tourism business—almost two-thirds of it from eco-tourists—is the country's largest source of income.

A concern is that without careful government control, the 1 million tourists visiting Costa Rica each year can degrade some of the protected areas. Increased tourism could also stimulate the building of too many hotels, resorts, and other potentially harmful forms of development.

To reduce deforestation the government has eliminated subsidies for converting forest to cattle grazing land. And it pays landowners to maintain or restore tree coverage. The goal is to make sustaining forests profitable. It has worked. Costa Rica has gone from having one of the world's highest deforestation rates to one of the lowest.

What Is Adaptive Ecosystem Management? Cooperation and Flexibility

People with competing interests can work together to develop adaptable plans for managing and sustaining nature reserves.

Managing and sustaining a nature reserve is difficult. One problem is that reserves are constantly changing in response to environmental changes. Another is that they are affected by a variety of biological, cultural, economic, and political factors. In addition, their size, shape, and biological makeup often are determined by

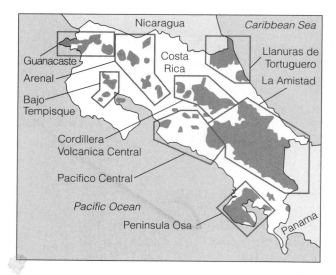

Figure 6-19 Solutions: Costa Rica has consolidated its parks and reserves into eight *megareserves* designed to sustain about 80% of the country's rich biodiversity.

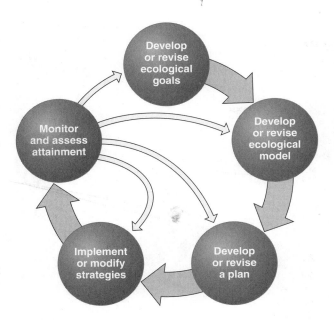

Figure 6-20 Solutions: the adaptive ecosystem management process.

political, legal, and economic factors that depend on land ownership and conflicting public demands rather than by ecological principles and considerations.

One way to deal with these uncertainties and conflicts is through *adaptive ecosystem management*. It is based on using four principles. *First*, integrate ecological, economic, and social principles to help maintain and restore the sustainability and biological diversity of reserves while supporting sustainable economies and communities.

Second, seek ways to get government agencies, private conservation organizations, scientists, business interests, and private landowners to reach a consensus on how to achieve common conservation objectives.

Third, view all decisions and strategies as scientific and social experiments and use failures as opportunities for learning and improvement. *Fourth*, emphasize continual information gathering, monitoring, reassessment, flexibility, adaptation, and innovation in the face of uncertainty and usually unpredictable change. Figure 6-20 summarizes the adaptive ecosystem management process.

What Areas Should Receive Top Priority for Establishing Reserves? Protecting Centers and Hot Spots

We can prevent or slow down losses of biodiversity by concentrating efforts on protecting hot spots where significant biodiversity is under immediate threat.

In reality, few countries are physically, politically, or financially able to set aside and protect large biodiversity reserves. To protect as much of the earth's remaining biodiversity as possible, conservation biologists use an *emergency action* strategy that identifies and quickly protects *biodiversity hot spots*. These "ecological arks" are areas especially rich in plant and animal species that are found nowhere else and are in great danger of extinction or serious ecological disruption.

Figure 6-21 (p. 120) shows 25 hot spots. They contain almost two-thirds of the earth's terrestrial biodiversity and are the only locations for more than one-third of the planet's known terrestrial plant and animal species. According to Norman Myers: "I can think of no other biodiversity initiative that could achieve so much at a comparatively small cost, as the hot spots strategy."

What Is Wilderness and Why Is It Important? Land Protected from Us

Wilderness is land legally set aside in a large enough area to prevent or minimize harm from human activities.

One way to protect undeveloped lands from human exploitation is by legally setting them aside as *wilderness*. According to the U.S. Wilderness Act of 1964, *wilderness* consists of areas "of undeveloped land affected primarily by the forces of nature, where man is a visitor who does not remain." U.S. president Theodore Roosevelt summarized what we should do with wilderness: "Leave it as it is. You cannot improve it."

The U.S. Wilderness Society estimates that a wilderness area should contain at least 4,000 square kilometers (1,500 square miles); otherwise, it can be affected by air, water, and noise pollution from nearby human activities.

Wilderness supporters cite several reasons for preserving wild places. One is that they are areas where people can experience the beauty of nature and observe natural biological diversity. Such areas can also enhance the mental and physical health of visitors by allowing them to get away from noise, stress, development, and large numbers of people. Wilderness preservationist John Muir advised us,

> Climb the mountains and get their good tidings. Nature's peace will flow into you as the sunshine into the trees. The winds will blow their freshness into you, and the storms their energy, while cares will drop off like autumn leaves.

Even those who never use the wilderness areas may want to know they are there, a feeling expressed by novelist Wallace Stegner:

> Save a piece of country . . . and it does not matter in the slightest that only a few people every year will go into it. This is precisely its value. . . . We simply need that wild country available to us, even if we never do more than drive to its edge and look in. For it can be a means of reassuring ourselves of our sanity as creatures, a part of the geography of hope.

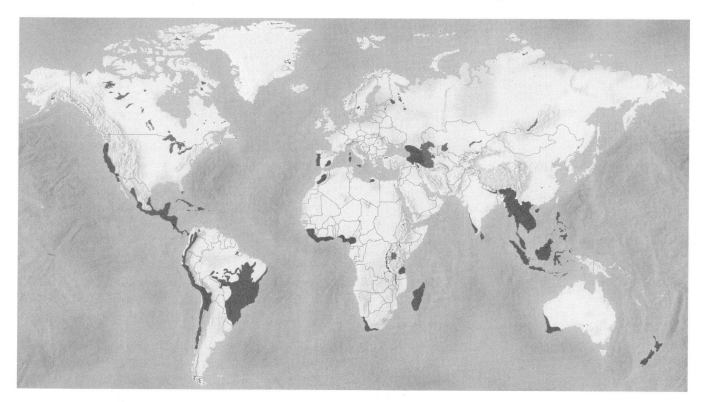

Figure 6-21 Endangered natural capital: twenty-five *hot spots* identified by ecologists as important but endangered centers of biodiversity that contain a large number of endemic plant and animal species found nowhere else. Research is adding new hot spots to this list. (Data from Center for Applied Biodiversity Science at Conservation International)

Some critics oppose protecting wilderness for its scenic and recreational value for a small number of people. They believe this is an outmoded concept that keeps some areas of the planet from being economically useful to humans.

Most biologists disagree. To them, the most important reasons for protecting wilderness and other areas from exploitation and degradation are to *preserve their biodiversity* as a vital part of the earth's natural capital and to *protect them as centers for evolution* in response to mostly unpredictable changes in environmental conditions. In other words, wilderness is a biodiversity and wildness bank and an eco-insurance policy.

Some analysts also believe wilderness should be preserved because the wild species it contains have a right to exist (or struggle to exist) and play their roles in the earth's ongoing saga of biological evolution and ecological processes, without human interference.

Case Study: How Much Wilderness Has Been Protected in the United States? Fighting For Crumbs and Losing

Only a small percentage of the land area of the United States has been set aside as wilderness.

In the United States, preservationists have been trying to save wild areas from development since 1900. Overall, they have fought a losing battle. Not until 1964 did Congress pass the Wilderness Act. It allowed the government to protect undeveloped tracts of public land from development as part of the National Wilderness Preservation System.

The area of protected wilderness in the United States increased tenfold between 1970 and 2000. Still, only about 4.6% of U.S. land is protected as wilderness—almost three-fourths of it in Alaska. Only 1.8% of the land area of the lower 48 states is protected, most of it in the West. In other words, Americans have reserved 98% of the continental United States to be used as they see fit and have protected only about 2% as wilderness. According to a 1999 study by the World Conservation Union, the United States ranks 42nd among nations in terms of terrestrial area protected as wilderness, and Canada is in 36th place.

In addition, only 4 of the 413 wilderness areas in the lower 48 states are larger than 4,000 square kilometers (1,500 square miles). Also, the system includes only 81 of the country's 233 distinct ecosystems. Most wilderness areas in the lower 48 states are threatened habitat islands in a sea of development.

What Can You Do?

Sustaining Terrestrial Biodiversity

- Plant trees and take care of them.

- Recycle paper and buy recycled paper products.

- Buy wood and wood products made from trees that have been grown sustainably.

- Help rehabilitate or restore a degraded area of forest or grassland near your home.

- When building a home, save as many trees and as much natural vegetation and soil as possible.

- Landscape your yard with a diversity of plants natural to the area instead of having a monoculture lawn.

Figure 6-22 What can you do? Ways to help sustain terrestrial biodiversity.

Almost 400,000 square kilometers (150,000 square miles) in scattered blocks of public lands could qualify for designation as wilderness—about 60% of it in the national forests. For two decades, these areas have been temporarily protected while they were evaluated for wilderness protection. Wilderness supporters would like to see all of these areas protected as part of the wilderness system.

This is unlikely because of the political strength of industries that see these areas as resources for increased profits and short-term economic growth. The political efforts of these industries paid off when, in 2003, the Bush administration ceased protecting areas under consideration for classification as wilderness. This opens up many of these lands to road building, mining, oil drilling, logging, and off-road vehicle use. Such activities would also disqualify these areas for wilderness protection in the future.

Figure 6-22 lists some ways you can help sustain the earth's terrestrial biodiversity.

6-8 ECOLOGICAL RESTORATION

How Can We Rehabilitate and Restore Damaged Ecosystems? Making Amends for Our Actions

Scientists have developed a number of techniques for rehabilitating and restoring degraded ecosystems and creating artificial ecosystems.

Almost every natural place on the earth has been affected or degraded to some degree by human activities. However, much of the environmental damage we have inflicted on nature is at least partially reversible through **ecological restoration:** the process of repairing damage caused by humans to the biodiversity and dynamics of natural ecosystems. Examples include replanting forests, restoring grasslands, restoring wetlands, reclaiming urban industrial areas (brownfields), reintroducing native species, removing invasive species, and freeing river flows by removing dams.

Farmer and philosopher Wendell Berry says we should try to answer three questions in deciding whether and how to modify or rehabilitate natural ecosystems. *First,* what is here? *Second,* what will nature permit us to do here? *Third,* what will nature help us do here? An important strategy is to mimic nature and natural processes and ideally let nature do most of the work, usually through secondary ecological succession.

By studying how natural ecosystems recover, scientists are learning how to speed up repair operations using a variety of approaches. They include the following:

- *Restoration:* trying to return a particular degraded habitat or ecosystem to a condition as similar as possible to its natural state. However, lack of knowledge about the previous composition of a degraded area can make it impossible to restore an area to its earlier state.

- *Rehabilitation:* attempts to turn a degraded ecosystem back into a functional or useful ecosystem without trying to restore it to its original condition.

- *Replacement:* replacing a degraded ecosystem with another type of ecosystem. For example, a productive pasture or tree farm may replace a degraded forest.

- *Creating artificial ecosystems:* an example is the creation of artificial wetlands.

Researchers have suggested five basic science-based principles for carrying out ecological restoration.

- Mimic nature and natural processes and ideally let nature do most of the work, usually through secondary ecological succession.

- Recreate important ecological niches that have been lost.

- Rely on pioneer species, keystone species, foundation species, and natural ecological succession to facilitate the restoration process.

- Control and remove harmful nonnative species.

- If necessary, reconnect small patches to form larger ones and create corridors where existing patches are isolated.

Some analysts worry that environmental restoration could encourage continuing environmental destruction and degradation by suggesting any ecological harm we do can be undone. Some go further and say that we do not understand the incredible complexity of ecosystems well enough to restore or manage damaged natural ecosystems.

Restorationists agree that restoration should not be used as an excuse for environmental destruction. But they point out that so far we have been able to protect or preserve no more than about 7% of nature from the effects of human activities. So ecological restoration is badly needed for much of the world's ecosystems that we have damaged. They also point out that if a restored ecosystem differs from the original system this is better than nothing and that increased experience will improve the effectiveness of ecological restoration.

✗ *How Would You Vote?* Should we mount a massive effort to restore ecosystems we have degraded even though this will be quite costly? Cast your vote online at http://biology .brookscole.com/miller7.

Case Study: Ecological Restoration of a Tropical Dry Forest in Costa Rica

A degraded tropical dry forest in Costa Rica is being restored in a cooperative venture between tropical ecologists and local people.

Costa Rica is the site of one of the world's largest *ecological restoration* projects. In the lowlands of the country's Guanacaste National Park (Figure 6-19), a small tropical dry deciduous forest has been burned, degraded, and fragmented by large-scale conversion to cattle ranches and farms.

Now it is being restored and relinked to the rain forest on adjacent mountain slopes. The goal is to eliminate damaging nonnative grass and cattle and reestablish a tropical dry forest ecosystem over the next 100–300 years.

Daniel Janzen, professor of biology at the University of Pennsylvania and a leader in the field of restoration ecology, has helped galvanize international support and has raised more than $10 million for this restoration project. He recognizes that ecological restoration and protection of the park will fail unless the people in the surrounding area believe they will benefit from such efforts. Janzen's vision is to make the nearly 40,000 people who live near the park an essential part of the restoration of the degraded forest, a concept he calls *biocultural restoration.*

By actively participating in the project, local residents reap educational, economic, and environmental benefits. Local farmers make money by sowing large areas with tree seeds and planting seedlings started in Janzen's lab. Local grade school, high school, and university students and citizens' groups study the ecology of the park and go on field trips to the park. The park's location near the Pan American Highway makes it an ideal area for eco-tourism, which stimulates the local economy.

The project also serves as a training ground in tropical forest restoration for scientists from all over the world. Research scientists working on the project give guest classroom lectures and lead some of the field trips.

Janzen recognizes that in a few decades today's children will be running the park and the local political system. If they understand the ecological importance of their local environment, they are more likely to protect and sustain its biological resources. He believes that education, awareness, and involvement—not guards and fences—are the best ways to restore degraded ecosystems and protect largely intact ecosystems from unsustainable use.

6-9 SUSTAINING AQUATIC BIODIVERSITY

What Do We Know about the Earth's Aquatic Biodiversity and Why Should We Care about It? Some but Not Nearly Enough

We know fairly little about the biodiversity of the world's marine and freshwater systems that provides us with important economic and ecological services.

Although we live on a water planet, we have explored only about 5% of the earth's global ocean and know fairly little about its biodiversity and how it works. We also know fairly little about freshwater biodiversity. Scientific investigation of poorly understood marine and freshwater aquatic systems is a *research frontier* whose study could result in immense ecological and economic benefits.

Despite the lack of knowledge about overall marine biodiversity, scientists have established three general patterns of marine biodiversity. *First,* the greatest marine biodiversity occurs in coral reefs, estuaries, and the deep-sea floor. *Second,* biodiversity is higher near coasts than in the open sea because of the greater variety of producers, habitats, and nursery areas in coastal areas. *Third,* biodiversity is higher in the bottom region of the ocean than in the surface region because of the greater variety of habitats and food sources on the ocean bottom.

Why should we care about aquatic biodiversity? The answer is that it helps keep us alive and supports our economies. Marine systems provide a variety of important ecological and economic services (Figure 3-15, p. 59), as do freshwater systems (Figure 3-18, p. 61).

What Are the Major Human Impacts on Aquatic Biodiversity? Our Large Aquatic Footprints

Human activities have destroyed or degraded a large proportion of the world's coastal wetlands, coral reefs, mangroves, and ocean bottom, and overfished many marine and freshwater species.

The greatest threat to the biodiversity of the world's oceans is loss and degradation of habitats, as summarized in Figure 3-18, p. 61. Revisit this figure. For example, more than one-fourth of the world's diverse coral reefs have been severely damaged, mostly by human activities, and another 70% of these reefs may be severely damaged or eliminated by 2050.

And many bottom habitats are being degraded and destroyed by dredging operations and trawler boats, which, like giant submerged bulldozers, drag huge nets weighted down with heavy chains and steel plates over ocean bottoms to harvest bottom fish and shellfish. Each year, thousands of trawlers scrape and disturb an area of ocean bottom equal to the combined size of Brazil and India and about 150 times larger than the area of forests clear-cut each year. In 2004, some 1,134 scientists signed a statement urging the United Nations to declare a moratorium on bottom trawling on the high seas.

Studies indicate that about three-fourths of the world's 200 commercially valuable marine fish species (40% in U.S. waters) are either overfished or fished to their estimated sustainable yield. One of the results of the increasingly efficient global hunt for fish is that big fish in many populations of commercially valuable species are becoming scarce. Smaller fish are next as the fishing industry has begun working its way down marine food webs. This can reduce the breeding stock needed for recovery of depleted species, unravel food webs, and disrupt marine ecosystems. And we throw away almost one-third of the fish we catch because they are caught unintentionally in our harvest of commercially valuable species.

According to marine biologists, at least 1,200 marine species have become extinct in the past few hundred years, and many thousands of additional marine species could disappear during this century. Indeed, *fish are threatened with extinction by human activities more than any other group of species.*

Also, according to the UN Food and Agriculture Organization and the World Wildlife Fund, at least a fifth of the world's 10,000 known freshwater fish species (37% in the United States) are threatened with extinction or have already become extinct. Indeed,

freshwater animals are disappearing five times faster than land animals.

Why Is It Difficult to Protect Marine Biodiversity? Out of Sight, Out of Mind

Coastal development, the invisibility and vastness of the world's oceans, and lack of legal jurisdiction hinder protection of marine biodiversity.

There are several reasons why protecting marine biodiversity is difficult. One is rapidly growing coastal development and the accompanying massive inputs of sediment and other wastes from land into coastal water. This harms shore-hugging species and threatens biologically diverse and highly productive coastal ecosystems such as coral reefs, marshes, and mangrove forest swamps.

Another problem is that much of the damage to the oceans and other bodies of water is not visible to most people. And many people incorrectly view the seas as an inexhaustible resource that can absorb an almost infinite amount of waste and pollution.

In addition, most of the world's ocean area lies outside the legal jurisdiction of any country. Thus, it is an open-access resource, subject to overexploitation because of the tragedy of the commons.

How Can We Protect and Sustain Marine Biodiversity? Prevention Is the Key.

Marine biodiversity can be sustained by protecting endangered species, establishing protected sanctuaries, managing coastal development, reducing water pollution, and preventing overfishing.

Here are some ways to protect and sustain marine biodiversity. We can protect endangered and threatened aquatic species, as discussed in Chapter 7.

We can *establish protected marine sanctuaries.* Since 1986, the World Conservation Union has helped establish a global system of *marine protected areas* (*MPAs*), mostly at the national level. And about 90 of the world's 350 biosphere reserves include coastal or marine habitats.

Scientific studies show that within fully protected marine reserves, fish populations double, fish size grows by almost a third, fish reproduction triples, and species diversity increases by almost one-fourth. Furthermore, this improvement happens within 2–4 years after strict protection begins and lasts for decades.

However, less than 0.01% of the world's ocean area consists of fully protected marine reserves. In the United States, the total area of fully protected marine habitat is only about 130 square kilometers (50 square miles). In other words, *we have failed to strictly protect*

Managing Fisheries

Fishery Regulations

Set catch limits well below the maximum sustainable yield

Improve monitoring and enforcement of regulations

Economic Approaches

Sharply reduce or eliminate fishing subsidies

Charge fees for harvesting fish and shellfish from publicly owned offshore waters

Certify sustainable fisheries

Protected Areas

Established no-fishing areas

Establish more marine protected areas

Rely more on integrated coastal management

Consumer Information

Label sustainably harvested fish

Publicize overfished and threatened species

Bycatch

Use wide-meshed nets to allow escape of smaller fish

Use net escape devices for seabirds and sea turtles

Ban throwing edible and marketable fish back into the sea

Aquaculture

Restrict coastal locations for fish farms

Control pollution more strictly

Depend more on herbivorous fish species

Nonative Invasions

Kill organisms in ship ballast water

Filter organisms from ship ballast water

Dump ballast water far at sea and replace with deep-sea water

Figure 6-23 Solutions: ways to manage fisheries more sustainably and protect marine biodiversity.

99.9% of the world's ocean area from human exploitation. Also, many current marine sanctuaries are too small to protect most of the species within them and do not provide adequate protection from pollution that flows into coastal waters as a result of land use.

In 1997, a group of international marine scientists called for governments to increase fully protected marine reserves to at least 20% of the ocean's surface by 2020. A 2003 study by the Pew Fisheries Commission recommended establishing many more protected marine reserves in U.S. coastal waters and connecting them with protected corridors so fish can move back and forth between such areas.

We can also establish *integrated coastal management* in which groups of fishers, scientists, conservationists, citizens, business interests, developers, and politicians work together to identify shared problems and goals. Then they attempt to develop workable and cost-effective solutions that preserve biodiversity and environmental quality while meeting economic and social needs, ideally using adaptive ecosystem management (Figure 6-20). Currently, more than 100 integrated

coastal management programs are being developed throughout the world, including the Chesapeake Bay in the United States.

Another important strategy is to protect existing coastal and inland wetlands from being destroyed or degraded. We can also manage marine fisheries to prevent overfishing. Figure 6-23 lists measures that analysts have suggested for managing global fisheries more sustainably and protecting marine biodiversity. Most of these approaches rely on some sort of government regulation. Finally, we can regulate and prevent aquatic pollution, as discussed in Chapter 9.

6-10 WHAT CAN WE DO?

What Should Be Our Priorities? An Eight-Step Program

Biodiversity expert Edward O. Wilson has proposed eight priorities for protecting most of the world's remaining ecosystems and species.

In 2002, Edward O. Wilson, considered to be one of the world's foremost experts on biodiversity, proposed the following priorities for protecting most of the world's remaining ecosystems and species:

- Take immediate action to preserve the world's biological hot spots (Figure 6-21).

- Keep intact the world's remaining old-growth forests and cease all logging of such forests.

- Complete the mapping of the world's terrestrial and aquatic biodiversity so we know what we have and can make conservation efforts more precise and cost-effective.

- Determine the world's *marine hot spots* and assign them the same priority for immediate action as for those on land.

- Concentrate on protecting and restoring everywhere the world's lakes and river systems, which are the most threatened ecosystems of all.

- Ensure that the full range of the earth's terrestrial and aquatic ecosystems is included in a global conservation strategy.

- Make conservation profitable. This involves finding ways to raise the income of people who live in or near nature reserves so they can become partners in their protection and sustainable use.

- Initiate ecological restoration products worldwide to heal some of the damage we have done and increase the share of the earth's land and water allotted to the rest of nature.

According to Wilson, such a conservation strategy would cost about $30 billion per year—an amount that could be provided by a tax of one-cent-per-cup of coffee.

This strategy for protecting the earth's precious biodiversity that we are completely dependent on will not be implemented without bottom-up political pressure on elected officials from individual citizens and groups. It will also require cooperation among key people in government, the private sector, scientists, and engineers using adaptive ecosystem management (Figure 6-20).

We abuse land because we regard it as a commodity belonging to us. When we see land as a community to which we belong, we may begin to use it with love and respect.

ALDO LEOPOLD

CRITICAL THINKING

1. Explain why you agree or disagree with **(a)** the four principles that biologists and some economists have suggested for using public land in the United States (p. 103) and **(b)** the nine suggestions made by developers and resource extractors for managing and using U.S. public land (p. 105).

2. Explain why you agree or disagree with each of the proposals for providing more sustainable use of forests throughout the world, listed in Figure 6-12, p. 109. Which two of these proposals do you believe are the most important? Explain.

3. Should developed countries provide most of the money to preserve remaining tropical forests in developing countries? Explain.

4. Should there be a ban on the use of off-road motorized vehicles and snowmobiles on all public lands? Explain.

5. Are you in favor of establishing more wilderness areas in the United States, especially in the lower 48 states (or in the country where you live)? Explain. What might be some drawbacks of doing this?

6. Congratulations! You are in charge of the world. List the three most important features of your policies for using and managing **(a)** forests, **(b)** parks, and **(c)** aquatic biodiversity.

LEARNING ONLINE

The website for this book contains helpful study aids and many ideas for further reading and research. They include a chapter summary, review questions for the entire chapter, flash cards for key terms and concepts, a multiple-choice practice quiz, interesting Internet sites, references, and a guide for accessing thousands of InfoTrac® College Edition articles. Log on to

http://biology.brookscole.com/miller7

Then click on the Chapter-by-Chapter area, choose Chapter 6, and select a learning resource.

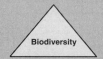
The last word in ignorance is the person who says of an animal or plant: "What good is it?" . . . If the land mechanism as a whole is good, then every part of it is good, whether we understand it or not. . . . Harmony with land is like harmony with a friend; you cannot cherish his right hand and chop off his left.

ALDO LEOPOLD

7-1 SPECIES EXTINCTION

What Are Three Types of Species Extinction? Local, Ecological, and Biological

Species can become extinct locally, ecologically, or globally.

Biologists distinguish among three levels of species extinction. One is *local extinction*. It occurs when a species is no longer found in an area it once inhabited but is still found elsewhere in the world. Most local extinctions involve losses of one or more populations of species.

The second type is *ecological extinction*. It occurs when so few members of a species are left that it can no longer play its ecological roles in the biological communities where it is found.

The third type is *biological extinction,* when a species is no longer found anywhere on the earth (Figure 7-1 and Case Study, below). Biological extinction is forever.

Case Study: The Passenger Pigeon: Gone Forever

The passenger pigeon, once the most common bird in North America, became extinct because of human activities, mostly overhunting and loss of habitat.

In 1813, bird expert John James Audubon saw a single flock of passenger pigeons that he estimated was 16 kilometers (10 miles) wide and hundreds of kilometers long and contained perhaps a billion birds. The flock took three days to fly past him and was so dense that it darkened the skies.

By 1914, the passenger pigeon (Figure 7-1, left) had disappeared forever. How could a species that was once the most common bird in North America and probably the world become extinct in only a few

decades? The answer is humans wiped them out. The main reasons for the extinction of this species were uncontrolled commercial hunting and loss of the bird's habitat and food supply as forests were cleared to make room for farms and cities.

Passenger pigeons were good to eat, their feathers made good filling for pillows, and their bones were widely used for fertilizer. They were easy to kill because they flew in gigantic flocks and nested in long, narrow colonies.

Beginning in 1858, passenger pigeon hunting became a big business. Shotguns, traps, artillery, and even dynamite were used. People burned grass or sulfur below their roosts to suffocate the birds. Shooting galleries used live birds as targets. In 1878, one professional pigeon trapper made $60,000 by killing 3 million birds at their nesting grounds near Petoskey, Michigan!

By the early 1880s, only a few thousand birds remained. At that point, recovery of the species was doomed because the females laid only one egg per nest. On March 24, 1900, a young boy in Ohio shot the last known wild passenger pigeon.

The last passenger pigeon on earth, a hen named Martha after Martha Washington, died in the Cincinnati Zoo in 1914. Her stuffed body is now on view at the National Museum of Natural History in Washington, D.C.

What Are Endangered and Threatened Species? Ecological Smoke Alarms

An endangered species could soon become extinct and a threatened species is likely to become extinct.

Biologists classify species heading toward biological extinction as either *endangered* or *threatened* (Figure 7-2, p. 128). An **endangered species** has so few individual survivors that the species could soon become extinct over all or most of its natural range. A **threatened,** or **vulnerable, species** is still abundant in its natural range but because of declining numbers is likely to become endangered in the near future.

Some species have characteristics that make them more vulnerable than others to ecological and biological extinction (Figure 7-3, p. 130). As biodiversity expert Edward O. Wilson puts it, the first animal species to go are the big, the slow, the tasty, and those with valuable parts such as tusks and skins."

Figure 7-1 Lost natural capital: some animal species that have become prematurely extinct largely because of human activities, mostly habitat destruction and overhunting.

Passenger pigeon   Great auk   Dodo   Dusky seaside sparrow   Aepyornis (Madagascar)

How Do Biologists Estimate Extinction Rates? Peering into a Cloudy Looking Glass

Scientists use measurements and models to estimate extinction rates.

Evolutionary biologists estimate that more than 99.9% of all species that have ever existed are now extinct because of a combination of background extinctions, mass extinctions, and mass depletions taking place over thousands to millions of years. Biologists also talk of an *extinction spasm,* wherein large numbers of species are lost over a period of a few centuries or at most 1,000 years.

Biologists trying to catalog extinctions have three problems. *First,* the extinction of a species typically takes such a long time that it is not easy to document. *Second,* we have identified only about 1.4–1.8 million of the world's estimated 5–100 million species. *Third,* we know little about most of the species we have identified.

The truth is that we do not know how many species are becoming extinct each year mostly because of our activities. But scientists do the best they can with the tools they have to estimate past and projected future extinction rates.

One approach is to study past records documenting the rate at which mammals and birds have become extinct since we arrived and compare that with the fossil records of such extinctions prior to our arrival. One way that biologists use to project future extinction rates is to observe how the number of species present increases with the size of an area. This *species–area relationship* suggests that on average a 90% loss of habitat causes the extinction of about 50% of the species living in that habitat. Another tool is to observe changes in species diversity at different latitudes.

Scientists also use models to estimate the risk that a particular species will become endangered or extinct within a certain time, based on factors such as trends in population size, changes in habitat availability, and interactions with other species.

Such estimates of extinction rates vary because of differing assumptions about the earth's total number of species, the proportion of these species that are found in tropical forests, the rate at which tropical forests are being cleared, and the reliability of the methods used to make these estimates.

How Are Human Activities Affecting Extinction Rates? Taking Out More Species

Biologists estimate that the current rate of extinction is at least 1,000 to 10,000 times the rate before we arrived.

In due time, all species become extinct but there is considerable evidence that we are hastening the final exit for a growing number of species. Before we came on the scene, the estimated extinction rate was roughly one species per million species annually. This amounted to an annual extinction rate of about 0.0001% per year.

Using the methods just described, biologists conservatively estimate that the current rate of extinction is at least 1,000 to 10,000 times the rate before we arrived. This amounts to an annual extinction rate of at least 0.1% to 1% per year.

So how many species are we losing prematurely each year? This depends on how many species are on the earth. Assuming that the extinction rate is 0.1%, each year we are losing 5,000 species per year if there are 5 million species, 14,000 if there are 14 million species (biologists' current best guess), and 100,000 if there are 100 million species.

Most biologists would consider the premature loss of 1 million species over 100–200 years an extinction crisis or spasm that, if kept up, would lead to a mass depletion or even a mass extinction. At an extinction

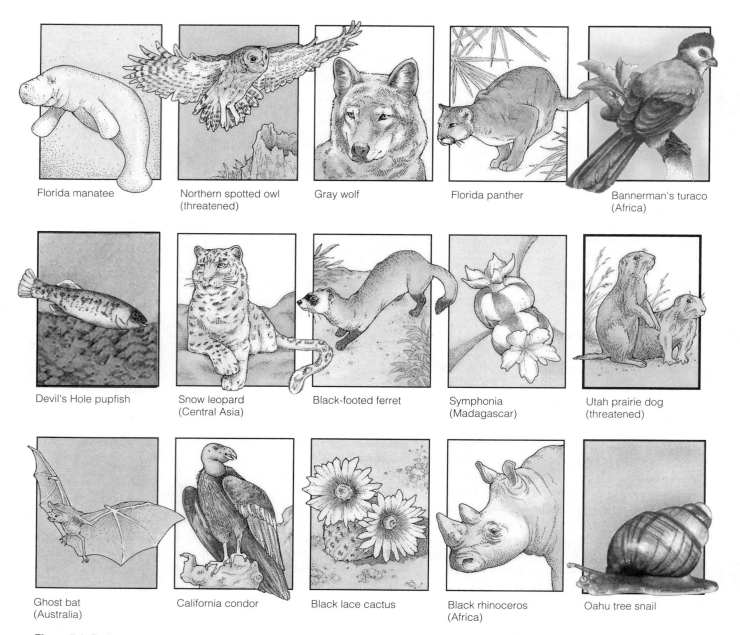

Figure 7-2 Endangered natural capital: species that are endangered or threatened with premature extinction largely because of human activities. Almost 30,000 of the world's species and 1,260 of those in the United States are officially listed as being in danger of becoming extinct. Most biologists believe the actual number of species at risk is much larger.

Florida manatee

Northern spotted owl (threatened)

Gray wolf

Florida panther

Bannerman's turaco (Africa)

Devil's Hole pupfish

Snow leopard (Central Asia)

Black-footed ferret

Symphonia (Madagascar)

Utah prairie dog (threatened)

Ghost bat (Australia)

California condor

Black lace cactus

Black rhinoceros (Africa)

Oahu tree snail

rate of 0.1% a year, the time it would take to lose 1 million species would be 200 years if there were a total of 5 million species, 71 years with a total of 14 million species, and 10 years with 100 million species. How many years would it take to lose 1 million species for each of these three species estimates if the extinction rate is 1% a year?

According to researchers Edward O. Wilson and Stuart Primm, at a 1% extinction rate, at least one-fifth of the world's current animal and plant species could be gone by 2030 and half could vanish by the end of this century. In the words of biodiversity expert Norman Myers, "Within just a few human generations, we shall—in the absence of greatly expanded conservation efforts—impoverish the biosphere to an extent that will persist for at least 200,000 human generations or twenty times longer than the period since humans emerged as a species."

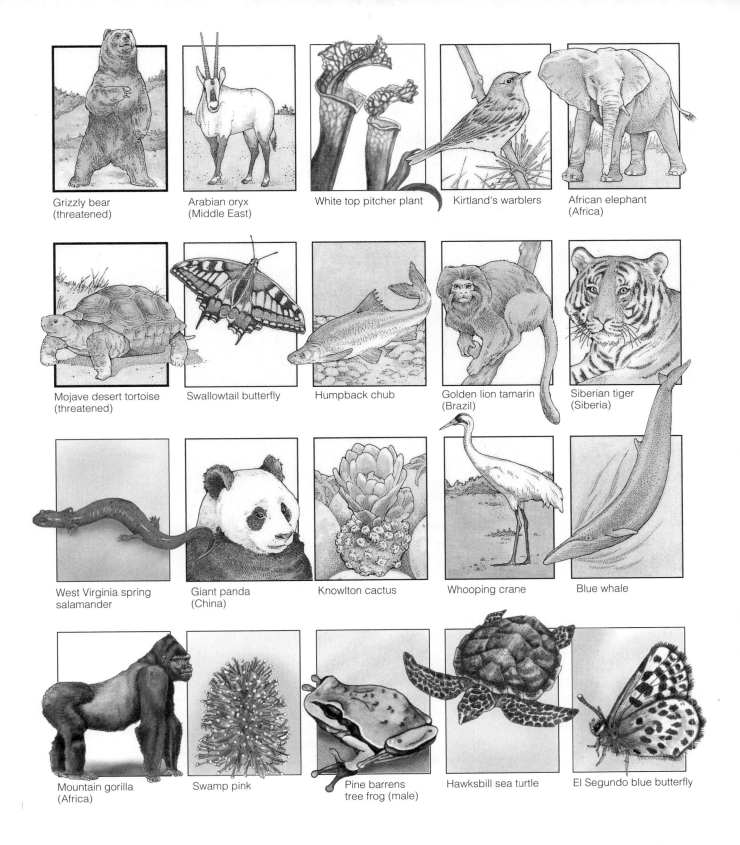

Grizzly bear
(threatened)

Arabian oryx
(Middle East)

White top pitcher plant

Kirtland's warblers

African elephant
(Africa)

Mojave desert tortoise
(threatened)

Swallowtail butterfly

Humpback chub

Golden lion tamarin
(Brazil)

Siberian tiger
(Siberia)

West Virginia spring
salamander

Giant panda
(China)

Knowlton cactus

Whooping crane

Blue whale

Mountain gorilla
(Africa)

Swamp pink

Pine barrens
tree frog (male)

Hawksbill sea turtle

El Segundo blue butterfly

Most biologists consider extinction rates of 0.1%–1% to be conservative estimates for several reasons. *First*, both the rate of species loss and the extent of biodiversity loss are likely to increase during the next 50–100 years because of the projected growth of the world's human population and resource use per person. In other words, the size of our already large ecological footprint (Figure 1-5, p. 9) is likely to increase.

Characteristic	Examples
Low reproductive rate	Blue whale, giant panda, rhinoceros
Specialized niche	Blue whale, giant panda, Everglades kite
Narrow distribution	Many island species, elephant seal, desert pupfish
Feeds at high trophic level	Bengal tiger, bald eagle, grizzly bear
Fixed migratory patterns	Blue whale, whooping crane, sea turtles
Rare	Many island species, African violet, some orchids
Commercially valuable	Snow leopard, tiger, elephant, rhinoceros, rare plants and birds
Large territories	California condor, grizzly bear, Florida panther

Figure 7-3 Characteristics of species that are prone to ecological and biological extinction.

Second, current and projected extinction rates are much higher than the global average in parts of the world that are endangered centers of the world's biodiversity. Conservation biologists urge us to focus our efforts on slowing the much higher rates of extinction in such hot spots (Figure 6-21, p. 120) as the best and quickest way to protect much of the earth's biodiversity from being lost prematurely.

Third, we are eliminating, degrading, and fragmenting many biologically diverse environments—such as tropical forests, tropical coral reefs, wetlands, and estuaries—that serve as potential colonization sites for the emergence of new species. Thus, in addition to increasing the rate of extinction, we may also be limiting long-term recovery of biodiversity by reducing the rate of speciation for some types of species. In other words, we are also creating a *speciation crisis.* See the Guest Essay by Norman Myers on this topic on the website for this chapter.

Some people, most of them not biologists, say the current estimated extinction rates are too high and are based on inadequate data and models. Researchers agree that their estimates of extinction rates are based on inadequate data and sampling. They continually strive to get better data and improve the models they use to estimate extinction rates.

However, they point to clear evidence that human activities have increased the rate of species extinction and that this rate is likely to rise. According to these biologists, arguing over the numbers and waiting to get better data and models should not be used as excuses for inaction. They call for us to implement a *precautionary strategy* now to help prevent a significant decrease in the earth's biodiversity as a result of our activities.

To these biologists, we are not heeding Aldo Leopold's warning about preserving biodiversity as we tinker with the earth: "To keep every cog and wheel is the first precaution of intelligent tinkering."

7-2 IMPORTANCE OF WILD SPECIES?

Why Should We Preserve Wild Species? They Have Value.

We should not cause the premature extinction of species because of the economic and ecological services they provide.

So what is all the fuss about? If all species eventually become extinct, why should we worry about losing a few more because of our activities? Does it matter that the passenger pigeon, the 80–100 remaining Florida panthers, or some unknown plant or insect in a tropical forest becomes prematurely extinct because of human activities?

We know that new species eventually evolve to take the place of ones lost through extinction spasms, mass depletions, or mass extinctions (Figure 3-4, p. 51). So why should we care if we speed up the extinction rate over the next 50–100 years? The answer is that *it will take at least 5 million years for speciation to rebuild the biodiversity we are likely to destroy during this century!*

Conservation biologists and ecologists say we should act now to prevent premature extinction of species because of their *instrumental value* based on their usefulness to us in the form of economic and ecological services (see top half of back cover). For example, species provide economic value in the form of food crops, fuelwood and lumber, paper, and medicine (Figure 7-4).

Another instrumental value is the *genetic information* in species that allows them to adapt to changing environmental conditions and to form new species. Genetic engineers use this information to produce new types of crops (Figure 3-5, p. 53) and foods and edible vaccines for viral diseases such as hepatitis B. Carelessly eliminating many of the species making up the world's vast genetic library is like burning books before we read them. Wild species also provide a way for us to learn how nature works and sustains itself.

The earth's wild plants and animals also provide us with *recreational pleasure.* Each year, Americans spend over three times as many hours watching wildlife—doing nature photography and bird watching, for example—as they spend on watching movies or professional sporting events.

Wildlife tourism, or *eco-tourism*, generates at least $500 billion per year worldwide, and perhaps twice as much. Conservation biologist Michael Soulé estimates that one male lion living to age 7 generates $515,000 in tourist dollars in Kenya but only $1,000 if killed for its skin. Similarly, over a lifetime of 60 years, a Kenyan elephant is worth about $1 million in eco-tourist revenue—many times more than its tusks are worth when sold illegally for their ivory.

Eco-tourism should not cause ecological damage but some of it does. The website for this chapter lists some guidelines for evaluating eco-tours.

Case Study: Why Should We Care about Bats? Ecological Allies

Because of the important ecological and economic roles bats play, we should view them as valuable allies, not as enemies to kill.

Bats—*the only mammals that can fly*—have two traits that make them vulnerable to extinction. *First,* they reproduce slowly. *Second,* many bat species live in huge colonies in caves and abandoned mines, which people sometimes block. This prevents them from leaving to get food and it can disturb their hibernation.

Bats play important ecological roles. About 70% of the world's 950 known bat species feed on crop-damaging nocturnal insects and other insect pest species such as mosquitoes. This makes them the major nighttime SWAT team for such insects.

In some tropical forests and on many tropical islands, *pollen-eating bats* pollinate flowers, and *fruit-eating bats* distribute plants throughout tropical forests by excreting undigested seeds.

As keystone species, such bats are vital for maintaining plant biodiversity and for regenerating large areas of tropical forest cleared by human activities. If you enjoy bananas, cashews, dates, figs, avocados, or mangos, you can thank bats.

Many people mistakenly view bats as fearsome, filthy, aggressive, rabies-carrying bloodsuckers. But most bat species are harmless to people, livestock, and crops. In the United States, only ten people have died of bat-transmitted disease in four decades of record keeping; more Americans die each year from falling coconuts.

Because of unwarranted fears of bats and lack of knowledge about their vital ecological roles, several

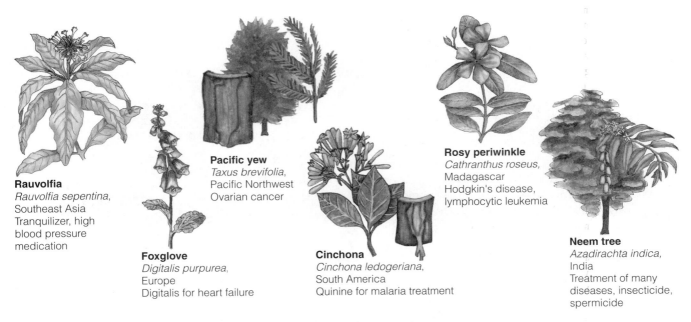

Rauvolfia
Rauvolfia sepentina,
Southeast Asia
Tranquilizer, high
blood pressure
medication

Foxglove
Digitalis purpurea,
Europe
Digitalis for heart failure

Pacific yew
Taxus brevifolia,
Pacific Northwest
Ovarian cancer

Cinchona
Cinchona ledogeriana,
South America
Quinine for malaria treatment

Rosy periwinkle
Cathranthus roseus,
Madagascar
Hodgkin's disease,
lymphocytic leukemia

Neem tree
Azadirachta indica,
India
Treatment of many
diseases, insecticide,
spermicide

Figure 7-4 Natural capital: *nature's pharmacy.* Parts of these and a number of other plants and animals (many of them found in tropical forests) are used to treat a variety of human ailments and diseases. Nine of the ten leading prescription drugs originally came from wild organisms. About 2,100 of the 3,000 plants identified by the National Cancer Institute as sources of cancer-fighting chemicals come from tropical forests. Despite their economic and health potential, fewer than 1% of the estimated 125,000 flowering plant species in tropical forests (and a mere 1,100 of the world's 260,000 known plant species) have been examined for their medicinal properties. Once the active ingredients in the plants have been identified, they can usually be produced synthetically. Many of these tropical plant species are likely to become extinct before we can study them.

bat species have been driven to extinction. Currently, about one-fourth of the world's bat species, including the ghost bat (Figure 7-2), are listed as endangered or threatened. Conservation biologists urge us to view bats as valuable allies, not as enemies.

What Is the Intrinsic Value of Wild Species? Existence Rights

Some believe that each wild species has an inherent right to exist.

Some people believe each wild species has *intrinsic* or *existence* value based on its inherent right to exist and play its ecological roles, regardless of its usefulness to us. According to this view, we have an ethical responsibility to protect species from becoming prematurely extinct as a result of human activities, and to prevent the degradation of the world's ecosystems and its overall biodiversity.

Some people distinguish between the survival rights of plants and those of animals, mostly for practical reasons. Poet Alan Watts once said he was a vegetarian "because cows scream louder than carrots."

Other people distinguish among various types of species. For example, they might think little about getting rid of the world's mosquitoes, cockroaches, rats, or disease-causing bacteria.

Some conservation biologists caution us not to focus primarily on protecting relatively big organisms—the plants and animals we can see and are familiar with. They remind us that the true foundation of the earth's ecosystems and ecological processes are the invisible bacteria, and the algae, fungi, and other *microorganisms* that decompose the bodies of larger organisms and recycle the nutrients needed by all life.

7-3 CAUSES OF PREMATURE EXTINCTION OF WILD SPECIES

What Is the Role of Habitat Loss and Degradation? Creating Homeless Species

The greatest threat to a species is the loss and degradation of the place where it lives.

Figure 7-5 shows the basic and secondary causes of the endangerment and premature extinction of wild species. Conservation biologists sometimes summarize the main secondary factors leading to premature extinction using the acronym **HIPPO** for **H**abitat destruction and fragmentation, **I**nvasive (alien) species, **P**opulation growth (too many people consuming too many resources), **P**ollution, and **O**verharvesting.

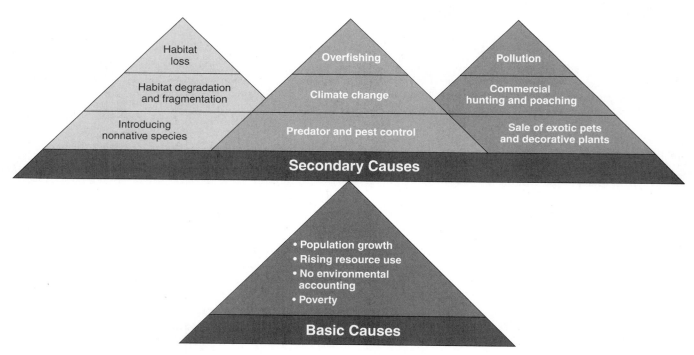

Figure 7-5 Degradation of natural capital: underlying and direct causes of depletion and premature extinction of wild species. The two biggest secondary causes of wildlife depletion and premature extinction are the loss or fragmentation and degradation of habitat, and deliberate or accidental introduction of nonnative species into ecosystems.

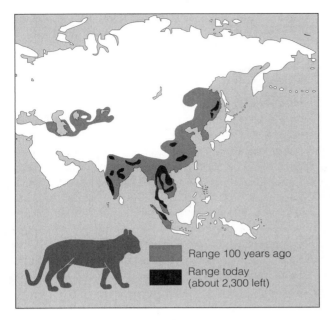

Indian Tiger

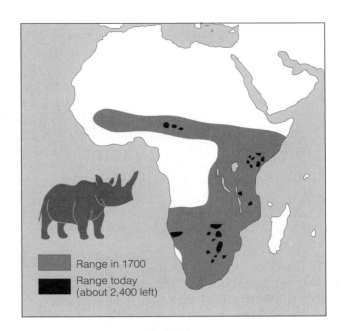

Black Rhino

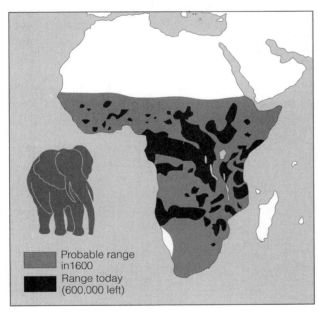

African Elephant

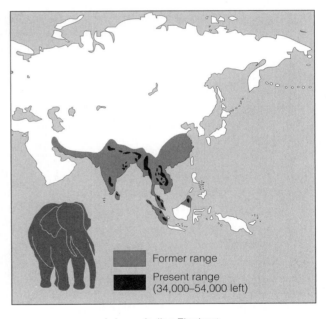

Asian or Indian Elephant

Figure 7-6 Degraded natural capital: reductions in the ranges of four wildlife species, mostly the result of habitat loss and hunting. What will happen to these and millions of other species when the world's human population doubles and per capita resource consumption rises sharply in the next few decades? (Data from International Union for the Conservation of Nature and World Wildlife Fund)

According to biodiversity researchers, the greatest threat to wild species is habitat loss (Figure 7-6), degradation, and fragmentation. In other words, many species have a hard time surviving when we take over their ecological "house" and their food supplies and make them homeless.

Deforestation of tropical forests is the greatest eliminator of species, followed by the destruction of coral reefs and wetlands, plowing of grasslands, and pollution of streams, lakes, and oceans. Globally, temperate biomes have been affected more by habitat loss and degradation than have tropical biomes because of widespread development in temperate countries over the past 200 years. But development is now shifting to many tropical biomes.

Island species, many of them *endemic species* found nowhere else on earth, are especially vulnerable to extinction when their habitats are destroyed, degraded, or fragmented. Any habitat surrounded by a different one can be viewed as a *habitat island* for most of the species that live there. Most national parks and other protected areas are habitat islands, many of them surrounded by potentially damaging logging, mining, energy extraction, and industrial activities. Freshwater lakes are also habitat islands that are especially vulnerable when nonnative species are introduced.

What Is the Role of Habitat Fragmentation? Isolating and Weakening Populations of Species

Species are more vulnerable to extinction when their habitats are divided into smaller, more isolated patches.

Habitat fragmentation occurs when a large, continuous area of habitat is reduced in area and divided into smaller, more scattered, and isolated patches or "habitat islands." This divides populations of a species into smaller and more isolated groups that are more vulnerable to predators, competitive species, disease, and catastrophic events such as a storm or fire. Also, it creates barriers that limit the abilities of some species to disperse and colonize new areas, get enough to eat, and find mates.

Case Study: How Do Human Activities Affect Bird Species? A Disturbing Message from the Birds

Our activities are causing serious declines in the populations of many bird species.

Approximately 70% of the world's 9,800 known bird species are declining in numbers, and about one of every six bird species are threatened with extinction, mostly because of habitat loss and fragmentation.

Conservation biologists view this decline of bird species with alarm. The reason is that birds are excellent *environmental indicators* because they live in every climate and biome, respond quickly to environmental changes in their habitats, and are easy to track and count.

Besides serving as indicator species, birds play important ecological roles. They include helping control populations of rodents and insects (which decimate many tree species), pollinating a variety of flowering plants, spreading plants throughout their habitats by consuming and excreting plant seeds, and scavenging dead animals. Conservation biologists urge us to listen more carefully to what birds are telling us about the state of the environment for them and for us.

What Is the Role of Deliberately Introduced Species? Good and Bad News

Many nonnative species provide us with food, medicine, and other benefits but a few can wipe out some native species, disrupt ecosystems, and cause large economic losses.

We depend heavily on nonnative organisms for ecosystem services, food, shelter, medicine, and aesthetic enjoyment. According to a 2000 study by ecologist David Pimentel, introduced species such as corn, wheat, rice, other food crops, cattle, poultry, and other livestock provide more than 98% of the U.S. food supply. Similarly, nonnative tree species are grown in about 85% of the world's tree plantations. Some deliberately introduced species have also helped control pests.

The problem is that some introduced species have no natural predators, competitors, parasites, or pathogens to help control their numbers in their new habitats. Such species can reduce or wipe out populations of many native species and trigger ecological disruptions. Figure 7-7 shows some of the estimated 50,000 nonnative species deliberately or accidentally introduced into the United States that have caused ecological and economic harm.

Nonnative species threaten almost half of the more than 1,260 endangered and threatened species in the United States and 95% of those in Hawaii, according to the U.S. Fish and Wildlife Service. They are also blamed for about two-thirds of fish extinctions in the United States between 1900 and 2000.

One example of a deliberately introduced nonnative species is the wild African honeybee (Figure 7-7), which was imported to Brazil in 1957 to help increase honey production. Instead, the bees displaced domestic honeybees and reduced the honey supply.

Since then, these nonnative bee species, popularly known as "killer bees," have moved northward into Central America. They have become established in Texas, Arizona, New Mexico, Puerto Rico, and California and are heading north at 240 kilometers (150 miles) per year. They should be stopped eventually by cold winters in the central United States unless they can adapt genetically to cold weather.

They are not the killer bees portrayed in some horror movies, but they are aggressive and unpredictable. They have killed thousands of domesticated animals and an estimated 1,000 people in the western hemisphere.

Figure 7-7 (facing page) **Threats to local and regional natural capital:** some nonnative species that have been deliberately or accidentally introduced into the United States.

Deliberately Introduced Species

Purple looselife

European starling

African honeybee ("Killer bee")

Nutria

Salt cedar (Tamarisk)

Marine toad (Giant toad)

Water hyacinth

Japanese beetle

Hydrilla

European wild boar (Feral pig)

Accidentally Introduced Species

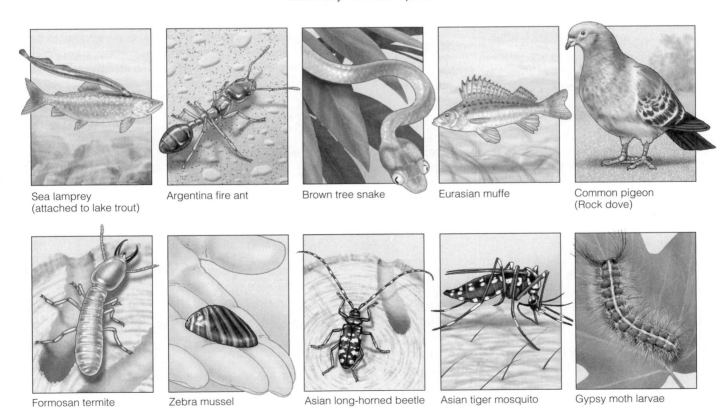

Sea lamprey (attached to lake trout)

Argentina fire ant

Brown tree snake

Eurasian muffe

Common pigeon (Rock dove)

Formosan termite

Zebra mussel

Asian long-horned beetle

Asian tiger mosquito

Gypsy moth larvae

What Is the Role of Accidentally Introduced Species? Aliens Taking Over

A growing number of accidentally introduced species are causing serious economic and ecological damage.

Many unwanted nonnative invaders arrive from other continents as stowaways on aircraft, in the ballast water of tankers and cargo ships, and as hitchhikers on imported products such as wooden packing crates. Cars and trucks can spread seeds of nonnative species imbedded in tire treads.

In the late 1930s, the extremely aggressive Argentina fire ant (Figure 7-7) was introduced accidentally into the United States in Mobile, Alabama. The ants may have arrived on shiploads of lumber or coffee imported from South America or by hitching a ride in the soil-containing ballast water of cargo ships.

Without natural predators, fire ants have spread rapidly by land and water (they can float) throughout the South, from Texas to Florida and as far north as Tennessee and Virginia (Figure 7-8). They are also found in Puerto Rico and recently have invaded California and New Mexico.

Wherever fire ants have gone, they have sharply reduced or wiped out up to 90% of native ant populations. Bother them, and up to 100,000 ants can swarm out of their nest to attack you with their painful and burning stings. They have killed deer fawns, birds, livestock, pets, and at least 80 people allergic to their venom. These ants have invaded cars and caused accidents by attacking drivers, damaged crops, disrupted phone service and electrical power, caused some fires by chewing through underground cables, and cost the United States an estimated $600 million per year. Their large mounds, which look like large boils on the land, can ruin cropland and their painful stings can make backyards uninhabitable.

Widespread pesticide spraying in the 1950s and 1960s temporarily reduced fire ant populations. But this chemical warfare actually hastened the advance of the rapidly multiplying fire ant by reducing populations of many native ant species. Worse, it promoted development of genetic resistance to pesticides in the fire ants through natural selection. In other words, we helped wipe out their competitors and make them genetically stronger.

Researchers at the U.S. Department of Agriculture are experimenting with the use of biological control agents to reduce fire ant populations. But before widespread use of these biological control agents begins, researchers must be sure they will not cause problems for native ant species or become pests themselves.

Fire ants are not all bad. They prey on some other insect pests, including ticks and horse fly larvae. Another harmful invader is the Formosan termite (Case Study, right).

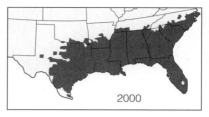

Figure 7-8 Natural capital degradation: expansion of the *Argentina fire ant* in southern states, 1918–2000. This invader is also found in Puerto Rico, New Mexico, and California. (Data from U.S. Department of Agriculture)

Solutions: How Can We Reduce Threats from Nonnative Species? Prevention Pays

Prevention is the best way to reduce the threats from harmful nonnative species because once they have arrived it is difficult and expensive to slow their spread.

Once a nonnative species gets established in an ecosystem, its wholesale removal is almost impossible—somewhat like trying to get smoke back into a chimney or trying to unscramble an egg. Thus, the best way to limit the harmful impacts of nonnative species is to prevent them from being introduced and becoming established.

There are several ways to do this. One is to identify both the major characteristics that allow species to become successful invaders and the types of ecosystems that are vulnerable to invaders (Figure 7-9). Such infor-

Characteristics of Successful Invader Species	Characteristics of Ecosystems Vulnerable to Invader Species
• High reproductive rate, short generation time	• Similar climate to habitat of invader
• Pioneer species	• Absence of predators on invading species
• Long lived	• Early successional species
• High dispersal rate	• Low diversity of native species
• Release growth-inhibiting chemicals into soil	• Absence of fire
• Generalists	• Disturbed by human activities
• High genetic variability	

Figure 7-9 Threats to natural capital: some general characteristics of successful invader species and ecosystems vulnerable to invading species.

The Termite from Hell

Forget killer bees and fire ants. The homeowner's nightmare is the Formosan termite (Figure 7-7). It is the most voracious, aggressive, and prolific of more than 2,000 known termite species.

These termites probably arrived on the U.S. mainland from Hawaii during or soon after World War II. They were stowaways in wooden packing materials on military cargo ships that docked in southern ports such as New Orleans, Louisiana, and Houston, Texas.

Formosan termites consume wood nine times faster than domestic termites. Their huge colonies can contain up to 73 million insects compared to about 1 million in the colonies of most native termites.

Domestic termites have to be in contact with soil, which can be chemically treated around the outside of a building to reduce infestation. But Formosan termites can establish a colony in an attic and in trees. This makes applying pesticides around the perimeter of a building virtually worthless in fighting these pests.

Over the past decade, the Formosan termite has caused more damage in New Orleans than hurricanes, floods, and tornadoes combined. Infestations affect as many as 90% of the houses and one-third of the oak trees in the city. The famous French Quarter has one of the world's most concentrated infestations.

Once confined to Louisiana, they have invaded at least a dozen other states, including Alabama, Florida, Mississippi, North and South Carolina, Texas, and California. They cause at least $1.1 billion in damage each year and these damages are increasing.

In New Orleans, the U.S. Department of Agriculture is using a variety of techniques all at once in an attempt to control the species in a heavily infested 15-block area of the French Quarter. They hope to develop other techniques for dealing with these invaders elsewhere.

One method is to bait the termites by putting out blocks of wood to detect their presence. Once the termites are found in a block of wood, it is replaced by another block that is bated with a pesticide toxic to termites. Termites feeding on this wood carry the pesticide back to their nest where it is spread to other members of the nest.

Scientists have also found a cottony mold that can kill 100% of the termites coming in contact with it within a week. They are working on a method for producing the mold and using it as part of the bait approach.

Critical Thinking

What important ecological roles do termites play in nature? If the Formosan termite and other termite species could be eradicated (a highly unlikely possibility), would you favor doing this? Explain.

---

mation can be used to screen out potentially harmful invaders.

We can also inspect imported goods that are likely to contain invader species. A third strategy is to identify major harmful invader species and pass international laws banning their transfer from one country to another, as is now done for endangered species.

We can also require cargo ships to discharge their ballast water and replace it with salt water at sea before entering ports, or require them to sterilize such water or pump nitrogen into the water to displace dissolved oxygen and kill most invader organisms.

How Serious Is the Illegal Taking or Killing of Wild Species? Making Big Money

Some protected species are illegally killed for their valuable parts or are sold live to collectors.

Organized crime has moved into illegal wildlife smuggling because of the huge profits involved—surpassed only by the illegal international trade in drugs and weapons. At least two-thirds of all live animals smuggled around the world die in transit.

Poverty can promote the illegal smuggling of wild species or their valuable parts. Some poor people struggling to survive in areas with rich stores of wildlife kill or trap such species in order to make enough money to survive and feed their families. Others are professional poachers.

To poachers, a live *mountain gorilla* is worth $150,000, a *panda* pelt $100,000 a *chimpanzee* $50,000, and an *Imperial Amazon macaw* $30,000. A *rhinoceros horn* is worth as much as $28,600 per kilogram ($13,000 per pound) because of its use in dagger handles in the Middle East and as a fever reducer and alleged aphrodisiac in China—the world's largest consumer of wildlife—and other parts of Asia.

In 1950, an estimated 100,000 tigers existed in the world. Despite international protection, today less than 7,500 tigers remain in the wild (about 4,000 of them in India), mostly because of habitat loss and poaching for their furs and bones. Bengal tigers are at risk because a tiger fur sells for $100,000 in Tokyo. With the body parts of a single tiger worth $5,000–20,000, it is not surprising that illegal hunting has skyrocketed, especially in India. Without emergency action, few if any tigers may be left in the wild within 20 years.

As commercially valuable species become endangered, their black market demand soars. This increases their chances of premature extinction from poaching. Most poachers are not caught. And the money they can make far outweighs the small risk of being caught, fined, and imprisoned.

What Are the Roles of Predators and Pest Control? If They Bother You, Kill Them.

Killing predators that bother us or cause economic losses threatens some species with premature extinction.

People try to exterminate species that compete with them for food and game animals. African farmers kill large numbers of elephants to keep them from trampling and eating food crops. Each year, U.S. government animal control agents shoot, poison, or trap thousands of coyotes, prairie dogs, wolves, bobcats, and other species that prey on livestock, on species prized by game hunters, on crops, or on fish raised in aquaculture ponds.

Since 1929, U.S. ranchers and government agencies have poisoned 99% of North America's prairie dogs because horses and cattle sometimes step into the burrows and break their legs. This has also nearly wiped out the endangered black-footed ferret (Figure 7-2; about 600 left in the wild), which preyed on the prairie dog.

What Is the Role of the Market for Exotic Pets and Decorative Plants? Are We Really Pet and Plant Lovers?

Legal and illegal trade in wildlife species used as pets or for decorative purposes threatens some species with extinction.

The global legal and illegal trade in wild species for use as pets is a huge and very profitable business. However, for every live animal captured and sold in the pet market, an estimated 50 others are killed.

About 25 million U.S. households have exotic birds as pets, 85% of them imported. More than 60 bird species, mostly parrots, are endangered or threatened because of this wild bird trade. According to the U.S. Fish and Wildlife Service, collectors of exotic birds may pay $10,000 for a threatened hyacinth macaw (Figure 7-2) smuggled out of Brazil; however, during its lifetime, a single macaw left in the wild might yield as much as $165,000 in tourist income. A 1992 study suggested that keeping a pet bird indoors for more than ten years doubles a person's chances of getting lung cancer from inhaling tiny particles of bird dander.

Other wild species whose populations are depleted because of the pet trade include amphibians, reptiles, mammals, and tropical fish (taken mostly from the coral reefs of Indonesia and the Philippines).

Divers catch tropical fish by using plastic squeeze bottles of cyanide to stun them. For each fish caught alive, many more die. In addition, the cyanide solution kills the coral animals that create the reef, which is a center for marine biodiversity.

Things don't have to be this way. Pilai Poonswad decided to do something about poachers taking hornbills—large, beautiful, and rare birds—from a rain forest in Thailand. She visited the poachers in their villages and showed them why the birds are worth more alive than dead. Now she has a number of ex-poachers earning much more money than they did before by taking eco-tourists into the forest to see these magnificent birds. Because of their vested financial interest in preserving the hornbills, they also help protect them from poachers.

Some exotic plants, especially orchids and cacti, are endangered because they are gathered (often illegally) and sold to collectors to decorate houses, offices, and landscapes. A collector may pay $5,000 for a single rare orchid, and a single, rare, mature crested saguaro cactus can earn cactus rustlers as much as $15,000.

In other words, collecting exotic pets and plants kills large numbers of them and endangers many of these species and others that depend on them. Are such collectors lovers or haters of the species they collect? Should we leave most exotic species in the wild?

What Are the Roles of Climate Change and Pollution? Speeding Up the Treadmill and Poisoning Species

Projected climate change and exposure to pollutants such as pesticides can threaten some species with premature extinction.

Most natural climate changes in the past have taken place over long periods of time. This gave species more time to adapt or evolve into new species in order to cope. But considerable evidence indicates that human activities such as greenhouse gas emissions and deforestation may bring about rapid climate change during this century. This could change the habitats of many species and accelerate the extinction of some species.

Pollution threatens populations and species in a number of ways. A major extinction threat is from the unintended effects of pesticides. According to the U.S. Fish and Wildlife Service, each year in the United States, pesticides kill about one-fifth of the country's beneficial honeybee colonies, more than 67 million birds, and 6–14 million fish. They also threaten about one-fifth of the country's endangered and threatened species.

During the 1950s and 1960s, populations of fish-eating birds such as the osprey, cormorant, brown pelican, and bald eagle plummeted. Research indicated that a chemical derived from DDT, when biologically magnified in food webs (Figure 7-10), made the birds'

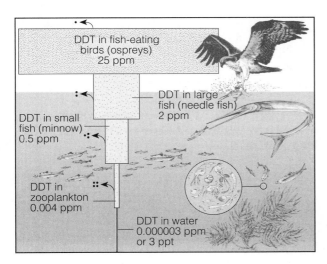

Figure 7-10 *Bioaccumulation* and *biomagnification.* DDT is a fat-soluble chemical that can accumulate in the fatty tissues of animals. In a food chain or web, the accumulated DDT can be biologically magnified in the bodies of animals at each higher trophic level. This diagram shows that the concentration of DDT in the fatty tissues of organisms was biomagnified about 10 million times in this food chain in an estuary near Long Island Sound in New York. If each phytoplankton organism takes up from the water and retains one unit of DDT, a small fish eating thousands of zooplankton (which feed on the phytoplankton) will store thousands of units of DDT in its fatty tissue. Then each large fish that eats ten of the smaller fish will ingest and store tens of thousands of units, and each bird (or human) that eats several large fish will ingest hundreds of thousands of units. Dots represent DDT, and arrows show small losses of DDT through respiration and excretion.

eggshells so fragile they could not reproduce successfully. Also hard hit were such predatory birds as the prairie falcon, sparrow hawk, and peregrine falcon, which help control rabbits, ground squirrels, and other crop eaters. Some *good news* is that since the U.S. ban on DDT in 1972, most of these species have made a comeback.

7-4 PROTECTING WILD SPECIES: THE LEGAL APPROACH

How Can International Treaties Help Protect Endangered Species? Some Success

International treaties have helped reduce the international trade of endangered and threatened species, but enforcement is difficult.

Several international treaties and conventions help protect endangered or threatened wild species. One of the most far reaching is the 1975 *Convention on International Trade in Endangered Species* (CITES). This treaty, now signed by 160 countries, lists some 900 species

that cannot be commercially traded as live specimens or wildlife products because they are in danger of extinction. The treaty also restricts international trade of 29,000 other species because they are at risk of becoming threatened.

CITES has helped reduce international trade in many threatened animals, including elephants, crocodiles, and chimpanzees. However, the effects of this treaty are limited because enforcement is difficult and varies from country to country, and convicted violators often pay only small fines. Also, member countries can exempt themselves from protecting any listed species, and much of the highly profitable illegal trade in wildlife and wildlife products goes on in countries that have not signed the treaty.

The Convention on Biological Diversity (CBD), ratified by 186 countries, legally binds signatory governments to reversing the global decline of biological diversity. However, its implementation has been slow because some key countries such as the United States have not ratified it. Also, it contains no severe penalties or other enforcement mechanisms.

How Can National Laws Help Protect Endangered Species? A Tough and Controversial Act in the United States

One of the world's most far-reaching and controversial environmental laws is the U.S. Endangered Species Act passed in 1973.

The United States controls imports and exports of endangered wildlife and wildlife products through two laws. One is the *Lacey Act of 1900.* It prohibits transporting live or dead wild animals or their parts across state borders without a federal permit.

The other is the *Endangered Species Act of 1973* (ESA), which was amended in 1982, 1985, and 1988. It was designed to identify and legally protect endangered species in the United States and abroad. This act is probably the most far-reaching environmental law ever adopted by any nation, which has made it controversial. Canada and a number of other countries have similar laws.

The National Marine Fisheries Service (NMFS) is responsible for identifying and listing endangered and threatened ocean species, and the U.S. Fish and Wildlife Services (USFWS) identifies and lists all other endangered and threatened species. Any decision by either agency to add or remove a species from the list must be based on biological factors alone, without consideration of economic or political factors. However, economic factors can be used in deciding whether and how to protect endangered habitat, and developing recovery plans for listed species.

The act also forbids federal agencies (except the Defense Department) to carry out, fund, or authorize projects that would jeopardize an endangered or threatened species or destroy or modify the critical habitat it needs to survive. On private lands, fines up to $100,000 and one year in prison can be imposed to ensure protection of the habitats of endangered species.

The act also makes it illegal for Americans to sell or buy any product made from an endangered or threatened species. These species cannot be hunted, killed, collected, or injured in the United States, and this protection has been extended to threatened and endangered foreign species.

In 2003 the Bush administration proposed eliminating protection of foreign species, causing an uproar by conservationists. With this rule change, American hunters, circuses, and the pet industry could pay individuals or governments to kill, capture, and import animals that are on the brink of extinction in other countries.

X *How Would You Vote?* Should the U.S. Endangered Species Act no longer protect threatened and endangered species in other countries? Cast your vote online at http://biology .brookscole.com/miller7.

Between 1973 and 2004, the number of U.S. species on the official endangered and threatened list increased from 92 to about 1,260 species—60% of them plants and 40% animals. According to a 2000 study by the Nature Conservancy, about one-third of the country's species are at risk of extinction, and 15% of all species are at high risk. This amounts to about 30,000 species, compared to the only 1,260 species currently protected under the ESA. The study also found that many of the country's rarest and most imperiled species are concentrated in a few hot spots (Figure 7-11).

What Are Critical Habitat Designations and Recovery Plans? How to Rebuild Populations

The Endangered Species Act requires protecting the critical habitat and developing a recovery plan for each listed species, but lack of funding and political opposition hinder these efforts.

The ESA generally requires the secretary of the interior to designate and protect the *critical habitat* needed for the survival and recovery of each listed species. So far, critical habitats have been established for only about one-third of the species on the ESA list, mostly because of political pressure and a lack of funds. Since 2001, the government stopped listing new species and designating critical habitat for listed species unless required to do so by court order.

Getting a species listed is only half the battle. Next, the USFWS or the NMFS is supposed to prepare a plan to help the species recover. By 2004, final recovery plans had been developed and approved for about 79% of the endangered or threatened U.S. species. However, about half of the plans exist only on paper, mostly because of political opposition and limited funds.

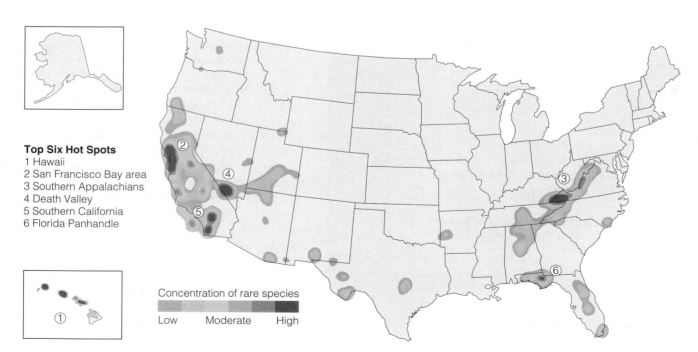

Top Six Hot Spots
1 Hawaii
2 San Francisco Bay area
3 Southern Appalachians
4 Death Valley
5 Southern California
6 Florida Panhandle

Concentration of rare species

Low   Moderate   High

Figure 7-11 Threatened natural capital: biodiversity hot spots in the United States. This map shows areas that contain the largest concentrations of rare and potentially endangered species. (Data from State Natural Heritage Programs, the Nature Conservancy, and Association for Biodiversity Information)

How Can We Encourage Private Landowners to Protect Endangered Species? Trying to Find Win-Win Compromises

Congress has amended the Endangered Species Act to help landowners protect endangered species on their land.

In 1982, Congress amended the ESA to allow the secretary of the interior to use *habitat conservation plans* (*HCP*). They are designed to strike a compromise between the interests of private landowners and those of endangered and threatened species, with the goal of not reducing the recovery chances of a protected species.

With an HCP, landowners, developers, or loggers are allowed to destroy some critical habitat in exchange for taking steps to protect members of the species. Such measures might include setting aside a part of the species' habitat as a protected area, paying to relocate the species to another suitable habitat, or paying money to have the government buy suitable habitat elsewhere. Once the plan is approved it cannot be changed, even if new data show that the plan is inadequate to protect a species and help it recover.

Some wildlife conservationists support this approach because it can reduce political pressure to seriously weaken or eliminate the ESA. However, there is growing concern that many of the plans have been approved without enough scientific evaluation of their effects on a species' recovery. Also, many of the plans are political compromises that do not protect the species or make inadequate provisions for its recovery.

Should the Endangered Species Act Be Weakened? One Side of the Story

Some believe that the Endangered Species Act should be weakened or repealed because it has been a failure and hinders the economic development of private land.

Since 1992, Congress has been debating its reauthorization of the ESA with proposals ranging from eliminating the act, weakening it, or strengthening it.

Opponents of the ESA contend that it puts the rights and welfare of endangered plants and animals above those of people, has not been effective in protecting endangered species, and has caused severe economic losses by hindering development on private land that contains endangered or threatened species. Since 1995, efforts to weaken the ESA have included the following suggested changes:

- Making protection of endangered species on private land voluntary.

- Having the government compensate landowners if it forces them to stop using part of their land to protect endangered species.

- Making it harder and more expensive to list newly endangered species by requiring government wildlife officials to navigate through a series of hearings and peer-review panels.

- Eliminating the need to designate critical habitats because developing and implementing a recovery plan is more important. Also, dealing with lawsuits for failure to develop critical habitats takes up most of the limited funds for carrying out the ESA.

- Allowing the secretary of the interior to permit a listed species to become extinct without trying to save it and to determine whether a species should be listed.

- Allowing the secretary of the interior to give any state, county, or landowner permanent exemption from the law, with no requirement for public notification or comment.

Other critics would go further and do away with this act. But since this is politically unpopular with the American public, most efforts are designed to weaken the act and reduce its meager funding.

Should the Endangered Species Act Be Strengthened? The Other Side of the Story

According to most conservation biologists, the Endangered Species Act should be strengthened and modified to develop a new system to protect and sustain the country's biodiversity.

Most conservation biologists and wildlife scientists agree that the ESA has some deficiencies and needs to be simplified and streamlined. But they contend that the ESA has not been a failure (Case Study, p. 142).

They also contest the charge that the ESA has caused severe economic losses. Government records show that since 1979 only about 0.05% of the almost 200,000 projects evaluated by the USFWS have been blocked or canceled as a result of the ESA.

A study by the U.S. National Academy of Sciences recommended three major changes to make the ESA more scientifically sound and effective.

- Greatly increase the meager funding for implementing the act.

- Develop recovery plans more quickly.

- When a species is first listed, establish a core of its survival habitat as a temporary emergency measure that could support the species for 25–50 years.

Most biologists and wildlife conservationists believe the United States should modify the act to emphasize protecting and sustaining biological diversity and ecological functioning rather than attempting to save individual species. This new ecosystems approach would follow three principles:

- Find out what species and ecosystems the country has.

- Locate and protect the most endangered ecosystems and species.

- Provide private landowners who agree to help protect specific endangered ecosystems with significant financial incentives (tax breaks and write-offs) and technical help.

X *How Would You Vote?* Should the Endangered Species Act be modified to protect the nation's overall biodiversity? Cast your vote online at http://biology.brookscole.com/miller7.

7-5 PROTECTING WILD SPECIES: THE SANCTUARY APPROACH

What Is the Role of Wildlife Refuges and Other Protected Areas? Protect the Homes of Species in Trouble

The United States has set aside 542 federal refuges for wildlife, but many refuges are suffering from environmental degradation.

In 1903, President Theodore Roosevelt established the first U.S. federal wildlife refuge at Pelican Island, Florida. Since then, the National Wildlife Refuge System has grown to 542 refuges. More than 35 million Americans visit these refuges each year to hunt, fish, hike, or watch birds and other wildlife.

More than three-fourths of the refuges serve as vital wetland sanctuaries for protecting migratory waterfowl. About one-fifth of U.S. endangered and threatened species have habitats in the refuge system, and some refuges have been set aside for specific endangered species. These have helped Florida's key deer, the brown pelican, and the trumpeter swan to recover.

Conservation biologists call for setting aside more refuges for endangered plants. They also urge Congress and state legislatures to allow abandoned military lands that contain significant wildlife habitat to become national or state wildlife refuges. However, according to a General Accounting Office study, activities considered harmful to wildlife occur in nearly 60% of the nation's wildlife refuges.

Can Gene Banks, Botanical Gardens, and Farms Help Save Endangered Species? Important but Limited Solutions

Establishing gene banks and botanical gardens, and using farms to raise threatened species can help protect species from extinction, but these options lack funding and storage space.

Gene or *seed banks* preserve genetic information and endangered plant species by storing their seeds in refrigerated, low-humidity environments. More than 100 seed banks around the world collectively hold about 3 million samples.

CASE STUDY

What Has the Endangered Species Act Accomplished?

Critics of the ESA call it an expensive failure because only 37 species have been removed from the endangered list. Most biologists agree that the act needs strengthening and modification. But they disagree that the act has been a failure, for four reasons.

First, species are listed only when they are in serious danger of extinction. This is like setting up a poorly funded hospital emergency room that takes only the most desperate cases, often with little hope for recovery, and saying it should be shut down because it has not saved enough patients.

Second, it takes decades for most species to become endangered or threatened. Thus, it usually takes

decades to bring a species in critical condition back to the point where it can be removed from the list. Expecting the ESA—which has been in existence only since 1973—to quickly repair the biological depletion of many decades is unrealistic.

Third, the most important measure of the law's success is that the conditions of almost 40% of the listed species are stable or improving. A hospital emergency room taking only the most desperate cases and then stabilizing or improving the condition of 40% of its patients would be considered an astounding success.

Fourth, the federal endangered species budget was only $58 million in 2005—about what the Department of Defense spends in a little more than an hour or 20¢ a year per U.S. citizen. To supporters of the

ESA, it is amazing that so much has been accomplished in stabilizing or improving the condition of almost 40% of the listed species on a shoestring budget.

Yes, the act can be improved and federal regulators have sometimes been too heavy handed in enforcing the act. But instead of gutting or doing away with this important act, biologists call for it to be strengthened and modified to help protect ecosystems and the nation's overall biodiversity.

Some critics say that only 20% of the endangered species are stable or improving. If correct, this is still an incredible bargain.

Critical Thinking

Should the budget for the Endangered Species Act be drastically increased? Explain.

Scientists urge the establishment of many more such banks, especially in developing countries. But some species cannot be preserved in gene banks. And the banks are expensive to operate and can be destroyed by accidents.

The world's 1,600 *botanical gardens* and *arboreta* contain living plants, representing almost one-third of the world's known plant species. However, they contain only about 3% of the world's rare and threatened plant species.

Botanical gardens also help educate an estimated 150 million visitors a year about the need for plant conservation. But these sanctuaries have too little storage capacity and too little funding to preserve most of the world's rare and threatened plants.

We can take pressure off some endangered or threatened species by raising individuals on *farms* for commercial sale. One example is the use of farms in Florida to raise alligators for their meat and hides. Another example is *butterfly farms* in Papua New Guinea, where many butterfly species are threatened by development activities.

What Role Can Zoos and Aquariums Play in Protecting Endangered Animal Species? Important but Expensive and Limited

Zoos and aquariums can help protect endangered animal species, but efforts lack funding and storage space.

Zoos, aquariums, game parks, and animal research centers are being used to preserve some individuals of critically endangered animal species, with the long-term goal of reintroducing the species into protected wild habitats.

Two techniques for preserving endangered terrestrial species are egg pulling and captive breeding. *Egg pulling* involves collecting wild eggs laid by critically endangered bird species and then hatching them in zoos or research centers. In *captive breeding*, some or all of the wild individuals of a critically endangered species are captured for breeding in captivity, with the aim of reintroducing the offspring into the wild.

Lack of space and money limits efforts to maintain populations of endangered species in zoos and research centers. The captive population of each species must number 100–500 individuals to avoid extinction through accident, disease, or loss of genetic diversity through inbreeding. Recent genetic research indicates that 10,000 or more individuals are needed for an endangered species to maintain its capacity for biological evolution.

However, zoos and research centers contain only about 3% of the world's rare and threatened plant species. Thus, the major conservation role of these facilities will be to help educate the public about the eco-

logical importance of the species they display and the need to protect habitat.

Public aquariums that exhibit unusual and attractive fish and some marine animals such as seals and dolphins also help educate the public about the need to protect such species. In the United States, more than 35 million people visit aquariums each year. However, unlike some zoos, public aquariums have not served as effective gene banks for endangered marine species, especially marine mammals that need large volumes of water.

Instead of seeing zoos and aquariums as sanctuaries, some critics see most of them as prisons for once-wild animals. They also contend that zoos and aquariums can foster the notion that we do not need to preserve large numbers of wild species in their natural habitats.

Some people criticize zoos and aquariums for putting on shows with animals wearing clothes, riding bicycles, or performing tricks. They see this as fostering the idea that the animals are there primarily to entertain us by doing things people do and, in the process, raising money for their keepers.

Conservation biologists point out that zoos, aquariums, and botanical gardens, regardless of their benefits and drawbacks, are not biologically or economically feasible solutions for most of the world's current endangered species and the much larger number of species expected to be threatened over the next few decades.

7-6 RECONCILIATION ECOLOGY

What Is Reconciliation Ecology? Rethinking Conservation Strategy

Reconciliation ecology involves finding ways to share the places that we dominate with other species.

In 2003, ecologist Michael L. Rosenburg wrote a book entitled *Win-Win Ecology: How Earth's Species Can Survive in the Midst of Human Enterprise* (Oxford University Press). He strongly supports the eight-point program of Edward O. Wilson to help save the earth's natural habitats by establishing and protecting nature reserves (p. 125). He also supports the species protection strategies discussed in this chapter.

But he contends that, in the long run, these approaches will fail for two reasons. *First,* current reserves are devoted to saving only about 7% of the world's terrestrial area. To Rosenberg, the real challenge is to help sustain wild species in the human-dominated portion of nature that makes up 97% of the planet's terrestrial ecological "cake."

Second, setting aside funds and refuges and passing laws to protect endangered and threatened species are essentially desperate attempts to save species that are in deep trouble. This can help a few species, but the real challenge is learning how keep more species from getting to such a point in the first place.

Rosenberg suggests that we develop a new form of conservation biology, called **reconciliation ecology.** It is the science of inventing, establishing, and maintaining new habitats to conserve species diversity in places where people live, work, or play. In other words, we need to learn how to share the spaces we dominate with other species.

How Can We Implement Reconciliation Ecology? Observe, Be Creative, and Cooperate with Your Neighbors

Some people are finding creative ways to practice reconciliation ecology in their neighborhoods and cities.

Practicing reconciliation ecology begins by looking at the habitats we prefer. Given a choice, most people prefer a grassy and fairly open habitat with a few scattered trees. We also like water and prefer to live near a stream, lake, river, or ocean. We also love flowers.

The problem is that most species do not like what we like or cannot survive in the habitats we prefer. No wonder so few of them live with us.

With restoration ecology, some of our monoculture yards can be replaced with diverse yards using plant species adapted to local climates that are selected to attract certain species. This would make neighborhoods more interesting, keep down insect pests, and require less use of noisy and polluting lawnmowers.

Communities could have contests and awards for people designing the most biodiverse and species-friendly yards and gardens. Signs could describe the type of ecosystem being mimicked and the species being protected as a way to educate and encourage experiments by other people. Who knows, some creative person might even be able to design more biologically diverse golf courses and cemeteries. People have worked together to help preserve bluebirds within human-dominated habitats (Case Study, right).

San Francisco's large Golden Gate Park is a 410-hectare (1,012-acre) oasis of gardens and trees in the midst of a large city. It is a good example of reconciliation ecology because it was designed and planted by humans who transformed it from a system of sand dunes.

The Department of Defense controls about 10 million hectares (25 million acres) of land in the United States. Rosenberg and other reconciliation ecologists believe that some of this land could serve as laboratories for developing and testing reconciliation ecology ideas.

Let me tell you a nice little story about bluebirds. *Bad news.* Populations of bluebirds in much of the eastern United States are declining.

There are two reasons. One is that these birds nest in tree holes of a certain size. Dead and dying trees once provided plenty of these holes. But today timber companies often cut down all of the trees, and many homeowners manicure their property by removing dead and dying trees.

A second reason is that two aggressive, abundant, and nonnative bird species—starlings and house sparrows—also like to nest in tree holes and take them away from bluebirds. To make matters worse, starlings eat blueberries that bluebirds need to survive during the winter.

Good news: People have come up with a creative way to help save the bluebird. They have designed nest boxes with holes large enough to accommodate bluebirds but too small for starlings. They also found that house sparrows like shallow boxes so they made the bluebird boxes deep enough to make them unattractive nesting sites for the sparrows.

In 1979, the North American Bluebird Society was founded to spread the word and encourage people to use the bluebird boxes on their properties and to keep house cats away from nesting bluebirds. Now bluebird numbers are building back up.

Restoration ecology works! Perhaps you might want to consider a career in this exciting new field.

Critical Thinking

See if you can come up with a reconciliation project to help protect threatened bird or other species in your neighborhood or school.

Some college campuses and schools might also serve as reconciliation ecology laboratories. How about your campus or school?

In this chapter, we have seen that protecting the species that make up part of the earth's biodiversity from premature extinction is a difficult, controversial, and challenging responsibility. Figure 7-12 lists some things you can do to help prevent the premature extinction of species.

We know what to do. Perhaps we will act in time.
EDWARD O. WILSON

What Can You Do?

Protecting Species

- Do not buy furs, ivory products, and other materials made from endangered or threatened animal species.

- Do not buy wood and paper products produced by cutting remaining old-growth forests in the tropics.

- Do not buy birds, snakes, turtles, tropical fish, and other animals that are taken from the wild.

- Do not buy orchids, cacti, and other plants that are taken from the wild.

Figure 7-12 What can you do? Ways to help prevent the premature extinction of species.

CRITICAL THINKING

1. Discuss your gut-level reaction to the following statement: "Eventually, all species become extinct. Thus, it does not really matter that the passenger pigeon is extinct, and that the whooping crane, and the world's remaining tiger species are endangered mostly because of human activities." Be honest about your reaction, and give arguments for your position.

2. Make a log of your own consumption of all products for a single day. Relate your level and types of consumption to the decline of wildlife species and the increased destruction and degradation of wildlife habitats in the United States (or the country where you live), in tropical forests, and in aquatic ecosystems.

3. Do you accept the ethical position that each *species* has the inherent right to survive without human interference, regardless of whether it serves any useful purpose for humans? Explain. Would you extend this right to the *Anopheles* mosquito, which transmits malaria, and to infectious bacteria? Explain.

4. Your lawn and house are invaded by fire ants, which can cause painful bites. What would you do?

5. Which of the following statements best describes your feelings toward wildlife: **(a)** As long as it stays in its space, wildlife is OK. **(b)** As long as I do not need its space, wildlife is OK. **(c)** I have the right to use wildlife habitat to meet my own needs. **(d)** When you have seen one redwood tree, fox, elephant, or some other form of wildlife, you have seen them all, so lock up a few of each species in a zoo or wildlife park and do not worry about protecting the rest. **(e)** Wildlife should be protected.

6. List your three favorite species. Examine why they are your favorites. Are they cute and cuddly looking, like the giant panda and the koala? Do they have humanlike qualities, like apes or penguins that walk upright? Are they large, like elephants or blue whales? Are they beautiful, like tigers and monarch butterflies? Are any of them plants? Are any of them species such as bats, sharks, snakes, or spiders that many people are afraid of? Are any of them microorganisms that help keep you alive? Reflect on what your choice of favorite species tells you about your attitudes toward most wildlife.

7. Environmental groups in a heavily forested state want to restrict logging in some areas to save the habitat of an endangered squirrel. Timber company officials argue that the well being of one type of squirrel is not as important as the well-being of the many families affected if the restriction causes the company to lay off hundreds of workers. If you had the power to decide this issue, what would you do and why? Can you come up with a compromise?

8. Congratulations! You are in charge of preventing the premature extinction of the world's existing species from human activities. What would be the three major components of your program to accomplish this goal?

LEARNING ONLINE

The website for this book contains helpful study aids and many ideas for further reading and research. They include a chapter summary, review questions for the entire chapter, flash cards for key terms and concepts, a multiple-choice practice quiz, interesting Internet sites, references, and a guide for accessing thousands of InfoTrac® College Edition articles. Log on to

http://biology.brookscole.com/miller7

Then click on the Chapter-by-Chapter area, choose Chapter 7, and select a learning resource.

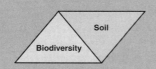

There are two spiritual dangers in not owning a farm. One is the danger of supposing that breakfast comes from the grocery, and the other that heat comes from the furnace.

ALDO LEOPOLD

8-1 HOW IS FOOD PRODUCED?

What Systems Provide Us with Food? The Challenges Ahead

Croplands, rangelands, and ocean fisheries supply most of our food and, since 1950, global food production from all three systems has increased dramatically.

Historically, humans have depended on three systems for their food supply: *Croplands* mostly produce grains, and provide about 77% of the world's food. *Rangelands* produce meat, mostly from grazing livestock, and supply about 16% of the world's food. *Oceanic fisheries* supply about 7% of the world's food.

Since 1950, there has been a staggering increase in global food production from all three systems. This occurred because of technological advances such as increased use of tractors and farm machinery and high-tech fishing boats and gear; inorganic chemical fertilizers; irrigation; pesticides; high-yield varieties of wheat, rice, and corn; densely populated feedlots and enclosed pens for raising cattle, pigs, and chickens; and aquaculture ponds and ocean cages for raising some types of fish and shellfish.

We face important challenges in increasing food production without causing serious environmental harm. To feed the world's 8.9 billion people projected to exist in 2050, we must produce and equitably distribute more food than has been produced since agriculture began about 10,000 years ago, and do it in an environmentally sustainable manner.

Can we do this? Some analysts say we can, mostly by using genetic engineering (Figure 3-5, p. 53). Others have doubts. They are concerned that environmental degradation, pollution, lack of water for irrigation, overgrazing by livestock, overfishing, and loss of vital ecological services may limit future food production. A key problem is that human activities continue to take over or degrade more of the planet's *net primary productivity*, which supports all life.

We also face the challenge of sharply reducing poverty because about one out of five people do not have enough land to grow their own food or enough money to buy enough food—regardless of how much

is available. This chapter analyzes the world's crop, meat, and fish production systems and how these systems can be made more sustainable.

What Plants and Animals Feed the World? Our Three Most Important Crops

Wheat, rice, and corn provide more than half of the calories in the food consumed by the world's people.

The earth has perhaps 30,000 plant species with parts that people can eat. However, only 14 plant and 8 terrestrial animal species supply an estimated 90% of our global intake of calories. Just three types of grain crops—*wheat, rice,* and *corn*—provide more than half the calories people consume.

Two-thirds of the world's people survive primarily on rice, wheat, and corn, mostly because they cannot afford meat. As incomes rise, most people consume more meat and other products of domesticated livestock, which in turn means more grain consumption by those animals.

Fish and shellfish are an important source of food for about 1 billion people, mostly in Asia and in coastal areas of developing countries. But on a global scale, fish and shellfish supply only 7% of the world's food and about 6% of the protein in the human diet.

What Are the Major Types of Food Production? High-Input and Low-Input Agriculture

About 80% of the world's food supply is produced by industrialized agriculture and 20% by subsistence agriculture.

There are two major types of agricultural systems: *industrialized* and *traditional*. **Industrialized agriculture,** or **high-input agriculture,** uses large amounts of fossil fuel energy, water, commercial fertilizers, and pesticides to produce single crops (monocultures) or livestock animals for sale. Practiced on about a fourth of all cropland, mostly in developed countries (Figure 8-1), high-input industrialized agriculture has spread since the mid-1960s to some developing countries.

Plantation agriculture is a form of industrialized agriculture used primarily in tropical developing countries. It involves growing *cash crops* (such as bananas, coffee, soybeans, sugarcane, cocoa, and vegetables) on large monoculture plantations, mostly for sale in developed countries.

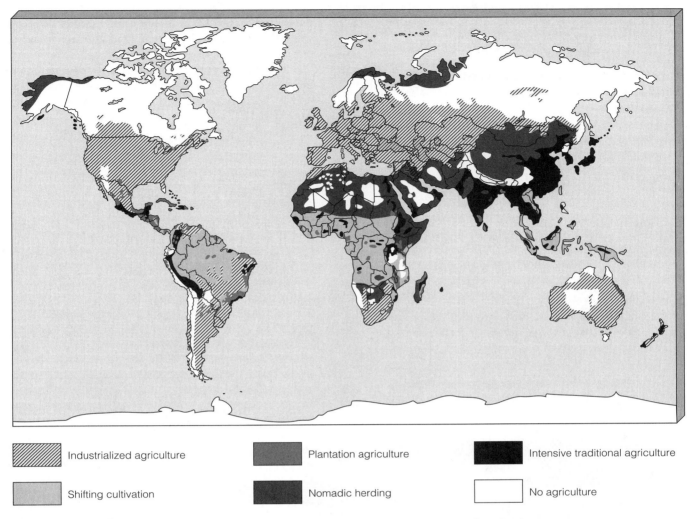

▨ Industrialized agriculture	▨ Plantation agriculture	■ Intensive traditional agriculture
▨ Shifting cultivation	■ Nomadic herding	□ No agriculture

Figure 8-1 Locations of the world's principal types of food production. Excluding Antarctica and Greenland, agricultural systems cover almost one-third of the earth's land surface.

An increasing amount of livestock production in developed countries is industrialized. Large numbers of cattle are brought to densely populated *feedlots*, where they are fattened up for about 4 months before slaughter. Most pigs and chickens in developed countries spend their lives in densely populated pens and cages and eat mostly grain grown on cropland.

Traditional agriculture consists of two main types, which together are practiced by about 2.7 billion people (42% of the world's people) in developing countries, and provide about a fifth of the world's food supply. **Traditional subsistence agriculture** typically uses mostly human labor and draft animals to produce only enough crops or livestock for a farm family's survival. In **traditional intensive agriculture,** farmers increase their inputs of human and draft-animal labor, fertilizer, and water to get a higher yield per area of cultivated land. They produce enough food to feed their families and to sell for income.

Croplands, like natural ecosystems, provide the ecological and economic services listed in Figure 8-2.

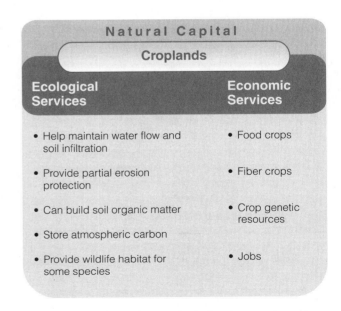

Figure 8-2 Natural capital: ecological and economic services provided by croplands.

Indeed, agriculture is the world's largest industry, providing a living for one of every five people.

How Have Green Revolutions Increased Food Production? High-Input Monocultures in Action

Since 1950, most of the increase in global food production has come from using high-input agriculture to produce more crops on each unit of land.

Farmers can produce more food by farming more land or getting higher yields per unit of area from existing cropland. Since 1950, most of the increase in global food production has come from increased yields per unit of area of cropland in a process called the **green revolution.**

The green revolution involves three steps. *First,* develop and plant monocultures of selectively bred or genetically engineered high-yield varieties of key crops such as rice, wheat, and corn. *Second,* produce high yields by using large inputs of fertilizer, pesti-

cides, and water. *Third,* increase the number of crops grown per year on a plot of land through *multiple cropping.*

This high-input approach dramatically increased crop yields in most developed countries between 1950 and 1970 in what is called the *first green revolution* (Figure 8-3, dark shading).

A *second green revolution* has been taking place since 1967. It involves introducing fast-growing dwarf varieties of rice and wheat (developed by Norman Bourlag, who later received a Nobel Peace Prize for his work), specially bred for tropical and subtropical climates, into several developing countries (Figure 8-3, lighter shading). Producing more food on less land is also an important way to protect biodiversity by saving large areas of forests, grasslands, wetlands, and easily eroded mountain terrain from being used to grow food.

Yield increases depend not only on fertile soil and ample water but also on high inputs of fossil fuels to run machinery, produce and apply inorganic fertilizers and pesticides, and pump water for irrigation. All told,

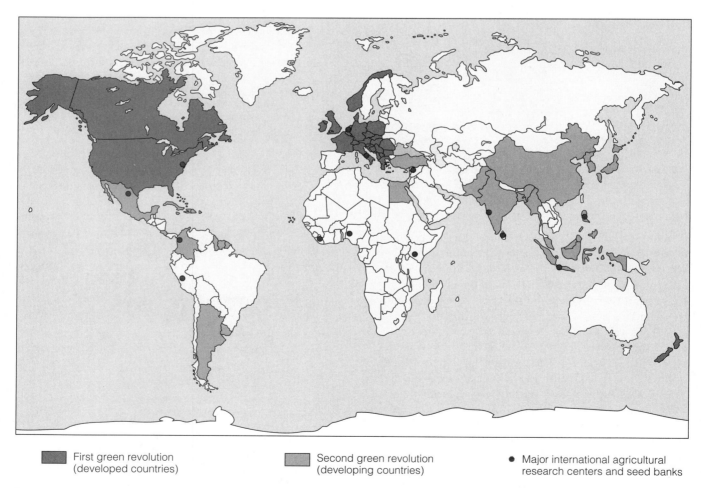

First green revolution (developed countries)

Second green revolution (developing countries)

● Major international agricultural research centers and seed banks

Figure 8-3 Countries whose crop yields per unit of land area increased during the two green revolutions. The first (dark shading) took place in developed countries between 1950 and 1970; the second (lighter shading) has occurred since 1967 in developing countries with enough rainfall or irrigation capacity. Several agricultural research centers and gene or seed banks (black dots) play a key role in developing high-yield crop varieties.

high-input green revolution agriculture uses about 8% of the world's oil output.

Case Study: Industrial Food Production in the United States: A Success Story

America's industrialized agricultural system produces about 17% of the world's grain but has a large environmental impact.

In the United States, industrialized farming has become *agribusiness* as big companies and larger family-owned farms have taken control of almost three-fourths of U.S. food production. According to environmental educator David Orr, "the U.S. food system is increasingly dominated by 'superfarms,' which are roughly to farming what WalMart is to retailing."

In total annual sales, agriculture is bigger than the automotive, steel, and housing industries combined. It generates about 18% of the country's gross domestic product and almost a fifth of all jobs in the private sector, employing more people than any other industry. With only 0.3% of the world's farm labor force, U.S. farms produce about 17% of the world's grain and nearly half of the world's grain exports.

Since 1950, U.S. farmers have used green revolution techniques to more than double the yield of key crops such as wheat, corn, and soybeans without cultivating more land. Such increases in the yield per hectare of key crops have kept large areas of forests, grasslands, wetlands, and easily erodible land from being converted to farmland.

In addition, the country's agricultural system has become increasingly efficient. While the U.S. output of crops, meat, and dairy products has been increasing steadily since 1975, the major inputs of labor and resources—with the exception of pesticides—to produce each unit of that output have fallen steadily since 1950.

This industrialization of agriculture has been made possible by the availability of cheap energy, most of it from oil. Putting food on the table consumes about 17% of all commercial energy used in the United States each year (Figure 8-4). The input of energy needed to produce a unit of food has fallen considerably and most plant crops in the United States provide more food energy than the energy used to grow them.

However, energy efficiency is much lower if we look at the whole U.S. food system. Considering the energy used to grow, store, process, package, transport, refrigerate, and cook all plant and animal food, *about 10 units of nonrenewable fossil fuel energy are needed to put 1 unit of food energy on the table.* By comparison, every unit of energy from human labor in traditional subsistence farming provides at least 1 unit of food energy and up to 10 units of food energy using traditional intensive farming.

What Growing Techniques Are Used in Traditional Agriculture? Low-Input Agrodiversity in Action

Many traditional farmers in developing countries use low-input agriculture to produce a variety of crops on each plot of land.

Traditional farmers in developing countries today grow about one-fifth of the world's food on about three-fourths of its cultivated land. Many traditional farmers simultaneously grow several crops on the same plot, a practice known as **interplanting.** Such crop diversity reduces the chance of losing most or all of the year's food supply to pests, bad weather, and other misfortunes.

Interplanting strategies vary. One type, **polyvarietal cultivation,** involves planting a plot with several varieties of the same crop. Another is **intercropping**—growing two or more different crops at the same time on a plot (for example, a carbohydrate-rich grain that uses soil nitrogen and a nitrogen-fixing plant that puts it back). A third type is **agroforestry,** or **alley cropping,** in which crops and trees are grown together.

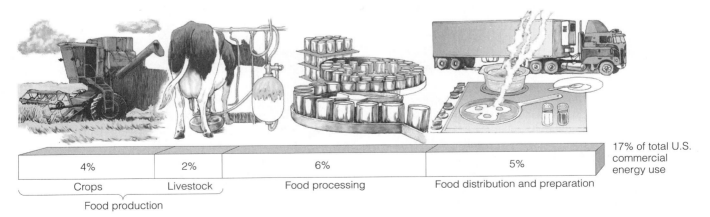

4%	2%	6%	5%	17% of total U.S. commercial energy use
Crops	Livestock	Food processing	Food distribution and preparation	

Food production

Figure 8-4 In the United States, industrialized agriculture uses about 17% of all commercial energy. In the United States, food travels an average 2,400 kilometers (1,500 miles) from farm to table. (Data from David Pimentel and Worldwatch Institute)

A fourth type is **polyculture** in which many different plants maturing at various times are planted together. Low-input polyculture has a number of advantages. There is less need for fertilizer and water because root systems at different depths in the soil capture nutrients and moisture efficiently. It provides more protection from wind and water erosion because the soil is covered with crops year-round. There is little or no need for insecticides because multiple habitats are created for natural predators of crop-eating insects. Also, there is little or no need for herbicides because weeds have trouble competing with the multitude of crop plants. The diversity of crops raised provides insurance against bad weather. This is a way of growing food by copying nature.

Recent ecological research found that on average, low-input polyculture produces higher yields per hectare of land than high-input monoculture. For example, a 2001 study by ecologists Peter Reich and David Tilman found that carefully controlled polyculture plots with 16 different species of plants consistently outproduced plots with 9, 4, or only 1 type of plant species.

8-2 SOIL EROSION AND DEGRADATION

What Causes Soil Erosion? Three Factors

Water, wind, and people cause soil erosion.

Most people in developed countries get their food from grocery stores, fast-food chains, and restaurants. But we need to remind ourselves that *all food comes from the earth or soil*—the base of life. This explains why preserving the world's topsoil (Figure 2-21, p. 35) is the key to producing enough food to feed the world's growing population.

Land degradation occurs when natural or human-induced processes decrease the future ability of land to support crops, livestock, or wild species. One type of land degradation is **soil erosion:** the movement of soil components, especially surface litter and topsoil, from one place to another. The two main agents of erosion are *flowing water* and *wind,* with water causing most soil erosion.

Some soil erosion is natural and some is caused by human activities. In undisturbed vegetated ecosystems, the roots of plants help anchor the soil, and usually soil is not lost faster than it forms. Soil becomes more vulnerable to erosion through human activities that destroy plant cover, including farming, logging, construction, overgrazing by livestock, off-road vehicle use, and deliberate burning of vegetation.

Soil erosion has two major harmful effects. One is *loss of soil fertility* through depletion of plant nutrients

in topsoil. The other harmful effect occurs when eroded soil ends up as sediment in nearby surface waters, where it can pollute water, kill fish and shellfish, and clog irrigation ditches, boat channels, reservoirs, and lakes.

Soil, especially topsoil, is classified as a renewable resource because natural processes regenerate it. However, if topsoil erodes faster than it forms on a piece of land, it eventually becomes a nonrenewable resource.

How Serious Is Global Soil Erosion? Mostly Bad News

Soil is eroding faster than it is forming on more than a third of the world's cropland, and much of this land also suffers from salt buildup and waterlogging.

A 1992 joint survey by the United Nations (UN) Environment Programme and the World Resources Institute estimated that topsoil is eroding faster than it forms on about 38% of the world's cropland (Figure 8-5). According to a 2000 study by the Consultative Group on International Agricultural Research, soil erosion and degradation has reduced food production on about 16% of the world's cropland. Study this figure to see which countries suffer the most from soil erosion and degradation. See the Guest Essay on soil erosion by David Pimentel on the website for this chapter.

Some analysts contend that erosion estimates are overstated because they underestimate the abilities of some local farmers to restore degraded land. The UN Food and Agriculture Organization (FAO) also points out that much of the eroded topsoil does not go far and is deposited further down a slope, valley, or plain. In some places, the loss in crop yields in one area could be offset by increased yields elsewhere.

Case Study: How Fast Is Soil Eroding in the United States? Some Hopeful News

Soil in the United States is eroding faster than it forms on most cropland, but since 1987, erosion has been cut by about two-thirds.

Bad news. According to the Natural Resources Conservation Service, soil on cultivated land in the United States is eroding about 16 times faster than it can form. Erosion rates are even higher in heavily farmed regions. An example is the Great Plains, which has lost one-third or more of its topsoil in the 150 years since it was first plowed.

Good news. Of the world's major food-producing nations, only the United States is sharply reducing some of its soil losses through a combination of planting crops without disturbing the soil and government-sponsored soil conservation programs.

These efforts to slow soil erosion are an important step and, since 1985, have cut soil losses on U.S. crop-

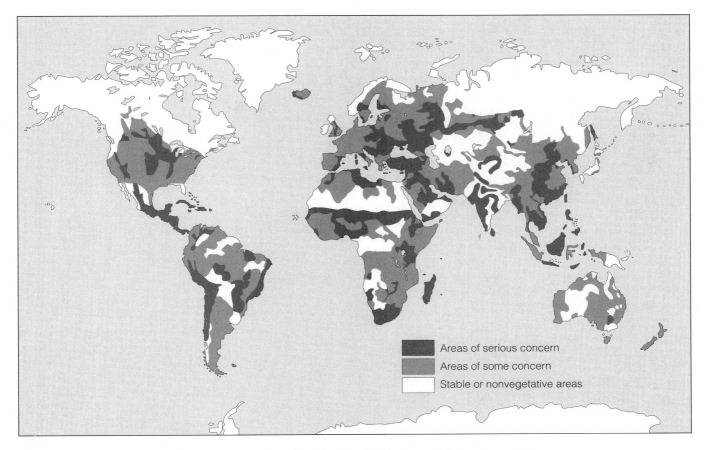

Figure 8-5 Natural capital degradation: global soil erosion. (Data from UN Environment Programme and the World Resources Institute)

---

land by about two-thirds. However, effective soil conservation is practiced today on only about half of all U.S. agricultural land and on about half of the country's most erodible cropland.

How Do Excess Salts and Water Degrade Soils? Crop Losses from Too Much Salt and Water

Repeated irrigation can reduce crop productivity by salt buildup in the soil and waterlogging of crop plants.

The one-fifth of the world's cropland that is irrigated produces almost 40% of the world's food. But irrigation has a downside. Most irrigation water is a dilute solution of various salts, picked up as the water flows over or through soil and rocks. Irrigation water not absorbed into the soil evaporates, leaving behind a thin crust of dissolved salts (such as sodium chloride) in the topsoil.

Repeated annual applications of irrigation water lead to the gradual accumulation of salts in the upper soil layers. This accumulation of salts is called **salinization** (Figure 8-6, p. 152). It stunts crop growth, low-

ers crop yields, and eventually kills plants and ruins the land.

According to a 1995 study, severe salinization has reduced yields on about a fifth of the world's irrigated cropland, and almost another third has been moderately salinized. The most severe salinization occurs in Asia, especially in China, India, and Pakistan.

Salinization affects almost one-fourth of irrigated cropland in the United States. But the proportion is much higher in some heavily irrigated western states.

We know how to prevent and deal with soil salinization, as summarized in Figure 8-7 (p. 152). But some of these remedies are expensive.

Another problem with irrigation is **waterlogging** (Figure 8-6). Farmers often apply large amounts of irrigation water to leach salts deeper into the soil. But without adequate drainage, water accumulates underground and gradually raises the water table. Saline water then envelops the deep roots of plants, lowering their productivity and killing them after prolonged exposure. At least one-tenth of the world's irrigated land suffers from waterlogging, and the problem is getting worse.

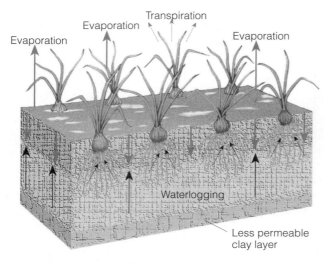

Evaporation   Evaporation   Transpiration   Evaporation

Waterlogging

Less permeable
clay layer

Salinization

1. Irrigation water contains small amounts of dissolved salts.

2. Evaporation and transpiration leave salts behind.

3. Salt builds up in soil.

Waterlogging

1. Precipitation and irrigation water percolate downward.

2. Water table rises.

Figure 8-6 Natural capital degradation: *salinization* and *waterlogging* of soil on irrigated land without adequate drainage can decrease crop yields.

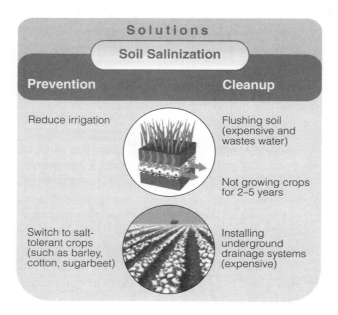

Solutions

Soil Salinization

Prevention

Reduce irrigation

Switch to salt-tolerant crops (such as barley, cotton, sugarbeet)

Cleanup

Flushing soil (expensive and wastes water)

Not growing crops for 2–5 years

Installing underground drainage systems (expensive)

Figure 8-7 Solutions: methods for preventing and cleaning up soil salinization.

8-3 SOIL CONSERVATION

How Can Conservation Tillage Reduce Soil Erosion? Do Not Disturb the Soil.

Modern farm machinery can plant crops without disturbing the soil.

Soil conservation involves using a variety of ways to reduce soil erosion and restore soil fertility, mostly by keeping the soil covered with vegetation.

Eliminating plowing and tilling is the key to reducing erosion and restoring healthy soil. Many U.S. farm-

ers do this by using **conservation-tillage farming** to disturb the soil as little as possible while planting crops.

With *minimum-tillage farming,* the soil is not disturbed over the winter. Then, at planting time, special tillers break up and loosen the subsurface soil without turning over the topsoil, previous crop residues, or any cover vegetation. In *no-till farming,* special planting machines inject seeds, fertilizers, and weed killers (herbicides) into thin slits made in the unplowed soil and then smooth over the cut.

In 2003, farmers used conservation tillage on about 45% of U.S. cropland. The USDA estimates that using conservation tillage on 80% of U.S. cropland would reduce soil erosion by at least half. Conservation tillage also has great potential to reduce soil erosion and raise crop yields in the Middle East and in Africa.

What Other Methods Can Reduce Soil Erosion? Several Tried and True Methods

Farmers have developed a number of ways to grow crops that reduce soil erosion.

Figure 8-8 shows some of the methods farmers have used to reduce soil erosion. One is **terracing,** which can reduce soil erosion on steep slopes by converting the land into a series of broad, nearly level terraces that run across the land's contours (Figure 8-8a). This retains water for crops at each level and reduces soil erosion by controlling runoff.

Another method is **contour farming,** which involves plowing and planting crops in rows across the slope of the land rather than up and down (Figure 8-8b). Each row acts as a small dam to help hold soil and to slow water runoff.

(a) Terracing

(b) Contour planting and strip cropping

(c) Alley cropping

(d) Windbreaks

Figure 8-8 Solutions: In addition to conservation tillage, soil conservation methods include **(a)** terracing, **(b)** contour planting and strip cropping, **(c)** alley cropping or agroforestry, and **(d)** windbreaks.

Farmers also use **strip cropping** to reduce soil erosion (Figure 8-8b). It involves planting alternating strips of a row crop (such as corn or cotton) and another crop that completely covers the soil (such as a grass or a grass and legume mixture). The cover crop traps soil that erodes from the row crop, catches and reduces water runoff, and helps prevent the spread of pests and plant diseases.

One way to reduce erosion is to leave crop residues on the land after the crops are harvested. Another is to plant **cover crops** such as alfalfa, clover, or rye immediately after harvest to help protect and hold the soil.

Another method for slowing erosion is **alley cropping** or **agroforestry,** in which several crops are planted together in strips or alleys between trees and shrubs that can provide fruit or fuelwood (Figure 8-8c). The trees or shrubs provide shade (which reduces water loss by evaporation) and help retain and slowly release soil moisture. They also can provide fruit, fuelwood, and trimmings that can be used as mulch (green manure) for the crops and as fodder for livestock.

Some farmers establish **windbreaks,** or **shelterbelts,** of trees (Figure 8-8d) to reduce wind erosion, help retain soil moisture, supply wood for fuel, and provide habitats for birds, pest-eating and pollinating insects, and other animals.

How Can We Maintain and Restore Soil Fertility? Conservation and Fertilizers

Soil conservation can reduce the loss of soil nutrients, and applying inorganic and organic fertilizers can help restore lost nutrients.

The best way to maintain soil fertility is through soil conservation. The next best thing to do is to restore some of the plant nutrients that have been washed, blown, or leached out of soil or removed by repeated crop harvesting.

Fertilizers are used to partially restore lost plant nutrients. Farmers can use **organic fertilizer** from plant and animal materials or **commercial inorganic fertilizer** produced from various minerals.

There are several types of *organic fertilizer*. One is **animal manure:** the dung and urine of cattle, horses, poultry, and other farm animals. It improves soil structure, adds organic nitrogen, and stimulates beneficial soil bacteria and fungi.

A second type of organic fertilizer called **green manure** consists of freshly cut or growing green vegetation plowed into the soil to increase the organic matter and humus available to the next crop. A third type is **compost,** produced when microorganisms in soil break down organic matter such as leaves, food wastes, paper, and wood in the presence of oxygen.

Crops such as corn, tobacco, and cotton can deplete nutrients (especially nitrogen) in the topsoil if planted on the same land several years in a row. One way to reduce such losses is **crop rotation.** Farmers plant areas or strips with nutrient-depleting crops one year. The next year, they plant the same areas with legumes whose root nodules add nitrogen to the soil. In addition to helping restore soil nutrients, this method reduces erosion by keeping the soil covered with vegetation.

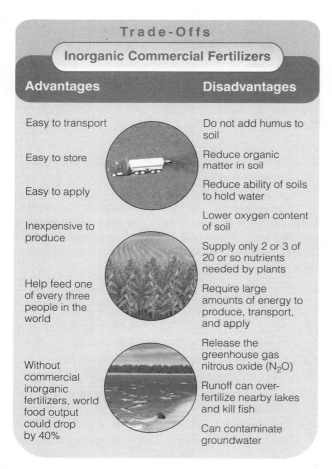

Figure 8-9 Trade-offs: advantages and disadvantages of using *inorganic commercial fertilizers* to enhance or restore soil fertility. Pick the single advantage and disadvantage that you think are the most important.

Can Inorganic Fertilizers Save the Soil? A Partial Solution

Inorganic fertilizers can help restore soil fertility if used with organic fertilizers and if their harmful environmental effects are controlled.

Many farmers (especially in developed countries) rely on *commercial inorganic fertilizers*. The active ingredients typically are inorganic compounds that contain *nitrogen, phosphorus,* and *potassium*. Other plant nutrients may also be present in low or trace amounts. These fertilizers account for about one-fourth of the world's crop yield.

Figure 8-9 lists the advantages and disadvantages of using inorganic fertilizers to enhance or restore soil fertility. Inorganic chemical fertilizers can replace depleted inorganic nutrients, but they do not replace organic matter. Thus for healthy soil, both inorganic and organic fertilizers should be used.

8-4 FOOD PRODUCTION, NUTRITION, AND ENVIRONMENTAL EFFECTS

How Much Has Food Production Increased? Impressive Gains That Are Slowing Down

After increasing significantly since 1950, global grain production has mostly leveled off since 1985, and per capita grain production has declined since 1978.

After almost tripling between 1950 and 1985, world grain production has essentially leveled off (Figure 8-10, left). And after rising by about 36% between 1950 and 1978, per capita food production has declined (Figure 8-10, right). The sharpest drops in per capita food production have occurred in Africa since 1970, in the former Soviet Union since 1990, and in China since 1998.

Good news. We produce more than enough food to meet the basic nutritional needs of every person on the earth. *Bad news:* one out of six people in developing countries are not getting enough to eat because food is not distributed equally among the world's people. This occurs because of differences in soil, climate, political and economic power, and average income per person.

Most agricultural experts agree that *the root causes of hunger and malnutrition are and will continue to be poverty and inequality,* which prevent poor people from growing or buying enough food regardless of how much is available. Other factors are war, corruption, and tariffs and subsidies that make it hard for poor people to have access to food that is produced.

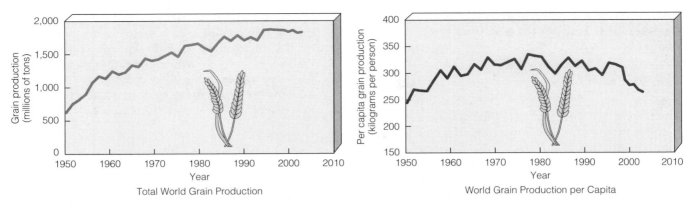

Figure 8-10 Total worldwide grain production of wheat, corn, and rice (left), and per capita grain production (right), 1950–2003. In order, the world's three largest grain-producing countries are China, the United States, and India. (Data from U.S. Department of Agriculture, Worldwatch Institute, UN Food and Agriculture Organization, and Earth Policy Institute)

How Serious Are Undernutrition and Malnutrition? Some Progress

Some people cannot grow or buy enough food to meet their basic energy needs, and others do not get enough protein and other key nutrients.

To maintain good health and resist disease, we need fairly large amounts of *macronutrients* (such as protein, carbohydrates, and fats), and smaller amounts of *micronutrients* consisting of various vitamins (such as A, C, and E) and minerals (such as iron, iodine, and calcium).

People who cannot grow or buy enough food to meet their basic energy needs suffer from **chronic undernutrition.** Chronically undernourished children are likely to suffer from mental retardation and stunted growth. They also are susceptible to infectious diseases such as measles and diarrhea that rarely kill children in developed countries.

Many of the world's poor can afford to live only on a low-protein, high-carbohydrate diet consisting only of grains such as wheat, rice, or corn. Many suffer from **malnutrition** resulting from deficiencies of protein and other key nutrients.

Good news. According to the UN Food and Agriculture Organization (FAO), the average daily food intake in calories per person in the world and in developing countries rose sharply between 1961 and 2000, and is projected to continue rising through 2030. Also, the estimated number of chronically undernourished or malnourished people fell from 918 million in 1970 to 825 million in 2001—about 95% of them in developing countries.

Bad news. About one of every six people in developing countries (including about one of every three children below age 5) is chronically undernourished or malnourished. The FAO estimates that at least 5.5 million people die prematurely from undernutrition, malnutrition, and increased susceptibility to normally nonfatal infectious diseases (such as measles and diarrhea) because of their weakened condition. This means that each day at least 15,100 people—80% of them children under age 5—die prematurely from these causes related to poverty.

Studies by the United Nations Children's Fund (UNICEF) indicate that one-half to two-thirds of childhood deaths from nutrition-related causes could be prevented at an average annual cost of $5–10 per child by the following measures:

- Immunizing children against childhood diseases such as measles

- Encouraging breast-feeding (except for mothers with AIDS)

- Preventing dehydration from diarrhea by giving infants a mixture of sugar and salt in a glass of water

- Preventing blindness by giving children a vitamin A capsule twice a year at a cost of about 75¢ per child, or fortifying common foods with vitamin A and other micronutrients at a cost of about 10¢ per child annually

- Providing family planning services to help mothers space births at least 2 years apart

- Increasing education for women, with emphasis on nutrition, drinking water sterilization, and child care

Some people in developed countries also suffer from lack of access to enough food for good health. In the United States, about 11 million people (half of them children under age 5) do not have access to enough food on a regular basis for good health.

How Serious Are Micronutrient Deficiencies? Important but Limited Progress

One of every three persons has a deficiency of one or more vitamins and minerals, especially vitamin A, iron, and iodine.

According to the World Health Organization (WHO), about one out of three people suffer from a deficiency of one or more vitamins and minerals. The most widespread micronutrient deficiencies in developing countries involve *vitamin A, iron,* and *iodine.*

According to the WHO, 120–140 million children in developing countries are deficient in vitamin A. Globally, about 250,000 children under age 6 go blind each year from a lack of vitamin A and up to 80% of them die within a year.

Other nutritional deficiency diseases are caused by the lack of minerals. Too little *iron*—a component of hemoglobin that transports oxygen in the blood—causes *anemia.* According to a 1999 survey by the WHO, one of every three people in the world, mostly women and children in tropical developing countries, suffers from iron deficiency. Iron deficiency causes fatigue, makes infection more likely, and increases a woman's chances of dying in childbirth and an infant's chances of dying of infection in its first year of life.

Elemental *iodine* is essential for proper functioning of the thyroid gland, which produces a hormone (thyroxine) that controls the body's rate of metabolism. Iodine is found in seafood and crops grown in iodine-rich soils. Chronic lack of iodine can cause stunted growth, mental retardation, and goiter—an abnormal enlargement of the thyroid gland that can lead to deafness. According to the United Nations, about 26 million children suffer brain damage each year from lack of iodine. And 600 million people—mostly in South and Southeast Asia—suffer from goiter.

How Serious Is Overnutrition? Bad and Getting Worse

After smoking, overnutrition is a major cause of preventable deaths.

Overnutrition occurs when food energy intake exceeds energy use and causes excess body fat. Too many calories, too little exercise, or both can cause overnutrition.

People who are underfed and underweight and those who are overfed and overweight face similar health problems: *lower life expectancy, greater susceptibility to disease and illness,* and *lower productivity and life quality.* We live in a world where 1 billion people have health problems because they do not get enough to eat and another 12 billion worry about health problems from eating too much. According to a 2004 study by the International Obesity Task Force, about one of every four people in the world is overweight and 5% are obese.

In developed countries, overnutrition is the second leading cause, after smoking of preventable deaths, mostly from heart disease, cancer, stroke, and diabetes. According to the Centers for Disease Control and Prevention, almost two-thirds of Americans adults are overweight or obese—the highest overnutrition rate of any developed country.

What Are the Environmental Effects of Producing Food? Agriculture Is Number One

Modern agriculture has a greater harmful environmental impact than any human activity, and these effects may limit future food production.

Modern agriculture has significant harmful effects on air, soil, water, and biodiversity, as Figure 8-11 shows. According to many analysts, agriculture has a greater harmful environmental impact than any human activity!

Some analysts believe these harmful environmental effects can be overcome and will not limit future food production. Other analysts disagree. For example, according to environmental expert Norman Myers, a combination of environmental factors may limit future food production. They include *soil erosion, salt buildup and waterlogging of soil on irrigated lands, water deficits and droughts,* and *loss of wild species* that provide the genetic resources for improved forms of foods.

According to a 2002 study by the UN Department for Economic and Social Affairs, close to 30% of the world's cropland has been degraded to some degree by soil erosion, salt buildup, and chemical pollution, and 17% has been seriously degraded. Such environmental factors may limit food production in India and China, the world's two most populous countries.

8-5 INCREASING CROP PRODUCTION

What Is the Gene Revolution? From Crossbreeding to Mixing Genes in a New Way

We can increase crop yields by using crossbreeding to mix the genes of similar types of organism and genetic engineering to mix those of different organisms.

For centuries, farmers and scientists have used *crossbreeding* to develop genetically improved varieties of crop strains. Such selective breeding has had amazing results. Ancient ears of corn were about the size of your little finger and wild tomatoes were once the size of a grape.

But traditional crossbreeding is a slow process, typically taking 15 years or more to produce a commercially valuable new variety, and can combine traits

Biodiversity Loss

Loss and degradation of habitat from clearing grasslands and forests and draining wetlands

Fish kills from pesticide runoff

Killing of wild predators to protect livestock

Loss of genetic diversity from replacing thousands of wild crop strains with a few monoculture strains

Soil

Erosion

Loss of fertility

Salinization

Waterlogging

Desertification

Air Pollution

Greenhouse gas emissions from fossil fuel use

Other air pollutants from fossil fuel use

Pollution from pesticide sprays

Water

Aquifer depletion

Increased runoff and flooding from land cleared to grow crops

Sediment pollution from erosion

Fish kills from pesticide runoff

Surface and groundwater pollution from pesticides and fertilizers

Overfertilization of lakes and slow-moving rivers from runoff of nitrates and phosphates from fertilizers, livestock wastes, and food processing wastes

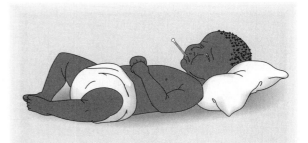

Human Health

Nitrates in drinking water

Pesticide residues in drinking water, food, and air

Contamination of drinking and swimming water with disease organisms from livestock wastes

Bacterial contamination of meat

Figure 8-11 Natural capital degradation: major environmental effects of food production. According to UN studies, land degradation reduced cumulative food production worldwide by about 13% on cropland and 4% on pastureland between 1950 and 2000.

only from species that are close to one another genetically. It also provides varieties that are useful for only about 5–10 years before pests and diseases reduce their effectiveness.

Scientists are creating a *third green revolution*—actually a *gene revolution*—by using *genetic engineering* to develop genetically improved strains of crops and livestock animals. It involves splicing a gene from one species and transplanting it into the DNA of another species (Figure 3-5, p. 53). Compared to traditional crossbreeding, gene splicing takes about half as long to develop a new crop, cuts costs, and allows the insertion of genes from almost any other organism into crop cells.

Ready or not, the world is entering the *age of genetic engineering*. More than two-thirds of the food products on U.S. supermarket shelves contain genetically engineered crops, and the proportion is increasing rapidly.

Despite the promise, there is considerable controversy over the use of *genetically modified food* (*GMF*) and other forms of genetic engineering. Such food is seen by its producers and investors as a potentially sustainable way to solve world food problems by its producers and investors, but some critics consider it potentially dangerous "Frankenfood." Figure 8-12 (p. 158) summarizes the projected advantages and disadvantages of this new technology. Study this figure carefully.

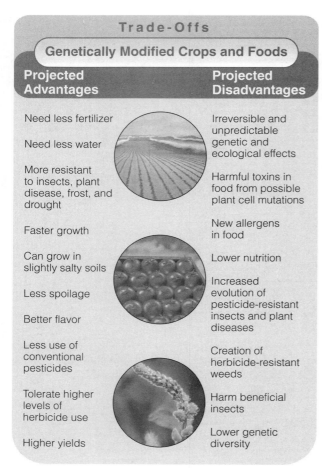

Trade-Offs

Genetically Modified Crops and Foods

Projected Advantages	Projected Disadvantages
Need less fertilizer	Irreversible and unpredictable genetic and ecological effects
Need less water	
More resistant to insects, plant disease, frost, and drought	Harmful toxins in food from possible plant cell mutations
Faster growth	New allergens in food
Can grow in slightly salty soils	Lower nutrition
Less spoilage	Increased evolution of pesticide-resistant insects and plant diseases
Better flavor	
Less use of conventional pesticides	Creation of herbicide-resistant weeds
Tolerate higher levels of herbicide use	Harm beneficial insects
Higher yields	Lower genetic diversity

Figure 8-12 Trade-offs: projected advantages and disadvantages of *genetically modified crops and foods*. Pick the single advantage and disadvantage that you think are the most important.

Critics recognize the potential benefits of genetically modified crops. But they warn that we know too little about the potential harm to human health and ecosystems from the widespread use of such crops. Also, genetically modified organisms cannot be recalled if they cause some unintended harmful genetic and ecological effects—as some scientists expect.

Most scientists and economists who have evaluated the genetic engineering of crops believe that its enormous potential benefits outweigh the much smaller risks. But critics call for more controlled field experiments, more research and long-term safety testing to better understand the risks, and stricter regulation of this rapidly growing technology. A 2004 study by the Ecological Society of America recommended more caution in releasing genetically engineered organisms into the environment.

X *How Would You Vote?* Do the advantages of genetically engineered foods outweigh their disadvantages? Cast your vote online at http://biology.brookscole.com/miller7.

Many analysts and consumer advocates believe governments should require mandatory labeling of genetically modified foods. This would provide consumers with information to help them make informed choices about the foods they buy. Such labeling is required in Japan, Europe, South Korea, Canada, Australia, and New Zealand and is favored by 81% of Americans polled in 1999.

Industry representatives and the U.S. Department of Agriculture oppose this because they claim that GM foods are not substantially different from foods developed by conventional crossbreeding methods. Also, they fear—probably correctly—that labeling such foods would hurt sales by arousing suspicion.

X *How Would You Vote?* Should all genetically engineered foods be so labeled? Cast your vote online at http://biology.brookscole.com/miller7.

Can We Continue Expanding the Green Revolution? Maybe, Maybe Not

Lack of resources such as water and fertile soil, and environmental factors may limit our ability to continue increasing crop yields.

Many analysts believe we can produce all the food we need in the future by spreading the use of existing high-yield green revolution crops and genetically engineered crops to more of the world.

Other analysts disagree. They point to several factors that have limited the success of the green and gene revolutions to date and may continue to do so. One problem is that without huge amounts of fertilizer and water, most green revolution crop varieties produce yields that are no higher (and are sometimes lower) than those from traditional strains. Another problem is that green revolution and genetically engineered crop strains and their high inputs of water, fertilizer, and pesticides cost too much for most subsistence farmers in developing countries.

Scientists also point out that continuing to increase fertilizer, water, and pesticide inputs eventually produces no additional increase in crop yields. For example, grain yields rose about 2.1% a year between 1950 and 1990 but dropped to 1.1% per year between 1990 and 2000 and to 0.5% between 1997 and 2002. No one knows whether this downward trend will continue.

There is also concern that crop yields in some areas may start dropping as soil erodes and loses fertility, irrigated soil becomes salty and waterlogged, underground and surface water supplies become depleted and polluted with pesticides and nitrates from fertilizers, and populations of rapidly breeding pests develop genetic immunity to widely used pesticides. We do not know how close we are to such environmental limits.

Also, according to Indian economist Vandana Shiva, overall gains in crop yields from new green and gene revolution varieties may be much lower than claimed. The yields are based on comparisons between the output per hectare of old and new *monoculture* varieties rather than between the even higher yields per hectare for *polyculture* cropping systems and the new monoculture varieties that often replace polyculture crops.

There is also concern that the projected increased loss of biodiversity can limit the genetic raw material needed for future green and gene revolutions. The UN Food and Agriculture Organization estimates that two-thirds of all seeds planted in developing countries are of uniform strains. Such genetic uniformity increases the vulnerability of food crops to pests, diseases, and harsh weather.

Will People Try New Foods? Changing Eating Habits Is Difficult.

A variety of plants and insects could be used as sources of food, but most consumers are reluctant to try new foods.

Some analysts recommend greatly increased cultivation of less widely known plants to supplement or replace such staples as wheat, rice, and corn. One of many possibilities is the *winged bean*, common in New Guinea and Southeast Asia. This fast-growing bean is a good source of protein and produces so many edible parts it has been called a supermarket on a stalk. It also needs little fertilizer because of nitrogen-fixing nodules in its roots.

Some edible insects—called *microlivestock*—are also important potential sources of protein, vitamins, and minerals in many parts of the world. There are about 1,500 edible insect species. Examples include black ant larvae (served in tacos in Mexico), giant waterbugs (crushed into vegetable dip in Thailand), emperor moth caterpillars (eaten in South Africa), cockroaches (eaten by Kalahari desert dwellers), lightly toasted butterflies (a favorite food in Bali), and fried ants (sold on the streets of Bogota, Colombia). Most of these insects are 58–78% protein by weight—three to four times as protein-rich as beef, fish, or eggs. One problem is getting farmers to take the financial risk of cultivating new types of food crops. Another is convincing consumers to try new foods. Would you try a bug soup?

Some plant scientists believe we should rely more on *polycultures of perennial crops*, which are better adapted to regional soil and climate conditions than most annual crops. Using perennials would also eliminate the need to till soil and replant seeds each year, greatly reducing energy use, saving water, and reducing soil erosion and water pollution from eroded sediment. Not surprisingly, large seed companies that make their money selling farmers seeds each year for annual crops generally oppose this idea.

Is Irrigating More Land the Answer? A Limited Solution

The amount of irrigated land per person has been falling since 1978 and is projected to fall much more during the next few decades.

About 40% of the world's food production comes from the 20% of the world's cropland that is irrigated. Between 1950 and 2003, the world's irrigated area tripled, with most of the growth occurring from 1950 to 1978.

However, the amount of irrigated land per person has been falling since 1978 and is projected to fall much more between 2004 and 2050. One reason is that since 1978, the world's population has grown faster than irrigated agriculture. Other factors are depletion of underground water supplies (aquifers), inefficient use of irrigation water, and salt buildup in soil on irrigated cropland. In addition, the majority of the world's farmers do not have enough money to irrigate their crops.

Is Cultivating More Land the Answer? Another Limited Solution

Significant expansion of cropland is unlikely over the next few decades because of poor soils, limited water, high costs, and harmful environmental effects.

Theoretically, clearing tropical forests and irrigating arid land could more than double the world's cropland. But much of this is *marginal land* with poor soil fertility, steep slopes, or both. Cultivation of such land is unlikely to be sustainable.

Much of the world's potentially cultivable land lies in dry areas, especially in Australia and Africa. Large-scale irrigation in these areas would require expensive dam projects, use large inputs of fossil fuel to pump water long distances, and deplete groundwater supplies by removing water faster than it is replenished. It would also require expensive efforts to prevent erosion, groundwater contamination, salinization, and waterlogging, all of which reduce crop productivity.

Furthermore, these potential increases in cropland would not offset the projected loss of almost one-third of today's cultivated cropland caused by erosion, overgrazing, waterlogging, salinization, and urbanization. Such cropland expansion would also reduce wildlife habitats and thus the world's biodiversity. Bottom line: *Many analysts believe that significant expansion of cropland is unlikely over the next few decades.*

Is Producing More Meat the Answer? More Protein at the Expense of the Environment

Meat and meat products are important sources of protein, but meat production has many harmful environmental effects.

Meat and meat products are good sources of high-quality protein. Between 1950 and 2003, world meat production increased more than fivefold, and per capita meat production more than doubled. It is likely to more than double by 2050 as affluence rises in middle-income developing countries such as China and people begin consuming more meat.

Some analysts expect most future increases in meat production to come from densely populated *feedlots,* where animals are fattened for slaughter by feeding on grain grown on cropland or meal produced from fish. Feedlots account for about 43% of the world's beef production, half of pork production, and almost three-fourths of poultry production.

Expanding feedlot production of meat will increase pressure on the world's grain supply because feedlot livestock consume grain produced on cropland instead of feeding on natural grasses. It will also increase pressure on the world's fish supply because about one-third of the world's fish catch is used to feed livestock. Livestock production also has an enormous environmental impact (Connections, below).

Can We Harvest More Fish and Shellfish? Vacuuming the Seas

After spectacular increases, the world's fish catch has leveled off.

The world's third major food-producing system consists of **fisheries:** concentrations of particular aquatic species suitable for commercial harvesting in a given ocean area or inland body of water. The world's commercial marine fishing industry is dominated by industrial fishing fleets using global satellite positioning equipment, sonar, huge nets and long fishing lines, spotter planes, and large factory ships that can process and freeze their catches.

About 55% of the annual commercial catch of fish and shellfish comes from the ocean using harvesting methods shown in Figure 8-13, mostly from plankton-rich coastal waters. Many commercially valuable

---

CONNECTIONS

Some Environmental Consequences of Meat Production

The meat-based diet of affluent people in developed and developing countries has a number of harmful environmental effects. More than half of the world's cropland (19% in the United States) is used to produce livestock feed grain (mostly field corn, sorghum, and soybeans). Livestock and fish raised for food also consume about 37% of the world's grain production and 70% in the United States.

Meat production uses more than half the water withdrawn from the world's rivers and aquifers each year. Most of this water is used to irrigate crops fed to livestock and to wash away animal wastes.

About 14% of U.S. topsoil loss is directly associated with livestock grazing. Cattle belch out about 16% of the methane (a greenhouse gas about 25 times more potent than carbon dioxide) released into the

atmosphere. Also, some of the nitrogen in commercial inorganic fertilizer used to grow livestock feed is converted to nitrous oxide, a greenhouse gas released from the soil into the atmosphere.

Livestock in the United States produce 20 times more waste (manure) than is produced by the country's human population. A single cow produces as much waste as 16 humans. Only about half of this nutrient-rich livestock waste is recycled into the soil. Manure washing off the land or leaking from lagoons used to store animal wastes can kill fish by depleting dissolved oxygen.

Producing meat can also endanger wildlife species. According to a 2002 report by the National Public Lands Grazing Campaign, livestock grazing in the United States has contributed to population declines of almost a fourth of the country's threatened and endangered species.

Some environmentalists have called for reducing livestock production (especially cattle) in order to reduce its environmental effects and to feed more people. This would decrease the environmental impact of livestock production, but it would not free up much land or grain to feed more of the world's hungry people.

Cattle and sheep that graze on rangeland use a resource (grass) that humans cannot eat, and most of this land is not suitable for growing crops. Moreover, because of poverty, insufficient economic aid, and the nature of global economic and food distribution systems, very little if any additional grain grown on land used to raise livestock or livestock feed would reach the world's hungry people.

Critical Thinking

Are you willing to eat less meat or not eat any meat? Explain.

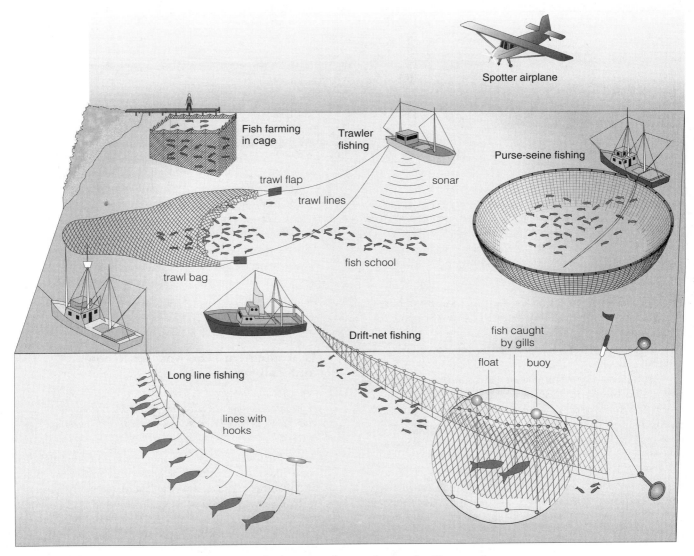

Figure 8-13 Major commercial fishing methods used to harvest various marine species. These methods have become so effective that many fish have become commercially extinct.

species are being overfished by these increasingly efficient methods that are "vacuuming" the seas of fish and shellfish. About one-third of the world's marine fish harvest is used as animal feed, fishmeal, and oils. The rest of the catch comes from using *aquaculture* to raise fish like livestock animals in feedlots in ponds and underwater cages, and from inland freshwater fishing from lakes, rivers, reservoirs, and ponds.

Figure 8-14 (p. 162) shows the effects of the global efforts to increase the seafood harvest. After increasing fourfold between 1960 and 1982, the annual commercial fish catch (marine plus freshwater harvest, but excluding aquaculture) has declined and leveled off (Figure 8-14, left). After doubling between 1950 and 1956, the per capita catch leveled off until 1980. Since then, it has been declining (Figure 8-14, right) and may continue to decline because of overfishing, pollution, habitat loss, and population growth.

How Are Overfishing and Habitat Degradation Affecting Fish Harvests? Dropping Yields

About three-fourths of the world's commercially valuable marine fish species are overfished or fished at their biological limit.

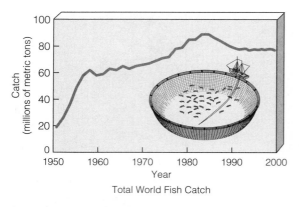

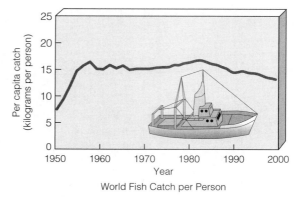

Figure 8-14 World fish catch (left) and world fish catch per person (right), 1950–2000. The total catch and per capita catches since 1990 may be about 10% lower than shown here because of the discovery in 2000 that since 1990, China had apparently been inflating its fish catches. (Data from UN Food and Agriculture Organization and Worldwatch Institute)

Fish are renewable resources as long as the annual harvest leaves enough breeding stock to renew the species for the next year. **Overfishing** is the taking of so many fish that too little breeding stock is left to maintain numbers.

Prolonged overfishing leads to *commercial extinction,* when the population of a species declines to the point at which it is no longer profitable to hunt for them. Fishing fleets then move to a new species or a new region, hoping eventually the overfished species will recover.

Overfishing is not new. Historical studies indicate that some species were overfished beginning centuries ago. However, overfishing has greatly accelerated with the expansion of today's large and efficient global fishing fleets.

According to the UN Food and Agriculture Organization, about three-fourths of the world's 200 commercially valuable marine fish species are either overfished or fished to their estimated maximum sustainable yield. According to the Ocean Conservancy, "we are spending the principal of our marine fish resources rather than living off the interest they provide." Analysts warn that some of these fisheries are so depleted that even if all fishing stopped immediately, it would take up to 20 years for stocks to recover.

In addition, degradation, destruction, and pollution of wetlands, estuaries, coral reefs, salt marshes, and mangrove swamps threaten populations of fish and shellfish.

Good news. In 1995, fisheries biologists studied population data for 128 depleted fish stocks and concluded that 125 of them could recover with careful management. This involves establishing fishing quotas, restricting use of certain types of fishing gear and methods, limiting the number of fishing boats, closing fisheries during spawning periods, and setting aside networks of no-take reserves.

Is Aquaculture the Answer? Good and Bad News

Aquaculture, the world's fastest growing type of food production, has advantages and disadvantages and scientists have proposed ways to reduce its harmful effects.

Aquaculture involves raising fish and shellfish for food like crops instead of going out in fishing boats and hunting and gathering them. It is the world's fastest growing type of food production and accounts for about one-third of the fish and shellfish we eat. China, the world leader, produces over two-thirds of the world's aquaculture output.

There are two basic types of aquaculture. One, called **fish farming,** involves cultivating fish in a controlled environment (often a coastal or inland pond, lake, reservoir, or rice paddy) and harvesting them when they reach the desired size.

The other is **fish ranching.** It involves holding anadromous species, such as salmon that live part of their lives in fresh water and part in salt water, in captivity for the first few years of their lives, usually in fenced-in areas or floating cages in coastal lagoons and estuaries. Then the fish are released, and adults are harvested when they return to spawn.

Figure 8-15 lists the major advantages and disadvantages of aquaculture. Some analysts project that freshwater and saltwater aquaculture production could provide at least half of the world's seafood by 2020. But other analysts warn that the harmful environmental effects of aquaculture (Figure 8-15, right) could limit future production.

Figure 8-16 lists some ways to make aquaculture more sustainable and to reduce its harmful environmental effects. However, even under the most optimistic projections, increasing both the wild catch and aquaculture will not increase world food supplies significantly. The reason is that currently fish and shellfish supply only about 1% of the calories and 6% of the protein in the human diet.

How Do Government Agricultural Policies Affect Food Production? To Interfere or Not to Interfere?

Governments can use price controls to keep food prices artificially low, give farmers subsidies to encourage food production, or eliminate food price controls and subsidies and let farmers and fishers respond to market demand.

Agriculture is a financially risky business. Whether farmers have a good year or a bad year depends on factors over which they have little control: weather, crop prices, crop pests and diseases, interest rates, and the global market. Because of the need for reliable food supplies despite fluctuations in these factors, most governments provide various forms of assistance to farmers and consumers.

Governments use three main approaches to do this. One is to use price controls to *keep food prices artificially low.* This makes consumers happy but means farmers may not be able to make a living.

Another is to *give farmers subsidies and tax breaks to keep them in business and encourage them to increase food production.* Globally, government price supports and other subsidies for agriculture total more than $300 billion per year (about $100 billion per year in the United States)—an average of more than half a million dollars per minute! If government subsidies are too generous and the weather is good, farmers may produce more food than can be sold. The resulting surplus depresses food prices, which reduces the financial incentive for farmers in developing countries to increase domestic food production—those connections again.

A third approach is to *eliminate most or all price controls and subsidies and let farmers and fishers respond to market demand without government interference.* However, some analysts urge that any phaseout of farm and fishery subsidies should be coupled with increased aid for the poor and the lower middle class, who would suffer the most from any increase in food prices. Many environmentalists say that instead of eliminating all subsidies, we should use them to reward farmers and ranchers who protect the soil, conserve water, reforest degraded land, protect and restore wetlands, conserve wildlife, and practice more sustainable agriculture and fishing.

Figure 8-15 Trade-offs: advantages and disadvantages of *aquaculture.* Pick the single advantage and disadvantage that you think are the most important.

Trade-Offs

Aquaculture

Advantages	Disadvantages
Highly efficient	Large inputs of land, feed, and water needed
High yield in small volume of water	Produces large and concentrated outputs of waste
Increased yields through crossbreeding and genetic engineering	Destroys mangrove forests
Can reduce overharvesting of conventional fisheries	Increased grain production needed to feed some species
Little use of fuel	Fish can be killed by pesticide runoff from nearby cropland
Profits not tied to price of oil	Dense populations vulnerable to disease
High profits	Tanks too contaminated to use after about 5 years

Solutions

More Sustainable Aquaculture

- Reduce use of fishmeal as a feed to reduce depletion of other fish

- Improve pollution management of aquaculture wastes

- Reduce escape of aquaculture species into the wild

- Restrict location of fish farms to reduce loss of mangrove forests and other threatened areas

- Farm some aquaculture species (such as salmon and cobia) in deeply submerged cages to protect them from wave action and predators and allow dilution of wastes into the ocean

- Set up a system for certifying sustainable forms of aquaculture

Figure 8-16 Solutions: Ways to make aquaculture more sustainable and reduce its harmful environmental effects.

8-6 PROTECTING FOOD RESOURCES: PEST MANAGEMENT

How Does Nature Keep Pest Populations under Control? Natural Enemies

Predators, parasites, and disease organisms found in nature control populations of most pest species as part of the earth's free ecological services.

A **pest** is any species that competes with us for food, invades lawns and gardens, destroys wood in houses, spreads disease, invades ecosystems, or is simply a nuisance. Worldwide, only about 100 species of plants (which we call weeds), animals (mostly insects), fungi, and microbes (which can infect crop plants and livestock animals) cause about 90% of the damage to the crops we grow.

In natural ecosystems and many polyculture agroecosystems, *natural enemies* (predators, parasites, and disease organisms) control the populations of about 98% of the potential pest species as part of the earth's free ecological services, and thus help keep any one species from taking over for very long.

When we clear forests and grasslands, plant monoculture crops, and douse fields with pesticides, we upset many of these natural population checks and balances. Then we must devise ways to protect our monoculture crops, tree plantations, and lawns from insects and other pests that nature once controlled at no charge.

What Are Pesticides, and How Are They Used? Repel or Kill Pests

We use chemicals to repel or kill pest organisms as plants have done for millions of years to defend themselves against hungry herbivores.

To help control pest organisms, we have developed a variety of **pesticides** or **biocides**—chemicals to kill or control populations of organisms we consider undesirable. Common types of pesticides include *insecticides* (insect killers), *herbicides* (weed killers), *fungicides* (fungus killers), and *rodenticides* (rat and mouse killers). *Biocide is* a more accurate name for these chemicals because most pesticides kill other organisms as well as their pest targets.

We did not invent the use of chemicals to repel or kill other species; plants have been producing chemicals to ward off, deceive, or poison herbivores that feed on them for about 225 million years. This is a never-ending, ever-changing coevolutionary process: herbivores overcome various plant defenses through natural selection; then new plant defenses are favored by natural selection in this ongoing cycle of evolutionary punch and counterpunch.

Since 1950, pesticide use has increased more than 50-fold, and most of today's pesticides are more than ten times as toxic as those used in the 1950s. About three-fourths of these chemicals are used in developed countries, but use in developing countries is soaring.

About one-fourth of pesticide use in the United States is for ridding houses, gardens, lawns, parks, playing fields, swimming pools, and golf courses of pests. According to the U.S. Environmental Protection Agency (EPA), the average lawn in the United States is doused with ten times more synthetic pesticides per hectare than U.S. cropland. Each year, more than 250,000 people in the United States become ill because of household pesticide use, and such pesticides are a major source of accidental poisonings and deaths for children under age 5.

Some pesticides, called *broad-spectrum agents,* are toxic to many species; others, called *selective,* or *narrow-spectrum agents,* are effective against a narrowly defined group of organisms. Pesticides vary in their *persistence,* the length of time they remain deadly in the environment. In 1962, biologist Rachel Carson warned against relying on synthetic organic chemicals to kill insects and other species we deem pests (Individuals Matter, right).

What Are the Advantages of Modern Synthetic Pesticides? Many Benefits

Modern pesticides save lives, increase food supplies, increase profits for farmers, work fast, and are safe if used properly.

Proponents of conventional chemical pesticides contend that their benefits outweigh their harmful effects. Conventional pesticides have a number of important benefits.

They save human lives. Since 1945, DDT and other chlorinated hydrocarbon and organophosphate insecticides probably have prevented the premature deaths of at least 7 million people (some say as many as 500 million) from insect-transmitted diseases such as malaria (carried by the *Anopheles* mosquito), bubonic plague (carried by rat fleas), and typhus (carried by body lice and fleas).

They increase food supplies. According to the UN Food and Agriculture Organization, about 55% of the world's potential human food supply is lost to pests—about two-thirds of that before harvest and the rest after. Without pesticides, these losses would be worse, and food prices would rise.

They increase profits for farmers. Pesticide companies estimate that every $1 spent on pesticides leads to an increase in U.S. crop yields worth approximately $4 (but studies have shown this benefit drops to about $2 if the harmful effects of pesticides are included).

Rachel Carson

Rachel Carson began her professional career as a biologist for the Bureau of U.S. Fisheries (later the U.S. Fish and Wildlife Service). In that capacity, she carried out research on oceanography and marine biology and wrote articles about the oceans and topics related to the environment.

In 1951, she wrote *The Sea Around Us,* which described in easily understandable terms the natural history of oceans and how human activities were harming them. This book sold more than 2 million copies, was translated into 32 languages, and won a National Book Award.

During the late 1940s and throughout the 1950s, DDT and related compounds were increasingly used to kill insects that ate food crops, attacked trees, bothered people, and transmitted diseases such as malaria.

In 1958, DDT was sprayed to control mosquitoes near the home and private bird sanctuary of one of Rachel Carson's friends. After the spraying, her friend witnessed the agonizing deaths of several birds. She begged Carson to find someone to investigate the effects of pesticides on birds and other wildlife.

Carson decided to look into the issue herself and found that independent research on the environmental effects of pesticides was almost nonexistent. As a well-trained scientist, she surveyed the scientific literature, became convinced that pesticides could harm wildlife and humans, and methodically developed information about the harmful effects of widespread use of pesticides.

In 1962, she published her findings in popular form in *Silent Spring,* an allusion to the silencing of "robins, catbirds, doves, jays, wrens, and scores of other bird voices" because of their exposure to pesticides. Many scientists, politicians, and policy makers read *Silent Spring,* and the public embraced it. But manufacturers of chemicals viewed the book as a serious threat to booming pesticide sales and mounted a campaign to discredit Carson. A parade of critical reviewers and industry scientists claimed her book was full of inaccuracies, made selective use of research findings, and failed to give a balanced account of the benefits of pesticides.

Some critics even claimed that, as a woman, she was incapable of understanding such a highly scientific and technical subject. Others charged that she was a hysterical woman and a radical nature lover trying to scare the public in order to sell books.

During these intense attacks, Carson was suffering from terminal cancer. Yet she strongly defended her research and countered her critics. She died in 1964—about 18 months after the publication of *Silent Spring*—without knowing that many historians consider her work an important contribution to the modern environmental movement then emerging in the United States.

They work faster and better than alternatives. Pesticides control most pests quickly and at a reasonable cost, have a long shelf life, are easily shipped and applied, and are safe when handled properly by farm workers. When genetic resistance occurs, farmers can use stronger doses or switch to other pesticides.

When used properly, their health risks are very low compared with their benefits. According to Elizabeth Whelan, director of the American Council on Science and Health (ACSH), which presents the position of the pesticide industry, "The reality is that pesticides, when used in the approved regulatory manner, pose no risk to either farm workers or consumers."

Newer pesticides are safer and more effective than many older pesticides. Greater use is being made of botanicals and microbotanicals. Derived originally from plants, they are safer to users and less damaging to the environment than many older pesticides. Genetic engineering is also being used to develop pest-resistant crop strains and genetically altered crops that produce pesticides.

Many new pesticides are used at much lower rates per unit area than older products. For example, application amounts per hectare for many new herbicides are 1/100 the rates for older ones, and genetically engineered crops could reduce the use of toxic insecticides.

What Is the Ideal Pesticide and Pest? An Ongoing Search

Scientists work to develop more effective and safer pesticides, but pests quickly develop genetic immunity to the pesticides we throw at them.

Scientists continue to search for the ideal pest-killing chemical, which would have these qualities:

- Kill only the target pest
- Not cause genetic resistance in the target organism
- Disappear or break down into harmless chemicals after doing its job
- Be more cost effective than doing nothing

The search continues, but so far no known natural or synthetic pesticide chemical meets all or even most of these criteria.

What Are the Disadvantages of Modern Synthetic Pesticides? Some Problems

Pesticides can promote genetic resistance to their effects, wipe out natural enemies of pest species, create new pest species, end up in the environment, and some can harm wildlife and people.

Opponents of widespread pesticide use believe their harmful effects outweigh their benefits. They site several serious problems with the use of conventional pesticides:

■ *They accelerate the development of genetic resistance to pesticides by pest organisms.* Insects breed rapidly, and within 5–10 years (much sooner in tropical areas) they can develop immunity to pesticides through natural selection and come back stronger than before. Weeds and plant disease organisms also develop genetic resistance, but more slowly. Since 1945, about 1,000 species of insects, mites, weed species, plant diseases, and rodents (mostly rats) have developed genetic resistance to one or more pesticides. Because of genetic resistance, many widely used insecticides (such as DDT) no longer do a good job of protecting people from insect-transmitted diseases in some parts of the world. Genetic resistance can also put farmers on a *pesticide treadmill*, whereby they pay more and more for a pest control program that often becomes less and less effective.

■ *Some insecticides kill natural predators and parasites that help control the populations of pest species.* Wiping out natural predators can unleash new pests, whose populations their predators had previously held in check, and cause other unexpected effects (Connections, p. 75). Of the 300 most destructive insect pests in the United States, 100 were once minor pests that became major pests after widespread use of insecticides. Mostly because of genetic resistance and reduction of natural predators, pesticide use has not reduced crop losses to pests in the United States (Spotlight, below).

■ *Pesticides do not stay put.* According to the U.S. Department of Agriculture (USDA), only 0.1–2% of the insecticide applied to crops by aerial spraying or ground spraying reaches the target pests. Also, less than 5% of herbicides applied to crops reach the target weeds. In other words, 98–99.9% of the pesticides and more than 95% of the herbicides we apply end up in the air, surface water, groundwater, bottom sediments, food, and nontarget organisms, including humans and wildlife (Figure 7-10, p. 139). Crops that have been genetically engineered to release small amounts of pesticides directly to pests can help overcome this problem. But this can promote genetic resistance to such pesticides.

■ *Some pesticides harm wildlife.* According to the USDA and the U.S. Fish and Wildlife Service, each year pesticides applied to cropland in the United States wipe out about 20% of U.S. honeybee colonies and damage another 15%. This costs farmers at least $200 million per year from reduced pollination of vital crops. Pesticides also kill more than 67 million birds and 6–14 million fish, and menace about one of every five endangered and threatened species in the United States.

How Successful Have Synthetic Pesticides Been in Reducing Crop Losses in the United States?

SPOTLIGHT

Studies indicate that pesticides have not been as effective in reducing crop losses in the United States as agricultural experts had hoped. David Pimentel, an expert in insect ecology, has evaluated data from more than 300 agricultural scientists and economists and come to the following conclusions:

■ Although the use of synthetic pesticides has increased 33-fold since 1942, more of the U.S. food supply is lost to pests today (an estimated 37%) than in the 1940s (31%). Losses attributed to insects almost doubled (from 7% to 13%) despite a tenfold increase in the use of synthetic insecticides.

■ The estimated environmental, health, and social costs of pesticide use in the United States range from $4 to $10 billion per year. The International Food Policy Research Institute puts the estimate much higher, at $100–200 billion per year, or $5–10 in damages for every dollar spent on pesticides.

■ Alternative pest control practices (right) could halve the use of chemical pesticides on 40 major U.S. crops without reducing crop yields.

Numerous studies and experience show that pesticide use can be reduced sharply without reducing yields, and in some cases yields even increase. Sweden has cut pesticide use in half with almost no decrease in crop yields. Campbell Soup uses no pesticides on tomatoes it grows in Mexico, and yields have not dropped. After a 65% cut in pesticide use on rice in Indonesia, yields increased by 15%.

Critical Thinking

Pesticide proponents argue that although crop losses to pests are higher today than in the past, without the widespread use of pesticides losses would be even higher. Explain why you agree or disagree with this argument.

■ *Some pesticides can threaten human health.* The World Health Organization (WHO) and the UN Environment Programme (UNEP) estimate that each year pesticides seriously poison at least 3 million agricultural workers in developing countries and at least 300,000 in the United States. This causes 20,000–40,000 deaths (about 25 in the United States) per year. Health officials believe the actual number of pesticide-related illnesses and deaths among the world's farm workers probably is greatly underestimated because of poor record-keeping, lack of doctors, inadequate reporting of illnesses, and faulty diagnoses.

According to studies by the National Academy of Sciences, exposure to legally allowed pesticide residues in food causes 4,000–20,000 cases of cancer per year in the United States and roughly half of these individuals die prematurely. Some scientists are becoming increasingly concerned about possible genetic mutations, birth defects, nervous system disorders (especially behavioral disorders), and effects on the immune and endocrine systems from long-term exposure to low levels of various pesticides. The pesticide industry disputes these claims.

✗ *How Would You Vote?* Do the advantages of using synthetic chemical pesticides outweigh their disadvantages? Cast your vote online at http://biology.brookscole.com/miller7.

Is the U.S. Public Adequately Protected from Exposure to Pesticides? Good and Bad News

Government regulation has banned a number of harmful pesticides but pesticide laws need to be strengthened.

How well the public in the United States is protected from the harmful effects of pesticides is controversial. The EPA banned or severely restricted the use of 56 active pesticide ingredients between 1972 and 2003. And the 1996 Food Quality Protection Act (FQPA) increased public protection from pesticides.

According to studies by the National Academy of Sciences, federal laws regulating pesticide use in the United States are inadequate and poorly enforced by the EPA, the Food and Drug Administration (FDA), and the USDA. Another study by the National Academy of Sciences found that up to 98% of the potential risk of developing cancer from pesticide residues on food grown in the United States would be eliminated if EPA standards were as strict for pre-1972 pesticides as they are for later ones. Another problem is that banned or unregistered pesticides may be manufactured in the United States and exported to other countries (Connections, right).

The pesticide industry disputes these findings and says that eating food grown by using pesticides for the past 50 years has never harmed anyone in the United

What Goes Around Can Come Around

CONNECTIONS

U.S. pesticide companies make and export to other countries pesticides that have been banned or severely restricted—or never even approved—in the United States. Other industrial countries also export banned and unapproved pesticides.

But what goes around can come around. In what environmentalists call a *circle of poison*, residues of some of these banned or unapproved chemicals exported to other countries can return to the exporting countries on imported food. Persistent pesticides such as DDT can also be carried by winds from other countries to the United States.

Environmentalists have urged Congress—without success—to ban such exports. Supporters of pesticide exports argue that such sales increase economic growth and provide jobs, and that banned pesticides are exported only with the consent of the importing countries. They also contend that if the United States did not export pesticides, other countries would.

In 1998, more than 50 countries met to finalize an international treaty that requires exporting countries to have informed consent from importing counties for exports of 22 pesticides and 5 industrial chemicals. In 2000, more than 100 countries developed an international agreement to ban or phase out the use of 12 especially hazardous persistent organic pollutants (POPs)—9 of them persistent hydrocarbon pesticides such as DDT and other chemically similar pesticides.

Critical Thinking

Should U.S. companies be allowed to export pesticides that have been banned, severely restricted, or not approved for use in the United States? Explain.

States. The industry also claims that the benefits of pesticides far outweigh their disadvantages.

What Are Other Ways to Control Pests? Copy Nature

A mix of cultivation practices and biological and ecological alternatives to conventional chemical pesticides can help control pests.

Many scientists believe we should greatly increase the use of biological, ecological, and other alternative methods for controlling pests and diseases that affect crops and human health. A number of methods are available.

One is the use of various *cultivation practices* to fake out pest species. Examples are rotating the types of crops planted in a field each year, adjusting planting times so major insect pests either starve or get eaten by their natural predators, and growing crops in areas where their major pests do not exist. Also, farmers can increase the use of polyculture, which uses plant diversity to reduce losses to pests. Homeowners can reduce weed invasions by cutting grass no lower than 8 centimeters (3 inches) high. This provides a dense enough cover to keep out crabgrass and many other undesirable weeds.

Genetic engineering can be used to speed up the development of pest- and disease-resistant crop strains. But there is controversy over whether the projected advantages of the increasing use of genetically modified plants and foods outweigh their projected disadvantages (Figure 8-12).

We can increase the use of *biological control*. It involves importing natural predators, parasites, and disease-causing bacteria and viruses to help regulate pest populations. This approach is nontoxic to other species, minimizes genetic resistance, and can save large amounts of money—about $25 for every $1 invested in controlling 70 pests in the United States. However, biological control agents cannot always be mass produced, are often slower acting and more difficult to apply than conventional pesticides, can sometimes multiply and become pests themselves, and must be protected from pesticides sprayed in nearby fields.

Sex attractants (called *pheromones*) can lure pests into traps or attract their natural predators into crop fields (usually the more effective approach). These chemicals attract only one species, work in trace amounts, have little chance of causing genetic resistance, and are not harmful to nontarget species. However, it is costly and time consuming to identify, isolate, and produce the specific sex attractant for each pest or predator.

Another approach is to use *hormones that disrupt an insect's normal life cycle* (Figure 8-17), and prevent it from reaching maturity and reproducing. Insect hormones have the same advantages as sex attractants. But they take weeks to kill an insect, often are ineffective with large infestations of insects, and sometimes break down before they can act. In addition, they must be applied at exactly the right time in the target insect's life cycle, can sometimes affect the target's predators and other nonpest species, and are difficult and costly to produce.

Some farmers controlled some insect pests by *spraying them with hot water.* This has worked well on cotton, alfalfa, and potato fields and in citrus groves in Florida, and the cost is roughly equal to that of using chemical pesticides.

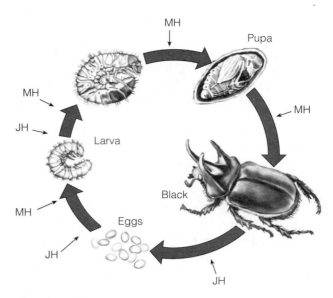

Figure 8-17 For normal insect growth, development, and reproduction to occur, certain juvenile hormones (JH) and molting hormones (MH) must be present at genetically determined stages in the insect's life cycle. If applied at the proper time, synthetic hormones disrupt the life cycles of insect pests and help control their populations.

Is Integrated Pest Management the Answer? A Combined Ecological Approach

An ecological approach to pest control uses an integrated mix of cultivation and biological methods, and small amounts of selected chemical pesticides as a last resort.

An increasing number of pest control experts and farmers believe the best way to control crop pests is a carefully designed **integrated pest management (IPM)** program. In this approach, each crop and its pests are evaluated as parts of an ecological system. Then farmers develop a control program that includes cultivation, biological, and chemical methods applied in proper sequence and with the proper timing.

The overall aim of IPM is not to eradicate pest populations but to reduce crop damage to an economically tolerable level. Fields are monitored carefully, and when an economically damaging level of pests has been reached, farmers first use biological methods (natural predators, parasites, and disease organisms) and cultivation controls, including vacuuming up harmful bugs. Small amounts of insecticides—mostly based on natural insecticides produced by plants—are applied only as a last resort. Also, different chemicals are used in order to slow the development of genetic resistance and to avoid killing predators of pest species.

In 1986, the Indonesian government banned the use of 57 of the 66 pesticides used on rice, and phased out pesticide subsidies over a 2-year period. It also launched a nationwide education program to help farmers switch to IPM. The results were dramatic. Between 1987 and 1992, pesticide use dropped by 65%, rice production rose by 15%, and more than 250,000 farmers were trained in IPM techniques. Sweden and Denmark have used IPM to cut their pesticide use in half.

The experiences of various countries show that a well-designed IPM program can reduce pesticide use and pest control costs by at least half, cut preharvest pest-induced crop losses by half, and improve crop yields. It can also reduce inputs of fertilizer and irrigation water, and slow the development of genetic resistance because pests are assaulted less often and with lower doses of pesticides. Thus IPM is an important form of *pollution prevention* that reduces risks to wildlife and human health.

Despite its promise, IPM, like any other form of pest control, has some disadvantages. It requires expert knowledge about each pest situation and is slower acting than conventional pesticides. In addition, methods developed for a crop in one area might not apply to areas with even slightly different growing conditions. Also, initial costs may be higher, although long-term costs typically are lower than those of using conventional pesticides.

Widespread use of IPM is hindered by government subsidies for conventional chemical pesticides and opposition by pesticide manufacturers, whose sales would drop sharply. There is also a lack of experts to help farmers shift to IPM.

A 1996 study by the National Academy of Sciences recommended that the United States shift from chemically based approaches to ecologically based pest management approaches. Within 5–10 years, such a shift could cut U.S. pesticide use in half, as it has in several other countries.

A growing number of scientists urge the USDA to use three strategies to promote IPM in the United States:

- Add a 2% sales tax on pesticides and use the revenue to fund IPM research and education.

- Set up a federally supported IPM demonstration project on at least one farm in every county.

- Train USDA field personnel and county farm agents in IPM so they can help farmers use this alternative.

The pesticide industry has successfully opposed such measures.

X HOW WOULD YOU VOTE? Should governments heavily subsidize a switch to integrated pest management? Cast your vote online at http://biology.brookscole.com/miller7.

Good news. Several UN agencies and the World Bank have joined together to establish an IPM facility. Its goal is to promote the use of IPM by disseminating information and establishing networks among researchers, farmers, and agricultural extension agents involved in IPM.

8-7 SOLUTIONS: SUSTAINABLE AGRICULTURE

What Is More Sustainable Agriculture? Learn from Nature

We can produce food more sustainably by reducing resource throughput and working with nature.

There are three main ways to reduce hunger and malnutrition and the harmful environmental effects of agriculture. One is to *slow population growth.* Another is to *reduce poverty* so that people can grow or buy enough food for their survival and good health.

The third is to develop and phase in systems of **sustainable** or **low-input agriculture**—also called **organic farming**—over the next few decades. Figure 8-18 (p. 170) lists the major components of more sustainable agriculture. Studies have shown that low-input organic farming produces roughly equivalent yields with lower carbon dioxide emissions, uses about half as much energy per unit of yield as conventional farming, improves soil fertility, and generally is more profitable for the farmer than high-input farming.

Most proponents of more sustainable agriculture are not opposed to high-yield agriculture. Instead, they see it as vital for protecting the earth's biodiversity by reducing the need to cultivate new and often marginal land. They call for using environmentally sustainable forms of both high-yield polyculture and high-yield monoculture for growing crops.

How Can We Make the Transition to More Sustainable Agriculture? Get Serious

More research, demonstration projects, government subsidies, and training can promote a shift to more sustainable agriculture.

Analysts suggest four major strategies to help farmers make the transition to more sustainable agriculture. *First,* greatly increase research on sustainable agriculture and improving human nutrition. *Second,* set up demonstration projects throughout each country so farmers can see how more sustainable agricultural systems work. *Third,* provide subsidies and increased foreign aid to encourage its use. *Fourth,* establish training programs in sustainable agriculture for farmers and government agricultural officials, and encourage the creation of college curricula in sustainable agriculture and human nutrition.

Phasing in more sustainable agriculture involves applying the four principles of sustainability (Figure 4-10, p. 74) to producing food. The goal is to feed

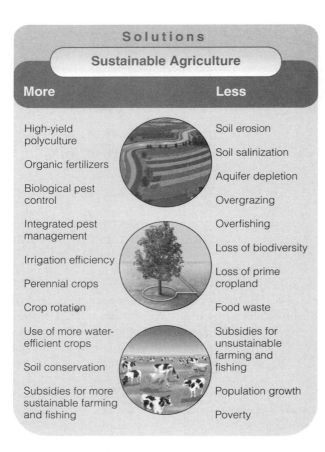

Solutions

Sustainable Agriculture

More	Less
High-yield polyculture	Soil erosion
Organic fertilizers	Soil salinization
Biological pest control	Aquifer depletion
Integrated pest management	Overgrazing
Irrigation efficiency	Overfishing
Perennial crops	Loss of biodiversity
Crop rotation	Loss of prime cropland
Use of more water-efficient crops	Food waste
Soil conservation	Subsidies for unsustainable farming and fishing
Subsidies for more sustainable farming and fishing	Population growth
	Poverty

Figure 8-18 Solutions: components of more sustainable, low-throughput agriculture.

the world's people while sustaining and restoring the earth's natural capital and living off the natural income it provides. This will not be easy but it can be done. Figure 8-19 lists some ways that you can promote more sustainable agriculture.

What Can You Do?

Sustainable Agriculture

- Waste less food
- Reduce or eliminate meat consumption
- Feed pets balanced grain foods instead of meat
- Use organic farming to grow some of your food
- Buy organic food
- Compost your food wastes

Figure 8-19 What can you do? Ways to promote more sustainable agriculture.

The sector of the economy that seems likely to unravel first is food. Eroding soils, deteriorating rangelands, collapsing fisheries, falling water tables, and rising temperatures are converging to make it difficult to expand food production fast enough to keep up with the demand.

LESTER R. BROWN

CRITICAL THINKING

1. Summarize the major economic and ecological advantages and limitations of each of the following proposals for increasing world food supplies and reducing hunger over the next 30 years: **(a)** cultivating more land by clearing tropical forests and irrigating arid lands, **(b)** catching more fish in the open sea, **(c)** producing more fish and shellfish with aquaculture, and **(d)** increasing the yield per area of cropland.

2. List five ways in which your lifestyle directly or indirectly contributes to soil erosion.

3. What could happen to energy-intensive agriculture in the United States and other industrialized countries if world oil prices rose sharply?

4. What are the three most important actions you would take to reduce hunger **(a)** in the country where you live and **(b)** in the world?

5. Should governments phase in agricultural tax breaks and subsidies to encourage farmers to switch to more sustainable farming? Explain your answer.

6. Explain why you support or oppose greatly increased use of **(a)** genetically modified food and **(b)** polyculture.

7. Explain how widespread use of a pesticide can **(a)** increase the damage done by a particular pest and **(b)** create new pest organisms.

8. Congratulations! You are in charge of the world. List the three most important features of your **(a)** your agricultural policy, **(b)** your policy to reduce soil erosion, and **(c)** your policy for pest management.

LEARNING ONLINE

The website for this book contains helpful study aids and many ideas for further reading and research. They include a chapter summary, review questions for the entire chapter, flash cards for key terms and concepts, a multiple-choice practice quiz, interesting Internet sites, references, and a guide for accessing thousands of InfoTrac® College Edition articles. Log on to

http://biology.brookscole.com/miller7

Then click on the Chapter-by-Chapter area, choose Chapter 8, and select a learning resource.

9 WATER RESOURCES AND WATER POLLUTION

Our liquid planet glows like a soft blue sapphire in the hard-edged darkness of space. There is nothing else like it in the solar system. It is because of water.

JOHN TODD

9-1 WATER'S IMPORTANCE, USE, AND RENEWAL

Why Is Water So Important and How Much Fresh Water is Available? Liquid Natural Capital

Water keeps us alive, moderates climate, sculpts the land, removes and dilutes wastes and pollutants, and is recycled by the hydrologic cycle.

We live on the water planet, with a precious film of water—most of it salt water—covering about 71% of the earth's surface. All organisms are made up of mostly water. Look in the mirror. What you see is about 60% water, most of it inside your cells.

You could survive for several weeks without food but only a few days without water. It takes huge amounts of water to supply you with food, shelter, and your other needs and wants. Water also plays a key role in sculpting the earth's surface, moderating climate, and removing and diluting water-soluble wastes and pollutants.

Despite its importance, water is one of our most poorly managed resources. We waste it and pollute it. We also charge too little for making it available. This encourages still greater waste and pollution of this renewable resource, for which we have no substitute. As Benjamin Franklin said many decades ago, "It is not until the well runs dry that we know the worth of water."

Only a tiny fraction of the planet's abundant water is readily available to us as fresh water (Figure 9-1). If the world's water supply were only 100 liters (26 gallons), our usable supply of fresh water would be only about 0.014 liter, or 2.5 teaspoons!

Good news: The world's fresh water supply is continuously collected, purified, recycled, and distributed in the solar-powered *hydrologic cycle* (Figure 2-25, p. 39). This magnificent water recycling and purification system works well as long as we do not overload water systems with slowly degradable and nondegradable wastes or withdraw water from underground supplies faster than it is replenished. *Bad news:* In parts of the world, we are doing both of these things.

Differences in average annual precipitation divide the world's countries and people into water *haves* and *have-nots*. For example, Canada, with only 0.5% of the world's population, has one-fifth of the world's fresh water, whereas China, with one-fifth of the world's people, has only 7% of the supply.

What Is Surface Water? Water on Top

Water that does not sink into the ground or evaporate into the air runs off into bodies of water.

One of our most precious resources is fresh water that flows across the earth's land surface and into the world's rivers, streams, lakes, wetlands, and estuaries. The precipitation that does not infiltrate the ground or return to the atmosphere by evaporation is called **surface runoff.** The region from which surface water drains into a river, lake, wetland, or other body of water is called its **watershed** or **drainage basin.**

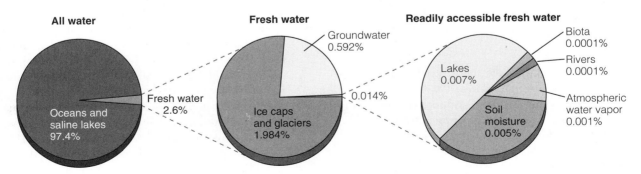

Figure 9-1 Natural capital: the planet's water budget. Only a tiny fraction by volume of the world's water supply is fresh water available for human use.

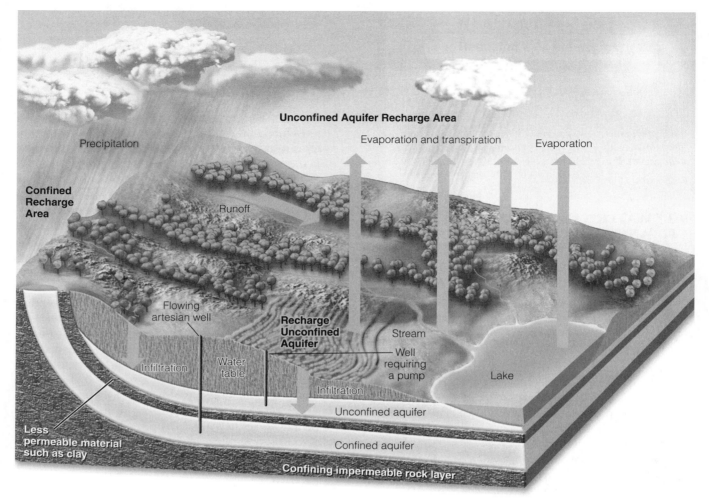

Figure 9-2 Natural capital: groundwater system. An *unconfined aquifer* is an aquifer with a water table. A *confined aquifer* is bounded above and below by less permeable beds of rock. Groundwater in this type of aquifer is confined under pressure. Some aquifers are replenished by precipitation and others are not.

About two-thirds of the world's annual runoff is lost by seasonal floods and is not available for human use. The remaining one-third is **reliable runoff:** the amount of runoff that we can generally count on as a stable source of fresh water from year to year.

What Is Groundwater? Water Down Below

Some precipitation infiltrates the ground and is stored in spaces in soil and rock.

Some precipitation infiltrates the ground and percolates downward through voids (pores, fractures, crevices, and other spaces) in soil and rock (Figure 9-2). The water in these spaces is called **groundwater**—one of our most important sources of fresh water.

Close to the surface, the spaces in soil and rock have little moisture in them. However, below a certain depth, in the **zone of saturation,** the spaces are completely filled with water. The **water table** is located at the top of the zone of saturation. It falls in dry weather

or when we remove groundwater faster than it is replenished and rises in wet weather.

Deeper down are geological layers called **aquifers:** porous, water-saturated layers of sand, gravel, or bedrock through which groundwater flows. They are like large elongated sponges through which groundwater seeps. Fairly watertight layers of rock or clay below an aquifer keep the water from seeping out. About one of every three people on the earth depends on water pumped out of aquifers for drinking and other uses.

Most aquifers are replenished naturally by precipitation that percolates downward through soil and rock in what is called **natural recharge.** But some are recharged from the side by *lateral recharge* from nearby streams. Most aquifers recharge extremely slowly.

Groundwater normally moves from points of high elevation and pressure to points of lower elevation and pressure. This movement is quite slow, typically only a meter or so (about 3 feet) per year and rarely more than 0.3 meter (1 foot) per day.

Some aquifers get very little, if any, recharge and on a human time scale are nonrenewable resources. They are found fairly deep underground and were formed tens of thousands of years ago. Withdrawals from them amount to *water mining* that, if kept up, will deplete these ancient deposits.

How Much of the World's Reliable Water Supply Are We Withdrawing? Taking Half Now and More Later

We are using more than half of the world's reliable runoff of surface water and could be using 70–90% by 2025.

During the last century, the human population tripled, global water withdrawal increased sevenfold, and per capita withdrawal quadrupled. As a result, we now withdraw about 34% of the world's reliable runoff. We leave another 20% of this runoff in streams to transport goods by boats, dilute pollution, and sustain fisheries and wildlife. Thus, *we directly or indirectly use about 54% of the world's reliable runoff of surface water.*

Because of increased population growth alone, global withdrawal rates of surface water could reach more than 70% of the reliable surface runoff by 2025 and 90% if per capita withdrawal of water continues rising at the current rate. This is a global average, with withdrawal rates already exceeding the reliable runoff in a growing number of areas.

How Do We Use the World's Fresh Water? Watering Crops Is Number One.

Irrigation is the biggest user of water (70%) followed by industries (20%), and cities and residences (10%).

Worldwide, we use about 70% of the water we withdraw each year from rivers, lakes, and aquifers to irrigate one-fifth of the world's cropland. This produces about 40% of the world's food, including two-thirds of the world's rice and wheat. Industry uses about 20% of the water withdrawn each year, and cities and residences use the remaining 10%. Agriculture and manufacturing use large amounts of water (Figure 9-3).

Case Study: Freshwater Resources in the United States

The United States has plenty of fresh water, but supplies vary in different areas depending on climate.

The United States has more than enough renewable fresh water. But much of it is in the wrong place at the wrong time or is contaminated by agricultural and industrial practices. The eastern states usually have ample precipitation, whereas many western states have too little.

In the East, the largest uses for water are for energy production, cooling, and manufacturing. The largest use by far in the West (85%) is for irrigation.

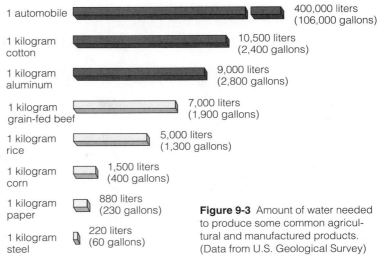

1 automobile	400,000 liters (106,000 gallons)
1 kilogram cotton	10,500 liters (2,400 gallons)
1 kilogram aluminum	9,000 liters (2,800 gallons)
1 kilogram grain-fed beef	7,000 liters (1,900 gallons)
1 kilogram rice	5,000 liters (1,300 gallons)
1 kilogram corn	1,500 liters (400 gallons)
1 kilogram paper	880 liters (230 gallons)
1 kilogram steel	220 liters (60 gallons)

Figure 9-3 Amount of water needed to produce some common agricultural and manufactured products. (Data from U.S. Geological Survey)

In many parts of the eastern United States, the most serious water problems are flooding, occasional urban shortages, and pollution. The major water problem in the arid and semiarid areas of the western half of the country is a shortage of runoff, caused by low precipitation, high evaporation, and recurring prolonged drought (Figure 9-4). Water tables in many areas are dropping rapidly as farmers and cities deplete aquifers faster than they are recharged. Many U.S. urban centers (especially in the West and Midwest) are located in areas that do not have enough water (Figure 9-4). In 2003, the U.S. Department of the Interior mapped out *water hot spots* in 17 western states (Figure 9-5, p. 174).

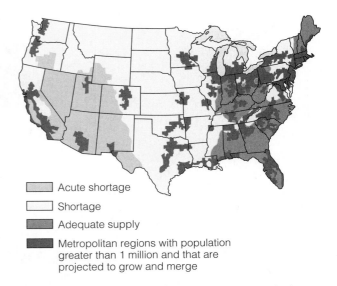

Acute shortage

Shortage

Adequate supply

Metropolitan regions with population greater than 1 million and that are projected to grow and merge

Figure 9-4 Water-deficit regions in the continental United States and their proximity to urban areas having populations greater than 1 million and that are merging to form megalopolises (dark areas) such as the one between Boston, Massachusetts and Washington, D.C., shown in Figure 5-11 (p. 88) (Data from U.S. Water Resources Council and U.S. Geological Survey)

Figure 9-5 Water *hot spot* areas in 17 western states that by 2025 could be faced with intense conflicts and "water wars" over competition for scarce water for urban growth, irrigation, recreation, and wildlife. Some analysts suggest that this is a map of places not to live over the next 25 years. (Data from U.S. Department of the Interior)

Legend:
- Highly likely conflict potential
- Substantial conflict potential
- Moderate conflict potential
- Unmet rural water needs

9-2 WATER RESOURCE PROBLEMS: TOO LITTLE WATER AND TOO MUCH WATER

What Causes Freshwater Shortages? Climate, Demand, and Poverty

About one out of six people does not have regular access to an adequate and affordable supply of clean water, and this could increase to at least one out of four people by 2050.

The two main factors causing water scarcity are a dry climate and too many people using the reliable supply of water. Figure 9-6 shows the degree of stress on the world's major river systems, based on a comparison of the amount of fresh water available with the amount used by humans.

A 2003 study by the United Nations, found that about one out of six people does not have regular access to an adequate and affordable supply of clean water and this could increase to at least one out of four people by 2050.

Poverty also plays a role in access to water. Even when a plentiful supply of water exists, most of the 1.1 billion poor people living on less than $1 a day cannot afford a safe supply of drinking water and must live in *hydrological poverty*. Most are cut off from municipal water supplies and must collect water from unsafe sources or buy water—often coming from polluted rivers—from private vendors at high prices. In water-

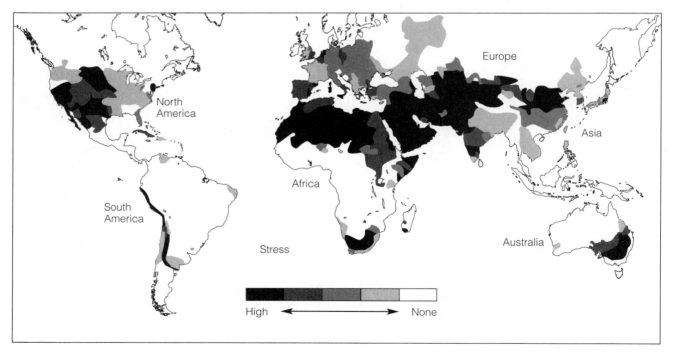

Figure 9-6 **Natural capital degradation:** stress on the world's major river basins, based on a comparison of the amount of water available with the amount used by humans. (Data from World Commission on Water Use in the 21st Century)

short rural areas in developing countries, many women and children must walk long distances each day, carrying heavy jars or cans, to get a meager and sometimes contaminated supply of water.

Case Study: Water Conflicts in the Middle East—A Growing Problem

Some 40% of the world's population, especially in the Middle East, clash over water supplies from shared river basins and the situation is expected to get worse.

In the near future, water-short countries in the Middle East are likely to engage in conflicts over access to fresh water resources. Most water in this dry region comes from three shared river basins: the Nile, the Jordan, and the Tigris-Euphrates (Figure 9-7). Three countries—Ethiopia, Sudan, and Egypt—use most of the water that flows in Africa's Nile River, with Egypt being last in line for the river's water.

To meet the water needs of its rapidly growing population, Ethiopia plans to divert more water from the Nile. So does Sudan. Such upstream diversions would reduce the amount of water available to water-short Egypt, which cannot exist without irrigation water from the Nile.

Egypt can go to war with Sudan and Ethiopia for more water, cut population growth, or improve irrigation efficiency. Other options are to import more grain to reduce the need for irrigation water, work out water-sharing agreements with other countries, or suffer the harsh human and economic consequences of hydrological poverty.

The Jordan Basin is by far the most water-short region, with fierce competition for its water among Jordan, Syria, Palestine (Gaza and the West Bank), and Israel. Syria plans to build dams and withdraw more water from the Jordan River, decreasing the downstream water supply for Jordan and Israel. Israel warns that it may destroy the largest dam that Syria plans to build.

Turkey, located at the headwaters of the Tigris and Euphrates Rivers, controls how much water flows downstream to Syria and Iraq before emptying into the Persian Gulf. Turkey is building 24 dams along the upper Tigris and Euphrates rivers to generate electricity and irrigate a large area of land.

If completed, these dams will reduce the flow of water downstream to Syria and Iraq by up to 35% in normal years and much more in dry years. Syria also plans to build a large dam along the Euphrates River to divert water arriving from Turkey. This will leave little water for Iraq and could lead to a war between it and Syria.

Resolving these water distribution problems will require a combination of regional cooperation in allo-

Figure 9-7 The Middle East, whose countries have some of the world's highest population growth rates. Because of the dry climate, food production depends heavily on irrigation. Existing conflicts between countries in this region over access to scarce water may soon overshadow both long-standing religious and ethnic clashes and take over valuable oil supplies.

cating water supplies, slowed population growth, improved efficiency in water use, higher water prices to help improve irrigation efficiency, and increased grain imports to reduce water needs. This will not be easy.

What Causes Flooding? Rain and People

Heavy rainfall, rapid snowmelt, removal of vegetation, and destruction of wetlands cause flooding.

Whereas some areas have too little water, others sometimes have too much because of natural flooding by streams, caused mostly by heavy rain or rapid melting of snow. A flood happens when water in a stream overflows its normal channel and spills into the adjacent area, called a **floodplain** (Figure 9-8, left, p. 176). Floodplains, which include highly productive wetlands, help provide natural flood and erosion control, maintain high water quality, and recharge groundwater.

People settle on floodplains because of their many advantages, including fertile soil, ample water for irrigation, availability of nearby rivers for transportation and recreation, and flat land suitable for crops, buildings, highways, and railroads.

Floods have several benefits. They provide the world's most productive farmland because of the nutrient-rich silt left after floodwaters recede. They also recharge groundwater and help refill wetlands.

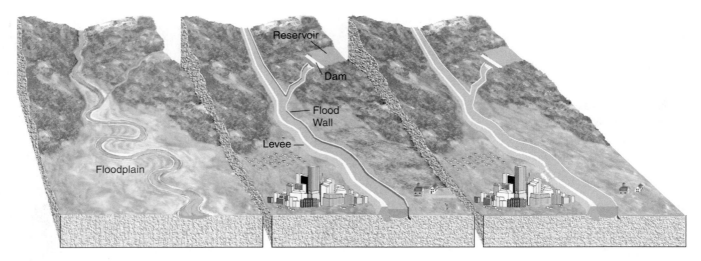

Figure 9-8 Land in a natural floodplain (left) often is flooded after prolonged rains. When the floodwaters recede, deposits of silt are left behind, creating a nutrient-rich soil. To reduce the threat of flooding and thus to allow people to live in floodplains, rivers have been narrowed and straightened (channelized), equipped with protective levees and walls, and dammed to create reservoirs that store and release water as needed. These alterations can give a false sense of security to floodplain dwellers. In the long run, such measures can greatly increase flood damage because they can be overwhelmed by prolonged rains (right), as happened along the Mississippi River in the midwestern United States during the summer of 1993.

Each year floods kill up to 25,000 thousands of people and cause tens of billions of dollars in property damage. In 2004 the United Nations warned that between 2003 and 2050, the number of people volnerable to floods will double from 1 billion to 2 billion.

Floods, like droughts, usually are considered natural disasters, but since the 1960s several types of human activities have contributed to the sharp rise in flood deaths and damages. One is *removal of water-absorbing vegetation,* especially on hillsides (Figure 9-9). Another is *draining wetlands* that absorb floodwaters and reduce the severity of flooding. *Living on floodplains* also increases the threat of damage from flooding. Flooding also increases when we *pave or build,* and replace water-absorbing vegetation, soil, and wetlands with highways, parking lots, and buildings that cannot absorb rainwater.

In developed countries, people deliberately settle on floodplains and then expect dams, levees, and other devices to protect them from floodwaters. However, when heavier-than-normal rains occur, these devices can be overwhelmed. In many developing countries, the poor have little choice but to try to survive in flood-prone areas (Case Study, below).

Case Study: Living on Floodplains in Bangladesh—Danger for the Poor

Bangladesh has experienced increased flooding because of upstream deforestation of Himalayan mountain slopes and the clearing of mangrove forests on its coastal floodplains.

Bangladesh is one of the world's most densely populated countries, with 141 million people (projected to reach 280 million by 2050) packed into an area roughly the size of Wisconsin. It is also one of the world's poorest countries.

The people of Bangladesh depend on moderate annual flooding during the summer monsoon season to grow rice and help maintain soil fertility in the delta basin. The annual floods deposit eroded Himalayan soil on the country's crop fields.

In the past, great floods occurred every 50 years or so. But since the 1970s they have come about every 4 years. Bangladesh's increased flood problems begin in the Himalayan watershed where several factors—rapid population growth, deforestation, overgrazing, and unsustainable farming on steep and easily erodible slopes—have greatly diminished the ability of its mountain soils to absorb water.

Think of a forest as a complex living sponge for catching, storing, using, and recycling water and releasing it in small amounts. Cut down the forests on the Himalayan mountains and what happens? Instead of being absorbed and released slowly, water from the monsoon rains runs off the denuded Himalayan foothills, carrying vital topsoil with it (Figure 9-9, right).

This increased runoff of soil, combined with heavier-than-normal monsoon rains, has increased the severity of flooding along Himalayan rivers and downstream in Bangladesh. A disastrous flood in 1998 covered two-thirds of Bangladesh's land area for 9 months, leveled 2 million homes, drowned at least 2,000 people, and left 30 million people homeless. It also destroyed

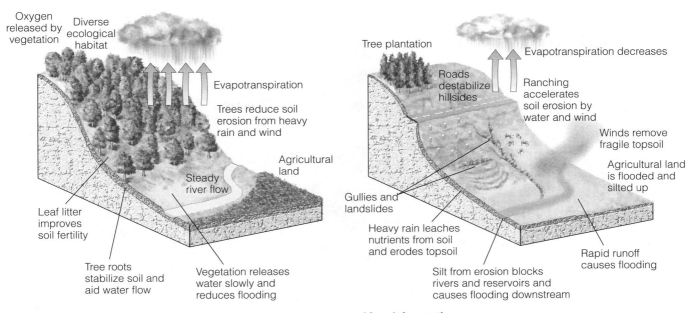

Forested hillside

Oxygen released by vegetation

Diverse ecological habitat

Evapotranspiration

Trees reduce soil erosion from heavy rain and wind

Agricultural land

Steady river flow

Leaf litter improves soil fertility

Tree roots stabilize soil and aid water flow

Vegetation releases water slowly and reduces flooding

After deforestation

Tree plantation

Roads destabilize hillsides

Evapotranspiration decreases

Ranching accelerates soil erosion by water and wind

Winds remove fragile topsoil

Agricultural land is flooded and silted up

Gullies and landslides

Heavy rain leaches nutrients from soil and erodes topsoil

Silt from erosion blocks rivers and reservoirs and causes flooding downstream

Rapid runoff causes flooding

Figure 9-9 Natural capital degradation: hillside before and after deforestation. Once a hillside has been deforested for timber and fuelwood, livestock grazing, or unsustainable farming, water from precipitation rushes down the denuded slopes, erodes precious topsoil, and floods downstream areas. A 3,000-year-old Chinese proverb says, "To protect your rivers, protect your mountains."

more than one-fourth of the country's crops, which caused thousands of people to die of starvation. In 2002, another flood left 5 million people homeless and flooded large areas of rice fields.

Living on Bangladesh's coastal floodplain also carries dangers from storm surges and cyclones. In 1970, as many as 1 million people drowned in one storm. And in 1991 another surge killed an estimated 139,000 people in 1991.

In their struggle to survive, the poor in Bangladesh have cleared many of the country's coastal mangrove forests for fuelwood, farming, and aquaculture ponds for raising shrimp. This has led to more severe flooding because these coastal wetlands shelter Bangladesh's low-lying coastal areas from storm surges and cyclones. Damages and deaths from cyclones in areas of Bangladesh still protected by mangrove forests have been much lower than in areas where the forests have been cleared.

Solutions: How Can We Reduce Flood Risks? Think about Where You Want to Live.

We can reduce flooding risks by controlling river water flows, preserving and restoring wetlands, identifying and managing flood-prone areas, and if possible, choosing not to live in such areas.

We can use several methods to reduce the risk from flooding. One is to *straighten and deepen streams*, a process called *channelization* (Figure 9-8, middle). Channelization can reduce upstream flooding, but it removes bank vegetation and increases stream velocity. This increased flow of water can promote upstream bank erosion, increase downstream flooding and sediment deposition, and reduce habitats for aquatic wildlife.

Another approach is to *build levees* or *floodwalls* along the sides of streams (Figure 9-8, middle). Levees contain and speed up stream flow, but this increases the water's capacity for doing damage downstream. They also do not protect against unusually high and powerful floodwaters, as occurred in 1993 when two-thirds of the levees built along the Mississippi River in the United States were damaged or destroyed.

Building dams can also reduce the threat of flooding by storing water in a reservoir and releasing it gradually. Dams have a number of advantages and disadvantages (Figure 9-10, p. 178). Another way to reduce flooding is to *preserve existing wetlands* and *restore degraded wetlands* to take advantage of the natural flood control provided by floodplains.

We can also *identify and manage floodplains* to get people out of flood-prone areas. This *prevention* or *precautionary* approach is based on thousands of years of experience that can be summed up in one idea: *Sooner or later the river (or the ocean) always wins.*

On a personal level, we can use the precautionary approach to *think carefully about where we live.* Many of the poor live in flood-prone areas because they have

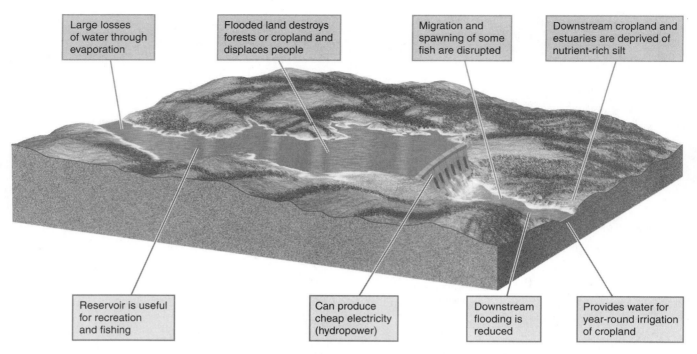

Large losses of water through evaporation

Flooded land destroys forests or cropland and displaces people

Migration and spawning of some fish are disrupted

Downstream cropland and estuaries are deprived of nutrient-rich silt

Reservoir is useful for recreation and fishing

Can produce cheap electricity (hydropower)

Downstream flooding is reduced

Provides water for year-round irrigation of cropland

Figure 9-10 Trade-offs: advantages (bottom boxes) and disadvantages (top boxes) of large dams and reservoirs. The world's 45,000 large dams (higher than 15 meters or 50 feet) capture and store about 14% of the world's runoff, provide water for about 45% of irrigated cropland, and supply more than half the electricity used by 65 countries. The United States has more than 70,000 large and small dams, capable of capturing and storing half of the country's entire river flow. Pick the single advantage and disadvantage that you think are the most important.

nowhere else to go. But most people can do some research and choose not to live in areas especially subject to flooding.

9-3 SUPPLYING MORE WATER

How Can We Increase Freshwater Supplies? Withdraw More and Waste Less

We can increase water supplies by building dams, bringing in water from somewhere else, withdrawing groundwater, converting salt water to fresh water, wasting less water, and importing food.

There are several ways to increase the supply of fresh water in a particular area. One is to build dams and reservoirs to store runoff for release as needed. Another is to bring in surface water from another area. We can also withdraw groundwater and convert salt water to fresh water (desalination).

Other strategies are to reduce water waste and import food to reduce water use in growing crops and raising livestock.

In *developed countries*, people tend to live where the climate is favorable and bring in water from another watershed. In *developing countries*, most people (especially the rural poor) must settle where the water is and try to capture and use the precipitation they need.

What Are the Advantages and Disadvantages of Large Dams and Reservoirs? Mixed Blessings

Large dams and reservoirs can produce cheap electricity, reduce downstream flooding, and provide year-round water for irrigating cropland, but they also displace people and disrupt aquatic systems.

Large dams and reservoirs have benefits and drawbacks (Figure 9-10). Their main purpose is to capture and store runoff and release it as needed to control floods, generate electricity, and supply water for irrigation and for towns and cities. Reservoirs also provide recreational activities such as swimming, fishing, and boating.

The more than 45,000 large dams built on the world's 227 largest rivers have increased the annual reliable runoff available for human use by nearly one-third. But a series of dams on a river, especially in arid areas, can reduce downstream flow to a trickle and prevent it from reaching the sea as a part of the hydrologic cycle. According to the World Commission on Water in the 21st Century, half of the world's major rivers are going dry part of the year because of flow reduction by dams.

This engineering approach to river management has displaced between 40 and 80 million people from their homes and has flooded an area of mostly productive land roughly equal to the area of California. In addition, this approach often impairs some of important ecological services rivers provide (Figure 9-11). In 2003, the World Resources Institute estimated that dams and reservoirs have strongly or moderately fragmented and disturbed 60% of the world's major river basins.

X *How Would You Vote?* Do the advantages of large dams outweigh their disadvantages? Cast your vote online at http://biology.brookscole.com/miller7.

Case Study: The California Water Transfer Project—Bringing Water to a Desert

The massive transfer of water from water-rich northern California to water-poor southern California has brought many benefits, but is also controversial.

Tunnels, aqueducts, and underground pipes can transfer stream runoff collected by dams and reservoirs from water-rich areas to water-poor areas. However, they also create environmental problems. Indeed, most of the world's dam projects and large-scale water transfers illustrate the important ecological principle that *you cannot do just one thing.* There are almost always a number of unintended environmental consequences.

Natural Capital

Ecological Services of Rivers

- Deliver nutrients to sea to help sustain coastal fisheries
- Deposit silt that maintains deltas
- Purify water
- Renew and renourish wetlands
- Provide habitats for wildlife

Figure 9-11 Natural capital: important ecological services provided by rivers. Currently, the services are given little or no monetary value when the costs and benefits of dam and reservoir projects are assessed. According to environmental economists, attaching even crudely estimated monetary values to these ecosystem services would help sustain them.

Figure 9-12 Solutions: California Water Project and the Central Arizona Project. These projects involve large-scale water transfers from one watershed in northern California to another in central and southern California. Arrows show the general direction of water flow.

One of the world's largest water transfer projects is the *California Water Project* (Figure 9-12). It uses a maze of giant dams, pumps, and aqueducts (cement-lined artificial rivers) to transport water from water-rich northern California to southern California's heavily populated arid and semiarid agricultural regions and cities. In effect, this project supplies massive amounts of water to areas that without such water would be mostly desert.

For decades, northern and southern Californians have feuded over how the state's water should be allocated under this project. Southern Californians say they need more water from the north to grow more crops and to support Los Angeles, San Diego, and other growing urban areas. Agriculture uses three-fourths of the water withdrawn in California, much of it used inefficiently for water-thirsty crops growing in desert-like conditions.

Opponents in the north say that sending more water south would degrade the Sacramento River, threaten fisheries, and reduce the flushing action that helps clean San Francisco Bay of pollutants. They also argue that much of the water sent south is wasted. They point to studies showing that making irrigation just 10% more efficient would provide enough water for domestic and industrial uses in southern California.

According to a 2002 joint study by a group of scientists and engineers, projected global warming will sharply reduce water availability in California (especially southern California) and other water-short states in the western United States (Figure 9-5) even under the study's best-case scenario. Some analysts project that sometime during this century, many of the people living in arid southern California cities (such as Los Angeles and San Diego) and farmers in this area, will have to move somewhere else because of a lack of water.

Pumping out more groundwater is not the answer because groundwater is already being withdrawn faster than it is replenished throughout much of California and desalinating ocean water is too expensive for irrigation. To most analysts, quicker and cheaper solutions are to improve irrigation efficiency, stop growing water-thirsty crops in a desert climate, and allow farmers to sell cities their legal rights to withdraw certain amounts of water from rivers.

Case Study: The Aral Sea Disaster—A Glaring Example of Unintended Consequences

Diverting water from the Aral Sea and its two feeder rivers mostly for irrigation has created a major ecological, economic, and health disaster.

The shrinking of the Aral Sea is a result of a large-scale water transfer project in an area of the former Soviet Union with the driest climate in central Asia. Since 1960, enormous amounts of irrigation water have been diverted from the inland Aral Sea and its two feeder rivers to create one of the world's largest irrigated areas, mostly for raising cotton and rice. The irrigation canal, the world's longest, stretches over 1,300 kilometers (800 miles).

This large-scale water diversion project, coupled with droughts and high evaporation rates in the area's hot and dry climate, has caused a regional ecological, economic, and health disaster. Since 1960, the sea's salinity has tripled, its surface area has decreased by 58%, and it has lost 83% of its volume of water. Water withdrawal for agriculture has reduced the sea's two supply rivers to mere trickles.

About 85% of the area's wetlands have been eliminated and roughly half the area's bird and mammal species have disappeared. In addition, a huge area of former lake bottom has been converted to a human-made desert covered with glistening white salt. The increased salt concentration caused the presumed extinction of 20 of the area's 24 native fish species. This has devastated the area's fishing industry, which once provided work for more than 60,000 people. Fishing villages and boats once on the sea's coastline now are abandoned in the middle of a salt desert.

Winds pick up the salty dust that encrusts the lake's now-exposed bed and blow it onto fields as far as 300 kilometers (190 miles) away. As the salt spreads, it pollutes water and kills wildlife, crops, and other vegetation. Aral Sea dust settling on glaciers in the Himalayas is causing them to melt at a faster than normal rate—another example of connections and unintended consequences.

To raise yields, farmers have increased inputs of herbicides, insecticides, fertilizers, and irrigation water on some crops. Many of these chemicals have percolated downward and accumulated to dangerous levels in the groundwater, from which most of the region's drinking water comes.

Shrinkage of the Aral Sea has altered the area's climate. The once-huge sea acted as a thermal buffer that moderated the heat of summer and the extreme cold of winter. Now there is less rain, summers are hotter and drier, winters are colder, and the growing season is shorter. The combination of such climate change and severe salinization have has reduced crop yields by 20–50% on almost a third of the area's cropland.

Finally, there have been increasing health problems from a combination of toxic dust, salt, and contaminated water for a growing number of the 58 million people living in the Aral Sea's watershed.

Can the Aral Sea be saved, and can the area's serious ecological and human health problems be reduced? *Encouraging news.* Since 1999, the United Nations and the World Bank have spent about $600 million to purify drinking water and upgrade irrigation and drainage systems, which improves irrigation efficiency and flushes salts from croplands. In addition, some artificial wetlands and lakes have been constructed to help restore aquatic vegetation, wildlife, and fisheries.

The five countries surrounding the lake and its two feeder rivers have worked to improve irrigation efficiency and to partially replace crops such as rice and cotton that have high water requirements with other crops requiring less irrigation water. As a result, the total annual volume of water in the Aral Sea basin has stabilized. But experts expect the largest portion of the Aral Sea to the south to continue shrinking.

What Are the Advantages and Disadvantages of Withdrawing Groundwater? Avoid Too Many Straws in the Glass.

Most aquifers are renewable sources unless water is removed faster than it is replenished or becomes contaminated.

Aquifers provide drinking water for about one-fourth of the world's people. In the United States, water pumped from aquifers supplies almost all of the drinking water in rural areas, one-fifth of that in urban areas, and 43% of irrigation water.

Relying more on groundwater has advantages and disadvantages (Figure 9-13). The *good news* is that aquifers are widely available and are renewable sources of water as long as the water is not withdrawn faster than it is replaced and as long as the aquifers do not become contaminated.

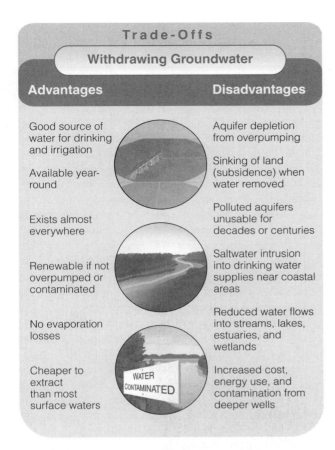

Trade-Offs

Withdrawing Groundwater

Advantages	Disadvantages
Good source of water for drinking and irrigation	Aquifer depletion from overpumping
Available year-round	Sinking of land (subsidence) when water removed
Exists almost everywhere	Polluted aquifers unusable for decades or centuries
Renewable if not overpumped or contaminated	Saltwater intrusion into drinking water supplies near coastal areas
No evaporation losses	Reduced water flows into streams, lakes, estuaries, and wetlands
Cheaper to extract than most surface waters	Increased cost, energy use, and contamination from deeper wells

WATER CONTAMINATED

Figure 9-13 Trade-offs: advantages and disadvantages of withdrawing groundwater. Pick the single advantage and disadvantage that you think are the most important.

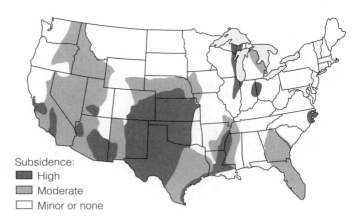

Subsidence:
- High
- Moderate
- Minor or none

Figure 9-14 Natural capital degradation: areas of greatest aquifer depletion from groundwater overdraft in the continental United States. Aquifer depletion is also high in Hawaii and Puerto Rico (not shown on map). It also causes the land below the aquifer to subside or sink in most of these areas. (Data from U.S. Water Resources Council and U.S. Geological Survey)

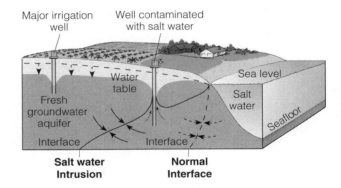

Major irrigation well · Well contaminated with salt water · Sea level · Water table · Salt water · Fresh groundwater aquifer · Seafloor · Interface · Interface

Salt water Intrusion · **Normal Interface**

Figure 9-15 Natural capital degradation: *saltwater intrusion* along a coastal region. When the water table is lowered, the normal interface (dashed line) between fresh and saline groundwater moves inland (solid line), making groundwater drinking supplies unusable.

The *bad news* is that water tables are falling in many areas of the world as the rate of pumping out water (mostly to irrigate crops) exceeds the rate of natural recharge from precipitation. The world's three largest grain-producing countries—China, India, and the United States—are overpumping many of their aquifers. Aquifer depletion also is a problem in Saudi Arabia, northern Africa (especially Libya and Tunisia), southern Europe, the Middle East, and parts of Mexico, Thailand, and Pakistan.

In the United States, groundwater is being withdrawn at four times its replacement rate. The most serious overdrafts are occurring in parts of the huge Ogallala Aquifer, extending from southern South Dakota to central Texas, and in parts of the arid southwestern United States (Figure 9-14). Serious groundwater depletion is also taking place in California's water-short Central Valley, which supplies about half the country's vegetables and fruits. Do you live in any of the shaded areas in Figure 9-14 or in a similar area in another country where groundwater is being withdrawn faster than it is replenished?

Groundwater overdrafts near coastal areas can contaminate groundwater supplies by causing the in-

trusion of salt water into fresh water aquifers used to supply water for irrigation and domestic purposes (Figure 9-15). This is an especially serious problem in many coastal areas in Florida, California, South Carolina, and Texas.

Figure 9-16 (p. 182) lists ways to prevent or slow the problem of groundwater depletion. Study this figure carefully.

With global water shortages looming, scientists are evaluating deep aquifers—found at depths of 0.8 kilometer (0.5 mile) or more—as future water sources. Preliminary results suggest that some of these aquifers hold enough water to support billions of people for centuries. They also indicate that the quality of water in these aquifers is much higher than the water in most of the world's rivers and lakes.

There are two major concerns about tapping these mostly one-time deposits of water. *First*, little is known

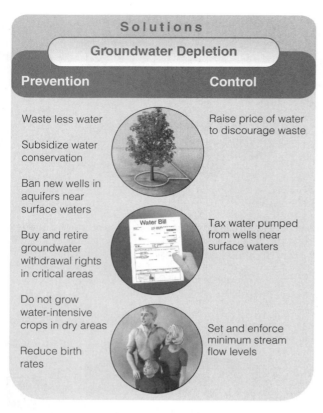

Solutions

Groundwater Depletion

Prevention	Control
Waste less water	Raise price of water to discourage waste
Subsidize water conservation	
Ban new wells in aquifers near surface waters	
Buy and retire groundwater withdrawal rights in critical areas	Tax water pumped from wells near surface waters
Do not grow water-intensive crops in dry areas	
Reduce birth rates	Set and enforce minimum stream flow levels

Figure 9-16 Solutions: ways to prevent or slow groundwater depletion. Which two of these solutions do you believe are the most important?

about the geological and ecological impacts of pumping water from deep aquifers. *Second*, we have no international water treaties governing rights to, and ownership of, water found under several different countries. Without such treaties, there could be legal and physical conflicts over who has the right to tap into and use these valuable resources.

How Useful Is Desalination? A Costly Option

Removing salt from seawater will probably not be done widely because of high costs and what to do with the resulting salt.

Desalination involves removing dissolved salts from ocean water or from brackish (slightly salty) water in aquifers or lakes. It is another way to increase supplies of fresh water.

One method for desalinating water is *distillation*—heating salt water until it evaporates, leaves behind salts in solid form, and condenses as fresh water. Another method is *reverse osmosis*—pumping salt water at high pressure through a thin membrane with pores that allow water molecules, but not most dissolved salts, to pass through. In effect, high pressure is used to push fresh water out of salt water.

There are about 13,500 desalination plants in 120 countries, especially in the arid, desert nations of the Middle East, North Africa, the Caribbean, and the Mediterranean. These plants meet less than 0.3% of the world's water needs.

There are two major problems with the widespread use of desalination. One is the high cost, because it takes a lot of energy to desalinate water. Currently, desalinating water costs two to three times as much as the conventional purification of fresh water, although recent advances in reverse osmosis have brought the energy costs down somewhat.

The second problem is that desalination produces large quantities of briny wastewater that contains lots of salt and other minerals. Dumping concentrated brine into a nearby ocean increases the salinity of the ocean water, which threatens food resources and aquatic life in the vicinity. Dumping it on land could contaminate groundwater and surface water.

Bottom line: Currently, significant desalination is practical only for wealthy and water-short countries and cities that can afford its high cost.

Scientists are working to develop new membranes for reverse osmosis that can separate water from salt more efficiently and under less pressure. If successful, this strategy could bring down the cost of using desalination to produce drinking water. Even so, desalinated water probably will not be cheap enough to irrigate conventional crops or meet much of the world's demand for fresh water unless scientists can figure out how to use solar energy or other means to desalinate seawater cheaply and how to safely dispose of the salt left behind.

9-4 REDUCING WATER WASTE

What Are the Benefits of Reducing Water Waste? A Win-Win Solution

We waste about two-thirds of the water we use, but we could reduce such waste to about 15%.

Mohamed El-Ashry of the World Resources Institute estimates that *65–70% of the water people use throughout the world is lost through evaporation, leaks, and other losses.* The United States, the world's largest user of water, does slightly better but still loses about half of the water it withdraws. El-Ashry believes it is economically and technically feasible to reduce such water losses to 15%, thereby meeting most of the world's water needs for the foreseeable future.

This win-win solution will also decrease the burden on wastewater plants and reduce the need for expensive dams and water transfer projects that destroy wildlife habitats and displace people. It will also slow depletion of groundwater aquifers and save energy and money.

According to water resource experts, the main cause of water waste is that *we charge too little for water.* Such *underpricing* is mostly the result of government subsidies that provide irrigation water, electricity, and diesel fuel for farmers to pump water from rivers and aquifers at below-market prices.

Subsidies keep the price of water so low that users have little or no financial incentive to invest in well-known water-saving technologies. According to water resource expert Sandra Postel, "By heavily subsidizing water, governments give out the false message that it is abundant and can afford to be wasted—even as rivers are drying up, aquifers are being depleted, fisheries are collapsing, and species are going extinct."

But farmers, industries, and others benefiting from government water subsidies argue that they promote settlement and agricultural production in arid and semiarid areas, stimulate local economies, and help lower prices of food, manufactured goods, and electricity for consumers.

Most water resource experts believe that in this century's era of water scarcity in many areas, governments will have to make the unpopular decision to raise water prices. China did this in 2002 because it faced water shortages in most of its major cities, rivers running dry, and falling water tables in key agricultural areas.

Higher water prices encourage water conservation but make it harder for low-income farmers and city dwellers to buy enough water to meet their needs. On the other hand, when South Africa raised water prices it established *lifeline* rates that give each household a set amount of water at a low price to meet basic needs. When users exceed this amount the price rises.

The second major cause of water waste is *lack of government subsidies for improving the efficiency of water use.* A basic rule of economics is that you get more of what you reward. Subsidies for efficient water use would sharply reduce water waste.

X *HOW WOULD YOU VOTE?* Should water prices be raised sharply to help reduce water waste? Cast your vote online at http://biology.brookscole.com/miller7.

How Can We Waste Less Water in Irrigation? Deliver Water Precisely When and Where It is Needed

About 60% of the world's irrigation water is wasted, but several improved irrigation techniques could reduce this waste to 5–20%.

About 60% of the irrigation water applied throughout the world does not reach the targeted crops and does not contribute to food production. Most irrigation systems obtain water from a groundwater well or a surface water source, which then flows by gravity through unlined ditches in crop fields so the crops can absorb the water (Figure 9-17, left). This *flood irrigation* method delivers far more water than needed for crop growth and typically loses 40% of the water through evaporation, seepage, and runoff.

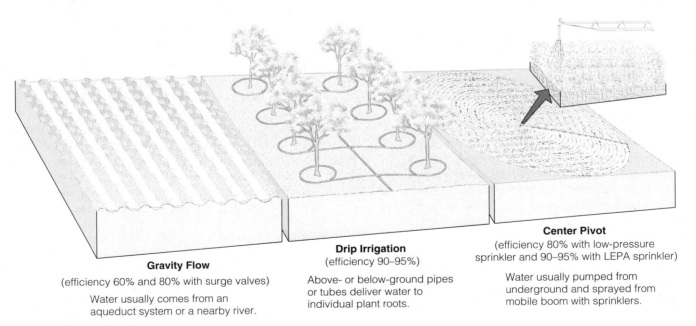

Gravity Flow
(efficiency 60% and 80% with surge valves)
Water usually comes from an aqueduct system or a nearby river.

Drip Irrigation
(efficiency 90–95%)
Above- or below-ground pipes or tubes deliver water to individual plant roots.

Center Pivot
(efficiency 80% with low-pressure sprinkler and 90–95% with LEPA sprinkler)
Water usually pumped from underground and sprayed from mobile boom with sprinklers.

Figure 9-17 Major *irrigation systems.* Because of high initial costs, center-pivot irrigation and drip irrigation are not widely used. This may change because of the development of new low-cost drip-irrigation systems.

More efficient and environmentally sound irrigation technologies exist that can greatly reduce water demands and waste on farms by delivering water more precisely to crops. One of these technologies is a *center-pivot low-pressure sprinkler* (Figure 9-17, right), which uses pumps to spray water on a crop. Typically, it allows 80% of the water to reach crops. Another method is *low-energy precision application (LEPA) sprinklers,* a form of center-pivot irrigation that puts 90–95% of the water where crops need it by spraying the water closer to the ground and in larger droplets than the center-pivot, low-pressure system. Another method is to use *soil moisture detectors* to water crops only when they need it.

Drip irrigation or *microirrigation systems* (shown in Figure 9-17, center) are the most efficient ways to deliver small amounts of water precisely to crops. These systems consist of a network of perforated plastic tubing installed at or below the ground level. Small holes or emitters in the tubing deliver drops of water at a slow and steady rate, close to the plant roots.

Drip irrigation is very efficient, with 90–95% of the water input reaching the crops. It is also adaptable because the flexible and lightweight tubing system can easily be fitted to match the patterns of crops in a field and left in place or moved around.

However, drip irrigation is used on just over 1% of the world's irrigated crop fields and 4% of those in the United States. But this percentage rises to 90% in Cyprus, 66% in Israel, and 13% in California. The main problem is that the capital cost of conventional drip irrigation systems is too high for most poor farmers and for use on low-value row crops.

Also, irrigation water is underpriced because of government subsidies. Raise water prices enough and drip irrigation would quickly be used to irrigate most of the world's crops. *Good news.* The capital cost of a new type of drip irrigation system is one-tenth as much per hectare as conventional drip systems.

Figure 9-18 lists other ways to reduce water waste in irrigating crops. Since 1950, water-short Israel has used many of these techniques to slash irrigation water waste by about 84% while irrigating 44% more land. Israel now treats and reuses 30% of its municipal sewage water for crop production and plans to increase this to 80% by 2025. The government also gradually removed most government water subsidies to raise the price of irrigation water to one of the highest in the world. Israelis also import most of their wheat and meat and concentrate on growing fruits, vegetables, and flowers that need less water.

Many of the world's poor farmers cannot afford to use most of the modern technological methods for increasing irrigation and irrigation efficiency. Instead, they use small-scale and low-cost traditional technologies.

Solutions

Reducing Irrigation Water Waste

- Lining canals bring water to irrigation ditches
- Leveling fields with lasers
- Irrigating at night to reduce evaporation
- Using soil and satellite sensors and computer systems to monitor soil moisture and add water only when necessary
- Polyculture
- Organic farming
- Growing water-efficient crops using drought-resistant and salt-tolerant crop varieties
- Irrigating with treated urban waste water
- Importing water-intensive crops and meat

Figure 9-18 Solutions: methods for reducing water waste in irrigation. Which two of these solutions do you believe are the most important?

Some (in Bangladesh, for example) use pedal-powered treadle pumps to move water through irrigation ditches. Others use buckets or small tanks with holes for drip irrigation.

How Can We Waste Less Water in Industry, Homes, and Businesses? Copy Nature, Raise Prices, Improve Efficiency

We can save water by changing to yard plants that need little water, using drip irrigation, raising water prices, fixing leaks, and using water-saving toilets and other appliances.

Figure 9-19 lists ways to use water more efficiently in industries, homes, and businesses. Many homeowners and businesses in water-short areas are replacing green lawns with vegetation adapted to a dry climate. This win-win approach, called *xeriscaping* (pronounced "ZER-i-scaping"), reduces water use by 30–85% and sharply reduces inputs of labor, fertilizer, and fuel. It also reduces polluted runoff, air pollution, and yard wastes.

About one-fifth of all U.S. public water systems do not have water meters and charge a single low rate for almost unlimited use of high-quality water. Many apartment dwellers have little incentive to conserve water because water use is included in their rent. In Boulder, Colorado, introducing water meters reduced water use by more than one-third.

Solutions

Reducing Water Waste

- Redesign manufacturing processes
- Landscape yards with plants that require little water
- Use drip irrigation
- Fix water leaks
- Use water meters and charge for all municipal water use
- Raise water prices
- Use waterless composting toilets
- Require water conservation in water-short cities
- Use water-saving toilets, showerheads, and front-loading clothes washers
- Collect and reuse household water to irrigate lawns and nonedible plants
- Purify and reuse water for houses, apartments, and office buildings

Figure 9-19 Solutions: methods of reducing water waste in industries, homes, and businesses. Which two of these solutions do you believe are the most important?

We can also save water by replacing the current system in which we use large amounts of water good enough to drink to dilute and wash or flush away industrial, animal, and household wastes with one that mimics the way nature deals with wastes. The FAO estimates that if current trends continue, within 40 years we will need the world's entire reliable flow of river water just to dilute and transport the wastes we produce.

One way to reduce the severity of this problem is to ban the discharge of industrial toxic wastes into municipal sewer systems. Another is to rely more on waterless composting toilets that convert human fecal matter to a small amount of dry and odorless soil-like humus material that can be removed from a composting chamber every year or so and returned to the soil as fertilizer. They work. I used one for almost two decades without any problems.

We can also return the nutrient-rich sludge produced by conventional waste treatment plants to the soil as a fertilizer. Banning the input of toxic industrial

How Can We Use Water More Sustainably? A Blue Revolution

We can use water more sustainably by cutting waste, raising water prices, preserving forests in water basins, and slowing population growth.

Sustainable water use is based on the commonsense principle stated in an old Inca proverb: "The frog does not drink up the pond in which it lives." Figure 9-20 lists ways to implement this principle.

Historically, our response to water scarcity has been to expand the supply by building more dams, transporting water from one area to another, and drilling more wells. These things will continue to some degree, but water resource experts project that the emphasis will begin shifting from increasing the water supply to reducing the water demand by using water more efficiently and stabilizing population.

The challenge in developing such a *blue revolution* is to implement a mix of strategies. One involves using

Solutions

Sustainable Water Use

- Not depleting aquifers
- Preserving ecological health of aquatic systems
- Preserving water quality
- Integrated watershed management
- Agreements among regions and countries sharing surface water resources
- Outside party mediation of water disputes between nations
- Marketing of water rights
- Raising water prices
- Wasting less water
- Decreasing government subsidies for supplying water
- Increasing government subsidies for reducing water waste
- Slowing population growth

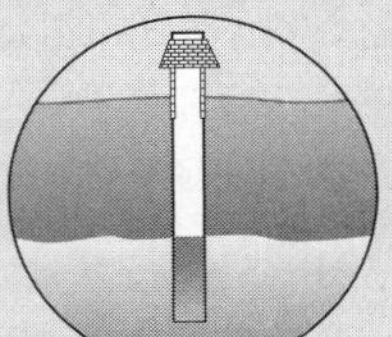

Figure 9-20 Solutions: methods for achieving more sustainable use of the earth's water resources.

technology to irrigate crops more efficiently and to save water in industries and homes. A second approach uses economic and political decisions to remove subsidies that cause water to be underpriced and thus wasted, while guaranteeing low prices for low-income consumers and adding subsidies that reward reduced water waste.

A third component is to switch to new waste treatment systems that accept only non-toxic wastes, use less or no water to treat wastes, return nutrients in plant and animal wastes to the soil, and mimic the ways that nature decomposes and recycles organic wastes. A fourth strategy is to leave enough water in rivers to protect wildlife, ecological processes, and the natural ecological services provided by rivers.

Each of us can help bring about this blue revolution by using and wasting less water. We can also support government policies that result in more sustainable use of the world's water and better ways to treat our industrial and household wastes. Figure 9-21 lists ways you can reduce your water use and waste.

What Can You Do?

Water Use and Waste

- Use water-saving toilets, showerheads, and faucet aerators.

- Shower instead of taking baths, and take short showers.

- Repair water leaks.

- Turn off sink faucets while brushing teeth, shaving, or washing.

- Wash only full loads of clothes or use the lowest possible water-level setting for smaller loads.

- Wash a car from a bucket of soapy water, and use the hose for rinsing only.

- If you use a commercial car wash, try to find one that recycles its water.

- Replace your lawn with native plants that need little if any watering.

- Water lawns and gardens in the early morning or evening.

- Use drip irrigation and mulch for gardens and flowerbeds.

- Use recycled (gray) water for watering lawns and houseplants and for washing cars.

Figure 9-21 What can you do? Ways you can reduce your use and waste of water. Which two of these practices do you believe are the most important?

9-5 TYPES, EFFECTS, AND SOURCES OF WATER POLLUTION

What Are the Major Types and Effects of Water Pollutants? Unseen Threats

Infectious bacteria, inorganic and organic chemicals, and excess heat pollute water.

Water pollution is any chemical, biological, or physical change in water quality that has a harmful effect on living organisms or makes water unsuitable for desired uses. Table 9-1 lists the major classes of water pollutants along with their human sources and their harmful effects.

What Are Point and Nonpoint Sources of Water Pollution? Concentrated and Diffuse Sources

Water pollution can come from single sources or a variety of dispersed sources.

Point sources discharge pollutants at specific locations through drain pipes, ditches, or sewer lines into bodies of surface water. Examples include factories, sewage treatment plants (which remove some but not all pollutants), underground mines, and oil tankers.

Because point sources are at specific places, they are fairly easy to identify, monitor, and regulate. Most developed countries control point-source discharges of many harmful chemicals into aquatic systems. But there is little control of such discharges in most developing countries.

Nonpoint sources are scattered and diffuse and cannot be traced to any single site of discharge. Examples include deposition from the atmosphere and runoff of chemicals into surface water from croplands, livestock feedlots, logged forests, urban streets, lawns, golf courses, and parking lots. They are usually large land areas or parts of the troposphere that pollute water by runoff, subsurface flow, or deposition from the atmosphere. There has been little progress in controlling nonpoint water pollution because of the difficulty and expense of identifying and controlling discharges from so many diffuse sources.

The leading sources of water pollution are agriculture, industries, and mining. *Agricultural activities* are by far the leading cause of water pollution. Sediment eroded from agricultural lands and overgrazed rangeland is the largest source. Other major agricultural pollutants include fertilizers and pesticides, bacteria from livestock and food processing wastes, and excess salt from soils of irrigated cropland.

Industrial facilities are another large source of water pollution. *Mining* is a third source. Surface mining disturbs the earth's surface, creating a major source of eroded sediments and runoff of toxic chemicals.

Table 9-1 Major Categories of Water Pollutants

INFECTIOUS AGENTS

Examples: Bacteria, viruses, protozoa, and parasitic worms

Major Human Sources: Human and animal wastes

Harmful Effects: Disease

OXYGEN-DEMANDING WASTES

Examples: Organic waste such as animal manure and plant debris that can be decomposed by aerobic (oxygen-requiring) bacteria

Major Human Sources: Sewage, animal feedlots, paper mills, and food processing facilities

Harmful Effects: Large populations of bacteria decomposing these wastes can degrade water quality by depleting water of dissolved oxygen. This causes fish and other forms of oxygen-consuming aquatic life to die.

INORGANIC CHEMICALS

Examples: Water-soluble (1) acids, (2) compounds of toxic metals such as lead (Pb), arsenic (As), and selenium (Se), and (3) salts such as sodium chloride (NaCl) in ocean water and fluorides (F^-) found in some soils

Major Human Sources: Surface runoff, industrial effluents, and household cleansers

Harmful Effects: Can (1) make fresh water unusable for drinking or irrigation, (2) cause skin cancers and crippling spinal and neck damage (F^-), (3) damage the nervous system, liver, and kidneys (Pb and As), (4) harm fish and other aquatic life, (5) lower crop yields, and (6) accelerate corrosion of metals exposed to such water.

ORGANIC CHEMICALS

Examples: Oil, gasoline, plastics, pesticides, cleaning solvents, detergents

Major Human Sources: Industrial effluents, household cleansers, surface runoff from farms and yards

Harmful Effects: Can (1) threaten human health by causing nervous system damage (some pesticides), reproductive disorders (some solvents), and some cancers (gasoline, oil, and some solvents) and (2) harm fish and wildlife.

PLANT NUTRIENTS

Examples: Water-soluble compounds containing nitrate (NO_3^-), phosphate (PO_4^{3-}), and ammonium (NH_4^+) ions

Major Human Sources: Sewage, manure, and runoff of agricultural and urban fertilizers

Harmful Effects: Can cause excessive growth of algae and other aquatic plants, which die, decay, deplete water of dissolved oxygen, and kill fish. Drinking water with excessive levels of nitrates lowers the oxygen-carrying capacity of the blood and can kill unborn children and infants ("blue-baby syndrome").

SEDIMENT

Examples: Soil, silt

Major Human Sources: Land erosion

Harmful Effects: Can (1) cloud water and reduce photosynthesis, (2) disrupt aquatic food webs, (3) carry pesticides, bacteria, and other harmful substances, (4) settle out and destroy feeding and spawning grounds of fish, and (5) clog and fill lakes, artificial reservoirs, stream channels, and harbors.

RADIOACTIVE MATERIALS

Examples: Radioactive isotopes of iodine, radon, uranium, cesium, and thorium

Major Human Sources: Nuclear and coal-burning power plants, mining and processing of uranium and other ores, nuclear weapons production, natural sources

Harmful Effects: Genetic mutations, miscarriages, birth defects, and certain cancers

HEAT (THERMAL POLLUTION)

Examples: Excessive heat

Major Human Sources: Water cooling of electric power plants and some types of industrial plants. Almost half of all water withdrawn in the United States each year is for cooling electric power plants.

Harmful Effects: Lowers dissolved oxygen levels and makes aquatic organisms more vulnerable to disease, parasites, and toxic chemicals. When a power plant first opens or shuts down for repair, fish and other organisms adapted to a particular temperature range can be killed by the abrupt change in water temperature—known as *thermal shock*.

Is the Water Safe to Drink? For Most but Not All

One of every five people in the world does not have access to safe drinking water.

Good news. About 95% of the people in developed countries and 74% of those in developing countries have access to clean drinking water.

Bad news. According to the WHO, about 1.4 billion people—about one in five—in developing countries do not have access to clean drinking water. As a result, each day about 9,300 people—most of them children under age 5—die prematurely from infectious diseases spread by contaminated water or the lack of water for adequate hygiene.

In Russia, half of all tap water is unfit to drink. In Africa, some 290 million people—about equal to the entire U.S. population—do not have access to safe drinking water.

The United Nations estimates it would cost about $23 billion a year over 8–10 years to bring low-cost safe water and sanitation to the 1.4 billion people who do not have it.

9-6 POLLUTION OF FRESHWATER STREAMS, LAKES, AND AQUIFERS

What Are the Water Pollution Problems of Streams? Pollution Overload and Low Flow

Flowing streams can recover from a moderate level of degradable water pollutants if their flows are not reduced.

Rivers and other flowing streams can recover rapidly from moderate levels of degradable, oxygen-demanding wastes and excess heat through a combination of dilution and biodegradation of such wastes by bacteria.

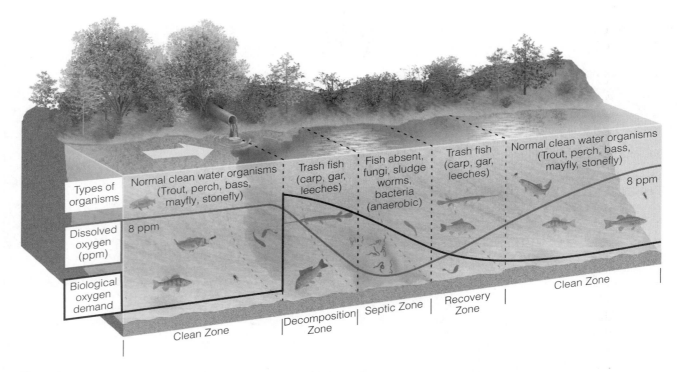

Figure 9-22 Natural capital: dilution and decay of degradable, oxygen-demanding wastes and heat in a stream, showing the oxygen sag curve and the curve of oxygen demand. Depending on flow rates and the amount of pollutants, streams recover from oxygen-demanding wastes and heat if they are given enough time and are not overloaded.

But this natural recovery process does not work if streams are overloaded with pollutants or when drought, damming, or water diversion for agriculture and industry reduce their flows. Also, these natural dilution and biodegradation processes do not eliminate slowly degradable and nondegradable pollutants.

In a flowing stream, the breakdown of degradable wastes by bacteria depletes dissolved oxygen and creates an *oxygen sag curve* (Figure 9-22). This reduces or eliminates populations of organisms with high oxygen requirements until the stream is cleansed of wastes. Similar oxygen sag curves can be plotted when heated water from industrial and power plants is discharged into streams.

What Have Developed Counties Done to Reduce Stream Pollution? Good and Bad News

Most developed countries have sharply reduced point-source pollution, but toxic chemicals and pollution from nonpoint sources are still problems.

Water pollution control laws enacted in the 1970s have greatly increased the number and quality of wastewater treatment plants in the United States and most other developed countries. Such laws also require industries to reduce or eliminate point-source discharges into surface waters.

These efforts have enabled the United States to hold the line against increased pollution by disease-causing agents and oxygen-demanding wastes in most of its streams. This is an impressive accomplishment given the rise in the country's economic activity, resource consumption, and population since passage of these laws.

One success story is the cleanup of Ohio's Cuyahoga River. It was so polluted that in 1959 and again in 1969, it caught fire and burned for several days as it flowed through Cleveland. The highly publicized image of this burning river prompted elected officials to enact laws limiting the discharge of industrial wastes into the river and sewage systems, and provide funds to upgrade sewage treatment facilities. Today the river is cleaner and is widely used by boaters and anglers. This accomplishment illustrates the power of bottom-up pressure by citizens to spur elected officials to change a severely polluted river into an economically and ecologically valuable public resource. Individuals matter!

There is also some *bad news.* Large fish kills and drinking water contamination still occur in parts of developed countries. Two causes of these problems are accidental or deliberate releases of toxic inorganic and organic chemicals by industries or mines and malfunctioning sewage treatment plants. A third cause is nonpoint runoff of pesticides and excess plant nutrients from cropland and animal feedlots.

What Have Developing Countries Done to Reduce Stream Pollution? Little Progress

Stream pollution in most developing countries is a serious and growing problem.

Available data indicate that stream pollution from discharges of untreated sewage and industrial wastes is a serious and growing problem in most developing countries. According to a 2003 report by the World Commission on Water in the 21st Century, half of the world's 500 rivers are heavily polluted, most of them running through developing countries. Most of these countries cannot afford to build waste treatment plants and do not have, or do not enforce, laws for controlling water pollution.

Industrial wastes and sewage pollute more than two-thirds of India's water resources and 54 of the 78 streams monitored in China. Only about 10% of the sewage produced in Chinese cities is treated. In Latin America and Africa, most streams passing through urban or industrial areas suffer from severe pollution.

What Are the Pollution Problems of Lakes? Too Little Flow and Mixing

Dilution of pollutants in lakes is less effective than in most streams because most lake water is not mixed well and has little flow.

In lakes and reservoirs, dilution of pollutants often is less effective than in streams for two reasons. One is that lakes and reservoirs often contain stratified layers that undergo little vertical mixing. The other is that they have little flow. The flushing and changing of water in lakes and large artificial reservoirs can take from 1 to 100 years, compared with several days to several weeks for streams.

This means that lakes and reservoirs are more vulnerable than streams to contamination by runoff or discharge of plant nutrients, oil, pesticides, and toxic substances such as lead, mercury, and selenium. These contaminants can kill bottom life and fish and birds that feed on contaminated aquatic organisms. Many toxic chemicals and acids also enter lakes and reservoirs from the atmosphere.

What Is Cultural Eutrophication and How Can It Be Reduced? Too Much of a Good Thing

Various human activities can overload lakes with plant nutrients, which decrease dissolved oxygen and kill some aquatic species.

Eutrophication is the name given to the natural nutrient enrichment of lakes, mostly from runoff of plant nutrients such as nitrates and phosphates from surrounding land. Over time, some lakes become more eutrophic, but others do not because of differences in the surrounding drainage basin.

Near urban or agricultural areas, human activities can greatly accelerate the input of plant nutrients to a lake. This process is called **cultural eutrophication.** It is mostly nitrate- and phosphate-containing effluents from various sources that cause such a change. These sources include runoff from farmland, animal feedlots, urban areas, and mining sites, and from the discharge of treated and untreated municipal sewage. Some nitrogen also reaches lakes by deposition from the atmosphere.

During hot weather or drought, this nutrient overload produces dense growths or "blooms" of organisms such as algae and cyanobacteria and thick growths of water hyacinths, duckweed, and other aquatic plants. These dense colonies of plant life can reduce lake productivity and fish growth by decreasing the input of solar energy needed for photosynthesis by the phytoplankton that support fish.

In addition, when the algae die, their decomposition by swelling populations of aerobic bacteria depletes dissolved oxygen in the surface layer of water near the shore and in the bottom layer. This oxygen depletion can kill fish and other aerobic aquatic animals. If excess nutrients continue to flow into a lake, anaerobic bacteria take over and produce gaseous decomposition products such as smelly, highly toxic hydrogen sulfide and flammable methane.

According to the U.S. Environmental Protection Agency (EPA), about one-third of the 100,000 medium to large lakes and about 85% of the large lakes near major population centers in the United States have some degree of cultural eutrophication. One-fourth of the lakes in China also suffer from cultural eutrophication.

There are several ways to *prevent* or *reduce* cultural eutrophication. They include using advanced (but expensive) waste treatment to remove nitrates and phosphates before wastewater enters lakes, banning or limiting the use of phosphates in household detergents and other cleaning agents, and using soil conservation and land-use control to reduce nutrient runoff.

There are also several ways to *clean up* lakes suffering from cultural eutrophication. Examples are mechanically removing excess weeds, controlling undesirable plant growth with herbicides and algicides, and pumping air through lakes and reservoirs to avoid oxygen depletion (an expensive and energy-intensive method).

As usual, pollution prevention is more effective and usually cheaper in the long run than cleanup. *Good news:* If excessive inputs of plant nutrients stop, a lake usually can return to its previous state (see Case Study that follows).

Case Study: Pollution in the Great Lakes—Hopeful Progress

Pollution of the Great Lakes has dropped significantly but there is a long way to go.

The five connected Great Lakes contain at least 95% of the fresh surface water in the United States and one-fifth of the world's fresh surface water. At least 38 million people in the United States and Canada obtain their drinking water from these lakes.

Despite their enormous size, these lakes are vulnerable to pollution from point and nonpoint sources. One reason is that less than 1% of the water entering the Great Lakes flows out to the St. Lawrence River each year. Another reason is that in addition to land runoff, these lakes get atmospheric deposition of large quantities of acids, pesticides, and other toxic chemicals, often blown in from hundreds or thousands of kilometers away.

By the 1960s, many areas of the Great Lakes were suffering from severe cultural eutrophication, huge fish kills, and contamination from bacteria and a variety of toxic industrial wastes. The impact on Lake Erie was particularly intense because it is the shallowest of the Great Lakes and has the highest concentrations of people and industrial activity along its shores. Many bathing beaches had to be closed, and by 1970 the lake had lost most of its native fish.

Since 1972, Canada and the United States have joined forces and spent more than $20 billion on a Great Lakes pollution control program. This program has decreased algal blooms and increased dissolved oxygen levels, thus increasing sport and commercial fishing catches in Lake Erie, and allowing most swimming beaches to reopen.

These improvements occurred mainly because of new or upgraded sewage treatment plants, better treatment of industrial wastes, and bans on the use of detergents, household cleaners, and water conditioners that contained phosphates.

Despite this important progress many problems remain. According to a 2000 survey by the EPA, more than three-fourths of the shoreline of the Great Lakes is not clean enough for swimming or for supplying drinking water. Nonpoint land runoff of pesticides and fertilizers from urban sprawl now surpasses industrial pollution as the greatest threat to the lakes. And sediments in 26 toxic hot spots remain heavily polluted. About half of the toxic compounds entering the lakes come from atmospheric deposition of pesticides, mercury from coal-burning plants, and other toxic chemicals from as far away as Mexico and Russia.

A recent survey by Wisconsin biologists found that one fish in four taken from the Great Lakes is unsafe for human consumption. Another problem has been an 80% drop in EPA funding for cleanup of the Great Lakes since 1992.

Some environmentalists call for banning the use of toxic chlorine compounds such as bleach in the pulp and paper industry around the Great Lakes. They would also ban new incinerators (which can release toxic chemicals into the atmosphere) in the area, and they would stop the discharge into the lakes of 70 toxic chemicals that threaten human health and wildlife. Officials of the industries involved have successfully opposed such bans.

Why Is Groundwater Pollution Such a Serious Problem? Not Easily Cleaned

Groundwater can become contaminated with a variety of chemicals because it cannot effectively cleanse itself and dilute and disperse pollutants.

According to many scientists, a serious threat to human health is the out-of-sight pollution of groundwater, a prime source of water for drinking and irrigation. Studies show that groundwater pollution comes from numerous sources (Figure 9-23). People who dump or spill gasoline, oil, and paint thinners and other organic solvents onto the ground also contaminate groundwater.

When groundwater becomes contaminated, it cannot cleanse itself of *degradable wastes* as flowing surface water does (Figure 9-22). One reason is that groundwater flows so slowly—usually less than 0.3 meter or 1 foot per day—that contaminants are not diluted and dispersed effectively. Another problem is that groundwater usually has much lower concentrations of dissolved oxygen (which helps decompose many contaminants) and smaller populations of decomposing bacteria. Also, the usually cold temperatures of groundwater slow down chemical reactions that decompose wastes.

Thus it can take hundreds to thousands of years for contaminated groundwater to cleanse itself of *degradable* wastes. On a human time scale, *nondegradable wastes* (such as toxic lead, arsenic, and fluoride) are there permanently.

What Is the Extent of Groundwater Pollution? Uncertain Overall but Serious in Some Areas

Leaks from chemical storage ponds, underground storage tanks, piping used to inject hazardous waste underground, and seepage of agricultural fertilizers can contaminate groundwater.

On a global scale, we do not know much about groundwater pollution because few countries go to the great expense of locating, tracking, and testing aquifers. But scientific studies in scattered parts of the world provide us with some *bad news*.

According to the EPA and the U.S. Geological Survey, one or more organic chemicals contaminate

Figure 9-23 Natural capital degradation:
principal sources of groundwater contamination in the United States. Another source is saltwaater intrusion from excessive groundwater withdrawal (Figure 9-15).

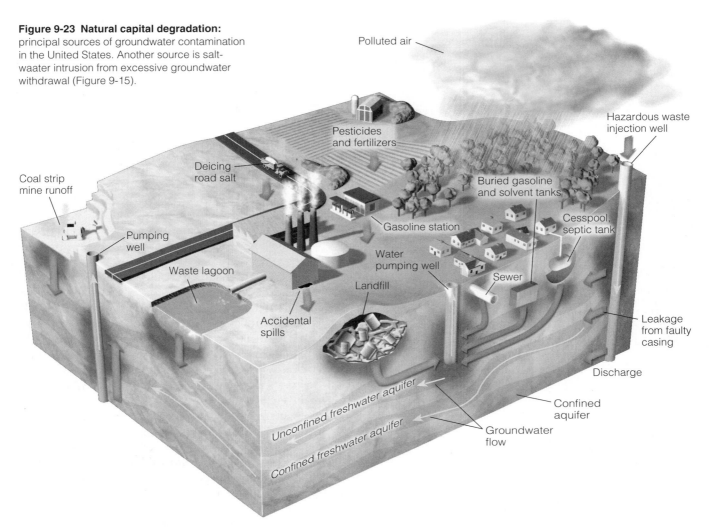

about 45% of *municipal* groundwater supplies in the United States. An EPA survey of 26,000 industrial waste ponds and lagoons in the United States found that one-third of them had no liners to prevent toxic liquid wastes from seeping into aquifers. One-third of these sites are within 1.6 kilometers (1 mile) of a drinking water well. The U.S. General Accounting Office estimated in 2002 that at least 76,000 underground tanks storing gasoline, diesel fuel, home heating oil, and toxic solvents were leaking their contents into groundwater in the United States.

During this century, scientists expect many of the millions of such tanks installed around the world in recent decades to corrode, leak, contaminate groundwater, and become a major global health problem. Determining the extent of a leak from a single underground tank can cost $25,000–250,000, and cleanup costs range from $10,000 to more than $250,000. If the chemical reaches an aquifer, effective cleanup is often not possible or is too costly. *Bottom line:* wastes we think we have thrown away or stored safely can escape and come back to haunt us.

Toxic *arsenic (As)* contaminates drinking water when a well is drilled into aquifers where soils and

rock are naturally rich in arsenic. According to the WHO, more than 112 million people are drinking water with arsenic levels 5–100 times the WHO standard, mostly in Bangladesh, China, and India's state of West Bengal.

There is also concern over arsenic levels in drinking water in parts of the United States. The U.S. plans to lower the acceptable level of arsenic in drinking water from 50 to 10 parts per billion (ppb). But according to the WHO and other scientists, even the 10-ppb standard is not safe. Many scientists call for lowering the standard to 3–5 ppb, but this would be very expensive.

Solutions: How Can We Protect Groundwater? Monitor and Say No

Prevention is the most effective and affordable way to protect groundwater from pollutants.

Figure 9-24 (p. 192) lists ways to prevent and clean up groundwater contamination. Pumping polluted groundwater to the surface, cleaning it up, and returning it to the aquifer is very expensive. Thus, *preventing contamination is the most effective and cheapest way to protect groundwater resources* (Figure 9-24, left).

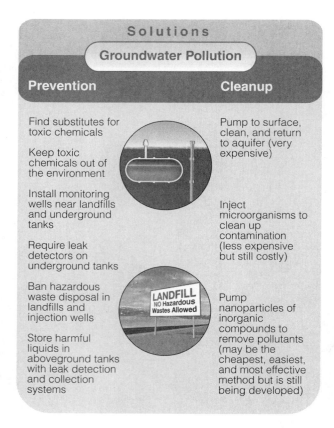

Solutions

Groundwater Pollution

Prevention	Cleanup
Find substitutes for toxic chemicals	Pump to surface, clean, and return to aquifer (very expensive)
Keep toxic chemicals out of the environment	
Install monitoring wells near landfills and underground tanks	Inject microorganisms to clean up contamination (less expensive but still costly)
Require leak detectors on underground tanks	
Ban hazardous waste disposal in landfills and injection wells	Pump nanoparticles of inorganic compounds to remove pollutants (may be the cheapest, easiest, and most effective method but is still being developed)
Store harmful liquids in aboveground tanks with leak detection and collection systems	

LANDFILL
NO Hazardous Wastes Allowed

Figure 9-24 Solutions: methods for preventing and cleaning up contamination of ground water. Which two of these solutions do you believe are the most important?

Underground tanks in the United States and a number of other developed countries are now strictly regulated. In the United States, thousands of old leaking tanks from gasoline stations and other facilities have been removed and the surrounding soil and groundwater have been treated to remove gasoline. This is expensive but there is little choice because the groundwater has already been contaminated.

9-7 OCEAN POLLUTION

How Much Pollution Can the Oceans Tolerate? We Do Not Know.

Oceans, if they are not overloaded, can disperse and break down large quantities of degradable pollutants.

The oceans can dilute, disperse, and degrade large amounts of raw sewage, sewage sludge, oil, and some types of degradable industrial waste, especially in deep-water areas. Also, some forms of marine life have been affected less by some pollutants than expected.

This has led some scientists to suggest it is safer to dump sewage sludge and most other harmful wastes into the deep ocean than to bury them on land or burn them in incinerators. Other scientists disagree, pointing out we know less about the deep ocean than we do about the moon. They add that dumping harmful wastes in the ocean would delay urgently needed pollution prevention and promote further degradation of this vital part of the earth's life-support system.

How Do Pollutants Affect Coastal Areas? More People and Development Equal More Pollution.

Pollution of coastal waters near heavily populated areas is a serious problem.

Coastal areas—especially wetlands and estuaries, coral reefs, and mangrove swamps—bear the brunt of our enormous inputs of wastes into the ocean (Figure 9-25). This is not surprising because about 40% of the world's population lives on or within 100 kilometers (62 miles) of the coast and 14 of the world's 15 largest metropolitan areas (each with 10 million people or more) are near coastal waters (Figure 5-12, p. 89).

In most coastal developing countries and in some coastal developed countries, municipal sewage and industrial wastes are dumped into the sea without treatment. For example, about 85% of the sewage from large cities along the Mediterranean Sea (with a coastal population of 200 million people during tourist season) is discharged into the sea untreated. This causes widespread beach pollution and shellfish contamination.

Recent studies of some U.S. coastal waters have found vast colonies of human viruses from raw sewage, effluents from sewage treatment plants (which do not remove viruses), and leaking septic tanks. According to one study, about one-fourth of the people using coastal beaches in the United States develop ear infections, sore throats, eye irritations, respiratory disease, or gastrointestinal disease.

Runoffs of sewage and agricultural wastes into coastal waters introduce large quantities of nitrate and phosphate plant nutrients, which can cause explosive growth of harmful algae. These *harmful algal blooms* (HABs) are called red, brown, or green toxic tides, depending on their color. They can release waterborne and airborne toxins that damage fisheries, kill some fish-eating birds, reduce tourism, and poison seafood.

Case Study: The Chesapeake Bay— An Estuary in Trouble

Pollutants from six states contaminate the shallow Chesapeake Bay estuary, but cooperative efforts have reduced some of the pollution inputs.

Since 1960, the Chesapeake Bay—America's largest estuary—has been in serious trouble from water pollution, mostly because of human activities. One problem is that between 1940 and 2004, the number of people

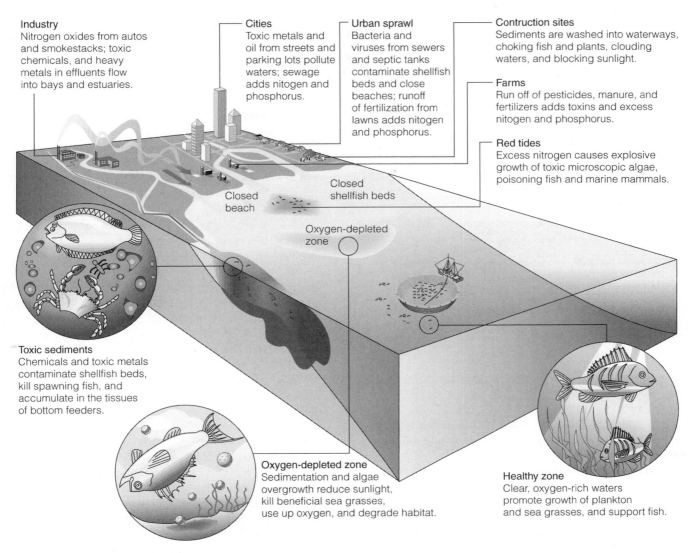

Industry
Nitrogen oxides from autos and smokestacks; toxic chemicals, and heavy metals in effluents flow into bays and estuaries.

Cities
Toxic metals and oil from streets and parking lots pollute waters; sewage adds nitogen and phosphorus.

Urban sprawl
Bacteria and viruses from sewers and septic tanks contaminate shellfish beds and close beaches; runoff of fertilization from lawns adds nitrogen and phosphorus.

Contruction sites
Sediments are washed into waterways, choking fish and plants, clouding waters, and blocking sunlight.

Farms
Run off of pesticides, manure, and fertilizers adds toxins and excess nitogen and phosphorus.

Red tides
Excess nitrogen causes explosive growth of toxic microscopic algae, poisoning fish and marine mammals.

Closed beach

Closed shellfish beds

Oxygen-depleted zone

Toxic sediments
Chemicals and toxic metals contaminate shellfish beds, kill spawning fish, and accumulate in the tissues of bottom feeders.

Oxygen-depleted zone
Sedimentation and algae overgrowth reduce sunlight, kill beneficial sea grasses, use up oxygen, and degrade habitat.

Healthy zone
Clear, oxygen-rich waters promote growth of plankton and sea grasses, and support fish.

Figure 9-25 Natural capital degradation: how residential areas, factories, and farms contribute to the pollution of coastal waters and bays.

living in the Chesapeake Bay area grew from 3.7 million to 17 million, and may soon reach 18 million.

Another problem is that the estuary receives wastes from point and nonpoint sources scattered throughout a huge drainage basin that includes 9 large rivers and 141 smaller streams and creeks in parts of six states (Figure 9-26, p. 194). Also, the bay has become a huge pollution sink because it is quite shallow, and only 1% of the waste entering it is flushed into the Atlantic Ocean.

Phosphate and nitrate levels have risen sharply in many parts of the bay, causing algal blooms and oxygen depletion. Commercial harvests of its once-abundant oysters, crabs, and several important fish have fallen sharply since 1960 because of a combination of pollution, overfishing, and disease.

Studies show that point sources, primarily sewage treatment plants and industrial plants (often in violation of their discharge permits), account for about 60% by weight of the phosphates. Nonpoint sources—mostly runoff of fertilizer and animal wastes from urban, suburban, and agricultural land and deposition from the atmosphere—account for about 60% by weight of the nitrates.

In 1983, the United States implemented the Chesapeake Bay Program. It is the country's most ambitious attempt at *integrated coastal management* in which citizens' groups, communities, state legislatures, and the federal government have worked together to reduce pollution inputs into the bay. Strategies include establishing land-use regulations in the bay's six watershed states to reduce agricultural and urban runoff, banning phosphate detergents, upgrading sewage treatment plants, and better monitoring of industrial discharges. In addition, wetlands are being restored and large areas of the bay are being

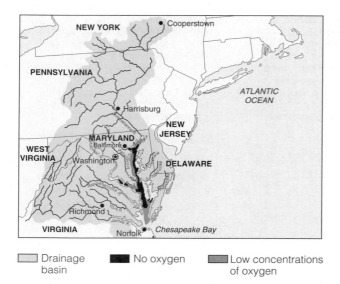

☐ Drainage basin ■ No oxygen ▨ Low concentrations of oxygen

Figure 9-26 Natural capital: *Chesapeake Bay,* the largest estuary in the United States, is severely degraded as a result of water pollution from point and nonpoint sources in six states and from deposition of air pollutants.

replanted with sea grasses to help filter out nutrients and other pollutants.

This hard work has paid off. Between 1985 and 2000, phosphorus levels declined 27%, nitrogen levels dropped 16%, and grasses growing on the bay's floor have made a comeback. This is a significant achievement given the increasing population in the watershed and the fact that nearly 40% of the nitrogen inputs come from the atmosphere.

However, there is still a long way to go, and a sharp drop in state and federal funding has slowed progress. But despite some setbacks, the Chesapeake Bay Program shows what can be done when diverse groups work together and use adaptive ecosystem management (Figure 6-20, p. 119) to achieve goals that benefit both wildlife and people.

What Are the Effects of Oil on Ocean Life? Land Activities, Not Tankers, Are the Major Culprit.

Most oil pollution comes from human activities on land, such as leaks and spills by industries and people changing and dumping their own motor oil.

Crude petroleum (oil as it comes out of the ground) and *refined petroleum* (fuel oil, gasoline, and other processed petroleum products) reach the ocean from a number of sources.

Tanker accidents and blowouts at offshore drilling rigs (when oil escapes under high pressure from a borehole in the ocean floor) get most of the publicity because of their high visibility. But *studies show that most*

ocean oil pollution comes from activities on land. According to a 2002 study by the Pew Oceans Commission, every 8 months an amount of oil equal to that spilled by the *Exxon Valdez* tanker drains from the land into the oceans. Almost half (some experts estimate 90%) of the oil reaching the oceans is waste oil dumped, spilled, or leaked onto the land or into sewers by cities, industries, and people changing their own motor oil.

Volatile organic hydrocarbons in oil immediately kill a number of aquatic organisms, especially in their vulnerable larval forms. Some other chemicals in oil form tar-like globs that float on the surface and coat the feathers of birds (especially diving birds) and the fur of marine mammals. This oil coating destroys their natural insulation and buoyancy, causing many of them to drown or die of exposure from loss of body heat.

Heavy oil components that sink to the ocean floor or wash into estuaries can smother bottom-dwelling organisms such as crabs, oysters, mussels, and clams or make them unfit for human consumption. Some oil spills have killed reef corals.

Research shows that most (but not all) forms of marine life recover from exposure to large amounts of *crude oil* within about 3 years. But recovery from exposure to *refined oil,* especially in estuaries and salt marshes, can take 10–15 years. Oil slicks that wash onto beaches can have a serious economic impact on coastal residents, who lose income from fishing and tourist activities.

If they are not too large, oil spills can be partially cleaned up by mechanical (floating booms, skimmer boats, and absorbent devices such as large pillows filled with feathers or hair), chemical, fire, and natural methods (such as using cocktails of natural bacteria to speed up oil decomposition).

But scientists estimate that current methods can recover no more than 15% of the oil from a major spill. This explains why preventing oil pollution is the most effective and, in the long run, the least costly approach.

Solutions: How Can We Protect Coastal Waters? Think Prevention

Preventing or reducing the flow of pollution from the land and from streams emptying into the ocean is the key to protecting the oceans.

Figure 9-27 list ways analysts have suggested for preventing and reducing excessive pollution of coastal waters. Study this figure carefully.

The key to protecting oceans is to reduce the flow of pollution from the land and from streams emptying into the ocean. Thus, ocean pollution control must be linked with land-use and air pollution policies because about one-third of all pollutants entering the ocean worldwide come from air emissions from land-based sources.

Solutions

Coastal Water Pollution

Prevention	Cleanup
Reduce input of toxic pollutants	Improve oil-spill cleanup capabilities
Separate sewage and storm lines	
Ban dumping of wastes and sewage by maritime and cruise ships in coastal waters	Sprinkle nanoparticles over an oil or sewage spill to dissolve the oil or sewage without creating harmful byproducts (still under development)
Ban ocean dumping of sludge and hazardous dredged material	
Protect sensitive areas from development, oil drilling, and oil shipping	Require at least secondary treatment of coastal sewage
Regulate coastal development	
Recycle used oil	Use wetlands, solar-aquatic, or other methods to treat sewage
Require double hulls for oil tankers	

Figure 9-27 Solutions: methods for preventing and cleaning up excessive pollution of coastal waters. Which two of these solutions do you believe are the most important?

9-8 PREVENTING AND REDUCING SURFACE WATER POLLUTION

How Can We Reduce Surface Water Pollution from Nonpoint Sources? Emphasize Prevention

The key to reducing nonpoint pollution, most of it from agriculture, is to prevent it from reaching bodies of surface water.

There are a number of ways to reduce nonpoint water pollution, most of it from agriculture. Farmers can reduce soil erosion, especially by keeping cropland covered with vegetation and by reforesting critical watersheds. They can also reduce fertilizer running off into surface waters and leaching into aquifers by using slow-release fertilizer, using none on steeply sloped land, and planting buffer zones of vegetation between cultivated fields and nearby surface water.

Applying pesticides only when needed and relying more on integrated pest control (p. 168) can reduce pesticide runoff. Farmers can control runoff and infiltration of manure from animal feedlots by planting buffers and locating feedlots and animal waste sites away from steeply sloped land, surface water, and flood zones.

✗ *HOW WOULD YOU VOTE?* Should we greatly increase efforts to reduce water pollution from nonpoint sources? Cast your vote online at http://biology.brookscole.com/miller7.

How Can We Reduce Water Pollution from Point Sources? Legal and Market Approaches

Most developed countries use laws to set water pollution standards, but in most developing countries such laws do not exist or are poorly enforced.

The Federal Water Pollution Control Act of 1972 (renamed the Clean Water Act when it was amended in 1977) and the 1987 Water Quality Act form the basis of U.S. efforts to control pollution of the country's surface waters. The Clean Water Act sets standards for allowed levels of key water pollutants and requires polluters to get permits specifying how much of various pollutants they can discharge into aquatic systems.

The EPA is also experimenting with a *discharge trading policy* that uses market forces to reduce water pollution in the United States. Under this program, a water pollution source is allowed to pollute at higher levels than allowed in its permits by buying credits from permit holders with pollution levels below their allowed levels.

Some environmentalists support discharge trading. But they warn that such a system is no better than the caps set for total pollution levels in various areas, and call for careful scrutiny of the cap levels. They also warn that discharge trading could allow pollutants to build up to dangerous levels in areas where credits are bought. In addition, they call for gradually lowering the caps to encourage prevention of water pollution and development of better technology for controlling water pollution, neither of which is a part of the current EPA discharge trading system.

According to Sandra Postel, director of the Global Water Policy Project, most cities in developing countries discharge 80–90% of their untreated sewage directly into rivers, streams, and lakes, which are used for drinking water, bathing, and washing clothes.

How Can We Reduce Water Pollution from Point Sources? The Technological Approach

Septic tanks and various levels of sewage treatment can reduce point-source water pollution.

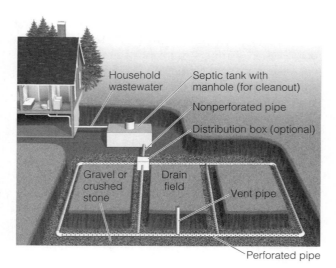

Figure 9-28 Solutions: *septic tank system* used for disposal of domestic sewage and wastewater in rural and suburban areas. This system separates solids from liquids, digests organic matter and large solids, and discharges the liquid wastes in a network of buried pipes with holes located over a large drainage or absorption field. As these wastes drain from the pipes and percolate downward, the soil filters out some potential pollutants, and soil bacteria decompose biodegradable materials. To be effective, septic tank systems must be properly installed in soils with adequate drainage, not placed too close together or too near well sites, and pumped out when the settling tank becomes full.

In rural and suburban areas with suitable soils, sewage from each house usually is discharged into a **septic tank** (Figure 9-28). About one-fourth of all homes in the United States are served by septic tanks.

In U.S. urban areas, most waterborne wastes from homes, businesses, factories, and storm runoff flow through a network of sewer pipes to *wastewater* or *sewage treatment plants.* Raw sewage reaching a treatment plant typically undergoes one or both of two levels of wastewater treatment. One is **primary sewage treatment.** It is a *physical* process that uses screens and a grit tank to remove large floating objects and solids such as sand and rock, and a settling tank that allows suspended solids to settle out as sludge (Figure 9-29). A second level is called **secondary sewage** treatment. It is a *biological* process in which aerobic bacteria remove up to 90% of dissolved and biodegradable, oxygen-demanding organic wastes.

Because of the Clean Water Act, most U.S. cities have combined primary and secondary sewage treatment plants. According to the EPA, however, at least two-thirds of these plants have at times violated water pollution regulations. Also, 500 cities have failed to meet federal standards for sewage treatment plants, and 34 East Coast cities simply screen out large floating objects from their sewage before discharging it into coastal waters.

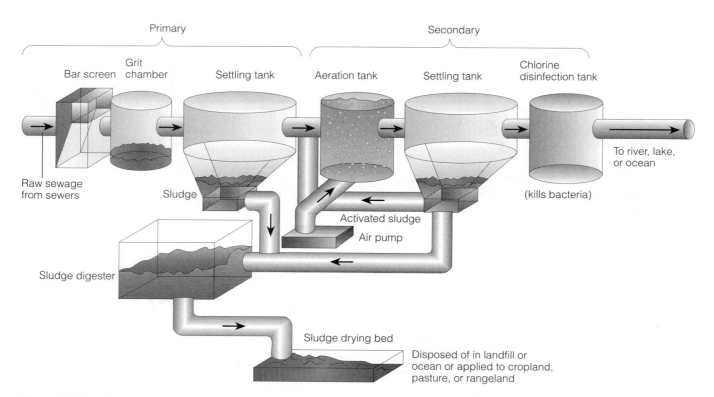

Figure 9-29 Solutions: primary and secondary sewage treatment.

Before discharge, water from primary, secondary, or more advanced treatment undergoes *bleaching* to remove water coloration and *disinfection* to kill disease-carrying bacteria and some but not all viruses. The usual method for doing this is *chlorination.* But chlorine can react with organic materials in water to form small amounts of chlorinated hydrocarbons. Some of these chemicals cause cancers in test animals and may damage the human nervous, immune, and endocrine systems.

Use of other disinfectants, such as ozone and ultraviolet light, is increasing but they cost more and their effects do not last as long as chlorination.

How Can We Improve Sewage Treatment? Eliminate Toxic Chemicals

Preventing toxic chemicals from reaching sewage treatment plants would eliminate such chemicals from the sludge and water discharged from such plants.

Environmental scientist Peter Montague calls for redesigning the sewage treatment system. The idea is to prevent toxic and hazardous chemicals from reaching sewage treatment plants and thus from getting into sludge and the water discharged from such plants.

He suggests several ways to do this. One is to require industries and businesses to remove toxic and hazardous wastes from water sent to municipal sewage treatment plants. Another is to encourage industries to reduce or eliminate the use and waste of toxic chemicals.

Another suggestion is to have more households, apartment buildings, and offices eliminate sewage outputs by switching to waterless *composting toilet systems* that are installed, maintained, and managed by professionals. Such systems would be cheaper to install and maintain than current sewage systems because they do not require vast systems of underground pipes connected to centralized sewage treatment plants. They also save large amounts of water.

Solutions: How Can We Treat Sewage by Working with Nature? Ecological Purification

Natural and artificial wetlands and other ecological systems can be used to treat sewage.

John Todd has developed an ecological approach to treating sewage, which he calls *living machines.* This ecological purification process begins when sewage flows into a passive solar greenhouse or outdoor sites containing rows of large open tanks populated by an increasingly complex series of organisms. In the first set of tanks, algae and microorganisms decompose organic wastes, with sunlight speeding up the process. Water

hyacinths, cattails, bulrushes, and other aquatic plants growing in the tanks take up the resulting nutrients.

After flowing though several of these natural purification tanks, the water passes through an artificial marsh of sand, gravel, and bulrushes to filter out algae and remaining organic waste. Some of the plants also absorb (sequester) toxic metals such as lead and mercury and secrete natural antibiotic compounds that kill pathogens.

Next, the water flows into aquarium tanks. Snails and zooplankton in these tanks consume microorganisms and are in turn consumed by crayfish, tilapia, and other fish that can be eaten or sold as bait. After ten days, the clear water flows into a second artificial marsh for final filtering and cleansing.

The water can be made pure enough to drink by using ultraviolet light or by passing the water through an ozone generator, usually immersed out of sight in an attractive pond or wetland habitat. Operating costs are about the same as for a conventional sewage treatment plant.

Some communities work with nature by using nearby natural wetlands to treat sewage, and others create artificial wetlands for such purposes (Solutions, p. 198).

Case Study: How Successful Has the United States Been in Reducing Water Pollution from Point Sources? Good and Bad News

Water pollution laws have significantly improved water quality in many U.S. streams and lakes, but there is a long way to go.

Great news. According to the EPA, the Clean Water Act of 1972 led to a number of improvements in U.S. water quality. Between 1992 and 2002, the number of Americans served by community water systems that met federal health standards increased from 79% to 94%. Also, between 1972 and 2002, the percentage of U.S. stream lengths found to be fishable and swimmable increased from 36% to 60% of those tested. And the amount of topsoil lost through agricultural runoff was cut by about 1.1 billion metric tons (1 billion tons) annually. In addition, between 1972 and 2002, the proportion of the U.S. population served by sewage treatment plants increased from 32% to 74%. And between 1974 and 2002, annual wetland losses decreased by 80%. These are impressive achievements given the increases in the U.S. population and per capita consumption of water and other resources since 1972.

Bad news. In 2000, the EPA found that 45% of the country's lakes and 40% of the streams surveyed were too polluted for swimming or fishing. Runoff of

Using Wetlands to Treat Sewage

SOLUTIONS

More than 150 cities and towns in the United States use natural and artificial wetlands to treat sewage as a low-tech, low-cost alternative to expensive waste treatment plants. For example, the coastal town of Arcata, California, created some 65 hectares (160 acres) of wetlands between the town and the adjacent Humboldt Bay. The marshes and ponds, developed on this land that was once a dump, act as an inexpensive natural waste treatment plant. The project cost less than half the estimated cost of a conventional treatment plant.

Here's how it works. First, sewage goes to sedimentation tanks, where the solids settle out as sludge that is removed and processed for use as fertilizer. Next, the liquid is pumped into oxidation ponds, where bacteria break down remaining wastes. After a month or so, the water is released into the artificial marshes, where plants and bacteria carry out further filtration and cleansing. Then the purified water flows into the Humboldt Bay with its abundant marine life.

The marshes and lagoons also serve as an Audubon Society bird sanctuary and provide habitats for thousands of otters, seabirds, and marine animals. The town even celebrates its natural sewage treatment system with an annual "Flush with Pride" festival.

Critical Thinking

List some possible drawbacks to creating artificial wetlands to treat sewage. Do these drawbacks outweigh the benefits? Explain.

animal wastes from hog, poultry, and cattle feedlots and meat processing facilities pollutes 7 of every 10 U.S. rivers. Most livestock wastes are not treated and are stored in lagoons that sometimes leak. They can also overflow or rupture as a result of excessive rainfall and spill their contents into nearby streams and rivers and sometimes into residential areas.

Fish caught in more than 1,400 different waterways and more than a fourth of the nation's lakes are unsafe to eat because of high levels of pesticides, mercury, and other toxic substances. And a 2003 internal study by the EPA found that at least half of the country's 6,600 largest industrial facilities and municipal wastewater treatment plants have illegally discharged toxic or biological wastes into waterways for years without government enforcement actions or fines.

Should the U.S. Clean Water Act Be Strengthened or Weakened? A Raging Controversy

Some want to strengthen the Clean Water Act while others want to weaken it.

Some environmentalists and a 2001 report by the EPA's inspector general call for the Clean Water Act to be strengthened. Suggested improvements include increased funding and authority to control nonpoint sources of pollution, upgrading the computer system for monitoring compliance with the law, and strengthening programs to prevent and control toxic water pollution.

Other suggestions include providing more funding and authority for integrated watershed and airshed planning to protect groundwater and surface water from contamination, and expanding the rights of citizens to bring lawsuits to ensure that water pollution laws are enforced. The National Academy of Sciences also calls for halting the loss of wetlands, higher standards for wetland restoration, and creating and evaluating new wetlands before filling any natural wetlands.

Many people oppose these proposals, contending that the Clean Water Act's regulations and government wetlands regulations are already too restrictive and costly. Farmers and developers see the law as a curb on their rights as property owners to fill in wetlands. They also believe they should be compensated for any property value losses because of federal wetland protection.

State and local officials want more discretion in testing for and meeting water quality standards. They argue that in many communities it is unnecessary and too expensive to test for all the water pollutants required by federal law.

X *HOW WOULD YOU VOTE?* Should the U.S. Clean Water Act be strengthened? Cast your vote online at http://biology .brookscole.com/miller7.

9-9 DRINKING WATER QUALITY

How Is Urban Drinking Water Purified? High-Tech and Low-Tech Approaches

Centralized water treatment plants can provide safe drinking water for city dwellers in developed countries and simpler and cheaper ways can be used to purify drinking water for individuals and villages in developing countries.

Most developed countries have laws establishing drinking water standards, but most developing countries do not have such laws or do not enforce them. Areas in developed countries that depend on surface water usually store it in a reservoir for several days. This improves clarity and taste by increasing dissolved oxygen content and allowing suspended matter to settle.

Next, the water is pumped to a purification plant and treated to meet government drinking water standards. In areas with very pure groundwater sources, little treatment except disinfection is necessary.

Ways to purify drinking water can be simple. In tropical countries without centralized water treatment systems, the WHO is urging people to purify drinking water by exposing a clear plastic bottle filled with contaminated water to intense sunlight. In the strong sunlight found in most tropical countries, heat and the sun's UV rays can kill infectious microbes in as little as 3 hours. Painting one side of the bottle black can improve heat absorption in this simple solar disinfection method. Where it has been used, incidences of dangerous childhood diarrhea have decreased by 30–40%.

In Bangladesh, households receive strips of cloth for filtering cholera-producing bacteria from drinking water. Villages using this approach have cut cholera cases in half.

Case Study: Is Bottled Water the Answer? Solution or Expensive Rip-off?

Some bottled water is not as pure as tap water and costs much more.

Despite some problems, experts say the United States has some of the world's cleanest drinking water. Yet about half of all Americans worry about getting sick from tap water contaminants, and many drink bottled water or install expensive water purification systems.

Studies reveal that in the United States bottled water is 240 times to 10,000 times more expensive than tap water, about one-fourth of it is tap water, and bacteria contaminate about one-third of such water. On the other hand, some countries must rely on bottled water because some of their tap water is too polluted to drink.

Before drinking expensive bottled water and buying costly home water purifiers, health officials suggest that consumers have their water tested by local health authorities or private labs (not companies trying to sell water purification equipment). The goals are to identify what contaminants (if any) must be removed and determine the type of purification needed to remove such contaminants. Independent experts contend that unless tests show otherwise, for most urban and suburban Americans served by large municipal drinking water systems, home water treatment systems are not worth the expense and maintenance hassles.

Buyers should check out companies selling water purification equipment and be wary of claims that the EPA has approved a treatment device. Although the EPA does *register* such devices, it neither tests nor approves them.

X | HOW WOULD YOU VOTE? Should pollution standards be established for bottled water? Cast your vote online at http://biology.brookscole.com/miller7.

How Can We Reduce Water Pollution? Individuals Matter

Shifting our priorities from controlling to preventing and reducing water pollution will require bottom-up political action by individuals and groups.

It is encouraging that since 1970 most of the world's developed countries have enacted laws and regulations that have significantly reduced point-source water pollution. Most of these improvements were the result of *bottom-up* political pressure on elected officials by individuals and organized groups. However, little has been done to reduce water pollution in most developing countries.

To environmentalists the next step is to increase efforts to reduce and prevent water pollution in developed and developing countries by asking the question: *"How can we not produce water pollutants in the first place?"* Figure 9-30 lists ways to do this over the next several decades.

This shift to *preventing water pollution* will not take place in developed countries without *bottom-up* political pressure on elected officials. It will also not occur in developing countries without similar pressure from

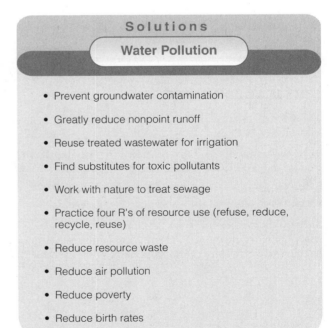

Solutions

Water Pollution

- Prevent groundwater contamination
- Greatly reduce nonpoint runoff
- Reuse treated wastewater for irrigation
- Find substitutes for toxic pollutants
- Work with nature to treat sewage
- Practice four R's of resource use (refuse, reduce, recycle, reuse)
- Reduce resource waste
- Reduce air pollution
- Reduce poverty
- Reduce birth rates

Figure 9-30 Solutions: methods for preventing and reducing water pollution. Which two of these solutions do you believe are the most important?

- Fertilize your garden and yard plants with manure or compost instead of commercial inorganic fertilizer.

- Minimize your use of pesticides.

- Never apply fertilizer or pesticides near a body of water.

- Grow or buy organic foods.

- Compost your food wastes.

- Do not use water fresheners in toilets.

- Do not flush unwanted medicines down the toilet.

- Do not pour pesticides, paints, solvents, oil, antifreeze, or other products containing harmful chemicals down the drain or onto the ground.

Figure 9-31 What can you do? Ways to help reduce water pollution. Which two of these practices do you believe are the most important.

citizens as well as financial and technical aid from developed countries. Figure 9-31 lists some actions you can take to help reduce water pollution.

It is a hard truth to swallow, but nature does not care if we live or die. We cannot survive without the oceans, for example, but they can do just fine without us.

ROGER ROSENBLATT

CRITICAL THINKING

1. How do human activities increase the harmful effects of prolonged drought? How can we reduce these effects?

2. How do human activities contribute to flooding and flood damage? How can these effects be reduced?

3. What role does population growth play in **(a)** water supply problems and **(b)** water pollution problems?

4. Should the prices of water for all uses be raised sharply to include more of its environmental costs and to encourage water conservation? Explain. What harmful and beneficial effects might this have on **(a)** business and jobs, **(b)** your lifestyle and the lifestyles of any children or grandchildren you might have, **(c)** the poor, and **(d)** the environment?

5. Explain why dilution is not always the solution to water pollution. Give examples and conditions for which this solution is or is not applicable.

6. For each of the eight categories of pollutants listed in Table 9-1, is it most likely to originate from **(a)** point sources or **(b)** nonpoint sources?

7. When you flush where does the wastewater go? Trace the actual flow of this wastewater in your community from your toilet through sewers to a wastewater treatment plant and from there to the environment. Try to visit a local sewage treatment plant to see what it does with your wastewater. Compare the processes it uses with those shown in Figure 9-29 (p. 196). What happens to the sludge produced by this plant? What improvements, if any, would you suggest for this plant?

8. Find out the price of tap water where you live. Then go to a grocery or other store and get prices per liter (or other volume unit) on all the available types of bottled water. Use these data to compare the price per liter of various brands of bottled water with the price of tap water.

9. Congratulations! You are in charge of the world. What are the three most important actions you would take to **(a)** manage the world's water resources, **(b)** sharply reduce point water pollution in developed countries, **(c)** sharply reduce nonpoint water pollution throughout the world, **(d)** sharply reduce groundwater pollution throughout the world, and **(e)** provide safe drinking water for the poor and other people in developing countries?

LEARNING ONLINE

The website for this book contains helpful study aids and many ideas for further reading and research. They include a chapter summary, review questions for the entire chapter, flash cards for key terms and concepts, a multiple-choice practice quiz, interesting Internet sites, references, and a guide for accessing thousands of InfoTrac® College Edition articles. Log on to

biology.brookscole.com/miller7

Then click on the Chapter-by-Chapter area, choose Chapter 9, and select a learning resource.

10 ENERGY

Energy

Energy

Typical citizens of advanced industrialized nations each consume as much energy in six months as typical citizens in developing countries consume during their entire life.

MAURICE STRONG

10-1 EVALUATING ENERGY RESOURCES

What Types of Energy Do We Use? Supplementing Free Solar Capital

About 99% of energy that heats the earth and our buildings comes from the sun, and the remaining 1% comes mostly from burning fossil fuels.

Almost all of the energy that heats the earth and our buildings comes from the sun at no cost to us. Without this essentially inexhaustible solar energy (*solar capital*), the earth's average temperature would be −240° C (−400° F), and life as we know it would not exist.

This direct input of solar energy also produces several other *indirect forms of renewable solar energy*. Examples are wind, falling and flowing water (hydropower), and biomass (solar energy converted to chemical en-

ergy and stored in the chemical bonds of organic compounds in trees and other plants).

Commercial energy sold in the marketplace makes up the remaining 1% of the energy we use to supplement the earth's direct input of solar energy. Most commercial energy comes from extracting and burning *nonrenewable mineral resources* obtained from the earth's crust, primarily carbon-containing fossil fuels—oil, natural gas, and coal—as shown in Figure 10-1

What Types of Commercial Energy Does the World Depend On? The Fossil Fuel Era

About 78% of the commercial energy we use comes from nonrenewable fossil fuels.

About 84% of the commercial energy consumed in the world comes from *nonrenewable* energy resources (78% from fossil fuels and 6% from nuclear power; Figure 10-2, p. 202, left). The remaining 16% comes from *renewable* energy resources—biomass (10%), hydropower (5%), and a combination of geothermal, wind, and solar energy (1%).

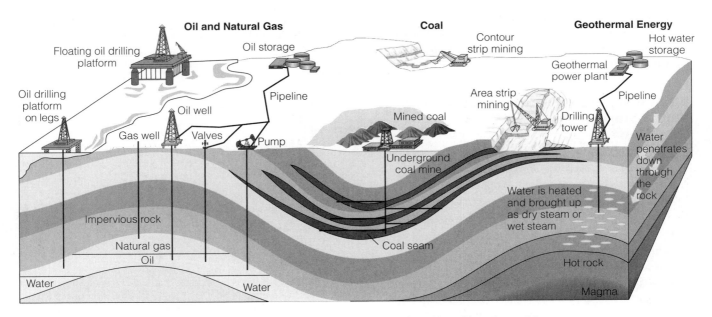

Figure 10-1 Natural capital: important nonrenewable energy resources that can be removed from the earth's crust are coal, oil, natural gas, and some forms of geothermal energy. Nonrenewable uranium ore is also extracted from the earth's crust and then processed to increase its concentration of uranium-235, which can be used as a fuel in nuclear reactors to produce electricity.

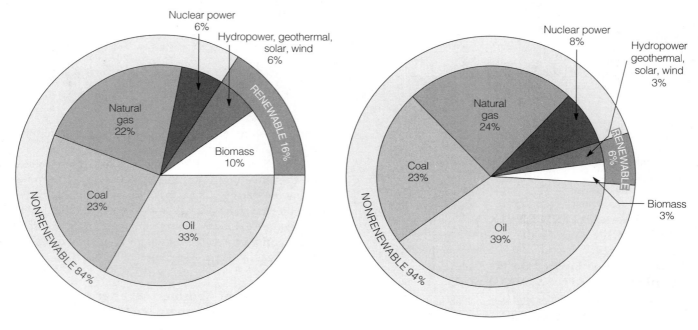

Figure 10-2 Commercial energy use by source for the world (left) and the United States (right) in 2002. Commercial energy amounts to only 1% of the energy used in the world; the other 99% is direct solar energy received from the sun and is not sold in the marketplace. (Data from U.S. Department of Energy, British Petroleum, Worldwatch Institute, and International Energy Agency)

Roughly half the world's people in developing countries burn wood and charcoal to heat their dwellings and cook their food. This *biomass energy* is renewable as long as wood supplies are not harvested faster than they are replenished.

Bad news: Many people in developing countries face a *fuelwood shortage* that is expected to get worse because of unsustainable harvesting of fuelwood. Also, people die prematurely from breathing particles emitted by burning wood indoors on open fires and in poorly designed primitive stoves.

What Is the Energy Future of the United States? Searching for Fossil Fuels Substitutes

There is debate over whether U.S. energy policy for this century should continue its dependence on oil and coal or depend more on natural gas, hydrogen, and solar cells.

The United States is the world's largest energy user, with the average American consuming as much energy in one day as a person in the poorest countries consumes in a year. In 2004, with only 4.6% of the population, it used 24% of the world's commercial energy. In contrast, India, with 16% of the world's people, used about 3% of the world's commercial energy.

About 94% of the commercial energy used in the United States comes from nonrenewable energy re-

sources (86% from fossil fuels and 8% from nuclear power; Figure 10-2, right). The remaining 6% comes mostly from renewable biomass and hydropower.

An important environmental, economic, and political issue is what energy resources the United States might be using by 2050 and 2100. Figure 10-3 shows the shifts in use of various commercial sources of energy in the United States since 1800, and one scenario projecting changes to a solar–hydrogen economy by

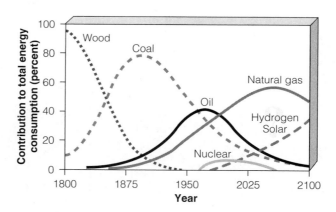

Figure 10-3 Shifts in the use of commercial energy resources in the United States since 1800, with projected changes to 2100. Shifts from wood to coal and then from coal to oil and natural gas have each taken about 50 years. The projected shift by 2100 is only one of many possible scenarios that depend on a variety of assumptions. (Data from U.S. Department of Energy)

2100. According to the U.S. Department of Energy and the Environmental Protection Agency, burning fossil fuels causes more than 80% of U.S. air pollution and 80% of U.S. carbon dioxide emissions. Many energy experts contend that the need to use cleaner and less climate-disrupting (noncarbon) energy resources—not the depletion of fossil fuels—is the driving force behind the projected transition to a solar–hydrogen energy age in the United States and other parts of the world before the end of this century.

Whether the shift shown in Figure 10-3, or some other scenario, occurs depends primarily on the energy resources the U.S. government decides to *promote* by use of subsidies and tax breaks. If we want energy alternatives such as solar energy and hydrogen to become main dishes instead of side orders on our energy menu, they must be nurtured by government subsidies and tax breaks.

A political problem is that the fossil fuel and nuclear power industries that have been receiving government subsidies for over 50 years understandably do not want to give them up. And they use their considerable political power to keep these subsidies even though they are mature and profitable industries that do not need such nurturing.

Thus the energy path of the United States (or any country) is primarily a political decision made by elected officials with pressure from officials of energy companies and from citizens. As a citizen, you can play an important role in deciding the energy future for yourself and any child you decide to have. Indeed, it is one of the most important political acts you can undertake. This explains why you should have an understanding of the advantages and disadvantages of our major energy options, as discussed in this chapter.

How Can We Decide Which Energy Resources to Use? Evaluating Alternative Resources

We need to answer several questions in deciding which energy resources to promote.

Energy policies need to be developed with the future in mind because experience shows that it usually takes at least 50 years and huge investments to phase in new energy alternatives to the point where they provide 10–20% of total energy use. Making projections such as those in Figure 10-3 and then converting such projections to energy policy involves trying to answer the following questions for *each* energy alternative:

- How much of the energy resource is likely to be available in the near future (the next 15–25 years) and the long term (the next 25–50 years)?

- What is the net energy yield for the resource?

- How much will it cost to develop, phase in, and use the resource?

- What government research and development subsidies and tax breaks will be used to help develop the resource?

- How will dependence on the resource affect national and global economic and military security?

- How vulnerable is the resource to terrorism?

- How will extracting, transporting, and using the resource affect the environment, human health, and the earth's climate? Should these harmful costs be included in the market price of each energy resource through a combination of taxes and phasing out environmentally harmful subsidies?

What Is Net Energy? The Only Energy That Really Counts

Net energy is the amount of high-quality usable energy available from a resource after subtracting the energy needed to make it available for use.

It takes energy to get energy. For example, before oil is useful to us it must be found, pumped up from beneath the ground or ocean floor, transferred to a refinery and converted to useful fuels (such as gasoline, diesel fuel, and heating oil), transported to users, and burned in furnaces and cars. Each of these steps uses high-quality energy. The second law of thermodynamics tells us that some of the high-quality energy we use in each step is wasted and degraded to lower-quality energy.

The usable amount of *high-quality energy* available from a given quantity of an energy resource is its **net energy.** It is the total amount of energy available from an energy resource minus the energy needed to find, extract, process, and get that energy to consumers. It is calculated by estimating the total energy available from the resource over its lifetime minus the amount of energy *used* (the first law of thermodynamics), *automatically wasted* (the second law of thermodynamics), and *unnecessarily wasted* in finding, processing, concentrating, and transporting the useful energy to users.

Net energy is like your net spendable income—your wages minus taxes and job-related expenses. For example, suppose for every 10 units of energy in oil in the ground we have to use and waste 8 units of energy to find, extract, process, and transport the oil to users. Then we have only 2 units of *useful energy* available from every 10 units of energy in the oil.

We can express net energy as the ratio of useful energy produced to the energy used to produce it. In the example just given, the *net energy ratio* would be 10/8, or approximately 1.25. The higher the ratio, the

Space Heating

Passive solar	5.8
Natural gas	4.9
Oil	4.5
Active solar	1.9
Coal gasification	1.5
Electric resistance heating (coal-fired plant)	0.4
Electric resistance heating (natural-gas-fired plant)	0.4
Electric resistance heating (nuclear plant)	0.3

High-Temperature Industrial Heat

Surface-mined coal	28.2
Underground-mined coal	25.8
Natural gas	4.9
Oil	4.7
Coal gasification	1.5
Direct solar (highly concentrated by mirrors, heliostats, or other devices)	0.9

Transportation

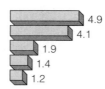

Natural gas	4.9
Gasoline (refined crude oil)	4.1
Biofuel (ethyl alcohol)	1.9
Coal liquefaction	1.4
Oil shale	1.2

Figure 10-4 *Net energy ratios* for various energy systems over their estimated lifetimes. The higher the net energy ratio, the greater the net energy available. (Data from U.S. Department of Energy and Colorado Energy Research Institute, *Net Energy Analysis*, 1976; and Howard T. Odum and Elisabeth C. Odum, *Energy Basis for Man and Nature*, 3rd ed., New York: McGraw-Hill, 1981)

greater the net energy. When the ratio is less than 1, there is a net energy loss.

Figure 10-4 shows estimated net energy ratios for various types of space heating, high-temperature heat for industrial processes, and transportation. Study this important figure carefully. In terms of net energy, how do the energy resources used to heat your home and propel your car (if you have one) stack up compared to other alternatives?

Currently, oil has a high net energy ratio because much of it comes from large, accessible, and cheap-to-extract deposits such as those in the Middle East. When those sources are depleted, the net energy ratio of oil will decline and prices will rise.

Conventional nuclear energy has a low net energy ratio because large amounts of energy are needed to extract and process uranium ore, convert it into nuclear fuel, build and operate nuclear power plants, dismantle the highly radioactive plants after their 15–60 years of useful life, and store the resulting highly radioactive wastes safely for 10,000–240,000 years, depending on the types of radioisotopes they contain. Each of these steps in what is called the *nuclear power fuel cycle* uses energy and costs money. Some analysts estimate that

ultimately the conventional nuclear power fuel cycle will lead to a net energy loss; we will have to put more energy into it than we will ever get out of it.

10-2 NONRENEWABLE FOSSIL FUELS

What Is Crude Oil, and How Is It Extracted and Processed? Gooey Stuff to Which We Are Addicted

Crude oil is a thick liquid containing hydrocarbons that we extract from underground deposits and separate into products such as gasoline, heating oil, and asphalt.

Petroleum, or **crude oil** (oil as it comes out of the ground), is a thick and gooey liquid consisting of hundreds of combustible hydrocarbons along with small amounts of sulfur, oxygen, and nitrogen impurities. This type of oil is also known as *conventional oil* or *light oil*.

Deposits of crude oil and natural gas often are trapped together under a dome deep within the earth's crust on land or under the seafloor (Figure 10-1). The crude oil is dispersed in pores and cracks in underground rock formations, somewhat like water saturating a sponge. To extract the oil, a well is drilled into the deposit. Then oil, drawn by gravity out of the rock pores and into the bottom of the well, is pumped to the surface.

Drilling for oil causes only moderate damage to the earth's land because the wells occupy fairly little land area. But drilling for oil and transporting it around the world involves some oil spills on land and in aquatic systems. In addition, harmful environmental effects are associated with the extraction, processing, and use of any nonrenewable resource from the earth's crust (Figure 1-6, p. 10).

After it is extracted, crude oil is transported to a *refinery* by pipeline, truck, or ship (oil tanker). There it is heated and distilled in gigantic columns to separate it into components with different boiling points (Figure 10-5)—a technological marvel based on complex chemistry and engineering. Some of the products of oil distillation, called **petrochemicals,** are used as raw materials in industrial organic chemicals, pesticides, plastics, synthetic fibers, paints, medicines, and many other products. Look at your clothes and the items in the room you are in and try to figure out how many of these things were made from chemicals produced by distilling oil.

Who Has the World's Oil Supplies? OPEC Rules

Eleven OPEC countries—most of them in the Middle East—have 78% of the world's proven oil reserves and most of the world's unproven reserves.

The oil industry is the world's largest business. Thus, control of the world's current and future oil reserves is the single greatest source of global economic and political power.

Oil *reserves* are identified deposits from which oil can be extracted profitably at current prices with current technology. The 11 countries that make up the Organization of Petroleum Exporting Countries (OPEC) have 78% of the world's crude oil reserves. This explains why OPEC is expected to have long-term control over the supplies and prices of the world's conventional oil. Today OPEC's members are Algeria, Indonesia, Iran, Iraq, Kuwait, Libya, Nigeria, Qatar, Saudi Arabia, the United Arab Emirates, and Venezuela.

Saudi Arabia, with 25%, has by far the largest proportion of the world's crude oil reserves. It is followed by Canada (15%) because its huge supply of oil sand was recently classified as a conventional source of oil. Other countries with large proven reserves are Iraq (11%), the United Arab Emirates (9.3%), Kuwait (9.2%),

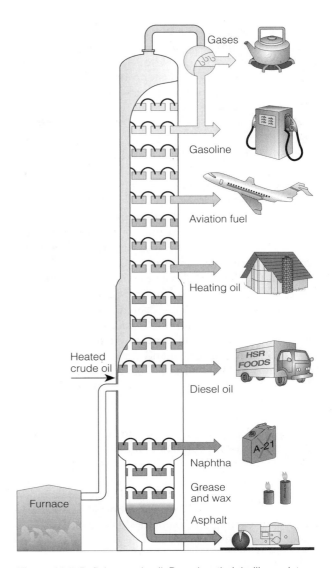

Figure 10-5 Refining crude oil. Based on their boiling points, components are removed at various levels in a giant distillation column. The most volatile components with the lowest boiling points are removed at the top of the column.

and Iran (8.6%). The United States—the world's largest oil user—has only 2.9% of the world's proven oil reserves and only a small percentage of its unproven reserves.

How Long Will Conventional Oil Supplies Last? The End of the Oil Era May Be in Sight

Known and projected global oil reserves should last for 42–93 years and U.S. reserves for 10–48 years, depending on how rapidly we use oil.

We are not yet running out of oil. But like all nonrenewable resources, the world's oil supplies are eventually expected to decline. At some point during this decline, prices will begin a steady rise as the world's consumers begin competing for the remaining oil reserves. In 1999, Mike Bowling, CEO of ARCO Oil, said at an energy conference in Houston, Texas, "We are embarked on the beginning of the last days of the Age of Oil." He then went on to discuss the need for the world to shift from a carbon-based to a hydrogen-based energy economy during this century (Figure 10-3). See Appendix 3 at the end of this book for a brief history of the Age of Oil.

According to geologists, known and projected global reserves of oil are expected to be 80% depleted within 42–93 years and U.S. reserves in *10–48 years,* depending on how rapidly we use oil. If these estimates are correct, oil should be reaching its sunset years sometime during this century.

Basically there are three options: look for more oil, use or waste less, or use something else. Some analysts—mostly economists—contend that rising oil prices (when oil consumption exceeds oil production) will stimulate exploration and lead to enough new reserves to meet future demand through the next century or longer. Many analysts argue that even if much more oil is somehow found, we are ignoring the consequences of the high (1–5% per year) growth rate in global oil consumption.

It is hard to get a grip on the incredible amount of oil we consume. Maybe this will help. *Stretched end to end, the number of barrels of oil we used in 2004 would circle the equator 636 times!*

Suppose we continue to use oil at the current rate with no increase in oil consumption—a highly unlikely assumption. Even under this conservative no growth estimate,

- Saudi Arabia, with the world's largest known crude oil reserves, could supply the world's entire oil needs for about 10 years

- The estimated reserves under Alaska's North Slope—the largest ever found in North America—would meet current world demand for only 6 months or U.S. demand for 3 years

- The estimated reserves in Alaska's Arctic National Wildlife Refuge would meet current oil demand for only 1–5 months and U.S. oil demand for 7–24 months

Thus, for the world just to keep using conventional oil at the current rate, we must discover global oil reserves that are the equivalent to a new Saudi Arabian supply every 10 years. According to most geologists, this is highly unlikely.

And many of developing countries such as China and India are rapidly expanding their use of oil. By 2025, China could be using as much oil as the United States, and the two countries could be competing to import dwindling supplies of increasingly expensive

oil. Indeed, if everyone in the world consumed as much oil as the average American, the world's proven oil reserves would be gone in a decade. Exponential growth is an incredibly powerful force.

Here is the problem in a nutshell: Oil is the most widely used energy resource in the world and the United States. Most people in developed countries are *oilaholics* and the world's largest suppliers for this addiction are Canada and the Middle Eastern countries of Saudi Arabia and Iraq. According to Philip E. Clapp, president of the National Environmental Trust, "The entire world economy is built on a bet of how long the House of Saud can continue."

There is talk of the United States reducing its dependence on imported oil and having more control over global oil prices by increasing domestic oil supplies. Most analysts consider this unrealistic because the United States has only 2.9% of the world's proven oil reserves and uses 26% of the world's annual oil production.

Despite intensive exploration, U.S. oil production peaked in 1975 and is expected to continue declining as its oil fields are gradually depleted. And the United States also produces most of its dwindling supply of oil at a high cost of $7.50–10 per barrel compared to production costs of about $2.50 in Saudi Arabia. Thus, opening all of the U.S. coastal waters, forests, and wild places to drilling would hardly put a dent in world oil prices or meet much of the U.S. demand for oil.

In 2003, the United States imported about 55% of the oil it used (up from 36% in 1973 when OPEC imposed an oil embargo against the U.S. and other nations). Reasons for this high dependence on imported oil are declining domestic oil reserves, higher production costs for domestic oil than for most sources of imported oil, and increased oil use. According to the Department of Energy (DOE), the United States could be importing 64–70% of the oil it uses by 2020.

Some analysts favor depending on oil imports, arguing that using up limited and declining domestic oil supplies is a drain-America-first policy that will increase future dependence on foreign oil supplies. What do you think?

What Are the Major Advantages and Disadvantages of Conventional Oil? A Difficult Choice

Conventional oil is a versatile fuel that can last for at least 50 years, but burning it produces air pollution and releases the greenhouse gas carbon dioxide.

Figure 10-6 lists the advantages and disadvantages of using conventional crude oil as an energy resource. A serious problem associated with the use of conventional crude oil is that burning oil or any carbon-containing fossil fuel releases CO_2 into the atmosphere

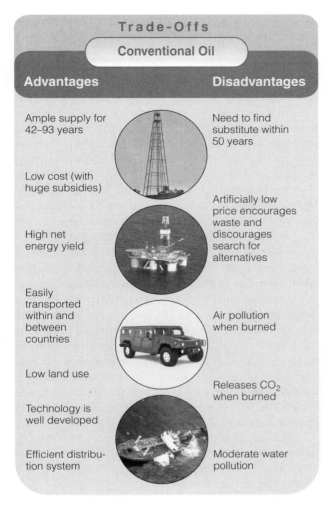

Trade-Offs

Conventional Oil

Advantages	Disadvantages
Ample supply for 42–93 years	Need to find substitute within 50 years
Low cost (with huge subsidies)	Artificially low price encourages waste and discourages search for alternatives
High net energy yield	
Easily transported within and between countries	Air pollution when burned
Low land use	
Technology is well developed	Releases CO_2 when burned
Efficient distribution system	Moderate water pollution

Figure 10-6 Trade-offs: advantages and disadvantages of using conventional crude oil as an energy resource. Pick the single advantage and disadvantage that you think are the most important.

and thus can help promote climate change from global warming. Currently, burning oil mostly as gasoline and diesel fuel for transportation accounts for 43% of global CO_2 emissions. Figure 10-7 compares the relative amounts of CO_2 emitted per unit of energy by the major fossil fuels and nuclear power.

How Useful Are Heavy Oils from Oil Sand and Oil Shale? Can Heavier Substitutes Save the Day?

Heavy oils from oil sand and oil shale could supplement conventional oil, but there are environmental problems.

Oil sand, or **tar sand,** is a mixture of clay, sand, water, and a combustible organic material called *bitumen*—a thick and sticky heavy oil with a high sulfur content. Oil sands nearest the earth's surface are dug up by gigantic electric shovels and mixed with hot water and steam to extract the bitumen, which is heated in huge

cookers to convert it into a low-sulfur synthetic crude oil suitable for refining.

Northeastern Alberta in Canada has about three-fourths of the world's oil sand resources, about a tenth of them close enough to the surface to be recovered by surface and underground mining. Improved technology may allow extraction of twice that amount.

Currently, these deposits supply about a fifth of Canada's oil needs and this proportion is expected to increase. Because of the dramatic reductions in development and production costs, in 2003 the oil industry began counting Canada's oil sands as reserves of conventional oil. This means that Canada has 15% of the world's oil reserves, second only to Saudi Arabia. If a pipeline is built to transfer some of this synthetic crude oil from western Canada to the northwestern United States, Canada could greatly reduce future U.S. dependence on oil imports from the Middle East and add to Canadian income.

Bad news: Extracting and processing oil sands has a severe impact on the land and produces much more water pollution, much more air pollution (especially sulfur dioxide), and more CO_2 per unit of energy than conventional crude oil.

Oily rocks are another potential supply of heavy oil. Such rocks, called *oil shales,* contain a solid combustible mixture of hydrocarbons called *kerogen.* It can be can be extracted from crushed oil shales by heating them in a large container, a process which yields a distillate called **shale oil.** Before the thick shale oil can be sent by pipeline to a refinery, it must be heated to increase its flow rate and processed to remove sulfur, nitrogen, and other impurities.

Estimated potential global supplies of shale oil are about 240 times larger than estimated global supplies of conventional oil. But most deposits are of such a low grade that with current oil prices and technology, it

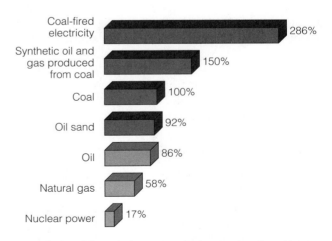

Coal-fired electricity	286%
Synthetic oil and gas produced from coal	150%
Coal	100%
Oil sand	92%
Oil	86%
Natural gas	58%
Nuclear power	17%

Figure 10-7 CO_2 emissions per unit of energy produced by various fuels, expressed as percentages of emissions produced by burning coal directly. (Data from U.S. Department of Energy)

Trade-Offs

Heavy Oils from Oil Shale and Oil Sand

Advantages	Disadvantages
Moderate cost (oil sand)	High cost (oil shale)
Large potential supplies, especially oil sands in Canada	Low net energy yield
Easily transported within and between countries	Large amount of water needed for processing
Efficient distribution system in place	Severe land disruption from surface mining
Technology is well developed	Water pollution from mining residues
	Air pollution when burned
	CO_2 emissions when burned

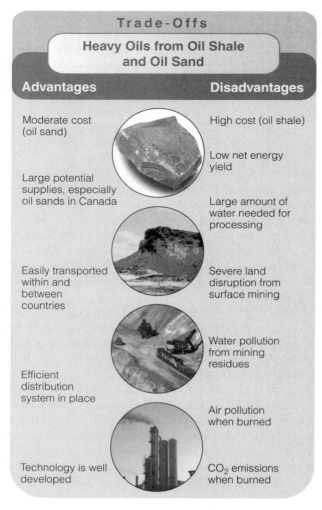

Figure 10-8 Trade-offs: advantages and disadvantages of using heavy oils from tar sand and oil shales as energy resources. Pick the single advantage and disadvantage that you think are the most important.

takes more energy and money to mine and convert the kerogen to crude oil than the resulting fuel is worth. Producing and using shale oil also has a much higher environmental impact than conventional oil.

Figure 10-8 lists the advantages and disadvantages of using heavy oil from oil sand and oil shales as energy resources. Overall, do you believe that the advantages outweigh the disadvantages? Explain.

What Is Natural Gas? Mostly Methane

Natural gas, consisting mostly of methane, is often found above reservoirs of crude oil.

In its underground gaseous state, **natural gas** is a mixture of 50–90% by volume of methane (CH_4), the simplest hydrocarbon. It also contains smaller amounts of heavier gaseous hydrocarbons such as ethane (C_2H_6),

propane (C_3H_8), and butane (C_4H_{10}), and small amounts of highly toxic hydrogen sulfide (H_2S).

Conventional natural gas lies above most reservoirs of crude oil (Figure 10-1). However, unless a natural gas pipeline has been built, deposits of natural gas found above oil deposits cannot be used. Indeed, the natural gas found above oil reservoirs in deep sea and remote land areas is often viewed as an unwanted byproduct and is burned off. This wastes a valuable energy resource and releases carbon dioxide into the atmosphere.

Unconventional natural gas is found by itself in other underground sources. So far, it costs too much to get natural gas from such unconventional sources, but the extraction technology is being developed rapidly.

When a natural gas field is tapped, propane and butane gases are liquefied and removed as **liquefied petroleum gas (LPG)**. LPG is stored in pressurized tanks for use mostly in rural areas not served by natural gas pipelines. The rest of the gas (mostly methane) is dried to remove water vapor, cleansed of poisonous hydrogen sulfide and other impurities, and pumped into pressurized pipelines for distribution. At a very low temperature, natural gas can be converted to **liquefied natural gas (LNG)**. This highly flammable liquid can then be shipped to other countries in refrigerated tanker ships.

Russia has about 31% of the world's proven natural gas reserves, followed by Iran (15%) and Qatar (9%). The United States has only 3% of the world's proven natural gas reserves.

The long-term global outlook for natural gas supplies is better than for conventional oil. At the current consumption rate, known reserves and undiscovered, potential reserves of conventional natural gas should last the world for 62–125 years and the United States for 55–80 years, depending on how rapidly it is used.

Geologists project that *conventional* and *unconventional* supplies of natural gas (the latter available at higher prices) should last at least 200 years at the current consumption rate and 80 years if usage rates rise 2% per year.

What Are the Advantages and Disadvantages of Natural Gas? The Best Fossil Fuel

Natural gas is a versatile and clean-burning fuel that should last for 62–125 years, but it releases the greenhouse gases carbon dioxide (when burned) and methane (from leaks) into the atmosphere.

Figure 10-9 lists the advantages and disadvantages of using conventional natural gas as an energy resource. Overall, do you believe that the advantages outweigh the disadvantages? Explain.

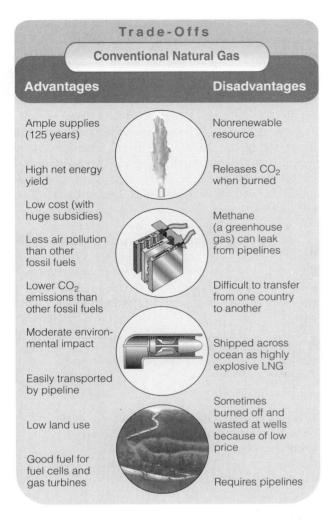

Trade-Offs

Conventional Natural Gas

Advantages	Disadvantages
Ample supplies (125 years)	Nonrenewable resource
High net energy yield	Releases CO_2 when burned
Low cost (with huge subsidies)	
Less air pollution than other fossil fuels	Methane (a greenhouse gas) can leak from pipelines
Lower CO_2 emissions than other fossil fuels	Difficult to transfer from one country to another
Moderate environmental impact	Shipped across ocean as highly explosive LNG
Easily transported by pipeline	
Low land use	Sometimes burned off and wasted at wells because of low price
Good fuel for fuel cells and gas turbines	Requires pipelines

Figure 10-9 Trade-offs: advantages and disadvantages of using conventional natural gas as an energy resource. Pick the single advantage and disadvantage that you think are the most important.

Because of its advantages over oil, coal, and nuclear energy, some analysts see natural gas as the best fuel to help make the transition to improved energy efficiency and greater use of solar energy and hydrogen over the next 50 years.

However, U.S. production of natural gas has been declining for a long time, and experts do not believe this situation will be reversed. More natural gas could be imported from Canada, but this will require building a major pipeline between the two countries. Also, natural gas production in Canada is expected to peak between 2020 and 2030. Then the United States and the rest of the world would have to rely increasingly on Russia and the Middle East for supplies of natural gas.

What Is Coal? A Mostly Carbon Fuel That Is Widely Used

Coal is an abundant energy resource that is burned mostly to produce electricity and steel.

Coal is a solid fossil fuel formed in several stages as the buried remains of land plants that lived 300–400 million years ago were subjected to intense heat and pressure over many millions of years (Figure 10-10). Coal is mostly carbon and contains small amounts of sulfur, released into the atmosphere as sulfur dioxide when the coal is burned. Burning coal also releases trace amounts of toxic mercury and radioactive materials. Coal is burned to generate 62% of the world's electricity (52% in the United States) and to make three-fourths of its steel.

Coal is the world's most abundant fossil fuel. According to the U.S. Geological survey, identified and unidentified supplies of coal could last the world for 214–1,125 years, depending on the rate of usage. The United States has one-fourth of the world's proven coal reserves. Russia has 16% and China 12%. In 2002, just over half of global coal consumption was split almost evenly between China and the United States.

China has enough proven coal reserves to last 300 years at its current rate of consumption. According to the U.S. Geological Survey, identified U.S. coal reserves should also

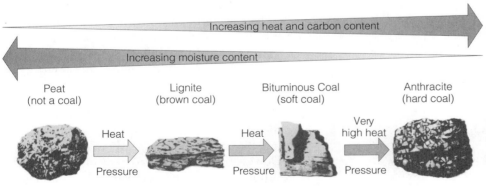

Increasing heat and carbon content →

← Increasing moisture content

Peat (not a coal)	Lignite (brown coal)	Bituminous Coal (soft coal)	Anthracite (hard coal)
Partially decayed plant and animal matter in swamps and bogs; low heat content	Heat → Pressure — Low heat content; low sulfur content; limited supplies in most areas	Heat → Pressure — Extensively used as a fuel because of its high heat content and large supplies; normally has a high sulfur content	Very high heat → Pressure — Highly desirable fuel because of its high heat content and low sulfur content; supplies are limited in most areas

Figure 10-10 Natural capital: stages in coal formation over millions of years. Peat is a soil material made of moist, partially decomposed organic matter.

last about 300 years at the current consumption rate, and unidentified U.S. coal resources could extend those supplies for perhaps another 100 years, at a higher cost. However, if U.S. coal use should increase by 4% a year—as the coal industry projects—the country's proven coal reserves would last only 64 years.

What Are the Advantages and Disadvantages of Coal? Plenty Around but Bad for the Environment and the Climate

Coal is the most abundant fossil fuel, but compared to oil and natural gas it is not as versatile, has a much higher environmental impact, and releases more carbon dioxide into the atmosphere.

Figure 10-11 lists the advantages and disadvantages of using coal as an energy resource. *Bottom line:* Coal is

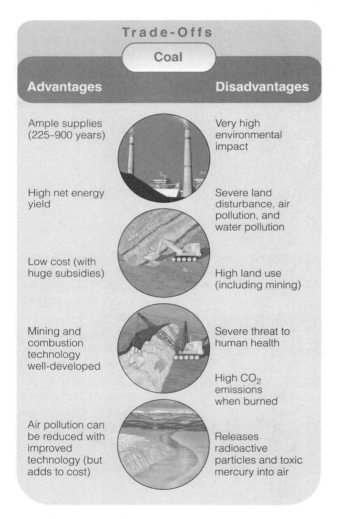

Figure 10-11 Trade-offs: advantages and disadvantages of using coal as an energy resource. Pick the single advantage and disadvantage that you think are the most important.

the world's most abundant fossil fuel but mining and burning coal has a severe environmental impact on the earth's air, water, and land, and accounts for over a third of the world's annual CO_2 emissions.

Each year in the United States alone, air pollutants from coal burning kill thousands of people prematurely (estimates range from 65,000 to 200,000), cause at least 50,000 cases of respiratory disease, and result in several billion dollars of property damage. Many people are unaware that burning coal is also responsible for about one-fourth of atmospheric mercury pollution in the United States, and it releases far more radioactive particles into the air than normally operating nuclear power plants.

Many analysts project a decline in coal use over the next 40–50 years because of its high CO_2 emissions (Figure 10-7) and harmful health effects, and the availability of safer and cheaper ways to produce electricity such as burning natural gas in gas turbines and wind energy.

X *HOW WOULD YOU VOTE?* Should coal use be phased out over the next 20 years? Cast your vote online at http://biology .brookscole.com/miller7.

What Are the Advantages and Disadvantages of Converting Solid Coal into Gaseous and Liquid Fuels? Better for the Air, Worse for the Climate

Coal can be converted to gaseous and liquid fuels that burn cleaner than coal, but the costs are high, and producing and burning them add more carbon dioxide to the atmosphere than burning coal.

Solid coal can be converted into **synthetic natural gas (SNG)** by **coal gasification** or into a liquid fuel such as methanol or synthetic gasoline by **coal liquefaction.** Figure 10-12 lists the advantages and disadvantages of using these *synfuels.* Overall, do you believe that the advantages outweigh the disadvantages? Explain.

Without huge government subsidies, most analysts expect synfuels to play a minor role as energy resources in the next 20–50 years. Compared to burning conventional coals, they require mining 50% more coal and their production and burning add 50% more carbon dioxide to the atmosphere. Also, they cost more to produce than coal.

However, the U.S. Department of Energy and a consortium of major oil companies are working on ways to reduce CO_2 emissions during the coal gasification process. If these efforts are successful, burning gasified coal could be a cheaper and cleaner way to produce electricity than by burning coal, oil, or natural gas. Stay tuned.

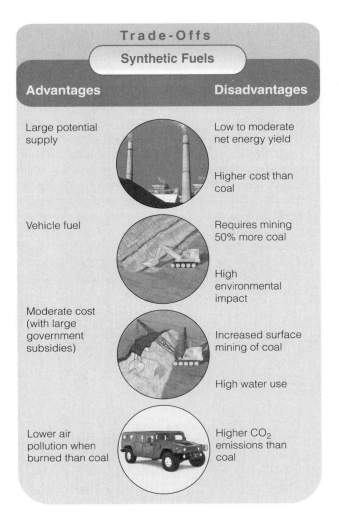

Trade-Offs

Synthetic Fuels

Advantages	Disadvantages
Large potential supply	Low to moderate net energy yield
Vehicle fuel	Higher cost than coal
	Requires mining 50% more coal
Moderate cost (with large government subsidies)	High environmental impact
	Increased surface mining of coal
	High water use
Lower air pollution when burned than coal	Higher CO_2 emissions than coal

Figure 10-12 Trade-offs: advantages and disadvantages of using synthetic natural gas (SNG) and liquid synfuels produced from coal. Pick the single advantage and disadvantage that you think are the most important.

10-3 NONRENEWABLE NUCLEAR ENERGY

What Is Nuclear Fission? Splitting Nuclei

Neutrons can split apart or fission the nuclei of certain isotopes with large mass numbers and release a large amount of energy.

The source of energy for nuclear power is **nuclear fission,** a nuclear change in which nuclei of certain isotopes with large mass numbers (such as uranium-235) are split apart into lighter nuclei when struck by neutrons.

Multiple fissions within a critical mass of nuclear fuel form a **chain reaction,** which releases an enormous amount of energy (Figure 10-13) that can be used to produce electricity at a nuclear power plant.

Nuclear fission produces radioactive fission fragments containing isotopes that spontaneously shoot out fast-moving particles (alpha and beta particles), gamma rays (a form of high-energy electromagnetic radiation; Figure 2-5, p. 22), or both. These unstable isotopes are called **radioactive isotopes,** or **radioisotopes.**

Exposure to alpha, beta, and gamma radiation and the high-speed neutrons emitted in a nuclear fission chain reaction and by the resulting radioactive wastes can harm human cells in two ways. *First,* harmful mutations of DNA molecules in genes and chromosomes can cause genetic defects in immediate offspring or several generations later. *Second,* tissue damage such as burns, miscarriages, eye cataracts, and cancers (bone, thyroid, breast, skin, and lung) can occur during the victim's lifetime.

How Does a Nuclear Fission Reactor Work? Splitting Nuclei to Produce Electricity

In a conventional nuclear reactor, isotopes of uranium and plutonium undergo controlled nuclear fission, and the resulting heat is used to produce steam that spins turbines to generate electricity.

To evaluate the advantages and disadvantages of nuclear power, we must know how a conventional nuclear power plant and its accompanying nuclear power fuel cycle work. In the reactor of a nuclear

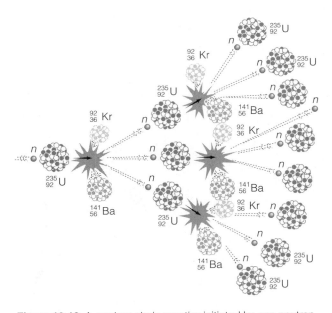

Figure 10-13 A *nuclear chain reaction* initiated by one neutron triggering fission in a single uranium-235 nucleus. This figure illustrates only a few of the trillions of fissions caused when a single uranium-235 nucleus is split within a critical mass of uranium-235 nuclei. The elements krypton (Kr) and barium (Ba), shown here as fission fragments, are only two of many possibilities.

power plant, the rate of fission in a nuclear chain reaction is controlled and the heat generated is used to produce high-pressure steam, which spins turbines that generate electricity.

Light-water reactors (LWRs) like the one diagrammed in Figure 10-14 produce about 85% of the world's nuclear-generated electricity (100% in the United States). *Control rods* are moved in and out of the reactor core to absorb neutrons and thus regulate the rate of fission and amount of power the reactor produces. A *coolant,* usually water, circulates through the reactor's core to remove heat to keep fuel rods and other materials from melting and to produce steam for generating electricity. The greatest danger in water-cooled reactors is a loss of coolant that would allow the nuclear fuel to quickly overheat, melt down, and possibly release radioactive materials to the environ-ment. An LWR reactor has an emergency core cooling system as a backup to help prevent such meltdowns.

As a further safety backup, a *containment vessel* with strong, thick walls surrounds the reactor core. It is designed to keep radioactive materials from escaping into the environment in case of an internal explosion or core meltdown within the reactor, and to protect the core from external threats such as a plane crash.

Water-filled pools or *dry casks* with thick steel walls are used for on-site storage of highly radioactive spent fuel rods removed when reactors are refueled. Spent-fuel pools or casks are in a separate building not nearly as well protected as the reactor core and are much more vulnerable to a head-on crash from a plane and to a terrorist attack. The long-term goal is to transport spent fuel rods and other long-lived radioactive wastes to an underground facility where they must be stored

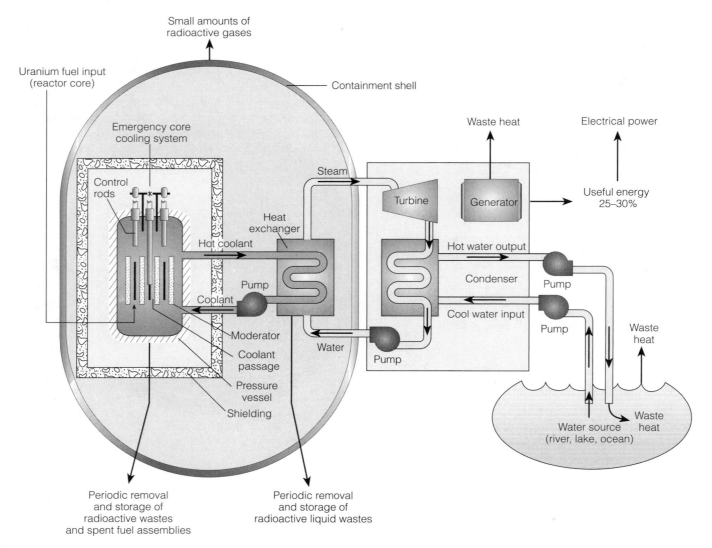

Figure 10-14 Light-water nuclear power plant with a pressurized water reactor. Some plants use huge cooling towers to transfer some of the waste heat to the atmosphere.

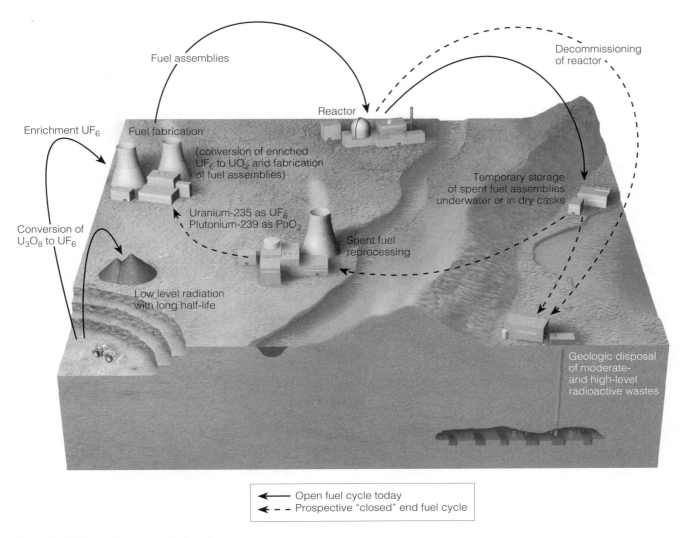

Enrichment UF$_6$

Fuel assemblies

Decommissioning
of reactor

Fuel fabrication

Reactor

(conversion of enriched
UF$_6$ to UO$_2$ and fabrication
of fuel assemblies)

Conversion of
U$_3$O$_8$ to UF$_6$

Uranium-235 as UF$_6$
Plutonium-239 as PuO$_2$

Temporary storage
of spent fuel assemblies
underwater or in dry casks

Spent fuel
reprocessing

Low level radiation
with long half-life

Geologic disposal
of moderate-
and high-level
radioactive wastes

Open fuel cycle today
Prospective "closed" end fuel cycle

Figure 10-15 The nuclear power fuel cycle.

safely for 10,000–240,000 years until their radioactivity falls to safe levels.

The overlapping and multiple safety features of a modern nuclear reactor greatly reduce the chance of a serious nuclear accident. But these safety features make nuclear power plants very expensive to build and maintain.

Nuclear power plants, each with one or more reactors, are only one part of the nuclear power fuel cycle (Figure 10-15). Study this figure carefully. Unlike other energy resources, nuclear energy produces highly radioactive materials that must be stored safely for 10,000–240,000 years. In addition, once a nuclear reactor comes to the end of its useful life after 40–60 years, it cannot just be shut down and abandoned like a coal-burning plant. It contains large quantities of intensely radioactive materials that must be kept out of the environment for many thousands of years.

In evaluating the safety and economic feasibility of nuclear power, energy experts and economists caution us to look at this entire cycle, not just the nuclear plant itself.

Case Study: What Happened to Nuclear Power? A Fading Dream

After more than 50 years of development and enormous government subsidies, nuclear power has not lived up to its promise.

In the 1950s, researchers predicted that by the year 2000 at least 1,800 nuclear power plants would supply 21% of the world's commercial energy (25% in the United States) and most of the world's electricity.

After almost 50 years of development, enormous government subsidies, and an investment of $2 trillion, these goals have not been met. Instead, by 2002,

442 commercial nuclear reactors in 30 countries were producing only 6% of the world's commercial energy and 19% of its electricity.

Since 1989, electricity production from nuclear power has increased only slightly and is now the world's slowest-growing energy source. According to the U.S. Department of Energy, the percentage of the world's electricity produced by nuclear power is projected to fall to 12% by 2025 because the retirement of aging existing plants is expected to exceed the construction of new ones.

No new nuclear power plants have been ordered in the United States since 1978, and all 120 plants ordered since 1973 have been canceled. In 2004, there were 103 licensed commercial nuclear power reactors in 31 states—most of them located in the eastern half of the country. These reactors generate about 21% of the country's electricity and 8% of its total energy. This percentage is expected to decline over the next two decades as existing plants wear out and are retired.

According to energy analysts and economists, there are several major reasons for the failure of nuclear power to grow as projected. They include multibillion-dollar construction cost overruns, higher operating costs and more malfunctions than expected, and poor management. Two other major setbacks have been public concerns about safety and stricter government safety regulations, especially after the accidents in 1979 at the Three Mile Island nuclear plant in Pennsylvania and in 1986 at the Chernobyl nuclear plant in Ukraine.

Another problem is investor concerns about the economic feasibility of nuclear power, taking into account the entire nuclear fuel cycle. At Three Mile Island, investors lost over a billion dollars in one hour from damaged equipment and repair, even though no human lives were lost. Also, concern has risen about the vulnerability of nuclear power plants to terrorist attack after destruction of the World Trade Center buildings in New York City on September 11, 2001, in the United States. Experts are especially concerned about the vulnerability of poorly protected and intensely radioactive spent fuel rods stored in water pools or casks outside of reactor buildings.

What Are the Advantages and Disadvantages of the Conventional Nuclear Power Fuel Cycle? Better than Coal but Much More Costly and Vulnerable to Terrorist Attack

The nuclear power fuel cycle has a fairly low environmental impact, an ample supply of fuel, and a very low risk of an accident, but costs are high, radioactive wastes must be stored safely for thousands of years, and facilities are vulnerable to terrorist attack.

Trade-Offs

Conventional Nuclear Fuel Cycle

Advantages	Disadvantages
Large fuel supply	High cost (even with large subsidies)
Low environmental impact (without accidents)	Low net energy yield
Emits 1/6 as much CO_2 as coal	High environmental impact (with major accidents)
Moderate land disruption and water pollution (without accidents)	Catastrophic accidents can happen (Chernobyl)
Moderate land use	No widely acceptable solution for long-term storage of radioactive wastes and decommissioning worn-out plants
Low risk of accidents because of multiple safety systems (except in 35 poorly designed and run reactors in former Soviet Union and eastern Europe)	Subject to terrorist attacks
	Spreads knowledge and technology for building nuclear weapons

Figure 10-16 Trade-offs: advantages and disadvantages of using the conventional nuclear power fuel cycle (Figure 10-15) to produce electricity. Pick the single advantage and disadvantage that you think are the most important.

Figure 10-16 lists the major advantages and disadvantages of nuclear power. Using nuclear power to produce electricity has some important advantages over coal-burning power plants (Figure 10-17).

Because of the built-in safety features, the risk of exposure to radioactivity from nuclear power plants in the United States and most other developed countries is extremely low. However, a partial or complete meltdown or explosion is possible, as accidents at the Chernobyl nuclear power plant in Ukraine (Case Study, p. 216) and the Three Mile Island plant in Pennsylvania taught us.

The U.S. Nuclear Regulatory Commission (NRC) estimates there is a 15–45% chance of a complete core meltdown at a U.S. reactor during the next 20 years. The NRC also found that 39 U.S. reactors have an 80% chance of containment shell failure from a meltdown or an explosion of gases inside the containment structures.

In the United States, there is widespread public distrust of the ability of the NRC and the Department of Energy (DOE) to enforce nuclear safety in commercial (NRC) and military (DOE) nuclear facilities. In 1996, George Galatis, a respected senior nuclear engineer, said, "I believe in nuclear power but after seeing the NRC in action I'm convinced a serious accident is not just likely, but inevitable. . . . They're asleep at the wheel."

The 2001 destruction of New York City's two World Trade Center towers by terrorist attack raised fears that a similar attack by a large plane loaded with fuel could break open a reactor's containment shell and set off a reactor meltdown that could create a major radioactive disaster.

Nuclear officials say such concerns are overblown and that U.S. nuclear plants could survive such an attack because of the thickness and strength of the containment walls. But a 2002 study by the Nuclear Control Institute found that the plants were not designed to withstand the crash of a large jet traveling at the impact speed of the two hijacked airliners that hit the World Trade Center.

An even greater concern is insufficient security at U.S. nuclear power plants against ground-level attacks by terrorists. During a series of ground-based security exercises made by the NRC between 1991 and 2001, mock attackers were able to simulate the destruction of enough equipment to cause a meltdown of nearly half of U.S. nuclear plants. A 2002 study also found that many security guards at nuclear power plants have low morale and are overworked, underpaid, undertrained, and not equipped with sufficient firepower to repel a serious ground attack by terrorists.

The NRC contends that the security weaknesses revealed by earlier mock tests have been corrected. But many nuclear power analysts are unconvinced and note that since September 2001 the NRC has stopped staging such tests.

According to critics, the problem is that the NRC is reluctant to require utilities to significantly upgrade plant security because this would increase the costs of nuclear power and make it less competitive in the marketplace.

Throughout the world, nuclear scientists and government officials urge the shutdown of 35 poorly designed and poorly operated nuclear reactors in some republics of the former Soviet Union and in eastern Europe. This is unlikely without economic aid from the world's developed countries.

Case Study: What Should We Do with High-Level Radioactive Waste? A Dangerous and Long-Lasting, Unintended Consequence

There is disagreement among scientists over methods for the long-term storage of high-level radioactive waste.

Each part of the nuclear power fuel cycle produces low-level and high-level solid, liquid, and gaseous radioactive wastes. *High-level radioactive wastes* give off large amounts of harmful radiation for a short time and small amounts for a long time. Such wastes must be stored safely for at least 10,000 years and about 240,000 years if plutonium-239 is not removed by reprocessing. They consist mainly of spent fuel rods from commercial nuclear power plants and assorted wastes from the production of nuclear weapons.

There is concern that some of the pools and casks used to stored spent fuel rods at various nuclear power plants in the United States are vulnerable to sabotage by terrorist attack. A spent-fuel pool typically holds five to ten times more long-lived radioactivity than the radioactive core inside a plant's reactor. Unlike the reactor core with its thick concrete protective dome, spent-fuel pools and dry casks have little protective cover.

An earthquake, a deliberate crash by a small airplane, or an attack by a group of suicidal terrorists could drain a pool or rupture a dry cask containing spent fuel rods. According to the NRC, this would release significant amounts of radioactive materials into

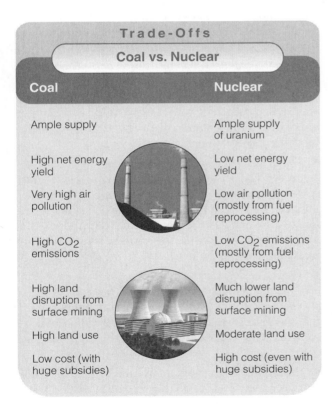

Trade-Offs

Coal vs. Nuclear

Coal	Nuclear
Ample supply	Ample supply of uranium
High net energy yield	Low net energy yield
Very high air pollution	Low air pollution (mostly from fuel reprocessing)
High CO_2 emissions	Low CO_2 emissions (mostly from fuel reprocessing)
High land disruption from surface mining	Much lower land disruption from surface mining
High land use	Moderate land use
Low cost (with huge subsidies)	High cost (even with huge subsidies)

Figure 10-17 Trade-offs: comparison of the risks of using conventional nuclear power (based on the nuclear power fuel cycle) and coal-burning plants to produce electricity. If you had to choose, would you rather live next door to a coal-fired power plant or a nuclear power plant?

Chernobyl is known around the globe as the site of the world's most serious nuclear power plant accident. On April 26, 1986, a series of explosions in one of the reactors in this nuclear power plant in Ukraine—then part of the Soviet Union—blew the massive roof off a reactor building and flung radioactive debris and dust high into the atmosphere. A huge radioactive cloud spread over much of Ukraine, Belarus, Russia, and other parts of eastern Europe, and eventually encircled the planet.

According to various UN studies, the disaster was caused by poor reactor design and human error.

Around 30 people near the accident site died from radiation exposure and nearly 2,000 children developed thyroid cancer. Health officials estimate that the accident exposed more than half a million people to dangerous levels of radioactivity, which caused from 8,000 to 15,000 premature deaths.

More than 100,000 people had to leave their homes. Most were not evacuated until at least 10 days after the accident. In 2003, Ukraine officials downgraded the 27-kilometer (17-mile) area surrounding the reactor to a "zone with high risk" to allow those willing to accept the health risk to return to their homes.

Chernobyl taught us that *a major nuclear accident anywhere has effects that reverberate throughout much of the world.*

Critical Thinking

After a nuclear accident or terrorist attack, exposed children and adults can take iodine tablets to help prevent thyroid cancer. The stable isotope of iodine in the tablets saturates the thyroid gland and blocks the uptake of radioactive iodine isotopes released by a nuclear accident. In 1997, French officials began distributing potassium iodide tablets to 600,000 people living within 10 kilometers (6 miles) of its 24 nuclear power installations. Has such action been taken in the country where you live? If not, why not?

---

the troposphere, contaminate large areas for decades, and create economic and psychological havoc.

U.S. nuclear power officials consider such events to be highly unlikely, worst-case scenarios and question some of the estimates. They also contend that nuclear power facilities are safe from attack. Critics are not convinced and call for constructing much more secure structures to protect spent-fuel storage sites. They accuse the NRC of failure to require this because it would impose additional costs on utility companies, raise the cost of nuclear power, and make it a less attractive energy alternative.

After more than 50 years of research, scientists still don't agree on whether there is a safe way for storing high-level radioactive waste. Some believe the long-term safe storage or disposal of high-level radioactive wastes is technically possible. Others disagree, pointing out that it is impossible to demonstrate that any method will work for 10,000–240,000 years.

Here are some of the proposed methods and their possible drawbacks:

- *Bury it deep underground.* This favored strategy is under study by all countries producing nuclear waste and is the option being pursued in the United States (Case Study, next page).

- *Shoot it into space or into the sun.* Costs would be very high, and a launch accident—like the explosion of the space shuttle *Challenger*—could disperse high-

level radioactive wastes over large areas of the earth's surface. This strategy has been abandoned for now.

- *Bury it under the Antarctic ice sheet or the Greenland ice cap.* The long-term stability of the ice sheets is not known. They could be destabilized by heat from the wastes, and retrieving the wastes would be difficult or impossible if the method failed. This strategy is prohibited by international law.

- *Dump it into descending zones of the earth's crust in the deep ocean.* But wastes eventually might be spewed out somewhere else by volcanic activity, and containers might leak and contaminate the ocean before being carried downward. Also, retrieval would be impossible if the method did not work. This strategy is prohibited by international law.

- *Bury it in thick deposits of mud on the deep-ocean floor in areas that tests show have been geologically stable for 65 million years.* The waste containers eventually would corrode and release their radioactive contents. This approach is prohibited by international law.

- *Change it into harmless, or less harmful, isotopes.* Currently, no way exists to do this. Even if a method was developed, costs probably would be very high, and the resulting toxic materials and low-level (but very long-lived) radioactive wastes would still need to be disposed of safely.

Case Study: What Should the United States Do with Its High-Level Radioactive Wastes? Controversy over Desert Burial

Scientists disagree over the decision to store high-level nuclear wastes at an underground storage site in Nevada.

In 1985, the U.S. Department of Energy (DOE) announced plans to build a repository for underground storage of high-level radioactive wastes from commercial nuclear reactors on federal land in the Yucca Mountain desert region, 160 kilometers (100 miles) northwest of Las Vegas, Nevada.

The proposed facility is expected to cost at least $58 billion to build (financed partly by a tax on nuclear power). It is scheduled to open by 2010, but may not open until 2015.

Some scientists argue that it should never be allowed to open, mostly because rock fractures and tiny cracks may allow water to leak into the site and eventually corrode radioactive waste storage casks. Geologists also point out a nearby active volcano and 32 active earthquake fault lines running through the site—an unusually high number. In 1998, Jerry Szymanski, formerly the DOE's top geologist at Yucca Mountain and now an outspoken opponent of the site, said that if water flooded the site it could cause an explosion so large that "Chernobyl would be small potatoes."

In 2002, the U.S. National Academy of Sciences, in collaboration with Harvard and University of Tokyo scientists, urged the U.S. government to slow down and rethink the nuclear waste storage process. These scientists contend that storing spent fuel rods in dry-storage casks in well-protected buildings at nuclear plant sites is an adequate solution for at least 100 years in terms of safety and national security. This would buy time to carry out more research on this complex problem and to evaluate other sites and storage methods that might be more acceptable scientifically and politically.

Opponents also contend that the Yucca Mountain waste site should not be opened because *it can decrease national (homeland) security.* The wastes would be put into specially designed casks and shipped by truck or rail cars to the Nevada site. This would require about 19,600 shipments of wastes over much of the country for the estimated 38 years before the site is filled. Critics contend that it is much more difficult to protect such a large number of shipments from a terrorist attack than to provide more secure ways to store such wastes at nuclear power plant sites.

The DOE and proponents of nuclear power say the risks of an accident or sabotage of nuclear waste shipments are negligible. Opponents believe the risks are underestimated, especially after the events of September 11, 2001.

Despite these and other objections from scientists and citizens, during the summer of 2002, Congress approved Yucca Mountain as the official site for storing the country's commercial nuclear wastes. Opponents want the law repealed.

X *HOW WOULD YOU VOTE?* Should highly radioactive spent fuel be stored in well-protected buildings at nuclear power plant sites instead of shipping them to a single site for underground burial? Cast your vote online at http://biology.brookscole.com/miller7.

What Can We Do with Worn-Out Nuclear Plants? A Costly Dilemma

When a nuclear reactor reaches the end of its useful life we have to keep its highly radioactive materials from reaching the environment for thousands of years.

When a nuclear power plant comes to the end of its useful life, it must be *decommissioned,* or retired—the last step in the nuclear power fuel cycle. Scientists have proposed three ways to do this.

One is to dismantle the plant after it closes and store its large volume of highly radioactive materials in a high-level, nuclear waste storage facility, whose safety is questioned by many scientists. A second approach is to put up a physical barrier around the plant and set up full-time security for 30–100 years before the plant is dismantled. A third option is to enclose the entire plant in a tomb that must last and be monitored for several thousand years. Regardless of the method chosen, decommissioning adds to the total costs of nuclear power as an energy option.

At least 228 large commercial reactors worldwide (20 in the United States) are scheduled for retirement by 2012. However, the U.S. Nuclear Regulatory Commission has approved extending the 40-year expected life expectancy of at least 40 U.S. reactors to 60 years. Opponents contend this could increase the risk of nuclear accidents in aging reactors.

What Are "Dirty" Bombs? A Serious Radioactive Threat

Terrorists could wrap conventional explosives around small amounts of various radioactive materials that are fairly easy to get, detonate such bombs, and contaminate fairly large areas with radioactivity for decades.

Since the terrorist attacks in the United States on September 11, 2001, there has been growing concern about threats from explosions of so-called *"dirty" bombs.* Such a bomb consists of an explosive such as dynamite mixed with or wrapped around some form of radioactive material—an amount that could fit in a coffee cup.

Such radioactive materials can be stolen from thousands of poorly guarded and difficult to protect sources or bought on the black market. Sources might be hospitals that use radioisotopes (such as cobalt-60) to treat cancer and diagnose various diseases. Other sources might include university research labs, and industries using radioisotopes to detect leaks in underground pipes, irradiate food, examine mail and other materials, and detect flaws in pipe welds and boilers.

Since 1986, the NRC has recorded 1,700 incidents in the United States in which radioactive materials used by industrial, medical, or research facilities have been stolen or lost. And since 1991, the International Atomic Energy Agency (IAEA) has detected 671 incidents of illicit trafficking in dirty-bomb materials.

Detonating a dirty bomb at street level or on a rooftop does not cause a nuclear blast. But it could kill up to 1,000 people in densely populated cities and spread radioactive material over several blocks. This would pose cancer risks for decades, and cause a lot of terror and panic—the primary objective of terrorists.

Can Nuclear Power Reduce Dependence on Imported Oil and Help Reduce Global Warming? Misleading Myths

Because so little oil is burned to produce electricity in the United States, building more nuclear power plants is not a way to reduce dependence on imported oil and is also not a major way to reduce carbon dioxide emissions.

Some proponents of nuclear power in the United States claim it will help reduce dependence on imported oil. But other analysts point out that use of nuclear power has little effect on U.S. oil use because burning oil typically produces only 2–3% of the electricity in the United States. Also, the major use for oil is in transportation, which would not be affected by increasing the use of nuclear power plants to produce electricity.

Nuclear power advocates also contend that increased use of nuclear power would reduce the threat of global warming by eliminating emissions of CO_2 compared to burning coal. Scientists point out this is only partially correct. Nuclear plants themselves are not emitters of CO_2. However, the nuclear fuel cycle does produce some CO_2, although it is much less than that produced by burning coal or natural gas to produce electricity (Figure 10-7). Environmentalists and many energy experts argue that reducing energy waste and increasing the use of wind turbines, solar cells, and hydrogen to produce electricity are better ways to reduce CO_2 emissions.

Can We Afford Nuclear Power? Burning Money

Even with massive government subsidies, the nuclear power fuel cycle is an expensive way to produce electricity compared to a number of other energy alternatives.

Experience has shown that the nuclear power fuel cycle is an expensive way to produce electricity, even when huge government subsidies partially shield it from free-market competition with other energy sources. In 1995, the World Bank said nuclear power is too costly and risky. *Forbes* business magazine has called the failure of the U.S. nuclear power program "the largest managerial disaster in U.S. business history, involving $1 trillion in wasted investment and $10 billion in direct losses to stockholders."

In recent years, the operating costs of many U.S. nuclear power plants have dropped, mostly because of less downtime. But environmentalists and economists point out that the cost of nuclear power must be based on the entire nuclear power fuel cycle, not merely the operating cost of individual plants. According to these analysts, when these costs (including nuclear waste disposal and decommissioning of worn-out plants) are included, the overall cost of nuclear power is very high (even with huge government subsidies) compared to many other energy alternatives.

Partly to address these concerns, the U.S. nuclear industry hopes to persuade the Congress and utility companies to build hundreds of smaller, second-generation plants using standardized designs, which they claim are safer and can be built more quickly (in 3–6 years).

These *advanced light-water reactors (ALWRs)* have built-in *passive safety features* designed to make explosions or the release of radioactive emissions almost impossible. However, according to *Nucleonics Week*, an important nuclear industry publication, "Experts are flatly unconvinced that safety has been achieved—or even substantially increased—by the new designs." In addition, these new designs do not eliminate the threats and the expense and hazards of long-term radioactive waste storage and power plant decommissioning.

Is Breeder Nuclear Fission a Feasible Alternative? A Failed Technology

Because of very high costs and bad safety experiences with several nuclear breeder reactors, this technology has essentially been abandoned.

Some nuclear power proponents urge the development and widespread use of **breeder nuclear fission reactors,** which generate more nuclear fuel than they consume by converting nonfissionable uranium-238 into fissionable plutonium-239. Because breeders would use more than 99% of the uranium in ore deposits, the world's known uranium reserves would last at least 1,000 years, and perhaps several thousand years.

However, if the safety system of a breeder reactor fails, the reactor could lose some of its liquid sodium coolant, which ignites when exposed to air and reacts explosively if it comes into contact with water. This could cause a runaway fission chain reaction and perhaps a nuclear explosion powerful enough to blast open the containment building and release a cloud of highly radioactive gases and particles into the atmosphere. Leaks of flammable liquid sodium can also cause fires, as has happened with all experimental breeder reactors built so far.

In addition, existing experimental breeder reactors produce plutonium so slowly that it would take 100–200 years for them to produce enough plutonium to fuel a significant number of other breeder reactors. In 1994, the United States ended government-supported research for breeder technology after providing about $9 billion in research and development funding.

In December 1986, France opened a commercial-size breeder reactor. It was so expensive to build and operate that after spending $13 billion the government spent another $2.75 billion to shut it down permanently in 1998. Because of this experience, other countries have abandoned their plans to build full-size commercial breeder reactors.

Is Nuclear Fusion a Feasible Alternative? Still at the Laboratory Stage

Nuclear fusion has a number of advantages, but after more than five decades of research and billions of dollars in government research and development subsidies, this technology is still at the laboratory stage.

Nuclear fusion is a nuclear change in which two isotopes of light elements, such as hydrogen, are forced together at extremely high temperatures until they fuse to form a heavier nucleus, releasing energy in the process. Scientists hope that controlled nuclear fusion will provide an almost limitless source of high-temperature heat and electricity. Research has focused on the D–T nuclear fusion reaction, in which two isotopes of hydrogen—deuterium (D) and tritium (T)—fuse at about 100 million degrees (Figure 10-18, top).

With nuclear fusion, there would be no risk of meltdown or release of large amounts of radioactive materials from a terrorist attack and little risk from additional proliferation of nuclear weapons because bomb-grade materials (such as enriched uranium-235 and plutonium-239) are not required for fusion energy. Fusion power might also be used to destroy toxic wastes, supply electricity for ordinary use, and decompose water to produce the hydrogen gas needed to run a hydrogen economy by the end of this century.

This sounds great. So what is holding up fusion energy? After more than 50 years of research and huge expenditures of mostly government funds, controlled nuclear fusion is still in the laboratory stage. None of the approaches tested so far have produced more energy than they use.

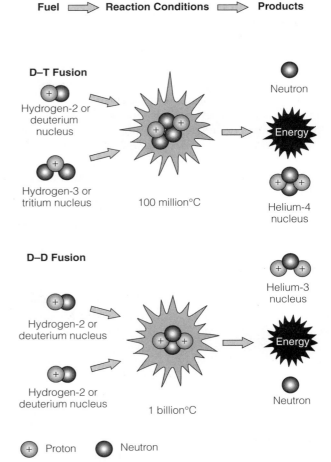

Fuel ⟹ Reaction Conditions ⟹ Products

Figure 10-18 The deuterium–tritium (D–T) and deuterium–deuterium (D–D) nuclear fusion reactions, which take place at extremely high temperatures.

If researchers can eventually get more energy out of nuclear fusion than they put in, the next step would be to build a small fusion reactor and then scale it up to commercial size. This is an extremely difficult engineering problem. Also, the estimated cost of a building and operating a commercial fusion reactor (even with huge government subsidies) is several times that of a comparable conventional fission reactor.

Proponents contend that with greatly increased federal funding, a commercial nuclear fusion power plant might be built by 2030 or perhaps by 2020. However, many energy experts do not expect nuclear fusion to be a significant energy source until 2100, if then.

What Should Be the Future of Nuclear Power in the United States? Phase Out or Keep Options Open

There is disagreement over whether the United States should phase out nuclear power or keep this option open in case other alternatives do not pan out.

Since 1948, nuclear energy (fission and fusion) has received about 58% of all federal energy research and development funds in the United States—compared to 22% for fossil fuels, 11% for renewable energy, and 8% for energy efficiency and conservation. Because the results of such a huge investment of taxpayer dollars have been disappointing, some analysts call for phasing out all or most government subsides and tax breaks for nuclear power and using the money to subsidize and accelerate the development of other more promising energy technologies.

To these analysts, nuclear power is a complex, expensive, inflexible, and centralized way to produce electricity that is too vulnerable to terrorist attack. They believe it is a technology whose time has passed in a world where electricity will increasingly be provided by small, decentralized, easily expandable power plants such as natural gas turbines, farms of wind turbines, arrays of solar cells, and hydrogen-powered fuel cells. According to investors and World Bank economic analysts, conventional nuclear power simply cannot compete in today's increasingly open, decentralized, and unregulated energy market unless it is artificially shielded from free-market competition by huge government subsidies.

Proponents of nuclear power argue that governments should continue funding research and development and pilot plant testing of potentially safer and cheaper reactor designs along with breeder fission and nuclear fusion. They say we need to keep these nuclear options available for use in the future if various renewable energy options fail to keep up with electricity demands and reduce CO_2 emissions to acceptable levels. Germany does not buy these arguments and has plans to phase out nuclear power over the next two decades.

X *How Would You Vote?* Should nuclear power be phased out in the country where you live over the next 20 to 30 years? Cast your vote online at http://biology.brookscole.com/miller7.

10-4 IMPROVING ENERGY EFFICIENCY

How Much Energy Do We Waste Unnecessarily? Throwing Money Away

About 43% of the energy used in the United States is wasted unnecessarily.

You may be surprised to learn that 84% of all commercial energy used in the United States is wasted (Figure 10-19). About 41% of this energy is wasted automatically because of the degradation of energy quality imposed by the second law of thermodynamics (p. 22). But about 43% is wasted unnecessarily. This waste is largely the result of using fuel-wasting motor vehicles,

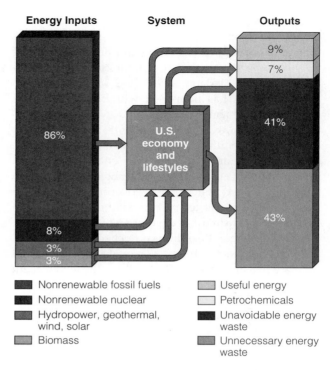

Figure 10-19 Flow of commercial energy through the U.S. economy. Note that only 16% of all commercial energy used in the United States ends up performing useful tasks or being converted to petrochemicals; the rest is either automatically and unavoidably wasted because of the second law of thermodynamics (41%) or wasted unnecessarily (43%). (Data from U.S. Department of Energy)

furnaces, and other devices, and living and working in leaky, poorly insulated, poorly designed buildings. See the Guest Essay on this topic by Amory Lovins on the website for this chapter. According to the U.S. Department of Energy (DOE), the United States unnecessarily wastes as much energy as two-thirds of the world's population consumes.

The way to reduce energy waste is to improve *energy efficiency* by using less energy to do more useful work. In other words, we learn how to do more with less. Reducing such energy waste has a number of economic and environmental advantages (Figure 10-20).

Improvements in energy efficiency since 1973 have cut U.S. energy bills by $275 billion a year. But unnecessary energy waste still costs the United States about $300 billion per year—an average of $570,000 per minute.

The energy conversion devices we use vary in their energy efficiencies (Figure 10-21). We can save energy and money by buying more energy-efficient cars, lighting, heating systems, water heaters, air conditioners, and appliances. Some energy-efficient models may cost more initially, but in the long run they usually save money by having a lower **life cycle cost:** initial cost plus lifetime operating costs.

Four widely used devices waste large amounts of energy. One is the *incandescent light bulb*, which wastes 95% of its energy input of electricity. In other words, it is a *heat bulb*. The second is a *nuclear power plant* producing electricity for space heating or water heating. Such a plant wastes about 86% of the energy in its nuclear fuel and probably 92% when we include the energy needed to deal with its radioactive wastes and to retire the plant. Third is a motor vehicle with an *internal combustion engine,* which wastes 75–80% of the energy in its fuel. The fourth is a *coal-burning power plant* in which two-thirds of the energy released by burning coal ends up as waste heat in the environment. Energy experts call for us to replace these devices or greatly improve their energy efficiency over the next few decades.

How Can We Save Energy and Money in Industry? Cogenerate, Buy New Motors, and Use Efficient Lighting

Industries can save energy and money by producing both heat and electricity from an energy source and by using more energy-efficient electric motors and lighting.

One way some industries save energy and money is to use **cogeneration,** or *combined heat and power (CHP)* systems. In such a system, two useful forms of energy (such as steam and electricity) are produced from the same fuel source. These systems have an efficiency of up to 80% (compared to about 30–40% for coal-fired boilers and nuclear power plants) and emit two-thirds less CO_2 per unit of energy produced than conventional coal-fired boilers.

Cogeneration has been widely used in western Europe for years. Its use in the United States (where it now produces 9% of the country's electricity) and China is growing.

Another way to save energy and money in industry is to *replace energy-wasting electric motors,* which consume about one-fourth of the electricity produced in the United States. Most of these motors are inefficient because they run only at full speed with their output throttled to match the task—somewhat like driving a car very fast with your foot on the brake pedal. Each year, a heavily used electric motor consumes 10 times its purchase cost in electricity—equivalent to using $200,000 worth of gasoline each year to fuel a $20,000 car! The costs of replacing such motors with new adjustable-speed drive motors would be paid back in about 1 year and save an amount of energy equal to that generated by 150 large (1,000-megawatt) power plants.

A third way to save energy is to *switch from low-efficiency incandescent lighting to higher-efficiency fluorescent lighting* (Figure 10-21).

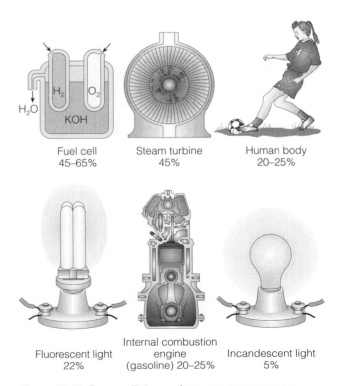

Solutions
Reducing Energy Waste

Prolongs fossil fuel supplies

Reduces oil imports

Very high net energy

Low cost

Reduces pollution and environmental degradation

Buys time to phase in renewable energy

Less need for military protection of Middle East oil resources

Improves local economy by reducing flow of money out to pay for energy

Creates local jobs

Figure 10-20 Solutions: advantages of reducing energy waste. Global improvements in energy efficiency could save the world about $1 trillion per year—an average of $114 million per hour!

Fuel cell 45–65%

Steam turbine 45%

Human body 20–25%

Fluorescent light 22%

Internal combustion engine (gasoline) 20–25%

Incandescent light 5%

Figure 10-21 Energy efficiency of some common energy conversion devices.

How Can We Save Energy in Transportation? Replace Gas Guzzlers with Gas Sippers

The best way to save energy in transportation is to increase the fuel efficiency of motor vehicles.

Good news: Between 1973 and 1985, the average fuel efficiency rose 37% for new cars sold in the United States because of government-mandated *Corporate Average Fuel Economy* (*CAFE*) standards. *Bad news:* Between 1985 and 2004, the average fuel efficiency of new cars sold in the United States leveled off or declined slightly.

Fuel-efficient cars are available but account for less than 1% of all car sales. One reason is that the inflation-adjusted price of gasoline today in the United States is low (Figure 10-22). A second reason is that two-thirds of U.S. consumers prefer SUVs, pickup trucks, minivans, and other large, inefficient vehicles. A third reason is the failure of elected officials to raise CAFE standards since 1985 because of opposition from automakers and oil companies.

Suppose that Congress required the average car in the United States get 17 kilometers per liter (kpl) [40 miles per gallon (mpg)] within 10 years. According to energy analysts, this would cut gasoline consumption in half, save more than three times the amount of oil in the nation's current proven oil reserves, and also save enough oil to eliminate all current oil imports to the United States from the Middle East. In 2003, China announced plans to impose much stricter fuel-efficiency standards than the United States to reduce the country's dependence on oil imports and reduce carbon dioxide emissions.

✗ HOW WOULD YOU VOTE? Should the government greatly increase fuel efficiency standards for all vehicles in the United States or the country where you live? Cast your vote online at http://biology.brookscole.com/miller7.

Case Study: Are Hybrid and Fuel-Cell Cars the Answer? New Options

Fuel-efficient, hybrid-electric vehicles are powered by a battery and a small internal combustion engine that recharges the battery, and electric vehicles are powered by fuel cells running on hydrogen

There is growing interest in developing *superefficient cars* that could eventually get 34–128 kilometers per liter (80–300 miles per gallon). Amory Lovins developed and promoted this concept in the 1980s. See his Guest Essay on the website for this chapter.

One type of energy-efficient car uses a *hybrid-electric internal combustion engine*. It runs on gasoline, diesel fuel, or natural gas and uses a small battery (recharged by the internal combustion engine) to provide the energy needed for acceleration and hill climbing.

Toyota introduced its first hybrid vehicle in 1997, and Honda and Nissan have been selling several models of hybrid vehicles in the United States since 2000. Carmakers plan to introduce up to 20 hybrid models, including cars, trucks, SUVs, and vans, in the next 4–5 years. Sales of hybrid motor vehicles are projected to grow rapidly and probably dominate motor vehicle sales between 2010 and 2030.

Another type of superefficient car is an electric vehicle that uses a *fuel cell*—a device that combines hydrogen gas (H_2) and oxygen gas (O_2) in the air to produce electrical energy to power the car and water vapor (H_2O), which is emitted into the atmosphere.

Fuel cells are at least twice as efficient as internal combustion engines, have no moving parts, require little maintenance, and produce little or no pollution, depending on how their hydrogen fuel is produced. Most major automobile companies have developed prototype fuel-cell cars and plan to market a variety of such vehicles by 2020 (with a few models available by 2010; Figure 10-23) and greatly increase their use by 2050. Until then, hybrids will probably have an advantage because they are available now and get their fuel from regular filling stations instead of having to depend on building a new network of hydrogen filling stations.

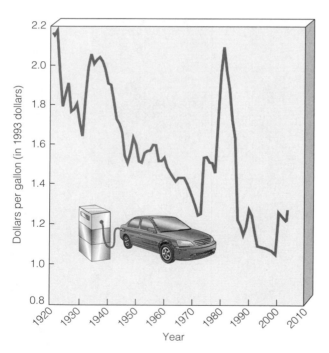

Figure 10-22 Real price of gasoline (in 1993 dollars) in the United States, 1920–2004. The 225 million motor vehicles in the United States use about 40% of the world's gasoline. Gasoline is one of the cheapest items American consumers buy and costs less per liter than bottled water. (Data from U.S. Department of Energy)

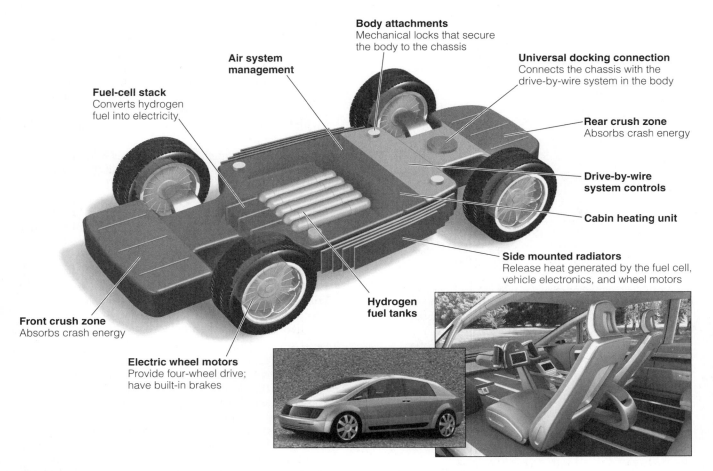

Figure 10-23 Prototype Hy-Wire (for hydrogen by wire) car of the future developed by General Motors. It combines a hydrogen fuel cell with drive-by-wire technology and should have a fuel efficiency equivalent of more than 43 kpl (100 mpg). It consists of a skateboard-like chassis and a variety of snap-on fiberglass bodies. It handles like a high-speed sports car, zips along with no engine noise, and emits only wisps of warm water vapor and heat—no smelly exhaust, no smog, no greenhouse gases. The company claims the car could be on the road within a decade, but some analysts believe that it will be 2020–2030 before a variety of such cars from various manufacturers will be mass-produced (Information from General Motors).

How Can We Design Buildings to Save Energy? Work with Nature

We can save energy in buildings by getting heat from the sun, superinsulating them, and using plant-covered ecoroofs.

Atlanta's 13-story Georgia Power Company building uses 60% less energy than conventional office buildings of the same size. The largest surface of the building faces south to capture solar energy. Each floor extends out over the one below it. This blocks out the higher summer sun to reduce air conditioning costs but allows warming by the lower winter sun. Energy-efficient lights focus on desks rather than illuminating entire rooms. In contrast, the conventional Sears Tower building in Chicago consumes more energy in a day than does a city of 150,000 people. Green architecture is be-

ginning to catch on in Europe, the United States, and Japan.

Another energy-efficient design is a *superinsulated house* (Figure 10-24, p. 224). Such houses typically cost 5% more to build than conventional houses of the same size. But this extra cost is paid back by energy savings within about 5 years and can save a homeowner $50,000–100,000 over a 40-year period. Superinsulated houses in Sweden use 90% less energy for heating and cooling that the typical American home.

Ecoroofs or *green roofs* covered with plants have been used in Germany, in other parts of Europe, and in Iceland for decades. With proper design, these plant-covered roof gardens provide good insulation, absorb storm water and release it slowly, outlast conventional roofs, and make a building or home more

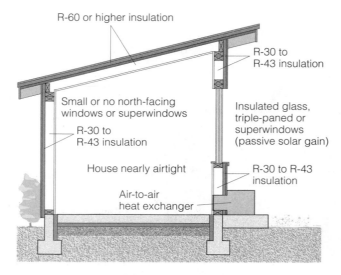

R-60 or higher insulation

R-30 to R-43 insulation

Small or no north-facing windows or superwindows

R-30 to R-43 insulation

Insulated glass, triple-paned or superwindows (passive solar gain)

House nearly airtight

R-30 to R-43 insulation

Air-to-air heat exchanger

Figure 10-24 Solutions: major features of a *superinsulated house.* Such a house is so heavily insulated and so airtight that heat from direct sunlight, appliances, and human bodies can warm it with little or no need for a backup heating system. An air-to-air heat exchanger prevents buildup of indoor air pollution.

energy efficient. Designing and installing such systems could be an interesting career.

How Can We Save Energy in Existing Buildings? Stop Leaks and Use Energy-Efficient Devices

We can save energy in existing buildings by insulating them, plugging leaks, and using energy-efficient heating and cooling systems, appliances, and lighting.

Here are some ways to save energy in existing buildings.

- *Insulate and plug leaks.* About one-third of heated air in U.S. homes and buildings escapes through closed windows and holes and cracks—roughly equal to the energy in all the oil flowing through the Alaska pipeline every year. During hot weather, these windows and cracks also let heat in, increasing the use of air conditioning. Although not very sexy, adding insulation and plugging leaks in a house are two of the quickest, cheapest, and best ways to save energy and money.

- *Use energy-efficient windows.* Replacing all windows in the United States with energy-efficient windows would cut expensive heat losses from houses by two-thirds, lessen cooling costs in the summer, and reduce CO_2 emissions. Widely available superinsulating windows insulate as well as 8–12 sheets of glass. Although they cost 10–15% more than double-glazed windows, this cost is paid back rapidly by the energy they save. Even better windows will reach the market soon.

- *Heat houses more efficiently.* In order, the most energy-efficient ways to heat space are superinsulation, a geothermal heat pump, passive solar heating, a conventional heat pump (in warm climates only), small cogenerating microturbines, and a high-efficiency (85–98%) natural gas furnace. The most wasteful and expensive way is to use electric resistance heating with the electricity produced by a coal-fired or nuclear power plant.

- *Heat water more efficiently.* One way to do this is to use a *tankless instant water heater* (about the size of a small suitcase) fired by natural gas or LPG but not by electricity. These devices, widely used in many parts of Europe, heat water instantly as it flows through a small burner chamber, provide hot water only when it is needed, and use less energy than traditional water heaters.*

- A well-insulated, conventional, natural gas or LPG water heater is fairly efficient. But all conventional natural gas and electric resistance heaters waste energy by keeping a large tank of water hot all day and night and can run out after a long shower or two.

- *Use energy-efficient appliances and lighting.* If all households in the United States used the most efficient frost-free refrigerator now available, 18 large (1,000-megawatt) power plants could close. Microwave ovens can cut electricity use for cooking by 25–50% (but not if used for defrosting food). Clothes dryers with moisture sensors cut energy use by 15%, and front-loading washers use 50% less energy than top-loading models but cost about the same. Replacing twenty-one energy-wasting incandescent light bulbs in a house or building with energy-efficient fluorescent bulbs typically saves $1,125. What a great investment payoff.

- *Set strict energy-efficiency standards for new buildings.* Building codes could require that new houses use 60–80% less energy than conventional houses of the same size, as has been done in Davis, California. Because of tough energy-efficiency standards, the average Swedish home consumes about one-third as much energy as the average American home of the same size.

Why Are We Still Wasting So Much Energy? We Get What We Reward

Low-priced fossil fuels and lack of government tax breaks for saving energy promote energy waste.

With such an impressive array of benefits (Figure 10-20), why is there so little emphasis on improving energy efficiency? One reason is a glut of low-cost gasoline (Figure 10-22) and other fossil fuels. As long as energy is

---

*They work great. I used them in a passive solar office and living space for 15 years. Models are available for $500–1,000 from companies such as Rinnai, Bosch, Takagi, and Envirotech. For information, visit foreverhotwater.com

artificially cheap because its market price does not include its harmful costs, people are more likely to waste it and not make investments in improving energy efficiency.

Another reason is a lack of sufficient tax breaks and other economic incentives for consumers and businesses to invest in improving energy efficiency.

Would you like to earn about 20% a year on your money, tax-free and risk free? Invest it in improving the energy efficiency of your home and in energy-efficient lights and appliances. You get your investment back in a few years and then make about 20% a year by having lower heating, cooling, and electricity bills. This is a win-win deal for you and the earth.

10-5 USING RENEWABLE ENERGY TO PROVIDE HEAT AND ELECTRICITY

What Are the Main Types of Renewable Energy? Solar Capital

Six types of renewable energy are solar, flowing water, wind, biomass, hydrogen, and geothermal.

One of the four keys to sustainability (Figure 4-10, p. 74), based on learning from nature, is to *rely mostly on renewable solar energy*. We can get renewable solar energy directly from the sun or indirectly from moving water, wind, and biomass. Two other forms of renewable energy are geothermal energy from the earth's interior and using renewable energy to produce hydrogen fuel from water. Like fossil fuels and nuclear power, each of these renewable energy alternatives has advantages and disadvantages, as discussed in the remainder of this chapter.

If renewable energy is so great why is it that it provides only 16% of the world's energy and 6% of the energy in the United States? One reason is that renewable energy resources have received and continue to receive much lower government tax breaks, subsidies, and research and development (R & D) funding than fossil fuels and nuclear power have received for decades. The other reason is that the prices we pay for fossil fuels and nuclear power do not include the costs of their harm to the environment and to human health.

In other words, the economic dice have been loaded against solar, wind, and other forms of renewable energy. If the economic playing field was made more even, energy analysts say that many of these forms of renewable energy would take over—another example of the *you-get-what-you-reward* economic principle in action.

How Can We Use Direct Solar Energy to Heat Houses and Water? Face the Sun and Store Its Heat

We can heat buildings by orienting them toward the sun (passive solar heating) or by pumping a liquid such as water through rooftop collectors (active solar heating).

Buildings and water can be heated by solar energy using two methods: passive and active (Figure 10-25). A **passive solar heating system** absorbs and stores heat from the sun directly within a structure (Figure 10-25, left). See the Guest Essay by Nancy Wicks on this topic on the website for this chapter. Energy-efficient windows and attached greenhouses face the sun to collect solar energy by direct gain. Walls and floors of concrete, adobe, brick, stone, salt-treated timber, and water in metal or plastic containers store much of the

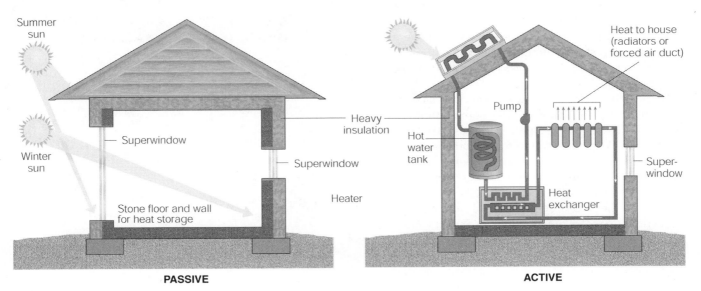

Figure 10-25 Solutions: passive and active solar heating for a home.

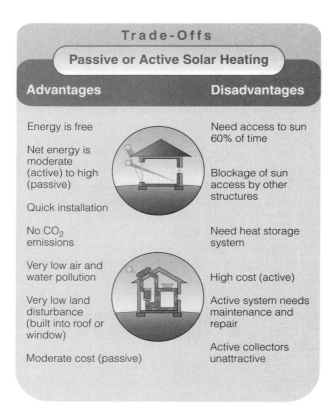

Trade-Offs

Passive or Active Solar Heating

Advantages	Disadvantages
Energy is free	Need access to sun 60% of time
Net energy is moderate (active) to high (passive)	Blockage of sun access by other structures
Quick installation	Need heat storage system
No CO_2 emissions	High cost (active)
Very low air and water pollution	Active system needs maintenance and repair
Very low land disturbance (built into roof or window)	Active collectors unattractive
Moderate cost (passive)	

Figure 10-26 Trade-offs: advantages and disadvantages of heating a house with passive or active solar energy. Pick the single advantage and disadvantage that you think are the most important.

collected solar energy as heat and release it slowly throughout the day and night. A small backup heating system such as a vented natural gas or propane heater may be used but is not necessary in many climates.

On a life cycle cost basis, good passive solar and superinsulated design is the cheapest way to heat a home or small building in regions with access to ample sunlight. Such a system usually adds 5–10% to the construction cost, but the life cycle cost of operating such a house is 30–40% lower. The typical payback time for passive solar features is 3–7 years.

An **active solar heating system** absorbs energy from the sun by pumping a heat-absorbing fluid (such as water or an antifreeze solution) through special collectors usually mounted on a roof or on special racks to face the sun (Figure 10-25, right). Some of the collected heat can be used directly. The rest can be stored in a large insulated container filled with gravel, water, clay, or a heat-absorbing chemical for release as needed. Active solar collectors can also supply hot water and are widely used in areas of the world with sunny climates.

Figure 10-26 lists the major advantages and disadvantages of using passive or active solar energy for heating buildings. Passive solar energy is great for new homes but cannot be used to heat existing homes and

buildings not oriented to receive sunlight or where other buildings or trees block access to sunlight. Active solar collectors are good for heating water in sunny areas. But most analysts do not expect widespread use of active solar collectors for heating houses because of high costs, maintenance requirements, and unappealing appearance.

How Can We Use Solar Energy to Generate High-Temperature Heat and Electricity? Desert Power

Large arrays of solar collectors in sunny deserts can produce high-temperature heat to spin turbines and produce electricity, but costs are high.

Several *solar thermal systems* can collect and transform radiant energy from the sun into high-temperature thermal energy (heat), which can be used directly or converted to electricity. These approaches are used mostly in desert areas with ample sunlight.

One method uses a *central receiver system*, called a *power tower*. Huge arrays of computer-controlled mirrors called *heliostats* track the sun and focus sunlight on a central heat collection tower (top drawing in Figure 10-27).

Another approach is a *solar thermal plant* in which sunlight is collected and focused on oil-filled pipes running through the middle of a large area of curved solar collectors (bottom drawing in Figure 10-27). This concentrated sunlight can generate temperatures high

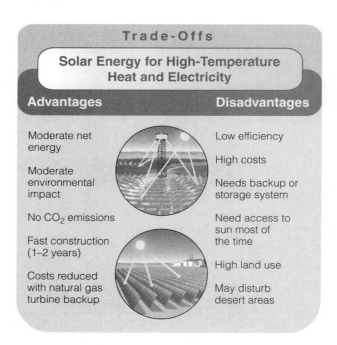

Trade-Offs

Solar Energy for High-Temperature Heat and Electricity

Advantages	Disadvantages
Moderate net energy	Low efficiency
Moderate environmental impact	High costs
No CO_2 emissions	Needs backup or storage system
Fast construction (1–2 years)	Need access to sun most of the time
Costs reduced with natural gas turbine backup	High land use
	May disturb desert areas

Figure 10-27 Trade-offs: advantages and disadvantages of using solar energy to generate high-temperature heat and electricity. Pick the single advantage and disadvantage that you think are the most important.

enough for producing steam to run turbines and generate electricity. At night or on cloudy days, high-efficiency, combined-cycle, natural gas turbines can supply backup electricity as needed.

On an individual scale, inexpensive *solar cookers* can focus and concentrate sunlight to cook food, especially in rural villages in sunny developing countries. They can be made by fitting an insulated box big enough to hold three or four pots with a transparent, removable top. Solar cookers reduce deforestation for fuelwood and the time and labor needed to collect firewood. They also reduce indoor air pollution from smoky fires.

Figure 10-27 lists the advantages and disadvantages of concentrating solar energy to produce high-temperature heat or electricity. Most analysts do not expect widespread use of such technologies over the next few decades because of high costs, limited suitable sites, and availability of much cheaper ways to produce electricity such as combined-cycle natural gas turbines and wind turbines.

How Can We Produce Electricity with Solar Cells? Use Your Roof, Outside Walls, or Windows As a Power Plant

Solar cells that convert sunlight to electricity can be incorporated into roofing materials or windows, and the high costs of doing this are expected to fall.

Solar energy can be converted directly into electrical energy by **photovoltaic (PV) cells,** commonly called **solar cells** (Figure 10-28). A typical solar cell is a transparent wafer that contains a semiconductor with a thickness ranging from less than that of a human hair to a sheet of paper. Sunlight energizes and causes electrons in the semiconductor to flow, creating an electrical current.

The semiconductor material used in solar cells can be made into paper-thin rigid or flexible sheets that can be incorporated into traditional-looking roofing materials (Figure 10-28). Glass walls and windows of buildings can also have built-in solar cells.

Easily expandable banks of solar cells can be used to provide electricity in developing countries for 1.7 billion people in rural villages without electricity. Such banks of cells can also produce electricity at a small power plant (see bottom drawing in Figure 10-29, p. 228), using combined-cycle, natural gas turbines to provide backup power when the sun is not shining. Another possibility is to use arrays of solar cells to convert water to hydrogen gas that can be distributed to energy users by pipeline, as natural gas is. With financing from the World Bank, India (the world's number-one market for solar cells) is installing solar-cell systems in 38,000 villages, and Zimbabwe is bringing solar electricity to 2,500 villages, mostly because they were long distances from power grids.

Figure 10-29 lists the advantages and disadvantages of solar cells. Current costs of producing electricity from solar cells are high but are expected to drop because of savings from mass production and new designs.

Currently, solar cells supply only about 0.05% of the world's electricity. But with increased government and private research and development, and greater government tax breaks and other subsidies, they could provide a quarter of the world's electricity by 2040. If such projections are correct, the production, sale, and

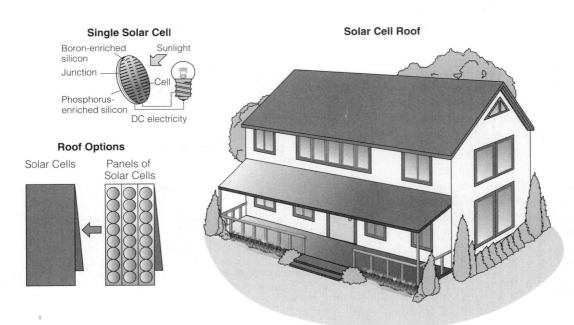

Single Solar Cell

Boron-enriched silicon
Junction
Phosphorus-enriched silicon
Sunlight
Cell
DC electricity

Roof Options

Solar Cells   Panels of Solar Cells

Solar Cell Roof

Figure 10-28 Solutions: photovoltaic (PV) (solar) cells can provide electricity for a house or building using new solar-cell roof shingles or PV-panel roof systems that look like metal roofs. Arrays of such cells can also produce electricity for a village or at a small power plant. New thin film solar cells can also be applied to windows and outside glass walls.

Trade-Offs

Solar Cells

Advantages	Disadvantages

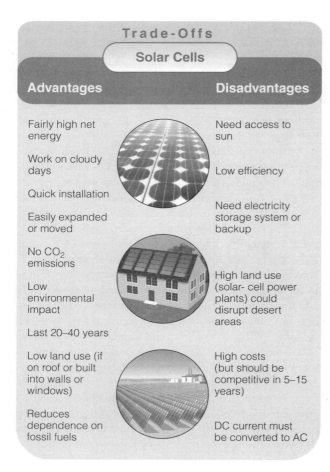

Advantages	Disadvantages
Fairly high net energy	Need access to sun
Work on cloudy days	Low efficiency
Quick installation	Need electricity storage system or backup
Easily expanded or moved	
No CO_2 emissions	High land use (solar-cell power plants) could disrupt desert areas
Low environmental impact	
Last 20–40 years	
Low land use (if on roof or built into walls or windows)	High costs (but should be competitive in 5–15 years)
Reduces dependence on fossil fuels	DC current must be converted to AC

Figure 10-29 Trade-offs: advantages and disadvantages of using solar cells to produce electricity. Pick the single advantage and disadvantage that you think are the most important.

installation of solar cells could become one of the world's largest and fastest-growing businesses. This is another exciting field to consider as a career choice.

How Can We Produce Electricity from Flowing Water? Renewable Hydropower

Water flowing in rivers and streams can be trapped in reservoirs behind dams and released as needed to spin turbines and produce electricity.

Solar energy evaporates water and deposits it as water and snow in other areas as part of the water cycle (Figure 2-25, p. 39). Water flowing from high elevations to lower elevations in rivers and streams can be controlled by dams and reservoirs and used to produce electricity (Figure 9-10, p. 178). This indirect form of renewable solar energy is called *hydropower*.

The most widely used approach is to build a high dam is across a large river to create a reservoir. Some of the water stored in the reservoir is allowed to flow

through huge pipes at controlled rates, spinning turbines and producing electricity. Smaller and lower dams can also be used.

In 2002, hydropower supplied about one-fifth of the world's electricity, 99% of the electricity in Norway, 75% in New Zealand, 25% in China, and 7% in the United States (but about 50% on the West Coast).

Figure 10-30 lists the advantages and disadvantages of using large-scale hydropower plants to produce electricity.

Because of increasing concern about the harmful environmental and social consequences of large dams, there has been growing pressure on the World Bank and other development agencies to stop funding new large-scale hydropower projects. Also, according to a 2000 study by the World Commission on Dams, hydropower in tropical countries is a major emitter of greenhouse gases. This occurs because the reservoirs that power the dams can trap rotting vegetation, which can emit greenhouse gases such as carbon dioxide and methane.

Trade-Offs

Large-Scale Hydropower

Advantages	Disadvantages

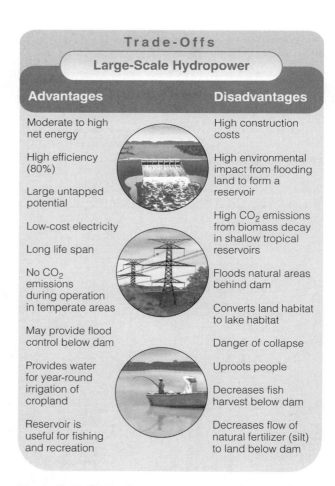

Advantages	Disadvantages
Moderate to high net energy	High construction costs
High efficiency (80%)	High environmental impact from flooding land to form a reservoir
Large untapped potential	
Low-cost electricity	High CO_2 emissions from biomass decay in shallow tropical reservoirs
Long life span	
No CO_2 emissions during operation in temperate areas	Floods natural areas behind dam
	Converts land habitat to lake habitat
	Danger of collapse
May provide flood control below dam	Uproots people
Provides water for year-round irrigation of cropland	Decreases fish harvest below dam
Reservoir is useful for fishing and recreation	Decreases flow of natural fertilizer (silt) to land below dam

Figure 10-30 Trade-offs: advantages and disadvantages of using large dams and reservoirs to produce electricity. Pick the single advantage and disadvantage that you think are the most important.

Small-scale hydropower projects eliminate most of the harmful environmental effects of large-scale projects. But their electrical output can vary with seasonal changes in stream flow.

We can also produce electricity from water flows by tapping into the energy from tides and waves. Most analysts expect these sources to make little contribution to world electricity production because of high costs and lack of enough areas with the right conditions.

What Is the Global Status of Wind Power? A Star Is Born.

Since 1995, the use of wind turbines to produce electricity has increased almost sevenfold.

The greater heating of the earth at the equator than at the poles and the earth's rotation set up flows of air called *wind*. This indirect form of solar energy can be captured by wind turbines and converted into electricity (Figure 10-31).

Since 1990, wind power has been by far the world's fastest-growing source of energy, with its use increasing almost sevenfold between 1995 and 2004. Europe is leading the way into the *age of wind energy*. About three-fourths of the world's wind power is produced in Europe in inland and offshore wind farms or parks. And European companies manufacture about 80% of the wind turbines sold in the global marketplace. Wind power also is being developed rapidly in India and to a lesser extent in China.

Much of the world's potential for wind power remains untapped. According to the 2003 Wind Force 12 report, wind parks on only one-tenth of the earth's land could produce twice the world's projected demand for electricity by 2020.

The DOE calls the Great Plains states of North Dakota, South Dakota, Nebraska, Kansas, Oklahoma, and Texas the "Saudi Arabia of wind," and points out that, in theory, they have enough wind resources to more than meet all the nation's electricity needs. According to the American Wind Energy Association, with increased and consistent government subsidies and tax breaks, wind power in the United States could produce almost a fourth of the country's electricity by 2025. This electricity can be passed through water to produce hydrogen gas, which can run fuel cells.

A growing number of U.S. farmers and ranchers make more money using their land for wind power production than by growing crops or raising cattle. In the 1980s, the United States was the world leader in wind technology, but lost that lead to Europe and Japan, mostly because of insufficient and irregular government subsidies and tax breaks for wind power compared to those received by fossil fuels and nuclear power.

Figure 10-32 (p. 230) lists the advantages and disadvantages of using wind to produce electricity. According to energy analysts, wind power has more advantages and fewer serious disadvantages than any other energy resource. Many governments and corporations are recognizing that wind is a vast, climate-benign, renewable energy resource that can supply both electricity and hydrogen fuel for running fuel cells at an affordable cost. They understand that *there is money in wind* and that our energy future may be *blowing in the wind.*

X *HOW WOULD YOU VOTE?* Should we greatly increase our dependence on wind power? Cast your vote online at http://biology.brookscole.com/miller7.

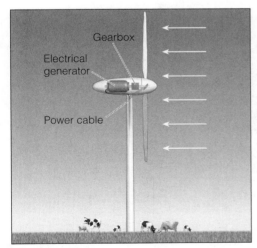

Wind turbine

Gearbox

Electrical generator

Power cable

Wind farm

Figure 10-31 Solutions: wind turbines can be used to produce electricity individually or in clusters, called wind farms or wind parks.

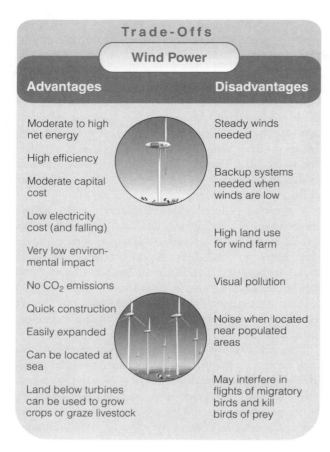

Trade-Offs
Wind Power

Advantages	Disadvantages
Moderate to high net energy	Steady winds needed
High efficiency	Backup systems needed when winds are low
Moderate capital cost	
Low electricity cost (and falling)	High land use for wind farm
Very low environmental impact	
No CO_2 emissions	Visual pollution
Quick construction	Noise when located near populated areas
Easily expanded	
Can be located at sea	
Land below turbines can be used to grow crops or graze livestock	May interfere in flights of migratory birds and kill birds of prey

Figure 10-32 Trade-offs: advantages and disadvantages of using wind to produce electricity. Wind power experts project that by 2020 wind power could supply more than 10% of the world's electricity and 10–25% of the electricity used in the United States. Pick the single advantage and disadvantage that you think are the most important.

How Useful Is Burning Solid Biomass? Burning Carbon Compounds

Plant materials and animal wastes can be burned to provide heat or electricity or converted into gaseous or liquid biofuels.

Biomass consists of plant materials (such as wood and agricultural waste) and animal wastes that can be burned directly as a solid fuel or converted into gaseous or liquid **biofuels.** Biomass is an indirect form of solar energy because it consists of combustible organic compounds produced by photosynthesis.

Most biomass is burned *directly* for heating, cooking, and industrial processes or *indirectly* to drive turbines and produce electricity. Burning wood and animal manure for heating and cooking supplies about 10% of the world's energy and about 30% of the energy used in developing countries.

Almost 70% of the people living in developing countries heat their homes and cook their food by burning wood or charcoal (derived from wood). How-

ever, about 2.7 billion people in these countries cannot find, or are too poor to buy, enough fuelwood to meet their needs.

One way to produce biomass fuel is to plant, harvest, and burn large numbers of fast-growing trees (such as cottonwoods, poplars, and sycamores), shrubs, perennial grasses (such as switchgrass), and water hyacinths in *biomass plantations.*

In agricultural areas, *crop residues* (such as sugarcane residues, rice husks, cotton stalks, and coconut shells) and *animal manure* can be collected and burned or converted into biofuels. Some ecologists argue that it makes more sense to use animal manure as a fertilizer and crop residues to feed livestock, retard soil erosion, and fertilize the soil.

Figure 10-33 lists the general advantages and disadvantages of burning solid biomass as a fuel. One problem is that burning biomass produces CO_2. However, if the rate of use of biomass does not exceed the rate at which it is replenished by new plant growth (which takes up CO_2), there is no net increase in CO_2

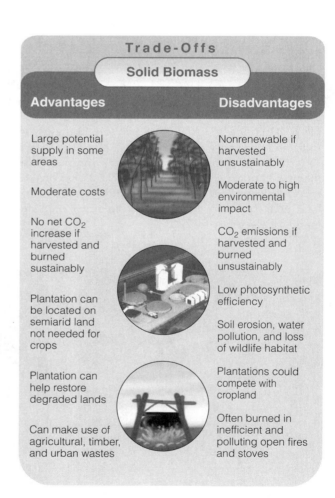

Trade-Offs
Solid Biomass

Advantages	Disadvantages
Large potential supply in some areas	Nonrenewable if harvested unsustainably
Moderate costs	Moderate to high environmental impact
No net CO_2 increase if harvested and burned sustainably	CO_2 emissions if harvested and burned unsustainably
	Low photosynthetic efficiency
Plantation can be located on semiarid land not needed for crops	Soil erosion, water pollution, and loss of wildlife habitat
Plantation can help restore degraded lands	Plantations could compete with cropland
Can make use of agricultural, timber, and urban wastes	Often burned in inefficient and polluting open fires and stoves

Figure 10-33 Trade-offs: general advantages and disadvantages of burning solid biomass as a fuel. Pick the single advantage and disadvantage that you think are the most important.

emissions. But repeated cycles of growing and harvesting biomass plantations can deplete the soil of key nutrients.

How Can Gaseous and Liquid Fuels Be Produced from Solid Biomass? Bacteria and Chemistry to the Rescue

Some forms of biomass can be converted into gaseous and liquid biofuels.

Bacteria and various chemical processes can convert some forms of biomass into gaseous and liquid biofuels. Examples include *biogas*, a mixture of 60% methane and 40% CO_2, *liquid ethanol*, and *liquid methanol*.

In rural China, anaerobic bacteria in more than 500,000 *biogas digesters* convert plant and animal wastes into methane gas that is used for heating and cooking. After the biogas has been removed, the almost odorless solid residue is used as fertilizer on food crops or on trees. When they work, biogas digesters are very efficient. But they are slow and unpredictable, a problem that could be corrected by developing more reliable models. They also add CO_2 to the atmosphere.

Some analysts believe liquid ethanol and methanol produced from biomass could replace gasoline and diesel fuel when oil becomes too scarce and expensive. *Ethanol* can be made from sugar and grain crops (sugarcane, sugar beets, sorghum, sunflowers, and corn) by fermentation and distillation. Gasoline mixed with 10–23% pure ethanol makes *gasohol*, which can be burned in conventional gasoline engines. Figure 10-34 lists the advantages and disadvantages of using ethanol as a vehicle fuel compared to gasoline. Do you believe that the advantages of using ethanol as a vehicle fuel outweigh the disadvantages? Explain.

Some analysts believe that liquid methanol produced from biomass could replace gasoline and diesel fuel when oil becomes too scarce or expensive. *Methanol* is made mostly from natural gas but also can be produced at a higher cost from carbon dioxide, coal, and biomass such as wood, wood wastes, agricultural wastes, sewage sludge, and garbage. Figure 10-35 lists the advantages and disadvantages of using methanol as a vehicle fuel compared to gasoline. Do you believe that the advantages of using methanol as a vehicle fuel outweigh the disadvantages? Explain.

Chemist George A. Olah believes that establishing a *methanol economy* is preferable to the highly publicized hydrogen economy. He points out that methanol can be produced chemically from carbon dioxide in the atmosphere, which could also help slow projected global warming. In addition, methanol can be converted to other hydrocarbon compounds that can be used to produce a variety of useful chemicals like many of the petrochemicals made from petroleum and natural gas.

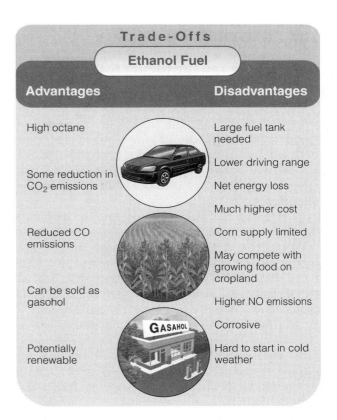

Figure 10-34   Trade-offs: advantages and disadvantages of using ethanol as a vehicle fuel compared to gasoline. Pick the single advantage and disadvantage that you think are the most important

Figure 10-35   Trade-offs: advantages and disadvantages of using methanol as a vehicle fuel compared to gasoline. Pick the single advantage and disadvantage that you think are the most important.

10-6 GEOTHERMAL ENERGY

How Can We Tap the Earth's Internal Heat? Going Underground

We can use geothermal energy stored in the earth's mantle to heat and cool buildings and to produce electricity.

Geothermal energy consists of heat stored in soil, underground rocks, and fluids in the earth's mantle. Scientists have developed several ways to tap into this stored energy to heat and cool buildings and to produce electricity.

Geothermal heat pumps can tap into the difference between underground and surface temperatures in most places and use a system of pipes and ducts to heat or cool a building. These devices extract heat from the earth in winter and in summer can store heat removed from a house in the earth. They are a very efficient and cost-effective way to heat or cool a space.

A related way to heat or cool a building is *geothermal exchange* or *geoexchange*. It involves using buried pipes filled with a fluid to move heat in or out of the ground, depending on the season and the heating or cooling requirements. In the winter, for example, heat is removed from fluid in pipes buried in the ground and blown through house ducts. In the summer, this process is reversed. According to the EPA, geothermal exchange is the most energy-efficient, cost-effective, and environmentally clean way to heat or cool a building.

We have also learned to tap into deeper and more concentrated underground reservoirs of geothermal energy. One type of reservoir contains *dry steam* with water vapor but no water droplets. Another consists of *wet steam*, a mixture of steam and water droplets. The third is *hot water* trapped in fractured or porous rock at various places in the earth's crust.

If such geothermal sites are close to the surface, wells can be drilled to extract the dry steam, wet steam, or hot water (Figure 10-1) and use it to heat homes and buildings or to spin turbines and produce electricity.

There are three other nearly nondepletable sources of geothermal energy. One is *molten rock* (magma). Another is *hot dry-rock zones*, where molten rock that has penetrated the earth's crust heats subsurface rock to high temperatures. A third source is low- to moderate-temperature *warm-rock reservoir deposits*. Heat from such deposits could be used to preheat water and run heat pumps for space heating and air conditioning. Hot dry-rock zones can be found almost anywhere about 8–10 kilometers (5–6 miles) below the earth's surface. Research is being carried out in several countries to see whether these zones can provide affordable geothermal energy.

Currently, about 22 countries (most of them in the developing world) are extracting energy from geothermal sites to produce about 1% of the world's electricity. Geothermal energy is used to heat about 85% of

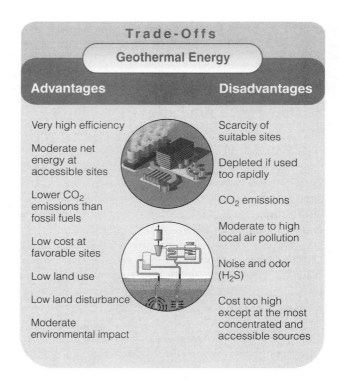

Figure 10-36 Trade-offs: advantages and disadvantages of using *geothermal energy* for space heating and to produce electricity or high-temperature heat for industrial processes. Pick the single advantage and disadvantage that you think are the most important.

Iceland's buildings, produce electricity, and grow most of its fruits and vegetables in greenhouses heated by geothermal energy.

Geothermal electricity meets the electricity needs of 6 million Americans and supplies 6% of California's electricity. The world's largest operating geothermal system, called *The Geysers*, extracts energy from a dry-steam reservoir north of San Francisco, California. But heat is being withdrawn from this geothermal site about 80 times faster than it is being replenished, converting this renewable resource to a nonrenewable source of energy. In 1999, Santa Monica, California, became the first city in the world to get all its electricity from geothermal energy.

Figure 10-36 lists the advantages and disadvantages of using geothermal energy. Do you believe that the advantages of geothermal energy outweigh its disadvantages? Explain.

10-7 HYDROGEN

Can Hydrogen Replace Oil? Goodbye Oil, Smog, and Carbon Dioxide Emissions, Hello Hydrogen

Some energy analysts view hydrogen gas as the best fuel to replace oil during the last half of this century.

When oil is gone or when what is left costs too much to use, how will we fuel vehicles, industry, and buildings? Many scientists and executives of major oil companies and automobile companies say the fuel of the future is hydrogen gas (H_2). Figure 10-37 lists the advantages and disadvantages of using hydrogen as an energy resource.

When hydrogen gas burns in air or in fuel cells, it combines with oxygen gas in the air and produces nonpolluting water vapor. Widespread use of hydrogen as a fuel would eliminate most of the air pollution problems we face today and greatly reduce the threats from global warming by emitting no CO_2—as long as the hydrogen is not produced from fossil fuels or other carbon-containing compounds.

So what is the catch? There are three problems in turning the vision of widespread use of hydrogen as a fuel into reality. *First,* hydrogen is chemically locked up in water and organic compounds such as methane and gasoline. *Second,* it takes energy and money to produce hydrogen from water and organic compounds. In other words, *hydrogen is not a source of energy. It is a fuel produced by using energy. Third,* fuel cells are the best way to use hydrogen to produce electricity, but current versions are expensive.

We could use electricity from coal-burning and conventional nuclear power plants to electrolyze water. But doing this is expensive and subjects us to the harmful environmental effects associated with using these fuels (Figures 10-11 and 10-16). We can also make hydrogen from coal and strip it from organic compounds found in fuels such as natural gas (methane), methanol, and gasoline. However, according to a 2002 study by physicist Marin Hoffer and a team of other scientists, producing hydrogen from organic compounds will add more CO_2 to the atmosphere per unit of heat generated than does burning these carbon-containing fuels directly. Thus using this approach could accelerate projected global warming unless we can develop affordable ways to store (sequester) the CO_2 underground or in the deep ocean.

Most proponents of hydrogen believe that if we are to get its very low pollution and low CO_2 benefits, the energy used to produce H_2 by decomposing water must come from low-polluting, renewable sources that also emit little or no CO_2. The most likely sources are electricity generated by wind farms, geothermal energy, solar cells (when their prices come down), or biological processes in bacteria and algae (Individuals Matter, p. 235).

In 1999 DaimlerChrysler, Royal Dutch Shell, Norsk Hydro, and Icelandic New Energy announced government-approved plans to turn the tiny country of Iceland into the world's first "hydrogen economy" by 2040.

Some analysts urge the United States to institute an Apollo-type program to spur the rapid development of a renewable-energy–hydrogen revolution that would be phased in during this century.

Once produced, hydrogen can be stored in a pressurized tank, as liquid hydrogen, and in solid metal hydride compounds, which when heated release hydrogen gas. Scientists are also evaluating ways to store H_2 by absorbing it onto the surfaces of activated charcoal or graphite nanofibers, which when heated release hydrogen gas. Another possibility is to store it inside tiny glass microspheres. Stay tuned for further developments from this important research.

Some *good news* is that metal hydrides, charcoal powders, graphite nanofibers, and glass microspheres containing hydrogen will not explode or burn if a vehicle's fuel tank or system is ruptured in an accident. This makes hydrogen a much safer fuel than gasoline, diesel fuel, methanol, and natural gas, which can explode or burn in an accident.

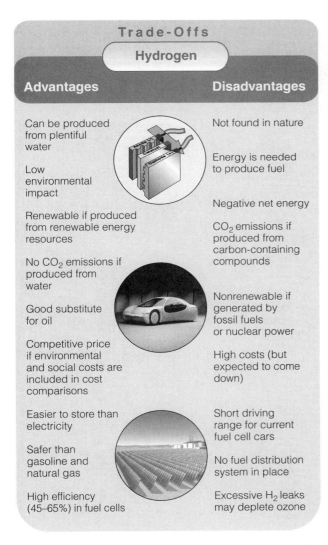

Trade-Offs

Hydrogen

Advantages	Disadvantages
Can be produced from plentiful water	Not found in nature
Low environmental impact	Energy is needed to produce fuel
Renewable if produced from renewable energy resources	Negative net energy
No CO_2 emissions if produced from water	CO_2 emissions if produced from carbon-containing compounds
Good substitute for oil	Nonrenewable if generated by fossil fuels or nuclear power
Competitive price if environmental and social costs are included in cost comparisons	High costs (but expected to come down)
Easier to store than electricity	Short driving range for current fuel cell cars
Safer than gasoline and natural gas	No fuel distribution system in place
High efficiency (45–65%) in fuel cells	Excessive H_2 leaks may deplete ozone

Figure 10-37 Trade-offs: advantages and disadvantages of using hydrogen as a fuel for vehicles and for providing heat and electricity. Pick the single advantage and disadvantage that you think are the most important.

10-8 A SUSTAINABLE ENERGY STRATEGY

What Are the Best Energy Alternatives? A New Vision

A more sustainable energy policy would improve energy efficiency, rely more on renewable energy, and reduce the harmful effects of using fossil fuels and nuclear energy.

Many scientists and energy experts who have evaluated energy alternatives have come to these three general conclusions:

First, *there will be a gradual shift from large, centralized macropower systems to smaller, decentralized micropower systems* such as small natural gas turbines for commercial buildings, wind turbines, fuel cells, and household solar panels and solar roofs (Figure 10-38).

This shift from centralized *macropower* to dispersed *micropower* is analogous to the computer industry's shift from large centralized mainframes to increasingly smaller, widely dispersed PCs, laptops, and handheld computers. Figure 10-39 lists some of the advantages of decentralized micropower systems over traditional macropower systems.

Second, *the best alternatives are a combination of improved energy efficiency and the use of natural gas as a fuel to make the transition to increased use of a variety of small-scale, decentralized, locally available, renewable energy resources and possibly nuclear fusion (if it proves feasible).*

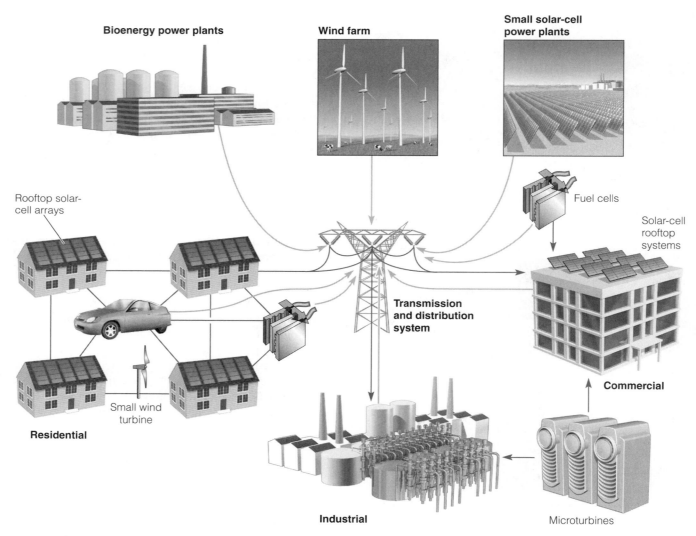

Figure 10-38 Solutions: *decentralized power system* in which electricity is produced by a large number of dispersed, small-scale *micropower systems.* Some would produce power on site and others would feed the power they produce into a conventional electrical distribution system. Over the next few decades, many energy and financial analysts expect a shift to this type of power system.

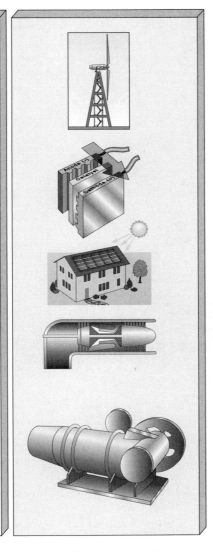

- Small modular units

- Fast factory production

- Fast installation (hours to days)

- Can add or remove modules as needed

- High energy efficiency (60–80%)

- Low or no CO_2 emissions

- Low air pollution emissions

- Reliable

- Easy to repair

- Much less vulnerable to power outages and terrorist attacks

- Useful anywhere

- Especially useful in rural areas in developing countries with no power

- Can use locally available renewable energy resources

- Easily financed (costs included in mortgage and commercial loan)

Figure 10-39 Solutions: advantages of micropower systems.

Figure 10-40 (p. 236) lists strategies for making the transition to a more sustainable energy future over the next 50 years. Carefully study this important figure.

Third, *over the next 50 years, the choice is not between using nonrenewable fossil fuels and various types of renewable energy.* Because of their supplies and low prices, fossil fuels will continue to be used in large quantities. The challenge is to find ways to reduce the harmful environmental impacts of widespread fossil fuel use, with special emphasis on reducing air pollution and emissions of greenhouse gases as less harmful alternatives are phased in.

What Role Will Economics Play in Energy Resource Use? Rewards Pay Off

Governments can use a combination of subsidies, tax breaks, and taxes to promote or discourage use of various energy alternatives.

To most analysts the key to making a shift to more sustainable energy resources and societies is economics and politics. Governments can use two basic economic and political strategies to help stimulate or dampen the short-term and long-term use of a particular energy resource.

One approach is to *keep energy prices artificially low to encourage use of selected energy resources.* This is done mostly by providing research and development subsidies and tax breaks, and by enacting regulations that help stimulate the development and use of energy resources receiving such support. For decades, this approach has been used to help stimulate the development and use of fossil fuels and nuclear power in the United States and in most other developed countries.

INDIVIDUALS MATTER

Producing Hydrogen from Green Algae Found in Pond Scum

In a few decades we may be able to use large-scale cultures of green algae to produce hydrogen gas. This simple plant grows all over the world and is commonly found in pond scum.

When living in air and sunlight, green algae carry out photosynthesis like other plants and produce carbohydrates and oxygen gas. However, in 2000, Tasios Melis, a researcher at the University of California, Berkeley, found a way to make these algae produce bubbles of hydrogen rather than oxygen.

First, he grew cultures of hundreds of billions of the algae in the normal way with plenty of sunlight, nutrients, and water. Then he cut off their supply of two key nutrients: sulfur and oxygen. Within 20 hours, the plant cells underwent a metabolic change and switched from an oxygen-producing to a hydrogen-producing metabolism, allowing the researcher to collect hydrogen gas bubbling from the culture.

Melis believes he can increase the efficiency of this hydrogen-producing process tenfold. If so, sometime in the future a *biological hydrogen plant* might cycle a mixture of algae and water through a system of clear tubes exposed to sunlight to produce hydrogen. The gene responsible for producing the hydrogen might even be transferred to other plants to produce hydrogen.

Critical Thinking

What might be some ecological problems related to the widespread use of this method for producing hydrogen?

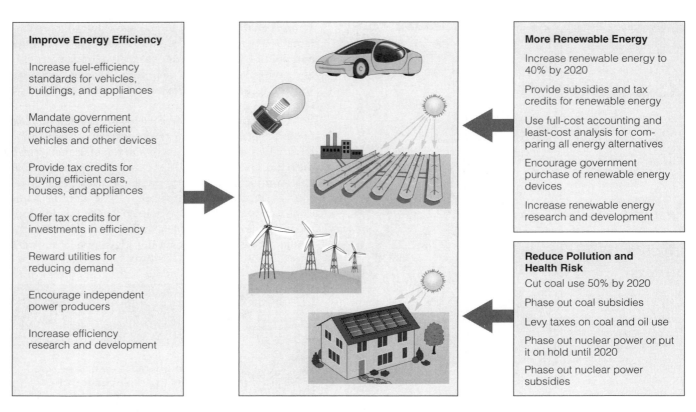

Improve Energy Efficiency

Increase fuel-efficiency standards for vehicles, buildings, and appliances

Mandate government purchases of efficient vehicles and other devices

Provide tax credits for buying efficient cars, houses, and appliances

Offer tax credits for investments in efficiency

Reward utilities for reducing demand

Encourage independent power producers

Increase efficiency research and development

More Renewable Energy

Increase renewable energy to 40% by 2020

Provide subsidies and tax credits for renewable energy

Use full-cost accounting and least-cost analysis for comparing all energy alternatives

Encourage government purchase of renewable energy devices

Increase renewable energy research and development

Reduce Pollution and Health Risk

Cut coal use 50% by 2020

Phase out coal subsidies

Levy taxes on coal and oil use

Phase out nuclear power or put it on hold until 2020

Phase out nuclear power subsidies

Figure 10-40 Solutions: suggestions of various analysts to help make the transition to a more sustainable energy future.

What Can You Do?

Energy Use and Waste

- Drive a car that gets at least 15 kilometers per liter (35 miles per gallon) and join a carpool.

- Use mass transit, walking, and bicycling.

- Superinsulate your house and plug all air leaks.

- Turn off lights, TV sets, computers, and other electronic equipment when they are not in use.

- Wash laundry in warm or cold water.

- Use passive solar heating.

- For cooling, open windows and use ceiling fans or whole-house attic or window fans.

- Turn thermostats down in winter and up in summer.

- Buy the most energy-efficient homes, lights, cars, and appliances available.

- Turn down the thermostat on water heaters to 43–49°C (110–120°F) and insulate hot water heaters and pipes.

Figure 10-41 What can you do? Ways to reduce your use and waste of energy.

This approach has created an uneven economic playing field that encourages energy waste and rapid depletion of nonrenewable energy resources. It also discourages improvements in energy efficiency and the development of renewable energy.

A second option is to *keep energy prices artificially high to discourage use of a resource.* Governments can raise the price of an energy resource by withdrawing existing tax breaks and other subsidies, enacting restrictive regulations, or adding taxes on its use. This increases government revenues, encourages improvements in energy efficiency, reduces dependence on imported energy, and decreases use of an energy resource that has a limited future supply.

X *HOW WOULD YOU VOTE?* Should the government increase taxes on fossil fuels and offset this by reducing income and payroll taxes and providing an energy safety net for the poor and the lower middle class? Cast your vote online at http://biology.brookscole.com/miller7.

We have the technology, creativity, and wealth to make the transition to a more sustainable energy future. Making this transition depends primarily on *politics*, which depends largely on the pressure individuals put on elected officials by voting with their ballots Figure 10-41 lists some ways you can contribute to making this transition by reducing the amount of energy you use and waste.

A transition to renewable energy is inevitable, not because fossil fuel supplies will run out—large reserves of oil, coal, and gas remain in the world—but because the costs and risks of using these supplies will continue to increase relative to renewable energy.

MOHAMED EL-ASHRY

CRITICAL THINKING

1. Just to continue using oil at the current rate (not the projected higher exponential increase in its annual use), we must discover and add to global oil reserves the equivalent of a new Saudi Arabian supply (the world's largest) *every 10 years*. Do you believe this is possible? If not, what effects might this have on your life and on the life of a child or grandchild you might have?

2. List five actions you can take to reduce your dependence on oil and the gasoline derived from it. Which of these things do you actually plan to do?

3. Explain why you are for or against continuing to increase oil imports in the United States or in the country where you live. If you favor reducing dependence on oil imports, list the three best ways to do this.

4. Explain why you agree or disagree with the following proposals by various energy analysts to solve U.S. energy problems: **(a)** find and develop more domestic supplies of oil, **(b)** place a heavy federal tax on gasoline and imported oil to help reduce the waste of oil resources, **(c)** increase dependence on nuclear power, and **(d)** phase out all nuclear power plants by 2025.

5. Would you favor having high-level nuclear waste transported by truck or train through the area where you live to a centralized underground storage site? Explain. What are the options?

6. Someone tells you that we can save energy by recycling it. How would you respond?

7. Explain why you agree or disagree with the following proposals by various energy analysts: **(a)** Government subsidies for all energy alternatives should be eliminated so all energy choices can compete in a true free-market system. **(b)** All government tax breaks and other subsidies for conventional fuels (oil, natural gas, and coal), synthetic natural gas and oil, and nuclear power (fission and fusion) should be phased out. They should be replaced with subsidies and tax breaks for improving energy efficiency and developing solar, wind, geothermal, hydrogen, and biomass energy alternatives. **(c)** Development of solar, wind, and hydrogen energy should be left to private enterprise and receive little or no help from the federal government, but nuclear energy and fossil fuels should continue to receive large federal subsidies.

8. Congratulations! You are in charge of the world. List the five most important features of your energy policy.

LEARNING ONLINE

The website for this book contains helpful study aids and many ideas for further reading and research. They include a chapter summary, review questions for the entire chapter, flash cards for key terms and concepts, a multiple-choice practice quiz, interesting Internet sites, references, and a guide for accessing thousands of InfoTrac® College Edition articles. Log on to

http://biology.brookscole.com/miller7

Then click on the Chapter-by-Chapter area, choose Chapter 10, and select a learning resource.

11 RISK, TOXICOLOGY, AND HUMAN HEALTH

The dose makes the poison.

PARACELSUS, 1540

11-1 RISKS AND HAZARDS

What Is Risk? The Chances We Take

Risk is a measure of the likelihood that you will suffer harm from a hazard.

Risk is the *possibility* of suffering harm from a hazard that can cause injury, disease, death, economic loss, or environmental damage. Risk is usually expressed in terms of *probability*: a mathematical statement about how likely it is to suffer harm from a hazard. Scientists often state probability in terms such as "The lifetime probability of developing lung cancer from smoking a pack of cigarettes a day is 1 in 250." This means that 1 of every 250 people who smoke a pack of cigarettes a day will develop lung cancer over a typical lifetime (usually considered 70 years).

It is important to distinguish between *possibility* and *probability*. When we say that it is *possible* that a smoker can get lung cancer we are saying that this event could happen. *Probability* gives us an estimate of the likelihood of such an event.

Risk assessment is the scientific process of estimating how much harm a particular hazard can cause to human health. **Risk management** involves deciding whether or how to reduce a particular risk to a certain level and at what cost. Figure 11-1 summarizes how risks are assessed and managed.

What Are the Major Types of Hazards? They Are All around Us

We can suffer harm from cultural hazards, physical hazards, chemical hazards, and biological hazards, but determining the risks involved is difficult.

We can suffer harm from four major types of hazards:

- *Cultural hazards* such as unsafe working conditions, smoking (Spotlight, at right), poor diet, drugs, drinking, driving, criminal assault, unsafe sex, and poverty.
- *Physical hazards* such as a fire, earthquake, volcanic eruption, flood, tornado, and hurricane.
- *Chemical hazards* from harmful chemicals in the air, water, soil, and food

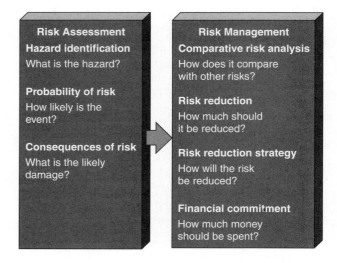

Figure 11-1 *Risk assessment* and *risk management*. These are important, difficult, and controversial processes.

- *Biological hazards* from pathogens (bacteria, viruses, and parasites)

Case Study: What Is the Biggest Killer? Puffing Life Away

Cigarette smoking is the world's most preventable major cause of suffering and premature death among adults.

What is roughly the diameter of a 30-caliber bullet, can be bought almost anywhere, is highly addictive, and kills about 13,700 people every day, 560 per hour, one every 6 seconds? It is a cigarette.

According to the World Health Organization (WHO), tobacco helped kill about 80 million people between 1950 and 2004. This is 2.6 times more than the 30 million people killed in battle in all wars during the 20th century!

The WHO estimates that each year tobacco contributes to the premature deaths of at least 5 million people (about half from developed countries and half from developing countries) from 34 illnesses including *heart disease, lung cancer, other cancers, bronchitis, emphysema,* and *stroke.* By 2030, the annual death toll from smoking-related diseases is projected to reach 10 million—an average of about 27,400 preventable deaths per day or 1 death every 3 seconds. About 70% of these deaths are expected to occur in developing countries.

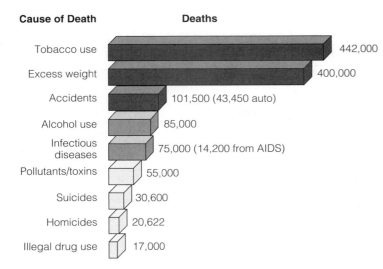

Cause of Death | **Deaths**

Tobacco use — 442,000
Excess weight — 400,000
Accidents — 101,500 (43,450 auto)
Alcohol use — 85,000
Infectious diseases — 75,000 (14,200 from AIDS)
Pollutants/toxins — 55,000
Suicides — 30,600
Homicides — 20,622
Illegal drug use — 17,000

Figure 11-2 Annual deaths in the United States from tobacco use and eight other causes in 2003. Smoking is by far the nation's leading cause of preventable death. (Data from U.S. National Center for Health Statistics and Centers for Disease Control and Prevention and U.S. Surgeon General)

According to a 2002 study by the Centers for Disease Control and Prevention, smoking kills about 442,000 Americans per year prematurely, an average of 1,210 deaths per day (Figure 11-2). This death toll is roughly equivalent to three fully loaded 400-passenger jumbo jets crashing accidentally *every day* with no survivors! Yet this ongoing major human tragedy rarely makes the news.

The overwhelming consensus in the scientific community is that the nicotine inhaled in tobacco smoke is highly addictive. Only 1 in 10 people who try to quit smoking succeed, about the same relapse rate as for recovering alcoholics and those addicted to heroin or crack cocaine. A British government study showed that adolescents who smoke more than one cigarette have an 85% chance of becoming smokers. People can also be exposed to secondhand smoke from others, called *passive smoking*.

According to a 2002 study by the Centers for Disease Control and Prevention, smoking in the United States costs about $158 billion a year for medical bills, increased insurance costs, disability, lost earnings and productivity because of illness, and property damage from smoking-caused fires. This is an average of about $7 per pack of cigarettes sold in the United States.

Many health experts urge that a $3–5 federal tax be added to the price of a pack of cigarettes in the United States. Such a tax would mean that the users of cigarettes (and other tobacco products), not the rest of society, would pay a much greater share of the health, economic, and social costs associated with their smoking.

Other suggestions that have been made for reducing the death toll and the health effects of smoking in the United States (and in other countries) include banning all cigarette advertising, prohibiting the sale of cigarettes and other tobacco products to anyone under 21 (with strict penalties for violators), and banning cigarette vending machines.

Analysts also call for classifying and regulating the use of nicotine as an addictive and dangerous drug, eliminating all federal subsidies and tax breaks to tobacco farmers and tobacco companies, and using cigarette tax income to finance an aggressive anti-tobacco advertising and education program. So far, the U.S. Congress has not enacted such reforms.

X *HOW WOULD YOU VOTE?* Do you favor classifying and regulating nicotine as an addictive and dangerous drug? Cast your vote online at http://biology.brookscole.com/miller7.

11-2 TOXICOLOGY: ASSESSING CHEMICAL HAZARDS

What Determines Whether a Chemical Is Harmful? How Much, How Often, and Genes

The harm caused by exposure to a chemical depends on the amount of exposure (dose), frequency of exposure, who is exposed, how well the body's detoxification systems work, and one's genetic makeup.

Toxicity measures how harmful a substance is in causing injury, illness, or death to a living organism. This depends on several factors. One is the **dose**, the amount of a substance a person has ingested, inhaled, or absorbed through the skin. Other factors are frequency of exposure, who is exposed (adult or child, for example), how well the body's detoxification systems (such as the liver, lungs, and kidneys) work, and the genetic makeup of individuals.

The type and amount of health damage that result from exposure to a chemical or other agent are called the **response.** An *acute effect* is an immediate or rapid harmful reaction to an exposure—ranging from dizziness to death. A *chronic effect* is a permanent or long-lasting consequence (kidney or liver damage, for example) from exposure to a single dose or to sublethal repeated doses of a harmful substance.

A basic concept of toxicology is that *any synthetic or natural chemical can be harmful if ingested in a large enough quantity.* For example, drinking 100 cups of strong coffee one after another would expose most people to a lethal dosage of caffeine. Similarly, downing 100 tablets of aspirin or 1 liter (1.1 quarts) of pure alcohol (ethanol) would kill most people.

The critical question is *how much exposure to a particular toxic chemical causes a harmful response?* This is the meaning of the chapter-opening quote by the German scientist Paracelsus about the dose making the poison.

Should We Be Concerned about Trace Levels of Toxic Chemicals in the Environment and in Our Bodies? It Depends on the Chemical.

Trace amounts of chemicals in the environment or your body may or may not be harmful.

Should we be concerned about trace amounts of various chemicals in air, water, food, and our bodies? The honest answer is that, in most cases, we do not know because of a lack of data and the difficulty of determining the effects of exposures to low levels of chemicals.

Some scientists think that trace levels of most chemicals are not harmful. They point to the dramatic increase in average life expectancy in the United States since 1950. Other scientists are not so sure and believe that much more research is needed to help us evaluate the possible long-term harm caused by exposure to low levels of the thousands of new synthetic chemicals that we have put into the environment during the past few decades.

Chemists are able to detect increasingly small amounts of potentially toxic chemicals in air, water, and food. This is good news, but it can give the false impression that dangers from toxic chemicals are increasing when, in some cases, all we are doing is uncovering levels of chemicals that have been around for a long time.

Some people also have the mistaken idea that natural chemicals are safe and synthetic chemicals are harmful. In fact, many synthetic chemicals are quite safe if used as intended, and many natural chemicals are deadly.

How Can We Estimate the Toxicity of a Chemical? Kill Half of the Animals in a Test Population

Chemicals vary widely in their toxicity to humans and other animals.

A **poison** or **toxin** is a chemical that adversely affects the health of a living human or animal by causing injury, illness, or death. One method for determining the relative toxicity of various chemicals is to measure their effects on test animals. A widely used method for estimating the toxicity of a chemical is to determine its *lethal dose (LD)*. This is often done by measuring a chemical's **median lethal dose** or **LD50:** the amount received in one dose that kills 50% of the animals (usually rats and mice) in a test population within a 14-day period.

Chemicals vary widely in their toxicity (Table 11-1). Some poisons can cause serious harm or death after a single acute exposure at very low dosages. Others cause such harm only at dosages so huge that it is nearly impossible to get enough into the body to cause injury or death. Most chemicals fall between these two extremes.

How Do Scientists Use Case Reports and Epidemiological Studies to Estimate Toxicity? Reports from the Front Lines and Controlled Experiments

We can estimate toxicity by using case reports about the harmful effects of chemicals on human health and by comparing the health of a group of people exposed to a chemical with that of a similar group not exposed to the chemical.

Table 11-1 Toxicity Ratings and Average Lethal Doses for Humans

Toxicity Ratings	LD50 (milligrams per kilogram of body weight)*	Average Lethal Dose†	Examples
Super toxic	Less than 0.01	Less than 1 drop	Nerve gases, botulism toxin, mushroom toxins, dioxin (TCDD)
Extremely toxic	Less than 5	Less than 7 drops	Potassium cyanide, heroin, atropine, parathion, nicotine
Very toxic	5–50	7 drops to 1 teaspoon	Mercury salts, morphine, codeine
Toxic	50–500	1 teaspoon to 1 ounce	Lead salts, DDT, sodium hydroxide, fluoride, sulfuric acid, caffeine, carbon tetrachloride
Moderately toxic	500–5,000	1 ounce to 1 pint	Methyl (wood) alcohol, ether, phenobarbitol, amphetamine, kerosene, aspirin
Slightly toxic	5,000–15,000	1 pint to 1 quart	Ethyl alcohol, Lysol, soaps
Essentially nontoxic	15,000 or greater	More than 1 quart	Water, glycerin, table sugar

*Dosage that kills 50% of individuals exposed.

†Amounts of substances that are liquids at room temperature when given to a 70.4-kilogram (155-pound) human.

Scientists use various methods to get information about the harmful effects of chemicals on human health. One is *case reports*, usually made by physicians. Case reports provide information about people suffering some adverse health effect or death after exposure to a chemical. Such information often involves accidental poisonings, drug overdoses, homicides, or suicide attempts.

Most case reports are not reliable sources for estimating toxicity because the actual dosage and the exposed person's health status are often not known. But such reports can provide clues about environmental hazards and suggest the need for laboratory investigations.

Another source of information is *epidemiological studies*. They involve comparing the health of people exposed to a particular chemical (the *experimental group*) with the health of another group of statistically similar people not exposed to the agent (the *control group*). The goal is to determine whether the statistical association between exposure to a toxic chemical and a health problem is strong, moderate, weak, or undetectable.

Three factors can limit the usefulness of epidemiological studies. One problem is that, in many cases, too few people have been exposed to high enough levels of a toxic agent to detect statistically significant differences. Another limitation is that conclusively linking an observed effect with exposure to a particular chemical is difficult because people are exposed to many different toxic agents throughout their lives. Another limitation is that we cannot use epidemiological studies to evaluate hazards from new technologies or chemicals to which people have not yet been exposed.

How Do Scientists Use Laboratory Experiments to Estimate Toxicity? Controversial Animal Testing

Exposing a population of live laboratory animals (especially mice and rats) to known amounts of a chemical is the most widely used method for determining the chemical's toxicity.

The most widely used method for determining toxicity is to expose a population of live laboratory animals (especially mice and rats) to measured doses of a specific substance under controlled conditions. Animal tests take 2–5 years and cost $200,000 to $2 million per substance tested. Such tests can be painful to the test animals and can kill or harm them. The goal is to develop data on the response of the test animals to various doses of a chemical (called a dose-response curve). But estimating the effects of low doses is difficult.

Animal welfare groups want to limit or ban the use of test animals or ensure that experimental ani-

mals are treated in the most humane manner possible. More humane methods for carrying out toxicity tests are available. They include computer simulations and using tissue cultures of cells and bacteria, chicken egg membranes, and measurements of changes in the electrical properties of individual animal cells.

These alternatives can greatly decrease the use of animals for testing toxicity. But scientists point out that some animal testing is needed because the alternative methods cannot adequately mimic the complex biochemical interactions of a live animal.

Acute toxicity tests are run to develop a **dose-response curve,** which shows the effects of various dosages of a toxic agent on a group of test organisms (Figure 11-3). Such tests are *controlled experiments* in which the effects of the chemical on a *test group* are compared with the responses of a *control group* of organisms not exposed to the chemical. Care is taken that organisms in each group are as identical as possible in age, health status, and genetic makeup, and that all are exposed to the same environmental conditions.

Fairly high dosages are used to reduce the number of test animals needed, obtain results quickly, and lower costs. Otherwise, tests would have to be run on millions of laboratory animals for many years, and manufacturers could not afford to test most chemicals.

For the same reasons, scientists usually use mathematical models to extrapolate the results of high-dose exposures to low-dose levels. Then they extrapolate the low-dose results from the test organisms to humans to estimate LD50 values for acute toxicity (Table 11-1).

According to the *nonthreshold dose-response model* (Figure 11-3, left), any dosage of a toxic chemical or ionizing radiation causes harm that increases with the

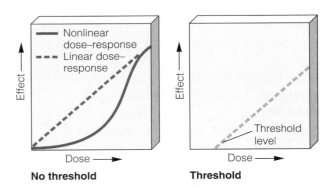

Figure 11-3 Two types of *dose-response curves*. The linear and nonlinear curves in the left graph apply if even the smallest dosage of a chemical or ionizing radiation has a harmful effect that increases with the dosage. The curve on the right applies if a harmful effect occurs only when the dosage exceeds a certain *threshold level*. Which model is better for a specific harmful agent is uncertain because of the difficulty in estimating the response to very low dosages.

dosage. With the *threshold dose-response model* (Figure 11-3, right), a threshold dosage must be reached before any detectable harmful effects occur, presumably because the body can repair the damage caused by low dosages of some substances.

Establishing which of these models applies at low dosages is extremely difficult and controversial. To be on the safe side, the nonthreshold dose-response model often is assumed.

Some scientists challenge the validity of extrapolating data from test animals to humans because human physiology and metabolism often differ from those of the test animals. Other scientists say that such tests and models work fairly well (especially for revealing cancer risks) when the correct experimental animal is chosen or a when chemical is toxic or harmful to several different test-animal species.

The problem of estimating toxicities gets worse. In real life, each of us is exposed to a variety of chemicals, some of which can interact in ways that decrease or enhance their individual effects over short and long times.

Toxicologists have great difficulty in estimating the toxicity of a single substance. But adding the problem of evaluating mixtures of potentially toxic substances, separating out which ones are the culprits, and determining how they can interact with one another is overwhelming from a scientific and economic standpoint. For example, just studying the interactions of three of the 500 most widely used industrial chemicals would take 20.7 million experiments—a physical and financial impossibility.

The effects of a particular chemical can also depend upon when exposure occurs. For example, people are likely to suffer more serious consequences from some chemicals as children than as adults and when their bodies' protective systems have been weakened from disease. As you can see, toxicologists have important but difficult jobs.

How Good Are Estimates of Toxicity? Taking Uncertainty into Account

Because all methods of estimating toxicity have serious limitations, allowed exposure levels are usually set well below the estimated harmful levels.

As we have seen, all methods for estimating toxicity levels and risks have serious limitations. But they are all we have. To take this uncertainty into account and minimize harm, scientists and regulators typically set allowed exposure levels to toxic substances and ionizing radiation at 1/100 or even 1/1,000 of the estimated harmful levels.

Despite their many limitations, carefully conducted and evaluated toxicity studies are important sources of information for understanding dose-response effects and estimating and setting exposure standards. But citizens, lawmakers, and regulatory officials must recognize the huge uncertainties and guesswork involved in all such studies.

11-3 CHEMICAL HAZARDS

What Are Toxic and Hazardous Chemicals? Causing Death and Harm

Toxic chemicals can kill and hazardous chemicals can cause various types of harm.

A **toxic chemical** can cause temporary or permanent harm or death to humans or animals. A **hazardous chemical** can harm humans or other animals because it is flammable or explosive or because it can irritate or damage the skin or lungs, interfere with oxygen uptake, or induce allergic reactions.

There are three major types of potentially toxic agents. One consists of **mutagens,** chemicals or ionizing radiation that cause random *mutations,* or changes, in the DNA molecules found in cells. An example is nitrous acid formed by the digestion of nitrite preservatives in foods. *Most mutations are harmless.* One reason is that organisms have biochemical repair mechanisms that can correct mistakes or changes in the DNA code.

But harmful mutations occurring in reproductive cells can be passed on to offspring and to future generations. It is generally accepted that there is no safe threshold for exposure to mutagens that can cause harmful mutations.

A second type consists of **teratogens,** chemicals that cause harm or birth defects to a fetus or embryo. Ethyl alcohol is an example of a teratogen. Drinking during pregnancy can lead to offspring with low birth weight and a number of physical, developmental, and mental problems. Thalidomide is also a potent teratogen.

The third group is **carcinogens,** chemicals or ionizing radiation that can cause or promote **cancer**—the growth of a malignant (cancerous) tumor, in which certain cells multiply uncontrollably. An example is benzene, a widely used chemical solvent. Many cancerous tumors spread by **metastasis,** when malignant cells break off from tumors and travel in body fluids to other parts of the body. There they start new tumors, making treatment much more difficult. Typically, 10–40 years may elapse between the initial exposure to a carcinogen and the appearance of detectable symptoms. Partly because of this time lag, many healthy teenagers and young adults have trouble believing their smoking, drinking, eating, and other lifestyle habits today could lead to some form of cancer before they reach age 50.

Case Study: What Effects Can Some Chemicals Have on Immune, Nervous, and Endocrine Systems? Possible Harm from Small Doses

Long-term exposure to some chemicals at low doses may disrupt the body's immune, nervous, and endocrine systems.

Since the 1970s, a growing body of research on wildlife and laboratory animals, along with some epidemiological studies of humans, indicate that long-term exposure to some chemicals in the environment can disrupt the body's immune, nervous, and endocrine systems.

The *immune system* consists of specialized cells and tissues that protect the body against disease and harmful substances by forming antibodies that make invading agents harmless. Ionizing radiation and some chemicals can weaken the human immune system and can leave the body wide open to attacks by allergens, infectious bacteria, viruses, and protozoans. Examples of some of these chemicals are arsenic and dioxins.

Some natural and synthetic chemicals in the environment can harm the human *nervous system* (brain, spinal cord, and peripheral nerves). For example, many poisons and the venom of poisonous snakes are neurotoxins, which inhibit, damage, or destroy nerve cells (neurons) that transmit electrochemical messages throughout the body. Effects can include behavioral changes, paralysis, and death. Examples of chemical neurotoxins are PCBs, mercury, and certain pesticides.

The *endocrine system* is a complex network of glands that releases very small amounts of *hormones* into the bloodstream of humans and other vertebrate animals. Low levels of these chemical messengers turn on and off bodily systems that control sexual reproduction, growth, development, learning ability, and behavior. Exposure to low levels of certain synthetic chemicals may disrupt the effects of natural hormones in animals, but more research is needed to verify the effects of these chemicals on humans.

Why Do We Know So Little about the Harmful Effects of Chemicals? Establishing Guilt Is Difficult.

Under existing laws, most chemicals are considered innocent until shown to be guilty, and estimating their toxicity to establish guilt is difficult, uncertain, and expensive.

According to risk assessment expert Joseph V. Rodricks, "Toxicologists know a great deal about a few chemicals, a little about many, and next to nothing about most." The U.S. National Academy of Sciences estimates that only about 10% of at least 80,000 chemicals in commercial use have been thoroughly screened for toxicity, and only 2% have been adequately tested to determine whether they are carcinogens, teratogens, or mutagens.

Hardly any of the chemicals in commercial use have been screened for possible damage to the nervous, endocrine, and immune systems.

Currently, federal and state governments do not regulate about 99.5% of the commercially used chemicals in the United States. There are several reasons for this lack of regulation. One is that under existing U.S. laws, most chemicals are *considered innocent until shown to be guilty.* Some analysts think this is the opposite of the way it should be. They ask why should chemicals have the same legal rights as people?

A second reason is that there are not enough funds, personnel, facilities, and test animals available to provide such information for more than a small fraction of the many individual chemicals we encounter in our daily lives. A third limitation is that analyzing the combined effects of multiple exposures to various chemicals and the possible interactions of such chemicals is too difficult and expensive.

Is Pollution Prevention the Answer? Taking Precautions

Preliminary but not conclusive evidence that a chemical causes significant harm should spur preventive action, some say.

So where does this leave us? We do not know a lot about the potentially toxic chemicals around us and inside of us, and estimating their effects is very difficult, time consuming, and expensive. Is there a way out of this dilemma?

Some scientists and health officials, especially those in European Union countries, are pushing for much greater emphasis on *pollution prevention.* They say we should not release into the environment chemicals that we know or suspect can cause significant harm. This means looking for harmless or less harmful substitutes for toxic and hazardous chemicals or recycling them within production processes so they do not reach the environment.

This prevention approach is based on the **precautionary principle:** When there is plausible but incomplete scientific evidence (frontier science evidence) of significant harm to humans or the environment from a proposed or existing chemical or technology, we should take action to prevent or reduce the risk instead of waiting for more conclusive (sound or consensus science) evidence. This principle is based on familiar axioms: "Look before you leap," "better safe than sorry," and "an ounce of prevention is worth a pound of cure."

Under this approach, those proposing to introduce a new chemical or technology would bear the burden of establishing its safety. This means two major changes in the way we evaluate risks. *First,* new chemicals and technologies would be assumed harmful until scientific studies can show otherwise. *Second,* existing chemicals and technologies that appear to have a strong chance of causing significant harm would be removed from the market until their safety can be established.

Some movement is being made in this direction, especially in the European Union. In 2000, negotiators agreed to a global treaty that would ban or phase out use of 12 of the most notorious *persistent organic pollutants (POPs),* also called the *dirty dozen.* The list included DDT and eight other persistent pesticides, PCBs, and dioxins and furans. New chemicals would be added to the list when the harm they could cause is seen as outweighing their usefulness. This treaty went into effect in 2004.

Manufacturers and businesses contend that widespread application of the precautionary principle would make it too expensive and almost impossible to introduce any new chemical or technology. They argue that we can never have a risk-free society.

On the other hand, proponents of increased reliance on the precautionary principle say that it will encourage innovation in developing less harmful alternative chemicals and technologies, and in finding ways to prevent as much pollution as possible instead of relying mostly on pollution control. It is true that we cannot have a risk-free society. But proponents believe we should make greater use of the precautionary principle effort to reduce many of the risks we face.

✗ HOW WOULD YOU VOTE? Should we rely more on the precautionary principle as a way to reduce the risks from chemicals and various technologies? Cast your vote online at http://biology.brookscole.com/miller7.

11-4 BIOLOGICAL HAZARDS: DISEASE IN DEVELOPED AND DEVELOPING COUNTRIES

What Are Nontransmissible and Transmissible Diseases? To Spread or Not to Spread

Diseases not caused by living organisms do not spread from one person to another, while those caused by living organisms such as bacteria and viruses can spread from person to person.

A **nontransmissible disease** is caused something other than a living organisms and does not spread from one person to another. Such diseases tend to develop slowly and have multiple causes. Examples are cardiovascular (heart and blood vessel) disorders, asthma, emphysema, and malnutrition.

A **transmissible disease** is caused by a living organism and can spread from one person to another. *Infectious agents* or *pathogens* (such as a bacterium, virus, protozoa, or parasite) (Figure 11-4) cause such diseases. These agents are spread by air, water, food, body fluids, and by some insects such as mosquitoes, fleas, and other nonhuman carriers. Figure 11-5 shows the annual death toll from the world's seven deadliest infectious diseases.

Great news: Since 1900, and especially since 1950, the incidence of infectious diseases and the death rates from such diseases have been greatly reduced. This has been achieved mostly by a combination of better health care, using antibiotics to treat infectious disease caused by bacteria, and developing vaccines to prevent the spread of some infectious viral diseases.

Bad news: Many disease-carrying bacteria have developed genetic immunity to widely used antibiotics (see Case Study below). Also, many disease-transmitting species of insects such as mosquitoes have become immune to widely used pesticides that once helped control their populations.

Case Study: Are We Losing Ground in Our Struggle against Infectious Bacteria? Growing Germ Resistance to Antibiotics

Rapidly producing infectious bacteria can undergo natural selection and become genetically resistant to widely used antibiotics.

We may be falling behind in our efforts to prevent infectious bacterial diseases because of the astounding reproductive rate of bacteria, some of which can produce 16,777,216 offspring in 24 hours. Their high reproductive rate allows them to become genetically resistant to an increasing number of antibiotics through natural selection. They can also transfer such resistance to nonresistant bacteria.

Other factors play a role in the serious rise in the incidence of some infectious bacterial diseases—such as tuberculosis (Case Study, at right)—once controlled by antibiotics. One is that harmful bacteria are spread around the globe by human travel and the trade of goods. Another is that overuse of pesticides increases populations of pesticide-resistant insects and other carriers of bacterial diseases.

An additional factor is overuse of antibiotics by doctors. According to a 2000 study by Richard Wenzel and Michael Edward, at least half of all antibiotics used to treat humans are prescribed unnecessarily. In many countries antibiotics are available without a prescription, which also promotes unnecessary use. Resistance to some antibiotics has also increased by the widespread use of antibiotics in livestock and diary animals to control disease and to promote growth.

The result of these factors acting together is that every major disease-causing bacterium now has strains that resist at least one of the roughly 160 antibiotics we use to treat bacterial infections, such as those from tuberculosis (Case Study, right).

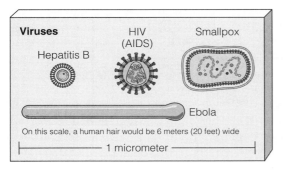
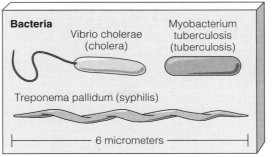
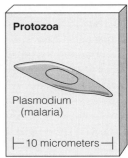

Figure 11-4 Examples of *pathogens* or agents that can cause transmissible diseases. A micrometer is one-millionth of a meter.

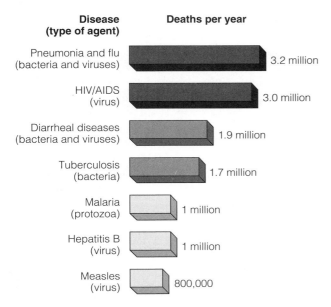

Figure 11-5 *Serious biological risks.* Each year the world's seven deadliest infectious diseases kill about 12.6 million people—most of them poor people in developing countries. Each day this amounts to about 34,500 mostly preventable deaths. (Data from World Health Organization)

Case Study: The Global Tuberculosis Epidemic—A Growing Threat

Tuberculosis (TB) kills about 1.7 million people a year and could kill 28 million people by 2020.

Since 1990, one of the world's most underreported stories has been the rapid spread of tuberculosis (TB). According to the World Health Organization, this highly infectious bacterial disease infects about 9 million people per year and kills about 1.7 million of them—most of them in developing countries. The WHO projects that between 2004 and 2020, about 28 million people will die of the disease unless current efforts and funding to control TB are greatly strengthened and expanded.

Many infected people do not appear to be sick, and about half of them do not know they are infected.

As a result, this serious health problem has been called a *silent global epidemic.*

Several factors account for the recent increase in TB. One is the lack of TB screening and control programs, especially in developing countries, where about 95% of the new cases occur. A second problem is that most strains of the TB bacterium have developed genetic resistance to almost all effective antibiotics.

Another factor is that increased population growth and urbanization increase contacts between people and spread TB, especially in areas where large numbers of the poor are crowded together. In addition, the spread of AIDS greatly weakens the immune system and allows TB bacteria to multiply in AIDS victims.

Slowing the spread of the disease involves early identification and treatment of people with active TB, especially those with a chronic cough. Treatment with a combination of four inexpensive drugs can cure 90% of those with active TB. However, to be effective, the drugs must be taken every day for 6–8 months. Because the symptoms disappear after a few weeks, many patients think they are cured and stop taking the drugs. This allows the disease to recur in a drug-resistant form. It then spreads to other people, and drug-resistant strains of TB bacteria develop.

How Serious Is the Threat from Viral Diseases? Watch Out for HIV, Flu, and Hepatitis B

HIV, flu, and hepatitis B viruses infect and kill many more people each year than the highly publicized Ebola, West Nile, and SARS viruses.

What are the world's three most widespread and dangerous viruses? The biggest killer is the *human immunodeficiency virus (HIV)* that is transmitted by unsafe sex, sharing of needles by drug users, infected mothers to offspring before or during birth, and exposure to infected blood. On a global scale, HIV infects at least 5 million new people a year (about 42,000 in the United States), and the resulting complications from AIDS kill about 3 million people a year (Case Study, p. 246).

The second biggest killer is the *influenza* or *flu* virus that is transmitted by the body fluids or airborne emissions of an infected person, and kills about 1 million people per year. The third largest killer is the *hepatitis B virus (HBV)* that damages the liver and kills about 1 million a year. Like HIV, it is transmitted by unsafe sex, sharing of needles by drug users, infected mothers to offspring before or during birth, and exposure to infected blood.

In recent years, three other viruses have received widespread coverage in the media. One is the *Ebola virus* transmitted by the blood or other body fluids of an infected person. Another is the *West Nile virus* that is transmitted by the bite of a common mosquito that became infected by feeding on birds that carry the virus. A third is the *severe acute respiratory syndrome (SARS) virus.*

Health officials are concerned about the emergence and spread of these three and other emerging viral diseases and are working hard to control their spread. But in terms of annual infection rates and deaths, the three most dangerous viruses by far are HIV, flu, and hepatitis B.

For example, the West Nile virus (spread by mosquitoes that bite infected birds) has spread throughout most of the continental United States, but the chances of being infected and killed by it are low (about 1 in 2,500). In 2003, the flu killed more Americans in two days than the West Nile virus killed during the entire year.

You can greatly reduce your chances of getting infectious diseases such as flu, the common cold, and SARS that spread from person to person by practicing good old-fashioned hygiene. Wash your hands thoroughly a lot and avoid touching your mouth, nose, and eyes.

The best weapons against viruses are *vaccines* that stimulate the body's immune system to produce antibodies to ward off viral infections. Immunization with vaccines has helped reduce the spread of viral diseases such as smallpox, polio, rabies, influenza, measles, and hepatitis B. But vaccines are not available for many viral diseases.

Case Study: How Serious Is the Global Threat from HIV and AIDS? A Rapidly Growing Health Threat

The spread of acquired immune deficiency syndrome (AIDS), caused by HIV, is one of the world's most serious and rapidly growing health threats.

The global spread of *acquired immune deficiency syndrome (AIDS)*, caused by the human immunodeficiency virus (HIV), is a serious and rapidly growing health threat. The virus itself is not deadly, but it kills immune cells and leaves the body defenseless against infectious bacteria and other viruses.

According to the WHO in 2004 some 38 million people worldwide (96% of them in developing countries, especially African countries south of the Sahara Desert) were infected with HIV. Every day about 14,000 more people—most of them between the ages of 15 and 24—get infected with HIV.

Within 7–10 years, at least half of those with HIV develop AIDS. This long incubation period means that infected people often spread the virus for several years without knowing they are infected. So far, *there is no vaccine to prevent HIV and no cure for AIDS. Once you get AIDS, you will eventually die,* although drugs may help some infected people live longer. However, only a tiny fraction of those suffering from AIDS can afford to use these costly drugs.

Between 1980 and 2004, more than 20 million people (460,000 in the United States) died of AIDS-related diseases. AIDS has caused the life expectancy of 700 million people living in sub-Saharan Africa to drop from 62 to 47 years. The premature deaths of teachers, health care workers, and other young productive adults in such countries leads to diminished education and health care, decreased food production and economic development, and disintegrating families. Such deaths drastically alter a country's age structure diagram (Figure 11-6). AIDS has also left 15 million orphans—roughly equal to every child under age 5 in America.

Between 2004 and 2020, the WHO estimates 60 million more deaths from AIDS and a death toll reaching as high as 5 million a year by 2020.

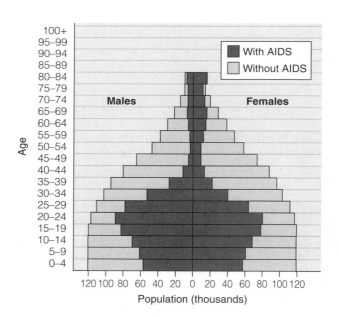

Figure 11-6 How AIDS can affect the age structure of a population. This figure shows the projected age structure of Botswana's population in 2020 with and without AIDS. (Data from U.S. Census Bureau)

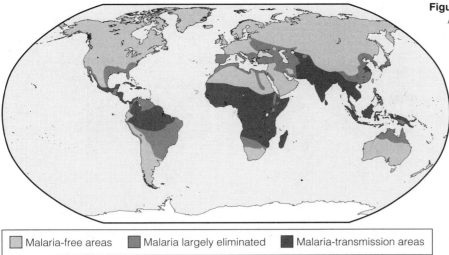

Figure 11-7 Worldwide distribution of *malaria*. About 40% of the world's current population lives in areas in which malaria is present, with the disease killing about 1 million people a year. (Data from the World Health Organization)

Malaria-free areas □ Malaria largely eliminated ■ Malaria-transmission areas

Case Study: Malaria: A Deadly Parasitic Disease That Is Making a Comeback

Malaria kills about 1 million people a year and has probably killed more people than all of the wars ever fought.

About one of every five people in the world—most of them living in poor African countries—is at risk from malaria (Figure 11-7). Worldwide, an estimated 300–500 million people are infected with the protozoan parasites that cause this disease, and 270–500 million new cases are reported each year. Malaria is not just a concern for the people living in the areas where it occurs, but also for anyone traveling to these areas—including many unsuspecting tourists—because there is no vaccine for this disease.

Malaria is caused by a parasite that is spread by the bites of certain mosquito species. It infects and destroys red blood cells, causing fever, chills, drenching sweats, anemia, severe abdominal pain and headaches, vomiting, extreme weakness, and greater susceptibility to other diseases. The disease kills about 1 million people each year (about 900,000 of them children under age 5—an average of 2,700 deaths per day. Many children who survive severe malaria have brain damage or impaired learning ability.

Malaria is caused by four species of protozoan parasites in the genus *Plasmodium*. Most cases of the disease are transmitted when an uninfected female of any of about 60 *Anopheles* mosquito species bites a person infected with *Plasmodium*, ingests blood that contains the parasite, and later bites an uninfected person (Figure 11-8). When this happens, *Plasmodium* parasites move out of the mosquito and into the human's bloodstream, multiply in the liver, and enter blood cells to continue multiplying. Malaria can also be transmitted by blood transfusions or by sharing needles.

The malaria cycle repeats itself until immunity develops, treatment is given, or the victim dies. *Over the course of human history, malarial protozoa probably have killed more people than all the wars ever fought.*

During the 1950s and 1960s, the spread of malaria was sharply curtailed by draining swamplands and marshes, spraying breeding areas with insecticides, and using drugs to kill the parasites in the bloodstream. Since 1970, malaria has come roaring back. Most species of the malaria-carrying *Anopheles* mosquito have become genetically resistant to most insecticides. Worse, the *Plasmodium* parasites have become genetically resistant to common antimalarial drugs.

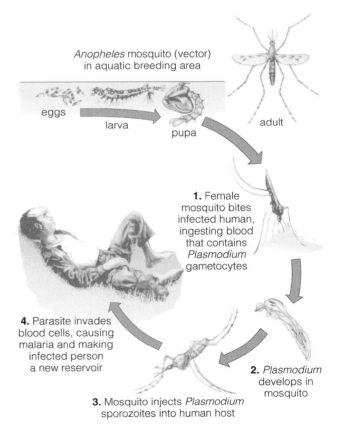

Anopheles mosquito (vector) in aquatic breeding area

eggs — larva — pupa — adult

1. Female mosquito bites infected human, ingesting blood that contains *Plasmodium* gametocytes

2. *Plasmodium* develops in mosquito

3. Mosquito injects *Plasmodium* sporozoites into human host

4. Parasite invades blood cells, causing malaria and making infected person a new reservoir

Figure 11-8 The life cycle of malaria. *Plasmodium* circulates from mosquito to human and back to mosquito.

Researchers are working to develop new anti-malarial drugs (such as artemisinins derived from the Chinese herbal remedy qinghaosu), vaccines, and biological controls for *Anopheles* mosquitoes. But such approaches receive too little funding and have proved more difficult than originally thought.

Other approaches include cultivating fish that feed on mosquito larvae (biological control), clearing vegetation around houses, planting trees that soak up water in low-lying marsh areas where mosquitoes thrive (a method that can degrade or destroy ecologically important wetlands), and using zinc and vitamin A supplements to boost resistance to malaria in children.

Spraying the inside of homes with low concentrations of DDT twice a year greatly reduces the number of malaria cases. But under an international treaty enacted in 2002, DDT and five of its chlorinated-hydrocarbon cousins are being phased out in developing countries. However, the treaty allows 25 countries to continue using DDT for malaria control until other alternatives are available. Health officials in developing countries call for much greater funding for research on finding ways to prevent and treat malaria.

Solutions: How Can We Reduce the Incidence of Infectious Diseases? More Money and Assistance

There are a number of ways to reduce the incidence of infectious diseases if the world is willing to provide the necessary funds and assistance.

Good news: According to the WHO, the global death rate from infectious diseases decreased by about two-thirds between 1970 and 2000 and is projected to continue dropping. Also, between 1971 and 2000, the percentage of children in developing countries immunized with vaccines to prevent tetanus, measles, diphtheria, typhoid fever, and polio increased from 10% to 84%—saving about 10 million lives a year.

Figure 11-9 lists measures health scientists and public health officials suggest to help prevent or reduce the incidence of infectious diseases that affect humanity (especially in developing countries). An important breakthrough has been the development of simple *oral rehydration therapy* to help prevent death from dehydration for victims of diarrheal diseases, which cause about one-fourth of all deaths of children under age 5. It involves administering a simple solution of boiled water, salt, and sugar or rice, at a cost of only a few cents per person. It has been the major factor in reducing the annual number of deaths from diarrhea from 4.6 million in 1980 to 1.9 million in 2002. Few investments have saved so many lives at such a low cost.

Bad news: The WHO estimates that only about 10% of global medical research and development money is

Solutions

Infectious Diseases

- Increase research on tropical diseases and vaccines
- Reduce poverty
- Decrease malnutrition
- Improve drinking water quality
- Reduce unnecessary use of antibiotics
- Educate people to take all of an antibiotic prescription
- Reduce antibiotic use to promote livestock growth
- Careful hand washing by all medical personnel
- Immunize children against major viral diseases
- Oral rehydration for diarrhea victims
- Global campaign to reduce HIV/AIDS

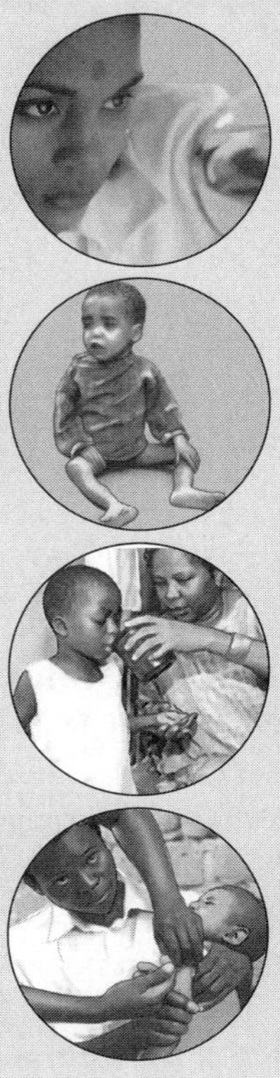

Figure 11-9 Solutions: ways to prevent or reduce the incidence of infectious diseases—especially in developing countries. Which two of these solutions do you believe are the most important?

spent on preventing infectious diseases in developing countries, even though more people worldwide suffer and die from these diseases than from all other diseases combined.

11-5 RISK ANALYSIS

How Can We Estimate Risks? Evaluate, Compare, Decide

Scientists have developed ways to evaluate and compare risks, decide how much risk is acceptable, and find affordable ways to reduce it.

Risk analysis involves identifying hazards and evaluating their associated risks (*risk assessment;* Figure 11-1, left), ranking risks (*comparative risk analysis*), determining options and making decisions about reducing or eliminating risks (*risk management;* Figure 11-1, right), and informing decision makers and the public about risks (*risk communication*).

Statistical probabilities based on past experience, animal testing and other tests, and epidemiological studies are used to estimate risks from older technologies and chemicals. To evaluate new technologies and products, risk evaluators use more uncertain statistical probabilities, based on models rather than actual experience and testing.

Figure 11-10 lists the results of a *comparative risk analysis*, summarizing the greatest ecological and health risks identified by a panel of scientists acting as advisers to the U.S. Environmental Protection Agency.

The greatest risks many people face today are rarely dramatic enough to make the daily news. In terms of the number of premature deaths per year (Figure 11-11, p. 250) and reduced life span (Figure 11-12, p. 251), *the greatest risk by far is poverty*. Its high death toll is a result of malnutrition, increased susceptibility to normally nonfatal infectious diseases, and often-fatal infectious diseases from lack of access to a safe water supply.

Thus, *the sharp reduction or elimination of poverty would do far more to improve longevity and human health than any other measure.* It would also greatly improve human rights, provide more people with income to stimulate economic development, and reduce environmental degradation and the threat of terrorism. Sharply reducing poverty is a win-win situation for people, economies, and the environment.

After the health risks associated with poverty and gender, the greatest risks of premature death are mostly the result of voluntary choices people make about their lifestyles (Figure 11-12). By far the best ways to reduce one's risk of premature death and serious health risks are to avoid smoking and exposure to smoke, lose excess weight, reduce consumption of foods containing cholesterol and saturated fats, eat a variety of fruits and vegetables, exercise regularly, not drink alcohol or drink no more than two drinks in a single day, avoid excess sunlight (which ages skin and causes skin cancer), and practice safe sex.

How Can We Estimate the Risks of Using Increasingly Complex Technology in Our Lives? A Difficult Task

Estimating risks from using certain technologies is difficult because of the unpredictability of human behavior, chance, and sabotage.

The more complex a technological system and the more people needed to design and run it, the more difficult it is to estimate the risks. The overall *reliability* or the probability (expressed as a percentage) that a person or device will complete a task without failing is the product of two factors:

$$\text{System reliability (\%)} = \text{Technology reliability} \times \text{Human reliability} \times 100$$

With careful design, quality control, maintenance, and monitoring, a highly complex system such as a nuclear power plant or space shuttle can achieve a high degree of technological reliability. But human reliability usually is much lower than technological reliability and almost impossible to predict: To err is human.

Suppose the technological reliability of a nuclear power plant is 95% (0.95) and human reliability is 75% (0.75). Then the overall system reliability is 71% (0.95 × 0.75 × 100 = 71%). Even if we could make the technology 100% reliable (1.0), the overall system reliability would still be only 75% (1.0 × 0.75 × 100 = 75%). The crucial dependence of even the most carefully designed systems on unpredictable human reliability

Comparative Risk Analysis

Most Serious Ecological and Health Problems

High-Risk Health Problems
- Indoor air pollution
- Outdoor air pollution
- Worker exposure to industrial or farm chemicals
- Pollutants in drinking water
- Pesticide residues on food
- Toxic chemicals in consumer products

High-Risk Ecological Problems
- Global climate change
- Stratospheric ozone depletion
- Wildlife habitat alteration and destruction
- Species extinction and loss of biodiversity

Medium-Risk Ecological Problems
- Acid deposition
- Pesticides
- Airborne toxic chemicals
- Toxic chemicals, nutrients, and sediment in surface waters

Low-Risk Ecological Problems
- Oil spills
- Groundwater pollution
- Radioactive isotopes
- Acid runoff to surface waters
- Thermal pollution

Figure 11-10 *Comparative risk analysis* of the most serious ecological and health problems according to scientists acting as advisers to the U.S. Environmental Protection Agency. Risks under each category are not listed in rank order. (Data from Science Advisory Board, *Reducing Risks*, Washington, D.C.: Environmental Protection Agency, 1990)

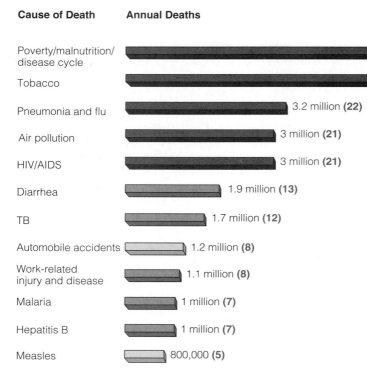

Cause of Death   **Annual Deaths**

Poverty/malnutrition/disease cycle — 11 million **(75)**

Tobacco — 5 million **(34)**

Pneumonia and flu — 3.2 million **(22)**

Air pollution — 3 million **(21)**

HIV/AIDS — 3 million **(21)**

Diarrhea — 1.9 million **(13)**

TB — 1.7 million **(12)**

Automobile accidents — 1.2 million **(8)**

Work-related injury and disease — 1.1 million **(8)**

Malaria — 1 million **(7)**

Hepatitis B — 1 million **(7)**

Measles — 800,000 **(5)**

Figure 11-11 Number of deaths per year in the world from various causes. Numbers in parentheses give these deaths in terms of the number of fully loaded 400-passenger jumbo jets crashing accidentally *every day of the year* with no survivors. Because of sensational media coverage, most people have a distorted view of the largest annual causes of death. (Data from United Nations and the World Health Organization)

helps explain essentially "impossible" tragedies such as the Chernobyl nuclear power plant accident and the *Challenger* and *Columbia* space shuttle accidents.

One way to make a system more foolproof or fail-safe is to move more of the potentially fallible elements from the human side to the technical side. However, chance events such as a lightning bolt can knock out an automatic control system, and no machine or computer program can completely replace human judgment. Also, the parts in any automated control system are manufactured, assembled, tested, certified, and maintained by fallible human beings. In addition, computer software programs used to monitor and control complex systems can also contain human error or can be deliberately modified by computer viruses to malfunction.

How Well Do We Perceive Risks? Most of Us Flunk

Most individuals are poor at evaluating the relative risks they face, mostly because of misleading information and irrational fears.

Most of us are not good at assessing the relative risks from the hazards that surround us. Also, many people deny or shrug off the high-risk chances of death (or injury) from voluntary activities they enjoy, such as *motorcycling* (1 death in 50 participants), *smoking* (1 in 300 participants by age 65 for a pack-a-day smoker), *hang gliding* (1 in 1,250), and *driving* (1 in 3,300 without a seatbelt and 1 in 6,070 with a seatbelt). Indeed, the most dangerous thing most people in many countries do each day is drive or ride in a car.

Yet some of these same people may be terrified about the possibility of being killed by a *gun* (1 in 28,000 in the United States), *flu* (1 in 130,000), *nuclear power plant accident* (1 in 200,000), *West Nile virus* (1 in 1 million), *lightning* (1 in 3 million), *commercial airplane crash* (1 in 9 million), *snakebite* (1 in 36 million), or *shark attack* (1 in 281 million).

What Factors Distort Our Perceptions of Risk? Irrational Fears and Perceptions Can Take Over.

Several factors can give people a distorted sense of risk.

Here are four factors that can cause people to see a technology or a product as being riskier than experts judge it to be. *First,* is the *degree of control* we have. Most of us have a greater fear of things that we do not have personal control over. For example, some individuals feel safer driving their own car for long distances through bad traffic than traveling the same distance on a plane. But look at the math. The risk of dying in a car accident while using your seatbelt is 1 in 6,070 whereas the risk of dying in a commercial airliner crash is 1 in 9 million. Can you think of another example?

Second, is the *fear of the unknown.* Most people have greater fear of a new, unknown product or technology than they do of an older and more familiar one. Examples include a greater fear of genetically modified food than of food produced by traditional plant-breeding techniques, and a greater fear of nuclear power plants than of more familiar coal-fired power plants. Can you think of another example?

Third, is whether or not we voluntarily take the risk. For example, we might perceive that the risk from driving, which is largely *voluntary,* is less than that from a nuclear power plant, which is mostly imposed on us whether we like it or not. Can you come up with another example?

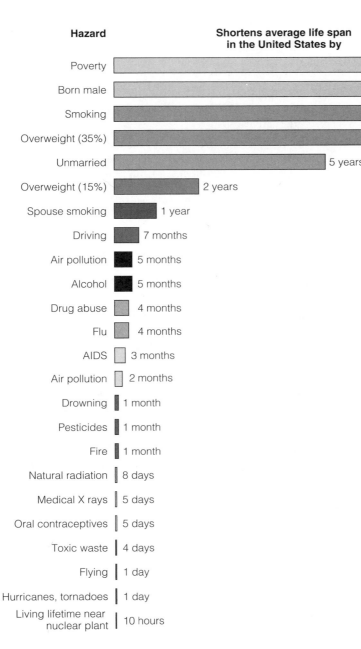

Hazard	Shortens average life span in the United States by
Poverty	7–10 years
Born male	7.5 years
Smoking	6 years
Overweight (35%)	6 years
Unmarried	5 years
Overweight (15%)	2 years
Spouse smoking	1 year
Driving	7 months
Air pollution	5 months
Alcohol	5 months
Drug abuse	4 months
Flu	4 months
AIDS	3 months
Air pollution	2 months
Drowning	1 month
Pesticides	1 month
Fire	1 month
Natural radiation	8 days
Medical X rays	5 days
Oral contraceptives	5 days
Toxic waste	4 days
Flying	1 day
Hurricanes, tornadoes	1 day
Living lifetime near nuclear plant	10 hours

Figure 11-12 Comparison of risks people face, expressed in terms of shorter average life span. After poverty and gender, the greatest risks people face are mostly from lifestyle choices they make. These are only generalized relative estimates. Individual response to some of these risks can vary with factors such as genetic variation, family medical history, emotional makeup, stress, and social ties and support. (Data from Bernard L. Cohen)

on careful risk analysis, it is usually seen as politics, not science. Residents will not be satisfied by estimates that the lifetime risks of cancer death from the facility are not greater than, say, 1 in 100,000. Instead, they point out that living near the facility means that they will have a much higher risk of dying from cancer than would people living further away.

How Can You Become Better at Risk Analysis? Analyze, Compare, and Evaluate Your Lifestyle

To become better at risk analysis you can carefully evaluate the barrage of bad news covered in the media, compare risks, and concentrate on reducing risks over which you have some control.

You can do three things to become better at estimating risks. *First,* carefully evaluate what the media presents. Recognize that the media often give an exaggerated view of risks to capture our interest and thus sell newspapers or gain TV viewers.

Second, compare risks. Do you risk getting cancer by eating a charcoal-broiled steak once or twice a week? Yes, because in theory anything can harm you. The question is whether this danger is great enough for you to worry about. In evaluating a risk the question is not, "Is it safe?" but rather, "How risky is it compared to other risks?"

Third, concentrate on the most serious risks to your life and health that you have some control over and stop worrying about smaller risks and those over which you have no control. When you worry about something, the most important question to ask is, "Do I have any control over this?"

You have control over major ways to reduce risks from heart attack, stroke, and many forms of cancer because you decide whether you smoke, what you eat, how much exercise you get, how much alcohol you consume, your exposure to the sun's ultraviolet rays, and whether or not you practice safe sex. Concentrate on evaluating these important choices, and you will have a much greater chance of living a longer, healthier, happier, and less fearful life.

Fourth, is whether a risk is *catastrophic, not chronic.* We usually have a much greater fear of a well-publicized death toll from a single catastrophic accident rather than the same or an even larger death toll spread out over a longer time. Examples include a severe nuclear power plant accident, an industrial explosion, or an accidental plane crash, as opposed to coal-burning power plants, automobiles, or smoking. Can you think of another example?

There is also concern over the *unfair distribution of risks* from the use of a technology or chemical. Citizens are outraged when government officials decide to put a hazardous waste landfill or incinerator in or near their neighborhood. Even when the decision is based

The burden of proof imposed on individuals, companies, and institutions should be to show that pollution prevention options have been thoroughly examined, evaluated, and used before lesser options are chosen.

JOEL HIRSCHORN

CRITICAL THINKING

1. Explain why you agree or disagree with the proposals for reducing the death toll and other harmful effects of smoking listed on p. 238. Do you believe that there should be a ban on smoking indoors in all public places? Explain.

2. Should we have zero pollution levels for all toxic and hazardous chemicals? Explain. What are the alternatives?

3. Do you believe health and safety standards in the workplace should be strengthened and enforced more vigorously, even if this causes a loss of jobs when companies transfer operations to countries with weaker standards? Explain.

4. Evaluate the following statements:
   a. We should not get worked up about exposure to toxic chemicals because almost any chemical can cause some harm at a large enough dosage.
   b. We should not worry so much about exposure to toxic chemicals because through genetic adaptation we can develop immunity to such chemicals.
   c. We should not worry so much about exposure to toxic chemicals because we can use genetic engineering to reduce or eliminate such problems.

5. How can changes in the age structure of a human population increase the spread of infectious diseases? How can the spread of infectious diseases affect the age structure of human populations?

6. Should laboratory-bred animals be used in laboratory experiments in toxicology? Explain. What are the alternatives?

7. What are the five major risks you face from **(a)** your lifestyle, **(b)** where you live, and **(c)** what you do for a living? Which of these risks are voluntary and which are involuntary? List the five most important things you can do to reduce these risks. Which of these things do you actually plan to do?

8. Congratulations! You are in charge of the world. List the three most important features of your program to reduce the risk from exposure to **(a)** toxic and hazardous chemicals, and **(b)** infectious disease organisms.

LEARNING ONLINE

The website for this book contains helpful study aids and many ideas for further reading and research. They include a chapter summary, review questions for the entire chapter, flash cards for key terms and concepts, a multiple-choice practice quiz, interesting Internet sites, references, and a guide for accessing thousands of InfoTrac® College Edition articles. Log on to

http://biology.brookscole.com/miller7

Then click on the Chapter-by-Chapter area, choose Chapter 11, and select a learning resource.

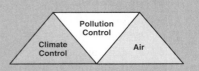

12 CLIMATE CHANGE, OZONE DEPLETION, AND AIR POLLUTION

We are embarked on the most colossal ecological experiment of all time—doubling the concentration in the atmosphere of an entire planet of one of its most important gases—and we really have little idea of what might happen.

PAUL A. COLINVAUX

12-1 ATMOSPHERE, WEATHER, AND CLIMATE

What Is the Troposphere? Weather Breeder

The innermost layer of the atmosphere is made up mostly of nitrogen and oxygen, with smaller amounts of water vapor and carbon dioxide.

We live at the bottom of a sea of air called the **atmosphere.** It is divided into several spherical layers (Figure 12-1), each layer characterized by abrupt changes in temperature as a result of differences in the absorption of incoming solar energy.

About 75–80% of the mass of the earth's air is found in the atmosphere's innermost layer, the **troposphere,** the atmospheric layer closest to the earth's surface. This layer extends only about 17 kilometers (11 miles) above sea level at the equator and about 8 kilometers (5 miles) over the poles. If the earth were the size of an apple, this lower layer containing the air we breathe would be no thicker than the apple's skin.

Take a deep breath. About 99% of the volume of the air you just inhaled consists of two gases: nitrogen (78%) and oxygen (21%). The remainder consists of water vapor (varying from 0.01% at the frigid poles to 4% in the humid tropics), slightly less than 1% argon (Ar), 0.038% carbon dioxide (CO_2), and trace amounts of several other gases.

The troposphere is also the layer of the atmosphere involved in the chemical cycling of many of the earth's vital nutrients. In addition, this thin and turbulent layer of rising and falling air currents and winds is largely responsible for the planet's short-term *weather* and long-term *climate.*

To biologist and environmental scientist David Suzuki, "Air is a matrix or universal glue that joins all life together. . . . Every breath is an affirmation of our connection with other living things, a renewal of our link with our ancestors, and a contribution to generations yet to come."

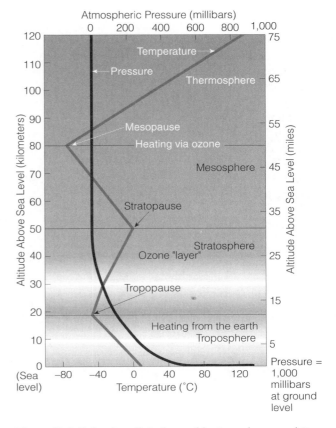

Figure 12-1 Natural capital: the earth's atmosphere consists of several layers. The average temperature of the atmosphere varies with altitude (light line). Most UV radiation from the sun is absorbed by ozone (O_3), found primarily in the stratosphere in the *ozone layer,* 17–26 kilometers (10–16 miles) above sea level.

What Is the Stratosphere? Earth's Global Sunscreen

Ozone in the atmosphere's second layer filters out most of the sun's UV radiation that is harmful to us and most other species.

The atmosphere's second layer is the **stratosphere,** which extends from about 17–48 kilometers (11–30 miles) above the earth's surface (Figure 12-1). Although the stratosphere contains less matter than the troposphere, its composition is similar, with two notable exceptions: its volume of water vapor is about

1/1,000 as much and its concentration of ozone (O_3) is much higher.

Stratospheric ozone is produced when some of the oxygen molecules there interact with ultraviolet (UV) radiation emitted by the sun. This "global sunscreen" of ozone in the stratosphere keeps about 95% of the sun's harmful UV radiation from reaching the earth's surface.

This UV filter of "good" ozone in the lower stratosphere allows humans and other forms of life to exist on land and helps protect us from sunburn, skin and eye cancers, cataracts, and damage to our immune systems. It also prevents much of the oxygen in the troposphere from being converted to ozone, a harmful air pollutant, in the lower troposphere.

Much evidence indicates that some human activities are *decreasing* the amount of beneficial or "good" ozone in the stratosphere and *increasing* the amount of harmful or "bad" ozone in the troposphere—especially in some urban areas.

How Does Weather Differ from Climate? Short- and Long-Term Atmospheric Conditions

Weather is an area's atmospheric conditions over short time periods and climate is the average of such conditions over long time periods.

Weather is an area's short-term atmospheric conditions—typically over hours or days. Examples of atmospheric conditions are temperature, pressure, moisture content, precipitation, sunshine, cloud cover, and wind direction and speed.

Climate is a region's long-term atmospheric conditions—typically over decades. *Average temperature* and *average precipitation* are the two main factors determining a region's climate and its effects on people. Figure 12-2 is a generalized map of the earth's major climate zones. In what type of climate zone do you live?

The temperature and precipitation patterns that lead to different climates are caused primarily by in-

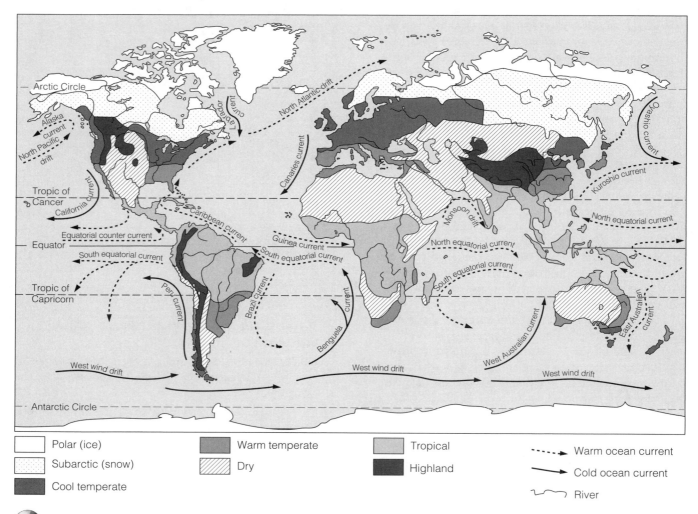

Figure 12-2 Natural capital: generalized map of the earth's current climate zones, showing the major contributing ocean currents and drifts.

coming solar energy and the way air circulates over the earth's surface. Solar energy heats the atmosphere, evaporates water, helps create seasons, and causes air to circulate.

Four major factors determine global air circulation patterns. One is the *uneven heating of the earth's surface.* Air is heated much more at the equator where the sun's rays strike directly than at the poles where sunlight strikes at an angle and thus is spread out over a much greater area. These differences in the amount of incoming solar energy help explain why tropical regions near the equator are hot, polar regions are cold, and temperate regions in between generally have intermediate average temperatures.

A second factor is *seasonal changes in temperature and precipitation.* The earth's axis—an imaginary line connecting the north and south poles—is tilted. As a result, various regions are tipped toward or away from the sun as the earth makes its yearlong revolution around the sun. This creates opposite seasons in the northern and southern hemispheres.

A third factor is *rotation of the earth on its axis.* The earth's rotation prevents air currents from moving due north and south from the equator. This results in the formation of six huge convection cells of swirling air masses, three north and three south of the equator, that transfer heat and water from one area to another and create different climate patterns in different parts of the earth (Figure 12-3).

Fourth, *properties of air, water, and land affect global air circulation.* Heat from the sun evaporates ocean water and transfers heat from the oceans to the atmosphere, especially near the hot equator. As this moisture-laden air rises, it cools and releases moisture as condensation because cold air holds less water vapor than warm air. The resulting cooler, drier air becomes denser, sinks (subsides), and creates an area of high pressure. As these air masses flow across the earth's surface, they pick up heat and moisture and begin to rise again. The resulting convection cells circulate air, heat, and moisture both vertically and from place to place in the troposphere.

If the earth's surface were uniform, its climate zones would occur in bands corresponding to latitude. One reason this does not happen is the presence of a mixture of continents and oceans, because continents heat and cool fairly quickly and oceans heat and cool more slowly. This alters wind and water flows and leads to an irregular distribution of climates and patterns of vegetation, as shown in Figure 12-3.

How Do Ocean Currents and Winds Affect Regional Climates? Moving Heat and Air Around

Ocean current and winds influence climate by redistributing heat received from the sun from one place to another.

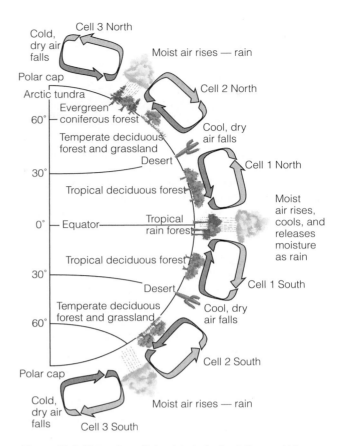

Figure 12-3 Natural capital: global air circulation and biomes. Heat and moisture are distributed over the earth's surface by vertical currents that form six large convection cells at different latitudes. The direction of airflow and the ascent and descent of air masses in these convection cells determine the earth's general climatic zones. The resulting uneven distribution of heat and moisture over the planet's surface leads to the forests, grasslands, and deserts that make up the earth's biomes.

The oceans absorb heat from the air circulation patterns just described, with the bulk of this heat absorbed near the warm tropical areas. This heat plus differences in water density create warm and cold ocean currents (Figure 12-2). These currents, driven by winds and the earth's rotation, redistribute heat received from the sun and thus influence climate and vegetation, especially near coastal areas. They also help mix ocean waters and distribute nutrients and dissolved oxygen needed by aquatic organisms.

How Do Gases Warm the Atmosphere and Affect Climate? The Natural Greenhouse Effect

Water vapor, carbon dioxide, and other gases influence climate by warming the lower troposphere and the earth's surface.

Small amounts of certain gases play a key role in determining the earth's average temperatures and thus its climates. These gases include water vapor (H_2O), carbon dioxide (CO_2), methane (CH_4), and nitrous oxide (N_2O).

Together, these gases, known as **greenhouse gases,** allow mostly visible light and some infrared radiation and ultraviolet (UV) radiation from the sun to pass through the troposphere. The earth's surface absorbs much of this solar energy and transforms it to longer-wavelength, infrared radiation, which rises into the troposphere.

Some of this infrared radiation escapes into space, and some is absorbed by molecules of greenhouse gases and emitted into the troposphere in all directions as even longer-wavelength infrared radiation, which warms the air. Some of this released energy is radiated into space and some warms the troposphere and the earth's surface. This natural warming effect of the air in the troposphere and surface is called the **greenhouse effect,** diagrammed earlier in Figure 2-10, p. 27.

The basic principle behind the natural greenhouse effect is well established. Indeed, without its current concentrations of greenhouse gases (especially water vapor, which is found in the largest concentration), the earth would be a cold and mostly lifeless planet.

12-2   CLIMATE CHANGE: HOW SERIOUS IS THE THREAT?

How Have the Earth's Temperature and Climate Changed in the Past? Climate Change Is Not New.

Temperature and climate have been changing throughout the earth's history.

Climate change is neither new nor unusual. The earth's average surface temperature and climate have been changing throughout the world's 4.7-billion-year history due to volcanic emissions, changes in solar input, continents moving as a result of shifting tectonic plate movements, strikes by large meteors, and other factors.

The troposphere's average temperature has changed gradually at some times (over hundreds to millions of years) and at other times fairly quickly (over a few decades to 100 years), as shown in the Figure 12-4 graphs.

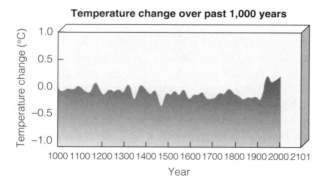

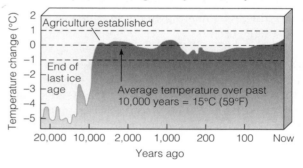

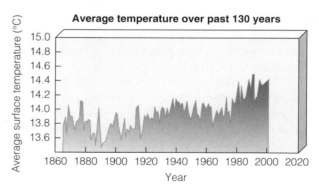

Figure 12-4 Estimated changes in the average global temperature of the atmosphere near the earth's surface over different periods of time. Past temperature changes are estimated by analysis of radioisotopes in rocks and fossils, plankton and radioisotopes in ocean sediments, ice cores from ancient glaciers, temperature measurements at different depths in boreholes drilled deep into the earth's surface, pollen from lake bottoms and bogs, tree rings, historical records, and temperature measurements (since 1861). (Data from Goddard Institute for Space Studies, Intergovernmental Panel on Climate Change, National Academy of Sciences, National Aeronautics and Space Agency, National Center for Atmospheric Research, and National Oceanic and Atmospheric Administration)

Over the past 900,000 years, the average temperature of the troposphere has undergone prolonged periods of *global cooling* and *global warming* (Figure 12-4, top left). These alternating cycles of freezing and thawing are known as *glacial and interglacial* (between ice ages) *periods*.

During each cold period, thick glacial ice covered much of the earth's surface for about 100,000 years. Most of it melted during a *warmer interglacial period* lasting 10,000–12,500 years that followed each glacial period, during which most of the ice melted. For roughly 12,000 years, we have had the good fortune to live in an interglacial period with a fairly stable climate and average global surface temperature (Figure 12-4, top right, and bottom left).

According to measurements of CO_2 in bubbles in glacial ice, estimated changes in tropospheric CO_2 levels correlate fairly closely with estimated variations in the atmosphere's average global temperature near the earth's surface during the past 160,000 years (Figure 12-5). Trace the curves in this figure.

Is the Earth's Troposphere Warming? Very Likely

There is considerable evidence that the earth's troposphere is warming, partly because of human activities.

In 1988, the United Nations and the World Meteorological Organization established the Intergovernmental Panel on Climate Change (IPCC) to document past climate change and project future climate change. The IPCC is a network of over 2,000 leading climate experts from 70 nations.

Panels of scientists from the U.S. National Academy of Sciences have also evaluated possible future climate changes. In addition, the U.S. Congress created the *U.S. Global Change Research Program* (*USGCRP*) in 1990 to project future climate changes and the potential impacts.

Figure 12-6 (p. 258) shows that since 1861, the concentrations of the greenhouse gases CO_2, CH_4, and N_2O in the troposphere have risen sharply, especially since 1950. According to studies by IPCC and the U.S. National Academy of Sciences, humans have increased concentrations of greenhouse gases in the troposphere by burning fossil fuels (which adds CO_2 and CH_4), clearing and burning forests and grasslands (which add CO_2 and N_2O), and planting rice and using inorganic fertilizers (both of which release N_2O into the troposphere)

In its 2001 report, the IPCC listed a number of findings indicating that *very likely* (90–99% probability) that the troposphere is getting warmer (Figure 12-4, bottom left). *First,* the 20th century was the hottest century in the past 1,000 years (Figure 12-4, bottom left). *Second,* since 1861 the average global temperature of the tropo-

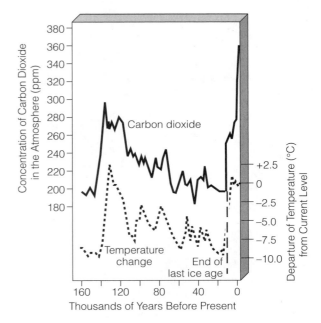

Figure 12-5 Atmospheric carbon dioxide levels and global temperature. Estimated long-term variations in average global temperature of the atmosphere near the earth's surface are graphed along with average tropospheric CO_2 levels over the past 160,000 years. The rough correlation between CO_2 levels in the troposphere and temperature, shown in these estimates based on ice core data, suggests a connection between these two variables, although no definitive causal link has been established. In 1999, the world's deepest ice core sample revealed a similar correlation between air temperatures and the greenhouse gases CO_2 and CH_4 going back 460,000 years. (Data from Intergovernmental Panel on Climate Change and National Center for Atmospheric Research)

sphere near the earth's surface has risen 0.6°C (1.1°F) over the entire globe and about 0.8°C (1.4°F) over the continents (Figure 12-4, bottom right). Most of this increase has taken place since 1980.

Third, the 16 warmest years on record have occurred since 1980 and the 10 warmest years since 1990 (Figure 12-4, bottom right). The hottest year was 1998, followed in order by 2002, 2001, and 2003. *Fourth,* glaciers and floating sea ice in some parts of the world are melting and shrinking. *Fifth,* during the last century the world's average sea level rose by 0.1–0.2 meters (4–8 inches), partly because of runoff of water from melting land-based ice and partly because ocean water expands when its temperature increases.

The global climate system is complex and still poorly understood. As a result, scientists cannot pin down the exact causes of such warming. They could be the result of natural climate fluctuations, changes in climate from human activities, or a combination of both factors—the most likely explanation according to most climate scientists. Regardless of the cause, it is virtually certain that significant climate change caused

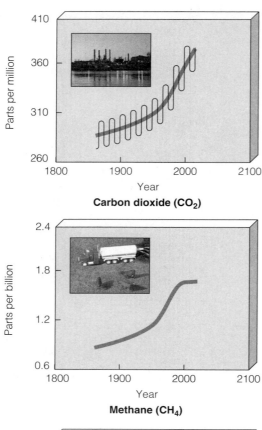

Carbon dioxide (CO$_2$)

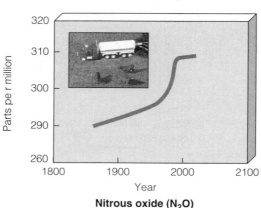

Methane (CH$_4$)

Nitrous oxide (N$_2$O)

Figure 12-6 Increases in average concentrations of the greenhouse gases, carbon dioxide, methane, and nitrous oxide in the troposphere between 1861 and 2003. The fluctuations in the CO$_2$ curve represent seasonal changes in photosynthetic activity that cause small differences between summer and winter concentrations of CO$_2$. (Data from Intergovernmental Panel on Climate Change, National Center for Atmospheric Research, and World Resources Institute)

by atmospheric warming (or cooling) over several decades to 100 years has important implications for human life, wildlife, and the world's economies.

Based on evidence about past changes in atmospheric CO$_2$ concentrations and atmospheric tempera-

tures, climate scientists believe that increased inputs of CO$_2$ and other greenhouse gases from human activities will enhance the earth's natural greenhouse effect and raise the average global temperature of the atmosphere near the earth's surface.

The terms global warming and global climate change often are used interchangeably, but they are not the same. **Global warming** refers to temperature increases in the troposphere, which in turn can cause climate change. **Global climate change** is a broader term that refers to changes in any aspects of the earth's climate, including temperature, precipitation, and storm intensity. It can involve global warming or cooling, but the focus in this chapter is on global warming. Global warming should not be confused with the different atmospheric problem of ozone depletion in the stratosphere (Table 12-1) that is discussed in Section 12-6.

What Is the Scientific Consensus about Future Climate Change and Its Effects? Change Is Under Way.

Human activities played a major role in recent climate changes and will very likely lead to significant climate change during this century.

To project the effects of increases in greenhouse gases on average global temperature, scientists develop complex *mathematical models* of how interactions among the earth's land, oceans, ice, and greenhouse gases determine the earth's average temperature. Then they run the models on supercomputers and they use the results to project future changes in the earth's average temperature. Figure 12-7 (p. 260) gives a greatly simplified summary of some of the interactions in the global climate system. Trace the flows and connections in this figure.

Such models provide scenarios of what is *very likely* (90–99% probability) or *likely* (66–90% probability) to happen to the average temperature of the troposphere based on various assumptions and data fed into the model. How well the results correspond to the real world depends on the assumptions of the model (based on current knowledge about the systems making up the earth, oceans, and atmosphere), and the accuracy of the data used.

In 1990, 1995, and 2001, the IPCC published reports that evaluated and are likely to change during this century (Figure 12-4). Here are three major findings of the 2001 report:

- Despite many uncertainties, the latest climate models match the records of global temperature changes since 1850 very closely.

- There is new and stronger evidence that most of the warming observed over the last 50 years is attributable to human activities.

Table 12-1 Major Characteristics of Global Warming and Ozone Depletion

Characteristic	Global Warming	Ozone Depletion
Region of atmosphere involved	Troposphere.	Stratosphere.
Major substances involved	CO_2, CH_4, N_2O (greenhouse gases).	O_3, O_2, chlorofluorocarbons (CFCs).
Interaction with radiation	Molecules of greenhouse gases absorb infared (IR) radiation from the earth's surface, vibrate, and release longer-wavelength IR radiation (heat) into the lower troposphere. This natural greenhouse effect helps warm the lower troposphere.	About 95% of incoming ultraviolet (UV) radiation from the sun is absorbed by O_3 molecules in the stratosphere and does not reach the earth's surface.
Nature of problem	There is a high (90–99%) probability that increasing concentrations of greenhouse gases in the troposphere from burning fossil fuels, deforestation, and agriculture are enhancing the natural greenhouse effect and raising the earth's average surface temperature (Figure 12-4, bottom right, and Figure 12-8, p. 261).	CFCs and other ozone-depleting chemicals released into the troposphere by human activities have made their way to the stratosphere, where they decrease O_3 concentration. This can allow more harmful UV radiation to reach the earth's surface.
Possible consequences	Changes in climate, agricultural productivity, water supplies, and sea level.	Increased incidence of skin cancer, eye cataracts, and immune system suppression and damage to crops and phytoplankton.
Possible responses	Decrease fossil fuel use and deforestation; prepare for climate change.	Eliminate or find acceptable substitutes CFCs and other ozone-depleting chemicals.

- It is *very likely* (90–99% probability) that the earth's mean surface temperature will increase by 1.4–5.8°C (2.5–10.4°F) between 2000 and 2100 (Figure 12-8, p. 261).

IPCC and other climate scientists holding the consensus or sound science view call for greatly increased research on how the earth's climate system works and for development of improved climate models to help narrow the range of uncertainty in projected temperature changes.

✗ *HOW WOULD YOU VOTE?* Do you believe that we will experience significant global warming during this century? Cast your vote online at http://biology.brookscole.com/miller7.

Why Should We Be Concerned about a Warmer Earth? The Speed of Change Is What Counts.

A rapid increase in the temperature of the troposphere during this century would give us little time to deal with its effects.

Climate scientists warn that the concern is not just a temperature change but how rapidly it occurs, regardless of cause. Past changes in the temperature of the troposphere often took place over thousands of years to a hundred thousand years (Figure 12-4, top left). The problem we face is a fairly sharp projected increase in temperature of the troposphere during this century (Figure 12-8, p. 261).

According to the IPPC, there is a 90–99% chance that this will be the fastest temperature change of the past 1,000 years. Such rapid temperature change can affect the availability of water resources by altering rates of evaporation and precipitation. It can also change wind patterns and weather; dry some areas; add moisture to others; alter ocean currents; shift areas where crops can be grown; increase average sea levels and flood some coastal wetlands, cities, and low-lying islands; and alter the structure and location of some of the world's biomes. These are major changes in the earth's atmospheric conditions. An increase in the earth's temperature within a few decades or a century gives us little time to deal with its effects.

In 2002, the U.S. National Academy of Sciences issued a study, which raised the possibility that the temperature of the troposphere could rise drastically in only a decade or two. The report lays out a nightmarish worst-case scenario in which ecosystems suddenly collapse, low-lying cities are flooded, forests are consumed in vast fires, grasslands die out and turn into dust bowls, wildlife disappears, and tropical waterborne and insect-transmitted infectious diseases spread rapidly beyond their current ranges.

These possibilities were affirmed by a 2003 analysis carried out by Peter Schwartz and Doug Randall for the Department of Defense. They concluded that global warming "must be viewed as a serious threat to global stability and should be elevated beyond a scientific debate to a U.S. national security concern."

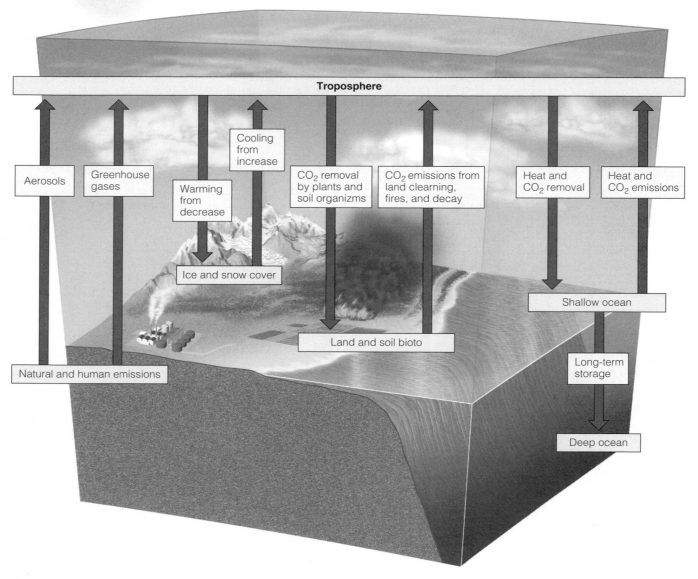

Figure 12-7 Natural capital: simplified model of some of the major processes that interact to determine the average temperature and greenhouse gas content of the troposphere and thus the earth's climate.

12-3 FACTORS AFFECTING THE EARTH'S TEMPERATURE

Scientists have identified a number of natural and human-influenced factors that might *amplify* or *dampen* projected changes in the average temperature of the troposphere. The fairly wide range of projected future temperature changes shown in Figure 12-8 results from including what is known about these factors in climate models. Let us examine some possible wild cards that could help or make matters worse or better during this century.

Can the Oceans Store More CO₂ and Heat? We Do Not Know.

There is uncertainty about how much CO_2 and heat the oceans can remove from the troposphere and how long they might remain in the oceans.

The oceans help moderate the earth's average surface temperature by removing about 29% of the excess CO_2 we pump into the atmosphere as part of the global carbon cycle. They also absorb heat from the atmosphere and slowly transfer some of it to the deep ocean, where it is removed from the climate system for long but unknown periods of time (Figure 12-7).

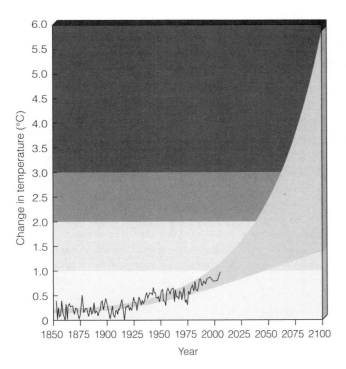

Figure 12-8 Comparison of measured changes in the average temperature of the atmosphere at the earth's surface between 1861 and 2003, and the projected range of temperature increase during the rest of this century. (Data from U.S. National Academy of Sciences, National Center for Atmospheric Research, and Intergovernmental Panel on Climate Change)

Ocean currents on the surface and deep down are connected and act like a gigantic conveyor belt to store CO_2 and heat in the deep sea and to transfer hot and cold water from the tropics to the poles (Figure 12-9). Scientists are concerned that in a warmer world an influx of fresh water from increased rain in the North Atlantic and thawing ice in the Arctic region might slow or disrupt this conveyor belt.

Evidence suggests large changes in the speed of this conveyor belt and its stopping and starting contributed to wild swings in northern hemisphere temperatures during past ice ages. Scientists are trying to learn more about how this belt operates to evaluate the likelihood of it slowing down or stalling during this century, and the effects this might have on regional and global atmospheric temperatures.

The large loop of shallow and deep ocean currents shown in Figure 12-9 helps keep much of the northern hemisphere fairly warm by pulling warm tropical water north, pushing cold water south, and releasing much of the heat stored in the water into the troposphere.

If this loop of currents should slow sharply or shut down, northern Europe and the northeast coast of North America would experience severe regional cooling. In other words, *global warming can lead to significant global cooling in some parts of the world*, with the climate of western Europe resembling that of Siberia.

How Might Changes in Cloud Cover Affect the Troposphere's Temperature? Another Uncertainty

Warmer temperatures create more clouds that could warm or cool the troposphere, but we do not know which effect might dominate.

One of the largest unknowns in global climate models is the effect of changes in the global distribution of clouds on the temperature of the troposphere. Warmer temperatures increase evaporation of surface water and create more clouds. These additional clouds can have a *warming effect,* by absorbing and releasing heat into the troposphere, or a *cooling effect* by reflecting more sunlight back into space.

The net result of these two opposing effects depends on several factors. One is how much water vapor

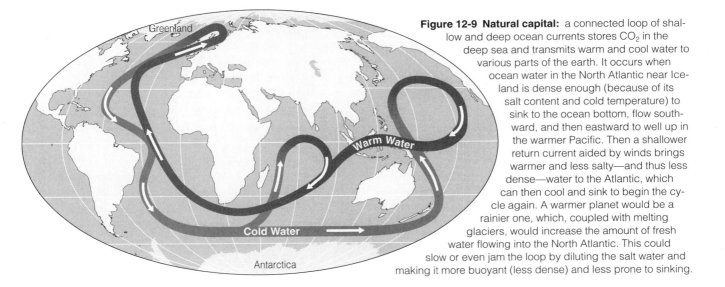

Figure 12-9 Natural capital: a connected loop of shallow and deep ocean currents stores CO_2 in the deep sea and transmits warm and cool water to various parts of the earth. It occurs when ocean water in the North Atlantic near Iceland is dense enough (because of its salt content and cold temperature) to sink to the ocean bottom, flow southward, and then eastward to well up in the warmer Pacific. Then a shallower return current aided by winds brings warmer and less salty—and thus less dense—water to the Atlantic, which can then cool and sink to begin the cycle again. A warmer planet would be a rainier one, which, coupled with melting glaciers, would increase the amount of fresh water flowing into the North Atlantic. This could slow or even jam the loop by diluting the salt water and making it more buoyant (less dense) and less prone to sinking.

will enter the troposphere as the earth's surface warms. An increase in thick and continuous clouds at low altitudes can *decrease* surface warming by reflecting and blocking more sunlight. However, an increase in thin and discontinuous cirrus clouds at high altitudes can warm the lower troposphere and *increase* surface warming.

In addition, infrared satellite images indicate that the wispy condensation trails (contrails) left behind by jet planes might have a greater impact on the temperature of the troposphere than scientists had thought. NASA scientists found that jet contrails expand and turn into large cirrus clouds that tend to release heat into the upper troposphere. If these preliminary results are confirmed, emissions from jet planes could be responsible for as much as half of the tropospheric warming in the northern hemisphere.

What climate scientists know about the effects of clouds has been included in the latest climate models, but much uncertainty remains.

How Might Outdoor Air Pollution Affect the Troposphere's Temperature? A Temporary Effect

Aerosol pollutants and soot produced by human activities can warm or cool the atmosphere, but such effects will decrease with any decline in outdoor air pollution.

Aerosols (microscopic droplets and solid particles) of various air pollutants are released or formed in the troposphere by volcanic eruptions and human activities and can either warm or cool the air depending on factors such as their size and the reflectivity of the underlying surface.

Most tropospheric aerosols, such as sulfate particles produced by fossil fuel combustion, tend to cool the atmosphere and thus can temporarily slow global warming. However, a recent study by Mark Jacobson of Stanford University indicated that tiny particles of *soot* or *black carbon aerosols*—produced mainly from incomplete combustion in coal burning, diesel engines, and open fires—may be responsible for 15–30% of global warming over the past 50 years. If so, such soot would be the second biggest contributor to global warming after the greenhouse gas CO_2.

Climate scientists do not expect aerosol pollutants to counteract or enhance projected global warming very much in the next 50 years for two reasons. One is that aerosols and soot fall back to the earth or are washed out of the lower atmosphere within weeks or months, whereas CO_2 and other greenhouse gases remain in the atmosphere for decades to several hundred years. The other is that aerosol inputs into the atmosphere are being reduced—especially in developed countries.

Can Increased CO_2 Levels Stimulate Photosynthesis and Remove More CO_2 from the Air? A Temporary and Limited Effect

Increased CO_2 in the troposphere could increase plant photosynthesis, but several factors can limit or offset this effect.

Some studies suggest that more CO_2 in the atmosphere could increase the rate of photosynthesis in some areas with adequate water and soil nutrients. This would remove more CO_2 from the troposphere and help slow global warming.

However, recent studies indicate that this CO_2 removal would be temporary for two reasons. One is that it would slow as the plants reach maturity and take up less CO_2 from the troposphere. The other is that carbon stored by the plants would be returned to the atmosphere as CO_2 when the plants die and decompose or burn.

How Might a Warmer Troposphere Affect Methane Emissions? Accelerated Warming

Warmer air can release methane gas stored in bogs, wetlands, and tundra soils and make the air even warmer.

Global warming could be accelerated by an increased release of methane (a potent greenhouse gas) from two major sources. One is bogs and other wetlands and the other is ice-like compounds called *methane hydrates* trapped beneath the arctic permafrost. Significant amounts of methane would be released into the troposphere if the permafrost in tundra and boreal forest soils melts, as is occurring in parts of Canada, Alaska, China, and Mongolia. The resulting tropospheric warming could lead to more methane release and still more warming.

12-4 POSSIBLE EFFECTS OF A WARMER WORLD

What Are Some Possible Effects of a Warmer Troposphere? Winners and Losers

A warmer climate would have beneficial and harmful effects, but poor nations in the tropics will suffer the most.

A warmer global climate could have a number of harmful and beneficial effects (Figures 12-10 and 12-11) for humans, other species, and ecosystems, depending mostly on location and on how rapidly the climate changes. Study these figures carefully. However, betting on living in an area with favorable climate change

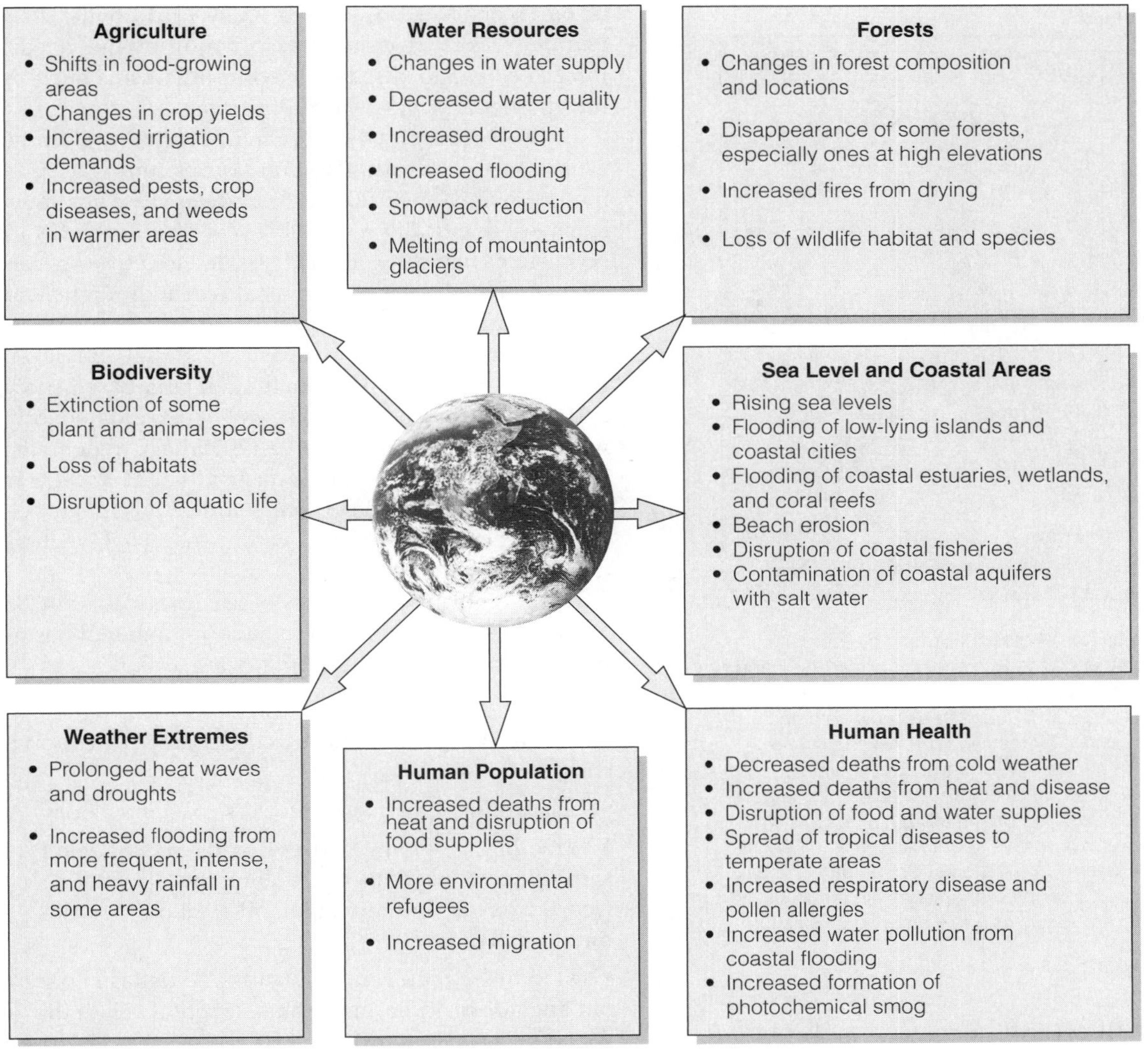

Figure 12-10
Winners and losers. Projected effects of a warmer atmosphere for the world and the United States. Most of these effects could be harmful or beneficial, depending on where one lives. Current models of the earth's climate cannot make reliable projections about where such effects might take place at a regional level and how long they might last. (Data from Intergovernmental Panel on Climate Change, U.S. Global Climate Change Research Program, U.S. National Academy of Science)

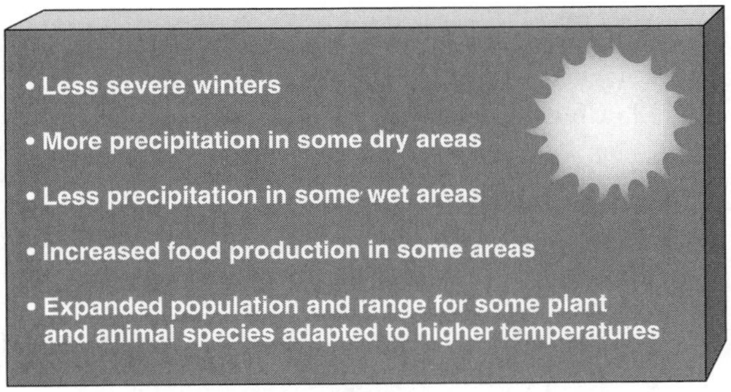

Agriculture
- Shifts in food-growing areas
- Changes in crop yields
- Increased irrigation demands
- Increased pests, crop diseases, and weeds in warmer areas

Water Resources
- Changes in water supply
- Decreased water quality
- Increased drought
- Increased flooding
- Snowpack reduction
- Melting of mountaintop glaciers

Forests
- Changes in forest composition and locations
- Disappearance of some forests, especially ones at high elevations
- Increased fires from drying
- Loss of wildlife habitat and species

Biodiversity
- Extinction of some plant and animal species
- Loss of habitats
- Disruption of aquatic life

Sea Level and Coastal Areas
- Rising sea levels
- Flooding of low-lying islands and coastal cities
- Flooding of coastal estuaries, wetlands, and coral reefs
- Beach erosion
- Disruption of coastal fisheries
- Contamination of coastal aquifers with salt water

Weather Extremes
- Prolonged heat waves and droughts
- Increased flooding from more frequent, intense, and heavy rainfall in some areas

Human Population
- Increased deaths from heat and disruption of food supplies
- More environmental refugees
- Increased migration

Human Health
- Decreased deaths from cold weather
- Increased deaths from heat and disease
- Disruption of food and water supplies
- Spread of tropical diseases to temperate areas
- Increased respiratory disease and pollen allergies
- Increased water pollution from coastal flooding
- Increased formation of photochemical smog

- Less severe winters
- More precipitation in some dry areas
- Less precipitation in some wet areas
- Increased food production in some areas
- Expanded population and range for some plant and animal species adapted to higher temperatures

Figure 12-11 *Winners.* Possible *beneficial effects* of a warmer atmosphere for some countries and people.

in the future is like playing a game of Russian roulette. Global climate models are improving, but so far scientists cannot make reliable projections of how the climates of particular regions are likely to change.

According to the IPCC, the largest burden of the harmful effects of moderate global warming will fall on people and economies in poorer tropical and subtropical nations without the economic and technological resources needed to adapt to those harmful effects. Some researchers estimate that by the end of this century the annual death toll from global warming could reach 6 million or more. Some analysts say these estimates are exaggerated. But even with a lower toll this is a serious and largely preventable human tragedy.

Here are a few of the effects projected by IPCC scientists based on current climate models. The largest temperature increases will take place at the earth's poles and are likely to cause melting of glaciers and floating ice. A warmer climate could expand ranges and populations of some plant and animal species that can adapt to warmer climates.

On the other hand, several factors could accelerate atmospheric warming by releasing large amounts of

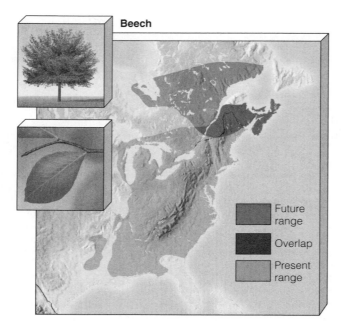

Beech

Future range

Overlap

Present range

Figure 12-12 Possible effects of global warming on the geographic range of beech trees based on ecological evidence and computer models. According to one projection, if CO_2 emissions doubled between 1990 and 2050, beech trees (now common throughout the eastern United States) would survive only in a greatly reduced range in northern Maine and southeastern Canada. This is only one of a number of tree species whose geographic ranges could be changed drastically by increased atmospheric warming. For example, native sugar maples are likely to disappear in the northeastern United States. On the other hand, ranges of some tree species adapted to a warm climate would spread. (Data from Margaret B. Davis and Catherine Zabinski, University of Minnesota)

CO_2 into the lower atmosphere. They include dieback of some forests that cannot spread fast enough to areas with more optimum temperatures (Figure 12-12), greatly increased wildfires in forest and grassland areas where the climate becomes drier, and tree deaths from increased disease and pest populations that would thrive in areas with a warmer climate.

For each 1°C (1.8°F) rise in the earth's average temperature, climate belts in middle latitude regions would shift toward the earth's poles by 100–150 kilometers (60–90 miles) or upward 150 meters (500 feet) in altitude. Such shifts could change areas where crops could be grown and affect the makeup and location of at least one-third of today's forests.

Changes in the structure and location of wildlife habitats could expand ranges and populations of some plant and animal species adapted to warmer climates. But such changes would threaten plant and animal species that could not migrate rapidly enough to new areas, and species with specialized niches. In addition, shifts in regional climate would threaten many parks, wildlife reserves, wilderness areas, wetlands, and coral reefs—thwarting some current efforts to stem the loss

of biodiversity. Also, species likely to do better in a warmer world are certain rapidly multiplying weeds, insect pests, and disease-carrying organisms such as mosquitoes and water-borne bacteria.

A 2004 report by the UN Environment Programme estimated that at least 1 million species could face premature extinction by 2050 unless greenhouse gas emissions are drastically reduced. According to the IPCC, ecosystems *most likely* to be disrupted and lose species are coral reefs, polar seas, coastal wetlands, arctic and alpine tundra, and high-elevation mountaintops.

Another problem is a rise in global sea level, caused by runoff from the melting of land-based snow and ice and by the fact that water expands slightly when heated. According to the 2001 IPCC report, the rate of sea level rise is now faster than at any other time during the past 1,000 years and IPCC climate scientists project a rise of 9–88 centimeters (4–35 inches) during this century.

The high projected rise in sea level of about 88 centimeters (35 inches) would have a number of harmful effects. They include the following:

- Threatening half of the world's coastal estuaries, wetlands (one-third of those in the United States), and coral reefs

- Disrupting many of the world's coastal fisheries

- Flooding low-lying barrier islands causing gently sloping coastlines (especially along the U.S. East Coast) to erode and retreat inland by about 1.3 kilometers (0.8 mile)

- Flooding agricultural lowlands and deltas in parts of Bangladesh, India, and China, where much of the world's rice is grown

- Contaminating freshwater coastal aquifers with saltwater

- Submerging some low-lying islands in the Pacific Ocean (the Marshall Islands) and the Indian Ocean (the Maldives, a chain of 1,200 small islands.

One comedian jokes that he plans to buy land in Kansas because it will probably become valuable beachfront property. Another boasts that she is not worried because she lives in a houseboat—the "Noah strategy."

12-5 DEALING WITH THE THREAT OF GLOBAL WARMING

What Are Our Options? The Great Climate Debate

There is disagreement over what we should do about the threat of global warming.

As we have seen, nearly all climate scientists agree that there is a 90–99% chance that the earth's temperature is very likely to increase during this century and that human activities play a part in this change. Despite this scientific consensus, there is debate among scientists over the causes of these changes (natural or human), how rapidly they might occur, the effects on humans and ecosystems, and how we should respond to this potentially serious long-term global threat. Global warming and climate change are difficult issues to deal with because they have many complex and still poorly understood causes, their effects are long-term and uneven, and there is controversy over how they should be addressed.

As a result of these scientific, economic, and political disagreements, there are three schools of thought concerning what we should do about projected global warming. One is to *do more research before acting.* With this *wait-and-see strategy,* many scientists and economists call for more research and a better understanding of the earth's climate system before making far-reaching and controversial economic and political decisions such as phasing out fossil fuels. This is the current position of the U.S. government.

A second and rapidly growing group of scientists, economists, business leaders, and political leaders (especially in the European Union) believe that we should *act now to reduce the risks from climate change brought about by global warming.* They argue that the potential for harmful economic, ecological, and social consequences is so great that global warming is a good candidate for applying the *precautionary principle.*

In 1997, more than 2,500 scientists from a variety of disciplines signed a Scientists' Statement on Global Climate Disruption and concluded, "We endorse those [IPCC] reports and observe that the further accumulation of greenhouse gases commits the earth irreversibly to further global climatic change and consequent ecological, economic, and social disruption. The risks associated with such changes justify preventive action through reductions in emissions of greenhouse gases." Also in 1997, 2,700 economists led by eight Nobel laureates declared, "As economists, we believe that global climate change carries with it significant environmental, economic, social, and geopolitical risks and that preventive steps are justified."

A third strategy is to *act now as part of a no-regrets strategy.* Scientists and economists supporting this approach say we should take the key actions needed to slow global warming—even if the threat does not materialize—because such actions lead to other important environmental, health, and economic benefits. For example, a reduction in the combustion of fossil fuels, especially coal, will lead to sharp reductions in air pollution that lowers food and timber productivity, decreases biodiversity, and prematurely kills large numbers of people. Reducing oil use would also decrease dependence on imported oil, which threatens economic and military security.

X *HOW WOULD YOU VOTE?* Should we act now to help slow global warming? Cast your vote online at http://biology.brookscole.com/miller7.

What Can We Do to Reduce the Threat? Conserve Energy, Use Renewable Energy, and Intercept Greenhouse Gas Emissions

We can improve energy efficiency, rely more on carbon-free renewable energy resources, and find ways to keep much of the CO_2 we produce out of the troposphere.

Figure 12-13 presents a variety of prevention and cleanup solutions that climate analysts have suggested for slowing the rate and degree of global warming. The solutions come down to three major strategies: *improve*

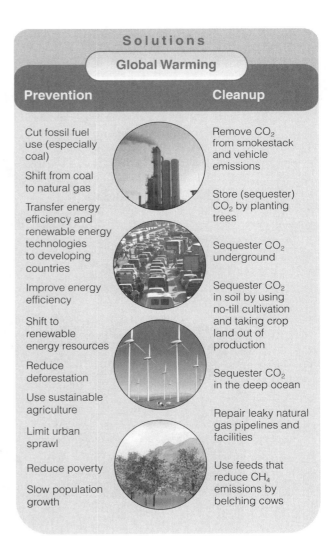

Figure 12-13 Solutions: methods for slowing global warming during this century.

energy efficiency to reduce fossil fuel use, shift from carbon-based fossil fuels to a mix of carbon-free renewable energy resources, and *sequester or store as much CO$_2$ as possible in soil, vegetation, underground, and in the deep ocean.* The effectiveness of these three strategies would be enhanced by *reducing population* to decrease the number of fossil fuel consumers and CO$_2$ emitters and by *reducing poverty* to decrease the need of the poor to clear more land for crops and fuelwood.

✗ *How Would You Vote?* Should we phase out the use of fossil fuels over the next fifty years? Cast your vote online at http://biology.brookscole.com/miller7.

Can We Remove and Store (Sequester) Enough CO$_2$ to Slow Global Warming? Are Output Approaches the Answer?

We can prevent some of the CO$_2$ we produce from circulating in the troposphere, but the costs may be high and the effectiveness of various approaches is unknown.

Figure 12-14 shows several potential techniques to remove CO$_2$ from the troposphere or from smokestacks, and store (sequester) it in other parts of the environment. One possible way to remove more CO$_2$ from the troposphere is to plant trees that store (sequester) it in biomass. But studies indicate that this is a temporary approach because trees release their stored CO$_2$ back into the atmosphere when they die and decompose or if they are burned (for example, by forest fires or to clear land for crops).

A second approach is *soil sequestration* in which plants such as switchgrass are used to remove CO$_2$ from the air and store it in the soil. But warmer temperatures can increase decomposition in soils and return some of the stored CO$_2$ to the atmosphere.

A third strategy is to *reduce the release of carbon dioxide and nitrous oxide from soil.* Ways to do this include *no-till cultivation* and setting aside depleted crop fields as conservation reserves.

A fourth approach is to remove CO$_2$ from smokestacks and *pump it deep underground* into unminable

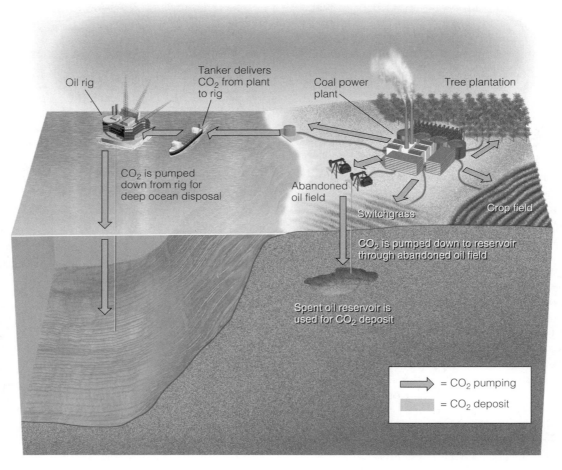

Figure 12-14 Solutions: methods for removing carbon dioxide from the atmosphere or from smokestacks and storing (sequestering) it in plants, soil, deep underground reservoirs, and the deep ocean.

coal seams and abandoned oil fields or *inject it into the deep ocean,* as shown in Figure 12-14. There are several problems with this strategy. One is that current methods can remove only about 30% of the CO_2 from smokestack emissions and would double or triple the cost of producing electricity by burning coal. The U.S. Department of Energy estimates that the cost of sequestering carbon dioxide in various underground and deep ocean repositories will have to be reduced at least 10-fold to make this approach economically feasible. In addition, injecting large quantities of CO_2 into the ocean could upset the global carbon cycle, seawater acidity, and some forms of deep-sea life in unpredictable ways.

How Can Governments Reduce the Threat of Climate Change? Use Sticks and Carrots

Governments can tax greenhouse gas emissions and energy use, increase subsidies and tax breaks for saving energy and using renewable energy, and decrease subsidies and tax breaks for fossil fuels.

Governments could use three major methods to promote the solutions to slowing global warming listed in Figure 12-13. One is to phase in *carbon taxes* on each unit of CO_2 emitted by fossil fuels (especially coal and gasoline) or *energy taxes* on each unit of fossil fuel (especially coal and gasoline) that is burned. Decreasing taxes on income, labor, and profits to offset increases in consumption taxes on carbon emissions or fossil fuel use could help make such a strategy more politically acceptable.

A second strategy is to *level the economic playing field* by greatly increasing government subsidies for energy-efficiency, carbon-free renewable-energy technologies, carbon sequestration, and more sustainable agriculture, and by phasing out subsidies and tax breaks for using fossil fuels.

The third strategy is *technology transfer.* Governments of developed countries could fund the transfer energy-efficiency, carbon-free renewable-energy, carbon-sequestration, and more sustainable agriculture technologies to developing countries. Increasing the current tax on each international currency transaction by a quarter of a penny could finance this technology transfer, which would then generate wealth for developing countries.

What Is Being Done to Reduce Greenhouse Gas Emissions? Some Thorny Problems

Getting countries to agree on reducing their greenhouse gas emissions requires dealing with some difficult issues.

In December 1997, more than 2,200 delegates from 161 nations met in Kyoto, Japan, to negotiate a treaty to help slow global warming. The resulting *Kyoto Protocol* would require 39 developed countries to cut emissions of CO_2, CH_4, and N_2O to an average of about 5.2% below 1990 levels by 2012. The initial steps in the protocol were directed at these 39 countries because they are responsible for about a majority of the world's CO_2 emissions (58% in 1999) and thus should take the lead in reducing their emissions.

The protocol would not require poorer developing countries to make cuts in their greenhouse gas emissions until a later version of the treaty. It also would allow greenhouse gas emissions trading among participating countries. For example, a country or business that reduced its CO_2 emissions or planted trees would receive a certain number of credits. It can use these credits to avoid having to reduce its emissions in other areas, bank them for future use, or sell them to other countries or businesses. By mid-2004, the Kyoto Protocol had been ratified by more than 120 countries.

Some analysts praise the Kyoto agreement as a small but important step in attempting to slow projected global warming and hope the conditions of the treaty will be strengthened in future negotiating sessions. But according to computer models, the 5.2% reduction goal of the Kyoto Protocol would shave only about 0.06°C (0.1°F) off the 0.7–1.7°C (1–3°F) temperature rise projected by 2060.

In 2001, President George W. Bush withdrew U.S. participation from the Kyoto Protocol because he argued that it was too expensive and did not require emissions reductions by developing countries such as China and India that have large and increasing emissions of greenhouse gases. This decision set off strong protests by many scientists, citizens, and leaders throughout most of the world who pointed out that strong leadership is needed by the United States because it has the highest total and per capita CO_2 emissions of any country. However, Scott Barnett, an expert on environmental treaties contends that the Kyoto Protocol is a badly thought out agreement that will not work.

X *HOW WOULD YOU VOTE?* Should the United States participate the Kyoto Protocol? Cast your vote online at http://biology.brookscole.com/miller7.

How Can We Move beyond the Kyoto Protocol Stalemate? Forging a New Strategy

Countries could work together to develop a new international approach to slowing global warming.

In 2004, Richard B. Stewart and Jonathan B. Wiener proposed that countries work together to develop a new strategy for slowing global warming. They conclude that the Kyoto Protocol will have little effect on future global warming without support and action by the world's largest greenhouse gas emitters—the United States and China, along with Russia and India.

They urge the development of a new climate treaty by the United States, China, India, Russia, and other major emitters among developing countries, and Australia and any other developed countries not participating in the Kyoto Protocol. The treaty would also include participation by other developing countries, create effective emissions trading program that includes developing countries omitted from such trading by the Kyoto Protocol, set achievable targets for reducing emissions for each 10 of the next 40 years, and evaluate global and national strategies for adapting to the harmful ecological and economic effects of global warming.

This or other alternative approaches would allow the United States to provide much-needed leadership on this important global issue instead of being seen as a spoiler. Such a parallel treaty could be used as a basis for overhauling the Kyoto Protocol. Or countries participating in the protocol could agree to join the new parallel treaty.

What Are Some Countries, States, Cities, and Businesses Doing to Help Delay Global Warming? Good News

Many countries, states, cities, and companies are reducing their greenhouse gas emissions, improving energy efficiency, and increasing their use of carbon-free renewable energy.

Many countries are reducing their greenhouse gas emissions. For example, by 2000 Great Britain had reduced its CO_2 emissions to its 1990 level, well ahead of its Kyoto target goal. It did this mostly by relying more on natural gas than on coal, improving energy efficiency in industry and homes, and reducing gasoline use by raising it's the tax on gasoline. Between 2000 and 2050, Great Britain aims to cut its CO_2 emissions by 60%, mostly by improving energy efficiency and relying on renewable resources for 20% of its energy by 2030.

According to a 2001 study by the Natural Resources Defense Council, China reduced its CO_2 emissions by 17% between 1997 and 2000, a period during which CO_2 emissions in the United States rose by 14%. The Chinese government did this by phasing out coal subsidies, shutting down inefficient coal-fired electric plants, stepping up its 20-year commitment to increase energy efficiency, and restructuring its economy to increase the use of renewable energy resources.

Since 1990, local governments in more than 500 cities around the world (including 110 in the United

What Can You Do?
Reducing CO_2 Emissions

- Drive a fuel-efficient car, walk, bike, carpool, and use mass transit
- Use energy-efficient windows
- Use energy-efficient appliances and lights
- Heavily insulate your house and seal all drafts
- Reduce garbage by recycling and reuse
- Insulate hot water heater
- Use compact fluorescent bulbs
- Plant trees to shade your house during summer
- Set water heater no higher than 49°C (120°F)
- Wash laundry in warm or cold water
- Use a low-flow shower head

Figure 12-15 What can you do? Ways to reduce your annual emissions of CO_2.

States) have established programs to reduce their greenhouse gas emissions. What is your community doing? In addition, a growing number of major global companies, such as Alcoa, DuPont, IBM, Toyota, BP Amoco, and Shell have established targets to reduce, by 2010, their greenhouse-gas emissions by 10–65% from 1990 levels.

Figure 12-15 lists some things you can do to cut CO_2 emissions. How many of these things are you doing?

How Can We Prepare for Global Warming? Get Ready for Change

A growing number of countries and cities are looking for ways to cope with the harmful effects of climate change.

According to the latest global climate models, the world needs to reduce emissions of greenhouse gases (not just CO_2) by at least 50% by 2018 to stabilize concentrations of CO_2 in the air at their present levels. Such a large reduction in emissions is extremely unlikely for political and economic reasons because it would require rapid, widespread changes in industrial processes, energy sources, transportation options, and individual lifestyles.

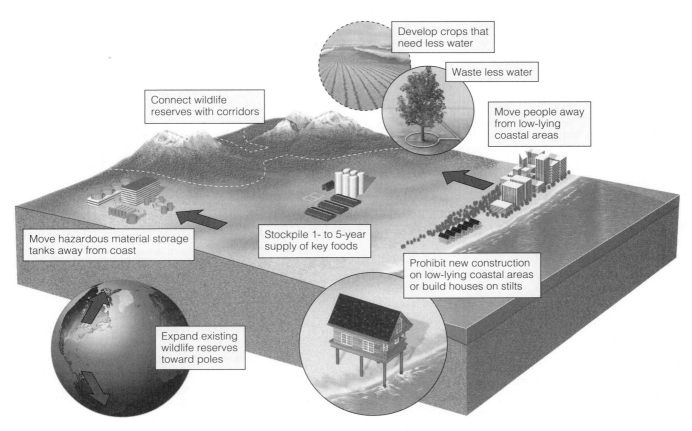

Figure 12-16 Solutions: ways to prepare for the possible long-term effects of climate change.

As a result, a growing number of analysts suggest we should also begin to prepare for the possible harmful effects of long-term atmospheric warming and climate change. Figure 12-16 shows some ways to implement this *adaptation strategy*.

12-6 OZONE DEPLETION IN THE STRATOSPHERE

What Is the Threat from Ozone Depletion? A Clear Danger

Less ozone in the stratosphere would allow more harmful UV radiation to reach the earth's surface.

A layer of ozone in the lower stratosphere (Figure 12-1) keeps about 95% of the sun's harmful ultraviolet (UV) radiation from reaching the earth's surface. Measuring instruments on balloons, aircraft, and satellites show considerable seasonal depletion (thinning) of ozone concentrations in the stratosphere above Antarctica and the Arctic. Similar measurements reveal a lower overall thinning everywhere except over the tropics.

Based on these measurements and mathematical and chemical models, the overwhelming consensus of researchers in this field is that ozone depletion (thinning) in the stratosphere is a serious threat to humans, other animals, and some of the sunlight-driven primary producers (mostly plants) that support the earth's food webs.

What Causes Ozone Depletion? From Dream Chemicals to Nightmare Chemicals

Widespread use of a number of useful and long-lived chemicals has reduced ozone levels in the stratosphere.

Thomas Midgley Jr., a General Motors chemist, discovered the first chlorofluorocarbon (CFC) in 1930, and chemists developed similar compounds to create a family of highly useful CFCs, known by their trade name as Freons.

These chemically stable (nonreactive), odorless, nonflammable, nontoxic, and noncorrosive compounds seemed to be dream chemicals. Inexpensive to manufacture, they became popular as coolants in air conditioners and refrigerators (replacing toxic sulfur dioxide and ammonia), propellants in aerosol spray cans, cleaners for electronic parts such as computer chips, fumigants for granaries and ship cargo holds, and bubbles in plastic foam used for insulation and packaging. Between 1960 and the early 1990s, CFC production rose sharply.

But it turned out that CFCs were too good to be true. In 1974, calculations by chemists Sherwood Rowland and Mario Molina at the University of California–Irvine indicated that CFCs were lowering the average concentration of ozone in the stratosphere. They shocked both the scientific community and the $28-billion-per-year CFC industry by calling for an immediate ban of CFCs in spray cans (for which substitutes were available).

Rowland and Molina's research led them to four major conclusions. *First*, CFCs remain in the troposphere because they are insoluble in water and chemically unreactive. *Second*, over 11–20 years these heavier-than-air compounds rise into the stratosphere mostly through convection, random drift, and the turbulent mixing of air in the troposphere.

Third, once they reach the stratosphere, the CFC molecules break down under the influence of high-energy UV radiation. This releases highly reactive chlorine atoms (Cl), as well as atoms of fluorine (F), bromine (Br), and iodine (I), which accelerate the breakdown of ozone (O_3) into O_2 and O in a cyclic chain of chemical reactions. This causes ozone in various parts of the stratosphere to be destroyed faster than it is formed.

Finally, each CFC molecule can last in the stratosphere for 65–385 years, depending on its type. During that time, each chlorine atom released from can convert hundreds molecules of O_3 to O_2.

Overall, according to Rowland and Molina's calculations and later models and atmospheric measurements of CFCs in the stratosphere, these dream molecules had turned into global ozone destroyers.

The CFC industry (led by DuPont), a powerful, well-funded adversary with a lot of profits and jobs at stake, attacked Rowland and Molina's calculations and conclusions. The two researchers held their ground, expanded their research, and explained the meaning of their calculations to other scientists, elected officials, and the media. After 14 years of delaying tactics, DuPont officials acknowledged in 1988 that CFCs were depleting the ozone layer and agreed to stop producing them once they found substitutes.

In 1995, Rowland and Molina received the Nobel Prize in chemistry for their work. In awarding the prize, the Royal Swedish Academy of Sciences said that they contributed to "our salvation from a global environmental problem that could have catastrophic consequences."

Measurements and models indicate that 75–85% of the observed ozone losses in the stratosphere since 1976 are the result of CFCs and other ozone-depleting chemicals (ODCs) released into the atmosphere by human activities beginning in the 1950s.

What Happens to Ozone Levels over the Earth's Poles during Part of the Year? It Drops Each Winter and Early Spring

During four months of each year, up to half of the ozone in the stratosphere over Antarctica is depleted.

In 1984, researchers analyzing satellite data discovered that 40–50% of the ozone in the upper stratosphere over Antarctica disappeared during September and November. This observed loss of ozone above Antarctica has been called an *ozone hole*. A more accurate term is *ozone thinning* because the ozone depletion varies with altitude and location. In 2003, the area of thinning was the largest ever.

Measurements indicate that CFCs and other ozone-depleting chemicals (ODCs) are the primary culprits. When partial sunlight returns in October, huge masses of ozone-depleted air above Antarctica flow northward and linger for a few weeks over parts of Australia, New Zealand, South America, and South Africa. This raises biologically damaging UV-B levels in these areas by 3–10% and in some years as much as 20%.

In 1988, scientists discovered that similar but usually less severe ozone thinning occurs over the Arctic from February–to June, with a typical ozone loss of 11–38% (compared to a typical 50% loss above Antarctica). When this mass of air above the Arctic breaks up each spring, large masses of ozone-depleted air flow south to linger over parts of Europe, North America, and Asia.

Models indicated that the Arctic is unlikely to develop the large-scale ozone thinning found over the Antarctic. But models also project ozone depletion over the Antarctic and Arctic will be at its worst between 2010 and 2019.

Why Should We Be Worried about Ozone Depletion? Life in the Ultraviolet Zone

Increased UV radiation reaching the earth's surface from ozone depletion in the stratosphere is harmful to human health, crops, forests, animals, and materials such as paints and plastics.

Why should we care about ozone loss? Figure 12-17 lists some of the expected effects of decreased levels of ozone in the stratosphere. From a human standpoint,

Natural Capital Degradation

Effects of Ozone Depletion

Human Health

- Worse sunburn
- More eye cataracts
- More skin cancers
- Immune system suppression

Food and Forests

- Reduced yields for some crops
- Reduced seafood supplies from reduced phytoplankton
- Decreased forest productivity for UV-sensitive tree species

Wildlife

- Increased eye cataracts in some species
- Decreased population of aquatic species sensitive to UV radiation
- Reduced population of surface phytoplankton
- Disrupted aquatic food webs from reduced phytoplankton

Air Pollution and Materials

- Increased acid deposition
- Increased photochemical smog
- Degradation of outdoor paints and plastics

Global Warming

- Accelerated warming because of decreased ocean uptake of CO_2 from atmosphere by phytoplankton and CFCs acting as greenhouse gases

Figure 12-17 Natural capital degradation: expected effects of decreased levels of ozone in the stratosphere.

the answer is that with less ozone in the stratosphere, more biologically damaging UV-A and UV-B radiation will reach the earth's surface. This will give humans worse sunburns, more eye cataracts (a clouding of the eye's lens that reduces vision and can cause blindness if not corrected), and more skin cancers (Connections, right).

Humans can make cultural adaptations to increased UV-B radiation by staying out of the sun, pro-tecting their skin with clothing, and applying sun-screens. However, plants and other animals that help support us and other forms of life cannot make such changes except through biological evolution, a process that can take a long time.

Connections: What Cancer Are You Most Likely to Get? Look in the Mirror

Exposure to UV radiation is a major cause of skin cancers.

Research indicates that years of exposure to UV-B ionizing radiation in sunlight is the primary cause of *squamous cell* and *basal cell skin cancers.* Together, these two types make up 95% of all skin cancers. Typically there is a 15- to 40-year lag between excessive exposure to UV-B and development of these cancers.

Caucasian children and adolescents who experience only a single severe sunburn double their chances of getting these two types of cancers. Some 90–95% of these types of skin cancer can be cured if detected early enough, although their removal may leave disfiguring scars. These cancers kill 1–2% of their victims, which amounts to about 2,300 deaths in the United States each year.

A third type of skin cancer, *malignant melanoma*, occurs in pigmented areas such as moles anywhere on the body. Within a few months, this type of cancer can spread to other organs.

It kills about one-fourth of its victims (most under age 40) within 5 years, despite surgery, chemotherapy, and radiation treatments. Each year it kills about 100,000 people (including 7,400 Americans), mostly Caucasians. It can be cured if detected early enough, but recent studies show that some melanoma survivors have a recurrence more than 15 years later.

A 2003 study found that women who used tanning parlors once a month or more increased their chance of developing malignant melanoma by 55%. The risk was highest for young adults. And a 2004 study at Dartmouth College found that people using tanning beds were more likely to develop basal cell and squamous cell skin cancers and 1.5 times more susceptible to squamous cell carcinoma.

Evidence indicates that people (especially Caucasians) who experience three or more blistering sunburns before age 20 are five times more likely to develop malignant melanoma than those who have never had severe sunburns. About one of every ten people who get malignant melanoma has an inherited gene that makes them especially susceptible to the disease. Figure 12-18 (p. 272) lists ways for you to protect yourself from harmful UV radiation.

What Can You Do?

Reducing Exposure to UV-Radiation

- Stay out of the sun, especially between 10 A.M. and 3 P.M.

- Do not use tanning parlors or sunlamps.

- When in the sun, wear protective clothing and sun–glasses that protect against UV-A and UV-B radiation.

- Be aware that overcast skies do not protect you.

- Do not expose yourself to the sun if you are taking antibiotics or birth control pills.

- Use a sunscreen with a protection factor of 15 or 25 if you have light skin.

- Examine your skin and scalp at least once a month for moles or warts that change in size, shape, or color or sores that keep oozing, bleeding, and crusting over. If you observe any of these signs, consult a doctor immediately.

Figure 12-18 What can you do? Ways to reduce your exposure to UV radiation.

12-7 PROTECTING THE OZONE LAYER

How Can We Protect the Ozone Layer? Say No

To reduce ozone depletion we must stop producing ozone-depleting chemicals.

The consensus of researchers in this field is that we should immediately stop producing all ozone-depleting chemicals. However, even with immediate action, models indicate it will take about 50 years for the ozone layer to return to 1980 levels and about 100 years for recovery to pre-1950 levels. *Good news*: Substitutes are available for most uses of CFCs, and others are being developed (Individuals Matter, right).

In 1987, representatives of 36 nations meeting in Montreal, Canada, developed a treaty commonly known as the *Montreal Protocol*. Its goal was to cut emissions of CFCs (but not other ozone depleters) into the atmosphere by about 35% between 1989 and 2000. After hearing more bad news about seasonal ozone thinning above Antarctica in 1989, representatives of 93 countries met in London in 1990 and in Copenhagen, Denmark (1992) and adopted the *Copenhagen Protocol,* an amendment which accelerated the phaseout of key ozone-depleting chemicals.

These landmark international agreements, now signed by 177 countries, are important examples of global cooperation in response to a serious global environmental problem. Without them, ozone depletion would be a much more serious threat, as shown in Figure 12-19. If nations continue to follow these treaties, ozone levels should return to 1980 levels by 2050 and 1950 levels by 2100.

The ozone protocols set an important precedent for global cooperation and action to avert potential global disaster by using *prevention* to solve a serious environmental problem. Nations and companies agreed to work together to solve this problem for three reasons. *First,* there was convincing and dramatic scientific evidence of a serious problem. *Second,* CFCs were produced by a small number of international companies. *Third,* the certainty that CFC sales would decline over a period of years unleashed the economic and creative resources of the private sector to find even more profitable substitute chemicals.

Ray Turner and His Refrigerator

INDIVIDUALS MATTER

Ray Turner, an aerospace manager at Hughes Aircraft in California, made an important low-tech, ozone-saving discovery by using his head—and his refrigerator. His concern for the environment led him to look for a cheap and simple substitute for the CFCs used as cleaning agents to remove films of oxidation from the electronic circuit boards manufactured at his plant.

He started by looking in his refrigerator. He decided to put drops of various substances on a corroded penny to see whether any of them removed the film of oxidation. Then he used his soldering gun to see whether solder would stick to the surface of the penny, indicating the film had been cleaned off.

First, he tried vinegar. No luck. Then he tried some ground-up lemon peel, also a failure. Next he tried a drop of lemon juice and watched as the solder took hold. The rest, as they say, is history.

Today, Hughes Aircraft uses inexpensive citrus-based solvents that are CFC-free to clean circuit boards. This new cleaning technique has reduced circuit board defects by about 75% at Hughes. And Turner got a hefty bonus. Now other companies, such as AT&T, clean computer boards and chips using acidic chemicals extracted from cantaloupes, peaches, and plums. Maybe you can find a solution to an environmental problem in your refrigerator, grocery store, drugstore, or backyard.

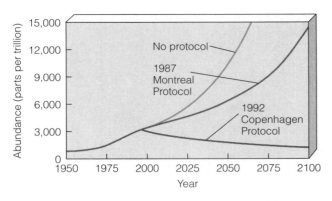

Figure 12-19 Solutions: projected concentrations of ozone-depleting chemicals (ODCs) in the stratosphere under three scenarios: no action, the 1987 Montreal Protocol, and the 1992 Copenhagen Protocol. Stratospheric ozone levels are not expected to return to 1980 levels until about 2050 and to 1950 levels by about 2100, assuming the international agreements are followed and any additional chemicals found to be depleting stratospheric ozone will also be phased out. (Data from World Meteorological Organization)

12-8 TYPES AND SOURCES OF AIR POLLUTION

What Are the Major Types and Sources of Outdoor Air Pollution? Burning Fossil Fuels Is the Major Culprit.

Outdoor air pollutants come mostly from natural sources and burning fossil fuels in motor vehicles and power and industrial plants.

Air pollution is the presence of one or more chemicals in the atmosphere in concentrations high enough to affect climate and harm organisms and materials. Air pollutants come from both natural sources such as volcanoes and hydrocarbons evaporating from some tree species, and human (anthropogenic) activities such as motor vehicles and coal-burning power plants.

Most natural sources of air pollution are spread out and, except for those from volcanic eruptions and some forest fires, rarely reach harmful levels. Most outdoor pollutants in urban areas enter the atmosphere from the burning of fossil fuels in power plants and factories (*stationary sources*) and in motor vehicles (*mobile sources*).

Scientists classify outdoor air pollutants into two categories. **Primary pollutants** are those emitted directly into the troposphere in a potentially harmful form. Examples are soot and carbon monoxide. While in the troposphere, some of these primary pollutants may react with one another or with the basic components of air to form new pollutants, called **secondary pollutants** (Figure 12-20, p. 274).

With their concentration of cars and factories, cities normally have higher outdoor air pollution levels than rural areas. However, prevailing winds can spread long-lived primary and secondary air pollutants emitted in urban and industrial areas to the countryside and to other urban areas.

Indoor pollutants come from infiltration of polluted outside air and various chemicals used or produced inside buildings. Experts in risk analysis rate indoor and outdoor air pollution as high-risk human health problems.

According to the World Health Organization (WHO), one out of every six people on the earth or more than 1.1 billion people live in urban areas where the outdoor air is unhealthy to breathe. Most of them live in densely populated cities in developing countries where air pollution control laws do not exist or are poorly enforced. In other words, poverty can mean poor air for the poor.

In the United States and most other developed countries, government-mandated standards set maximum allowable atmospheric concentrations, or criteria, for six *criteria or conventional air pollutants* commonly found in outdoor air. Table 12-2 (p. 275) lists the properties, sources, and effects of these six types of pollutants. Study this table carefully.

Most scientists would add two other chemicals to the six criteria pollutants shown in Table 12-2. One is *volatile organic compounds (VOCs)*, mostly hydrocarbons, because of their role in the formation of the photochemical smog that plagues many cities. The other is carbon dioxide (CO_2), because of its ability to change climate.

Most oil and coal companies do not want carbon dioxide classified as a pollutant and they lobby elected officials to prevent such classification. Otherwise, we might use less of these fuels and the companies would have to spend lots of money controlling CO_2 emissions. Such lobbying paid off. The EPA contended in 2003 that the Clean Air Act does not give the EPA the power to control gases that can cause global warming. Twelve states have sued the EPA for failure to regulate and some states are passing laws that regulate CO_2 emissions.

X *HOW WOULD YOU VOTE?* Should carbon dioxide be classified as an air pollutant? Cast your vote online at http://biology.brookscole.com/miller7.

What Is Photochemical Smog? Brown-Air Smog

Photochemical smog is a mixture of air pollutants formed by the reaction of nitrogen oxides and volatile organic hydrocarbons under the influence of sunlight.

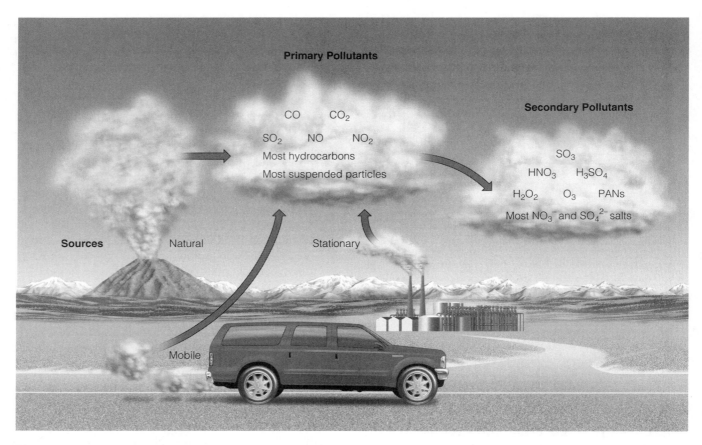

Figure 12-20 Natural capital degradation: Sources and types of air pollutants. Human inputs of air pollutants may come from *mobile sources* (such as cars) and *stationary sources* (such as industrial and power plants). Some *primary air pollutants* may react with one another or with other chemicals in the air to form *secondary air pollutants.*

A *photochemical reaction* is any chemical reaction activated by light. Air pollution known as **photochemical smog** is a mixture of primary and secondary pollutants formed under the influence of sunlight (Figure 12-21, p. 276). The resulting mixture of more than 100 chemicals is dominated by *photochemical ozone,* a highly reactive gas that harms most living organisms.

The formation of photochemical smog begins inside automobile engines and coal-burning power and industrial plants where high temperatures cause N_2 and O_2 in air to form colorless nitric oxide (NO). In the troposphere some of the NO is converted to nitrogen dioxide (NO_2), a yellowish-brown gas with a choking odor. It is the cause of the brownish haze that hangs over many cities during the afternoons of sunny days, explaining why photochemical smog is called *brown-air smog.* When exposed to UV radiation from the sun, the NO_2 engages in a complex series of reactions with hydrocarbons that produce photochemical smog (Figure 12-21, p. 276). Hotter days lead to higher levels of ozone and other components of photochemical smog. As traffic increases on a sunny day, photochemical smog (dominated by O_3) builds up to peak levels by early afternoon, irritating people's eyes and respiratory tracts.

All modern cities have some photochemical smog, but it is much more common in cities with sunny, warm, dry climates and lots of motor vehicles. Examples are Los Angeles, California; Denver, Colorado; and Salt Lake City, Utah, in the United States; Sydney, Australia; Mexico City, Mexico; São Paulo, Brazil; and Buenos Aires, Argentina. According to a 1999 study, if 400 million Chinese drive gasoline-powered cars by 2050 as projected, the resulting photochemical smog could cover the entire western Pacific in ozone, extending to the United States.

What Is Industrial Smog? Gray-Air Smog

Industrial smog is a mixture of sulfur dioxide, droplets of sulfuric acid, and a variety of suspended solid particles emitted by burning coal and oil.

Table 12-2 Major Outdoor Air Pollutants*

Carbon monoxide (CO)

Description: Colorless, odorless gas that is poisonous to air-breathing animals; forms during the incomplete combustion of carbon-containing fuels $(2\,C + O_2 \longrightarrow 2\,CO)$.

Major human sources: Cigarette smoking (Case Study, p. 238), incomplete burning of fossil fuels. About 77% (95% in cities) comes from motor vehicle exhaust.

Health effects: Reacts with hemoglobin in red blood cells and reduces the ability of blood to bring oxygen to body cells and tissues. This impairs perception and thinking; slows reflexes; causes headaches, drowsiness, dizziness, and nausea; can trigger heart attacks and angina; damages the development of fetuses and young children; and aggravates chronic bronchitis, emphysema, and anemia. At high levels it causes collapse, coma, irreversible brain cell damage, and death.

Nitrogen dioxide (NO₂)

Description: Reddish-brown irritating gas that gives photochemical smog its brownish color; in the atmosphere can be converted to nitric acid (HNO_3), a major component of acid deposition.

Major human sources: Fossil fuel burning in motor vehicles (49%) and power and industrial plants (49%).

Health effects: Lung irritation and damage; aggravates asthma and chronic bronchitis;

increases susceptibility to respiratory infections such as the flu and common colds (especially in young children and older adults).

Environmental effects: Reduces visibility; acid deposition of HNO_3 can damage trees, soils, and aquatic life in lakes.

Property damage: HNO_3 can corrode metals and eat away stone on buildings, statues, and monuments; NO_2 can damage fabrics.

Sulfur dioxide (SO₂)

Description: Colorless, irritating; forms mostly from the combustion of sulfur-containing fossil fuels such as coal and oil $(S + O_2 \longrightarrow SO_2)$; in the atmosphere can be converted to sulfuric acid (H_2SO_4), a major component of acid deposition.

Major human sources: Coal burning in power plants (88%) and industrial processes (10%).

Health effects: Breathing problems for healthy people; restriction of airways in people with asthma; chronic exposure can cause a permanent condition similar to bronchitis. According to the WHO, at least 625 million people are exposed to unsafe levels of sulfur dioxide from fossil fuel burning.

Environmental effects: Reduces visibility; acid deposition of H_2SO_4 can damage trees, soils, and aquatic life in lakes.

Property damage: SO_2 and H_2SO_4 can corrode metals and

eat away stone on buildings, statues, and monuments; SO_2 can damage paint, paper, and leather.

Suspended particulate matter (SPM)

Description: Variety of particles and droplets (aerosols) small and light enough to remain suspended in atmosphere for short periods (large particles) to long periods; cause smoke, dust, and haze.

Major human sources: Burning coal in power and industrial plants (40%), burning diesel and other fuels in vehicles (17%), agriculture (plowing, burning off fields), unpaved roads, construction.

Health effects: Nose and throat irritation, lung damage, and bronchitis; aggravates bronchitis and asthma; shortens life; toxic particulates (such as lead, cadmium, PCBs, and dioxins) can cause mutations, reproductive problems, cancer.

Environmental effects: Reduces visibility; acid deposition of H_2SO_4 droplets can damage trees, soils, and aquatic life in lakes.

Property damage: Corrodes metal; soils and discolors buildings, clothes, fabrics, and paints.

Ozone (O₃)

Description: Highly reactive, irritating gas with an unpleasant odor that forms in the troposphere as a major component of

photochemical smog (Figures 12-21).

Major human sources: Chemical reaction with volatile organic compounds (VOCs, emitted mostly by cars and industries) and nitrogen oxides to form photochemical smog (Figure 12-21).

Health effects: Breathing problems; coughing; eye, nose, and throat irritation; aggravates chronic diseases such as asthma, bronchitis, emphysema, and heart disease; reduces resistance to colds and pneumonia; may speed up lung tissue aging.

Environmental effects: Ozone can damage plants and trees; smog can reduce visibility.

Property damage: Damages rubber, fabrics, and paints.

Lead

Description: Solid toxic metal and its compounds, emitted into the atmosphere as particulate matter.

Major human sources: Paint (old houses), smelters (metal refineries), lead manufacture, storage batteries, leaded gasoline (being phased out in developed countries).

Health effects: Accumulates in the body; brain and other nervous system damage and mental retardation (especially in children); digestive and other health problems; some lead-containing chemicals cause cancer in test animals.

Environmental effects: Can harm wildlife.

*Data from U.S. Environmental Protection Agency.

Fifty years ago, cities such as London, England, and Chicago and Pittsburgh in the United States burned large amounts of heavy oil and coal (which contain sulfur impurities) in power plants and factories and for heating homes and cooking food. During the winter, people in such cities were exposed to **industrial smog,** consisting mostly of sulfur dioxide, aerosols containing suspended droplets of sulfuric acid, and a variety of suspended solid particles that give the re-

sulting smog a gray color, explaining why it is sometimes called *gray-air smog.*

Today, urban industrial smog is rarely a problem in most developed countries. This has happened because coal and heavy oil are burned in large boilers with reasonably good pollution controls or with tall smokestacks that transfer the pollutants to downwind rural areas. However, industrial smog is a problem in industrialized urban areas of China, India, Ukraine,

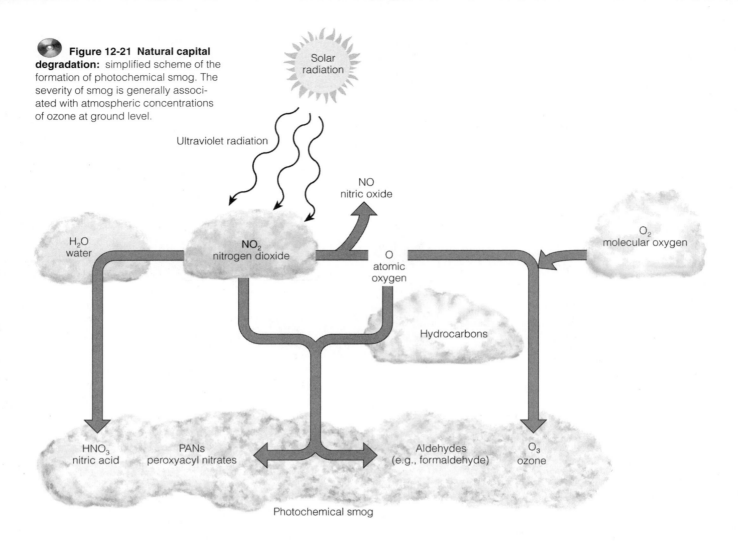

Figure 12-21 Natural capital degradation: simplified scheme of the formation of photochemical smog. The severity of smog is generally associated with atmospheric concentrations of ozone at ground level.

Solar radiation

Ultraviolet radiation

NO
nitric oxide

O_2
molecular oxygen

H_2O
water

NO_2
nitrogen dioxide

O
atomic oxygen

Hydrocarbons

HNO_3
nitric acid

PANs
peroxyacyl nitrates

Aldehydes
(e.g., formaldehyde)

O_3
ozone

Photochemical smog

and some eastern European countries, where large quantities of coal are burned in factories with inadequate pollution controls.

What Factors Influence the Formation of Photochemical and Industrial Smog? Rain, Wind, Buildings, Mountains, and Temperature

Outdoor air pollution can be reduced by precipitation, sea spray, and winds, and increased by urban buildings, mountains, and high temperatures.

Three natural factors help *reduce* air pollution. One is *rain and snow,* which help cleanse the air of pollutants. This helps explain why cities with dry climates are more prone to photochemical smog than cities with wet climates. A second factor is *salty sea spray from the oceans,* which can wash out much of the particulates and other water-soluble pollutants from air that flows from land over the oceans.

A third factor is *winds,* which can help sweep pollutants away, dilute them by mixing them with cleaner air, and bring in fresh air. However, these pollutants

are blown somewhere else or are deposited from the sky onto surface waters, soil, and buildings. There is no "away."

Three factors can *increase* air pollution. One is *urban buildings,* which can slow wind speed and reduce dilution and removal of pollutants. Another is *hills and mountains.* They can reduce the flow of air in valleys below them and allow pollutant levels to build up at ground level. In addition, *higher temperatures* found in most urban areas promote the chemical reactions leading to photochemical smog formation.

During daylight, the sun warms the air near the earth's surface. Normally, this warm air and most of the pollutants it contains rise to mix with the cooler air above it. This mixing of warm and cold air creates turbulence, which disperses the pollutants.

Under certain atmospheric conditions, however, a layer of warm air can lie atop a layer of cooler air nearer the ground, a situation known as a **temperature inversion.** Because the cooler air is denser than the warmer air above it, the air near the surface does not rise and mix with the air above it. Pollutants can concentrate in this stagnant layer of cool air near the ground.

Under certain conditions, temperature inversions can last for several days and allow pollutants to build up to dangerous concentrations.

What Is Acid Deposition and Where Does It Occur? Acids Falling on Your Head

Sulfur dioxide, nitrogen oxides, and particulates can react in the atmosphere to produce acidic chemicals that can travel long distances before returning to the earth's surface.

Most coal-burning power plants, ore smelters, and other industrial plants in developed countries use tall smokestacks to emit sulfur dioxide, suspended particles, and nitrogen oxides high into the troposphere where wind can mix, dilute, and disperse them.

These tall smokestacks reduce *local* air pollution, but they can increase *regional* air pollution downwind. This occurs because the primary pollutants, sulfur dioxide and nitrogen oxides, emitted high into the atmosphere are transported as much as 1,000 kilometers (600 miles) by prevailing winds. During their trip, they form secondary pollutants such as nitric acid vapor, droplets of sulfuric acid, and particles of acid-forming sulfate and nitrate salts.

These acidic substances remain in the atmosphere for 2–14 days, depending mostly on prevailing winds, precipitation, and other weather patterns. During this period they descend to the earth's surface in two forms. One is *wet deposition* as acidic rain, snow, fog, and cloud vapor. The other is *dry deposition* as acidic particles. The resulting mixture is called **acid deposition** (Figure 12-22), sometimes called *acid rain*. Its acidity usually is reported in terms of its pH (Figure 2-24, p. 38). Most dry deposition occurs within about 2–3 days, fairly near the emission sources, whereas most wet deposition takes place within 4–14 days in more distant downwind areas.

Acid deposition is a *regional* air pollution problem in most parts of the world that are downwind from coal-burning facilities and from urban areas with large numbers of cars. Figure 12-23 (p. 278) shows some of area throughout the world that are currently affected by acid deposition and areas likely to be affected in the future. Look at the map to see if you live in a major acid deposition area.

In some areas, soils contain basic compounds that can react with and neutralize, or *buffer,* some input of acids. The areas most sensitive to acid deposition are those containing thin, acidic soils without such natural buffering (Figure 12-23) and those in which the buffering capacity of soils has been depleted by decades of acid deposition.

Many acid-producing chemicals generated in one country are exported to other countries by prevailing winds. For example, acidic emissions from industrial-

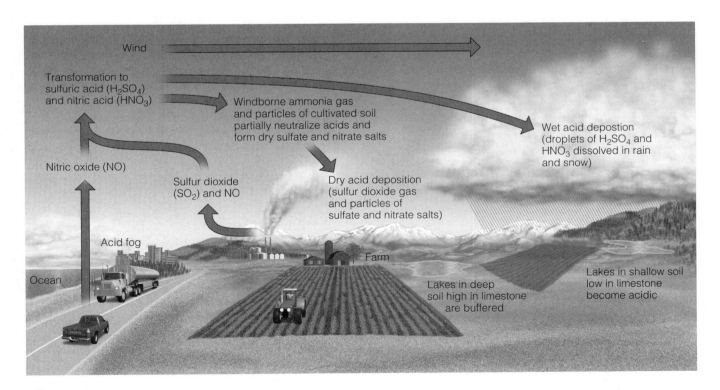

Wind

Transformation to sulfuric acid (H$_2$SO$_4$) and nitric acid (HNO$_3$)

Windborne ammonia gas and particles of cultivated soil partially neutralize acids and form dry sulfate and nitrate salts

Wet acid depostion (droplets of H$_2$SO$_4$ and HNO$_3$ dissolved in rain and snow)

Nitric oxide (NO)

Sulfur dioxide (SO$_2$) and NO

Dry acid deposition (sulfur dioxide gas and particles of sulfate and nitrate salts)

Acid fog

Ocean

Farm

Lakes in deep soil high in limestone are buffered

Lakes in shallow soil low in limestone become acidic

Figure 12-22 Natural capital degradation: *acid deposition,* which consists of rain, snow, dust, or gas with a pH lower than 5.6, is commonly called acid rain. Soils and lakes vary in their ability to buffer or remove excess acidity.

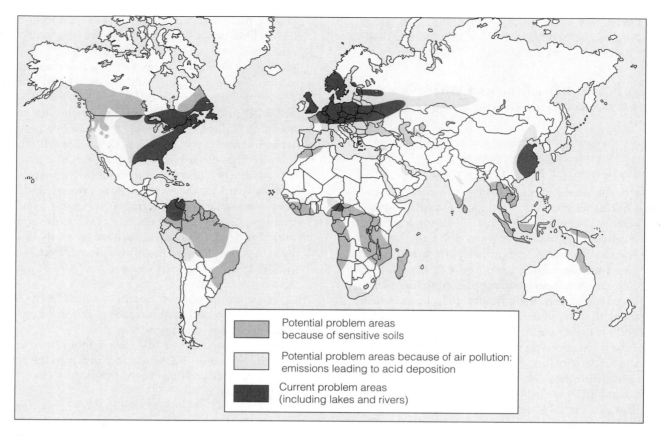

Figure 12-23 Natural capital degradation: regions where acid deposition is now a problem and regions with the potential to develop this problem. Such regions have large inputs of air pollution (mostly from power plants, industrial plants, and ore smelters), or are sensitive areas with soils and bedrock that cannot neutralize (buffer) inputs of acidic compounds. (Data from World Resources Institute and U.S. Environmental Protection Agency)

ized areas of western Europe (especially the United Kingdom and Germany) and eastern Europe blow into Norway, Switzerland, Austria, Sweden, the Netherlands, and Finland. Also, Some SO_2 and other emissions from coal-burning power and industrial plants in the Ohio Valley of the United States end up in Southeastern Canada. In addition, some acidic emissions in China end up in Japan and North and South Korea.

The worst acid deposition area is in Asia, especially China, which gets about 59% of its energy from burning coal. Scientists estimate that by 2025, China will emit more sulfur dioxide—a major culprit in acid deposition—than the United States, Canada, and Japan combined.

What Are Some Harmful Effects of Acid Deposition? Lung Disease, Corrosion, Haze, and Dead Fish

Acid deposition can cause or worsen respiratory disease, attack metallic and stone objects, decrease atmospheric visibility, and kill fish.

Acid deposition has a number of harmful effects. It contributes to human respiratory diseases such as bronchitis and asthma, and can leach toxic metals (such as lead and copper) from water pipes into drinking water. It also damages statues, national monuments, buildings, metals, and car finishes. Large amounts of money are spent each year to clean and repair monuments and buildings that have been attacked by acid deposition.

Acid deposition also decreases atmospheric visibility, mostly because of the sulfate particles it contains. On some days, the air in Grand Canyon National Park and in a number of other national parks is so smoggy from sulfate particles and other pollutants that visitors cannot see the magnificent vistas that are some of their main attractions.

Acid deposition has harmful ecological effects on aquatic systems. Below a pH of 4.5, most fish cannot survive. Acid deposition can also release aluminum ions (Al^{3+}) attached to minerals in nearby soil into lakes. These ions asphyxiate many kinds of fish by stimulating excessive mucus formation, which clogs their gills.

Because of excess acidity, several thousand lakes in Norway and Sweden contain no fish, and many more lakes there have lost most of their acid-neutralizing capacity. In Canada, at least 1,200 acidified lakes contain few if any fish, and some fish populations in many thousands more are declining because of increased acidity. In the United States, several hundred lakes (most in the Northeast) are threatened with excess acidity.

Acid deposition (often along with other air pollutants such as ozone) can harm forests and crops by leaching essential plant nutrients such as calcium and magnesium salts from soils. This reduces plant productivity and the ability of the soils to buffer or neutralize acidic inputs.

Acid deposition rarely kills trees directly, but can weaken them and make them more susceptible to other stresses such as severe cold, diseases, insect attacks, drought, and harmful mosses. Mountaintop forests are the terrestrial areas hardest hit by acid deposition. These areas tend to have thin soils without much buffering capacity, and trees on mountaintops (especially conifers such as red spruce and balsam fir that keep their leaves year-round) are bathed almost continuously in very acidic fog and clouds.

Note that *most of the world's lakes and forests are not being destroyed or seriously harmed by acid deposition.* It is a regional problem that can harm forests and lakes downwind from coal-burning facilities and from large car-dominated cities without adequate controls on emissions of nitrogen oxides from motor vehicles. Do you live in an area affected by acid deposition?

Figure 12-24 summarizes ways to reduce acid deposition. According to most scientists studying the problem, the best solutions are *prevention approaches* that reduce or eliminate emissions of sulfur dioxide, nitrogen oxides, and particulates.

Controlling acid deposition is a political hot potato. One problem is that the people and ecosystems it affects often are quite distant from the sources of the problem. Also, countries with large supplies of coal (such as China, India, Russia, and the United States) have a strong incentive to use it as a major energy resource. And owners of coal-burning power plants say the costs of adding equipment to reduce air pollution, using low-sulfur coal, or removing sulfur from coal are too high and would increase the cost of electricity for consumers.

Environmentalists respond that affordable and much cleaner ways are available to produce electricity—including wind turbines and burning natural gas in turbines. They also point out that the largely hidden health and environmental costs of burning coal are roughly twice its market cost. They urge inclusion of these costs in the price of producing electricity from coal. Then consumers would have more realistic knowledge about the harmful effects of burning coal.

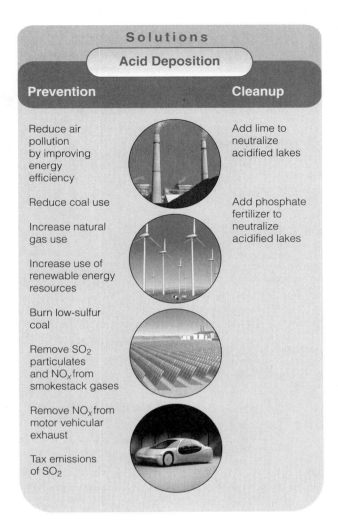

Figure 12-24 **Solutions:** methods for reducing acid deposition and its damage.

What Are the Types and Sources of Indoor Air Pollution? Being Indoors Can Be Hazardous to Your Health.

Indoor air pollution usually is a much greater threat to human health than outdoor air pollution.

If you are reading this book indoors, you may be inhaling more air pollutants with each breath than if you were outside. Figure 12-25 (p. 280) shows some typical sources of indoor air pollution. Which are you exposed to?

EPA studies have revealed some alarming facts about indoor air pollution in the United States. *First,* levels of 11 common pollutants generally are two to five times higher inside homes and commercial buildings than outdoors, and as much as 100 times higher in some cases. *Second,* pollution levels inside cars in traffic-clogged urban areas can be up to 18 times higher than outside. *Third,* the health risks from exposure to

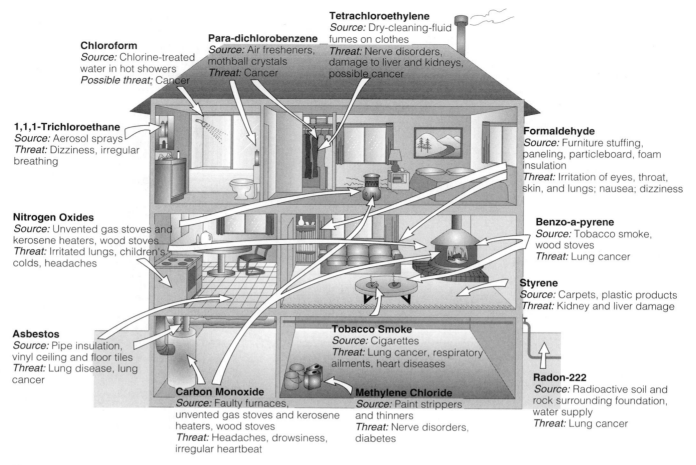

Chloroform
Source: Chlorine-treated water in hot showers
Possible threat: Cancer

Para-dichlorobenzene
Source: Air fresheners, mothball crystals
Threat: Cancer

Tetrachloroethylene
Source: Dry-cleaning-fluid fumes on clothes
Threat: Nerve disorders, damage to liver and kidneys, possible cancer

1,1,1-Trichloroethane
Source: Aerosol sprays
Threat: Dizziness, irregular breathing

Formaldehyde
Source: Furniture stuffing, paneling, particleboard, foam insulation
Threat: Irritation of eyes, throat, skin, and lungs; nausea; dizziness

Nitrogen Oxides
Source: Unvented gas stoves and kerosene heaters, wood stoves
Threat: Irritated lungs, children's colds, headaches

Benzo-a-pyrene
Source: Tobacco smoke, wood stoves
Threat: Lung cancer

Styrene
Source: Carpets, plastic products
Threat: Kidney and liver damage

Asbestos
Source: Pipe insulation, vinyl ceiling and floor tiles
Threat: Lung disease, lung cancer

Tobacco Smoke
Source: Cigarettes
Threat: Lung cancer, respiratory ailments, heart diseases

Radon-222
Source: Radioactive soil and rock surrounding foundation, water supply
Threat: Lung cancer

Carbon Monoxide
Source: Faulty furnaces, unvented gas stoves and kerosene heaters, wood stoves
Threat: Headaches, drowsiness, irregular heartbeat

Methylene Chloride
Source: Paint strippers and thinners
Threat: Nerve disorders, diabetes

Figure 12-25 Some important indoor air pollutants. (Data from U.S. Environmental Protection Agency)

such chemicals are magnified because people typically spend 70–98% of their time indoors or inside vehicles.

As a result of these studies, in 1990 the EPA placed indoor air pollution at the top of the list of 18 sources of cancer risk—causing as many as 6,000 premature cancer deaths per year. At greatest risk are smokers, infants and children under age 5, the old, the sick, pregnant women, people with respiratory or heart problems, and factory workers.

Danish and U.S. EPA studies have linked pollutants found in buildings to dizziness, headaches, coughing, sneezing, shortness of breath, nausea, burning eyes, chronic fatigue, irritability, skin dryness and irritation, and flu-like symptoms, known as the *sick-building syndrome.* New buildings are more commonly "sick" than old ones because of reduced air exchange (to save energy) and chemicals released from new carpeting and furniture. EPA studies indicate that almost one in five of the 4 million commercial buildings in the United States are considered "sick" (including the EPA headquarters).

According to the EPA and public health officials, the four most dangerous indoor air pollutants in de-

veloped countries are *cigarette smoke, formaldehyde, radioactive radon-222 gas* (see Case Study, at right), and *very fine and ultrafine particles.*

The chemical that causes most people in developed countries difficulty is *formaldehyde*, a colorless, extremely irritating gas widely used to manufacture common household materials. According to the EPA and the American Lung Association, 20–40 million Americans suffer from chronic breathing problems, dizziness, rash, headaches, sore throat, sinus and eye irritation, wheezing, and nausea caused by daily exposure to low levels of formaldehyde emitted from common household materials. Are you one of these people?

There are many sources of formaldehyde. They include building materials (such as plywood, particleboard, paneling, and high-gloss wood used in floors and cabinets), furniture, drapes, upholstery, adhesives in carpeting and wallpaper, urethane-formaldehyde insulation, fingernail hardener, and wrinkle-free coating on permanent-press clothing (Figure 12-25). The EPA estimates that as many as 1 of every 5,000 people who live in manufactured homes for more than 10 years will develop cancer from formaldehyde exposure.

In developing countries, the indoor burning of wood, charcoal, dung, crop residues, and coal in open fires or in unvented or poorly vented stoves for cooking and heating exposes inhabitants to high levels of particulate air pollution. According to the World Bank, as many as 2.8 million people (most of them women and children) in developing countries die prematurely each year from breathing elevated levels of such indoor smoke. *Thus, indoor air pollution for the poor is by far the world's most serious air pollution problem*—a glaring example of the relationship between poverty and environmental quality.

Case Study: Are You Being Exposed to Radioactive Radon Gas? Test the Air in Your House

Radon-222, a radioactive gas found in some soil and rocks, can seep into some houses and increase the risk of lung cancer.

Radon-222—a naturally occurring radioactive gas that you cannot see, taste, or smell—is produced by the radioactive decay of uranium-238. Most soil and rock contain small amounts of uranium-238. But this isotope is much more concentrated in underground deposits of minerals such as uranium, phosphate, granite, and shale.

When radon gas from such deposits seeps upward through the soil and is released outdoors, it disperses quickly in the atmosphere and decays to harmless levels. However, in buildings above such deposits radon gas can enter through cracks in foundations and walls, openings around sump pumps and drains, and hollow concrete blocks. Once inside, it can build up to high levels, especially in unventilated lower levels of homes and buildings.

Radon-222 gas quickly decays into solid particles of other radioactive elements such as polonium-210 that, if inhaled, expose lung tissue to a large amount of ionizing radiation from alpha particles. This exposure can damage lung tissue and lead to lung cancer over the course of a 70-year lifetime. Such exposure makes radon the second leading cause of lung cancer after smoking in the United States. Your chances of getting lung cancer from radon depend mostly on how much radon is in your home, how much time you spend in your home, and whether you are a smoker or have ever smoked.

Ideally, radon levels should be monitored continuously in the main living areas (not basements or crawl spaces) for 2 months to a year. By 2003, only about 6% of U.S. households had conducted radon tests (most lasting only 2–7 days and costing $20–100 per home). In 2003, the EPA urged Americans to test their homes for indoor radon gas.

For information about radon testing visit the EPA website at http://www.epa.gov/iaq/radon. According to the EPA, radon control could add $350–500 to the cost of a new home, and correcting a radon problem in an existing house could run $800–2,500. Remedies include sealing cracks in the foundation and walls, increasing ventilation by cracking a window or installing vents, and using a fan to create cross ventilation.

12-9   HARMFUL EFFECTS OF AIR POLLUTION

How Does Your Respiratory System Help Protect You from Air Pollution? Your Air Pollution Security System

Your respiratory system has several ways to help protect you from air pollution, but some air pollutants can overcome these defenses.

Your respiratory system has a number of mechanisms that help protect you from much air pollution. Hairs in your nose filter out large particles. Sticky mucus in the lining of your upper respiratory tract captures smaller (but not the smallest) particles and dissolves some gaseous pollutants. Sneezing and coughing expel contaminated air and mucus when pollutants irritate your respiratory system.

In addition, hundreds of thousands of tiny mucus-coated hair-like structures called *cilia* line your upper respiratory tract. They continually wave back and forth and transport mucus that traps the pollutants to your throat (where they are swallowed or expelled).

But prolonged or acute exposure to air pollutants including tobacco smoke can overload or break down these natural defenses. This can cause or contribute to various respiratory diseases. One that can start early in life is *asthma*, typically an allergic reaction causing sudden episodes of muscle spasms in the bronchial walls, which results in acute shortness of breath.

Years of smoking and breathing air pollutants can lead to other respiratory disorders including *lung cancer* and *chronic bronchitis*, which involves persistent inflammation and damage to the cells lining the bronchi and bronchioles in your lungs. The results are mucus buildup, painful coughing, and shortness of breath. Damage deeper in the lung can cause *emphysema*, which is irreversible damage to air sacs, or alveoli, leading to loss of lung elasticity, and acute shortness of breath.

People with respiratory diseases are especially vulnerable to air pollution, as are older adults, infants, pregnant women, and people with heart disease.

How Many People Die Prematurely from Air Pollution? A Major Killer

Each year, air pollution prematurely kills about 3 million people, mostly from indoor air pollution in developing countries.

Table 12-2 (p. 275) lists some of the harmful health effects from prolonged or chronic exposure to six major air pollutants. According to the World Health Organization, at least 3 million people worldwide (most of them in Asia) die prematurely each year from the effects of air pollution—an average of 8,200 deaths per day. About 2.8 million of these deaths (93%) are from *indoor* air pollution, mostly from burning wood or coal inside dwellings in developing countries. *This explains why the World Health Organization and the World Bank consider indoor air pollution one of the world's most crucial environmental problems.*

In the United States, the EPA estimates that annual deaths related to indoor and outdoor air pollution range from 150,000 to 350,000 people—equivalent to one to two fully loaded 400-passenger jumbo jets crashing *each day* with no survivors. Millions more become ill and lose work time. Most of these deaths are related to inhalation of fine and ultrafine particulates in indoor air.

According to recent studies by the EPA, each year more than 125,000 Americans get cancer from breathing soot-laden diesel fumes from buses trucks, tractors, bulldozers and other construction equipment, and portable generators. The EPA says that in one year a large diesel-powered bulldozer produces as much air pollution as 26 cars.

12-10 PREVENTING AND REDUCING AIR POLLUTION

How Have Laws Helped Reduce Air Pollution in the United States? Clean Air Acts to the Rescue

The Clean Air Acts in the United States have greatly reduced outdoor air pollution from six major pollutants.

The U.S. Congress passed Clean Air Acts in 1970, 1977, and 1990. With these acts, the federal government established air pollution regulations for key pollutants that are enforced by each state and by major cities.

Great news: According to a 2003 EPA report, combined *emissions* of the six criteria air pollutants decreased by 48% between 1970 and 2002, even with significant increases in gross domestic product, vehicle miles traveled, energy consumption, and population.

Bad news: After dropping in the 1980s, smog levels did not drop between 1993 and 2003, mostly because reducing smog requires much bigger cuts in emissions of nitrogen oxides from power and industrial plants and motor vehicles. Also, according to the EPA, in 2003 more than 170 million people lived in areas where air is unhealthy to breathe during part of the year because of high levels of air pollutants—primarily ozone and

fine particles. However, for most urban areas such conditions exist for only a few days a year.

How Can U.S. Air Pollution Laws Be Improved? We Can Do Better.

Environmentalists applaud the success of U. S. air pollution control laws, but have suggested several ways to make them more effective.

The reduction of outdoor air pollution in the United States since 1970 has been a remarkable success story. This occurred because of two factors. *First,* U.S. citizens insisted that laws be passed and enforced to improve air quality. *Second,* the country was affluent enough to afford such controls and improvements.

But more can be done. Environmentalists point to several deficiencies in the Clean Air Acts. One is *continuing to rely mostly on pollution cleanup rather than prevention.* An example of the power of prevention is that in the United States, the air pollutant with the largest drop (98% between 1970 and 2002) in its atmospheric level was lead, which was largely banned in gasoline. This is viewed as one the greatest environmental success stories in the country's history.

Second is the *failure of Congress to increase fuel-efficiency standards for cars, sport utility vehicles (SUVs), and light trucks.* According to environmental scientists, increased fuel efficiency would reduce air pollution from motor vehicles more quickly and effectively than any other method, reduce CO_2 emissions, save energy, and save consumers enormous amounts of money.

Third, there has been *inadequate regulation of emissions from inefficient two-cycle gasoline engines.* These engines are used in lawn mowers, leaf blowers, chain saws, jet skis, outboard motors, and snowmobiles. According to the California Air Resources Board, a 1-hour ride on a typical jet ski creates more air pollution than the average U.S. car does in a year, and operating a 100-horsepower marine engine for 7 hours emits more air pollutants than a new car driven 160,000 kilometers (100,000 miles). In 2001, the EPA announced plans to reduce emissions from most of these sources by 2007. But manufacturers push for extending these deadlines.

Fourth is the *failure of the acts to do much about reducing emissions of carbon dioxide and other greenhouse gases.* Fifth, the acts have *failed to deal seriously with indoor air pollution* even though it is by far the most serious air pollution problem in terms of poorer health, premature death, and economic losses from lost work time and increased health costs.

Finally, there is a need for *better enforcement of the Clean Air Acts.* According to a 2002 government study, doing this would save about 6,000 lives and prevent 140,000 asthma attacks each year in the United States.

Executives of companies affected by implementing such policies claim that implementing this would cost too much, harm economic growth, and cost jobs. Proponents contend that history has shown that almost all industry cost estimates of implementing various air pollution control standards in the United States were many times higher than the actual cost. In addition, implementing such standards has helped increase economic growth and create jobs by stimulating companies to develop new technologies for reducing air pollution emissions. Many of these technologies are sold in the international marketplace.

X HOW WOULD YOU VOTE? Should the 1990 U.S. Clean Air Act be strengthened? Cast your vote online at http://biology.brookscole.com/miller7.

Case Study: Should We Use the Marketplace to Reduce Pollution? Emissions Trading

Allowing producers of air pollutants to buy and sell government air pollution allotments in the marketplace can help reduce emissions.

To help reduce SO_2 emissions, the Clean Air Act of 1990 allows an *emissions trading policy,* which enables the 110 most polluting power plants in 21 states (primarily in the Midwest and East) to buy and sell SO_2 pollution rights.

Each year, a coal-burning power plant is given a certain number of pollution credits, or rights, that allow it to emit a certain amount of SO_2. A utility that emits less SO_2 than its limit has a surplus of pollution credits. It can use these credits to avoid reductions in SO_2 emissions at another of its plants, keep them for future plant expansions, or sell them to other utilities, private citizens, or environmental groups.

Proponents argue that this system allows the marketplace to determine the cheapest, most efficient way to get the job done instead of having the government dictate how to control pollution. Some environmentalists see this *cap-and-trade* market approach as an improvement over the regulatory *command-and-control* approach as long as it achieves net reduction in SO_2 pollution. This would be accomplished by limiting the total number of credits and gradually lowering the *emissions cap,* or annual number of credits, as has been done since 2000.

One of the neat things about the SO_2 emissions market is that anyone can participate. Environmental groups can buy up such rights to pollute and not use them. You could personally reduce air pollution by buying a certificate allowing you to add 0.9 metric ton (1 ton) of SO_2 to the atmosphere and hanging it on the wall. You could purchase these certificates and give them away as birthday or holiday gifts. See www.epa.gov/airmarkets/ for a list of brokers and other sellers of SO_2 permits.

Some environmentalists criticize the cap-and-trade program. They contend that it allows utilities with older, dirtier power plants to buy their way out of reducing pollution and keep on emitting unacceptable levels of SO_2. This could lead (as it has) to continuing high levels of air pollution in certain areas or "hot spots."

This approach also creates incentives to cheat because air quality regulation is based largely on self-reporting of emissions. Environmentalists call for unannounced spot monitoring by the government and large fines for cheaters.

In addition, the success of any emissions trading approach depends on how low the initial cap is set and then on how much it is reduced annually to promote continuing innovation in air pollution prevention and control. Without these elements, these critics say that emissions trading programs mostly move air pollutants from one area to another without achieving an overall reduction in air quality.

Good news: Between 1980 and 2002, the emissions trading system helped reduce SO_2 emissions from electric power plants in the United States by 40%. And the cost of doing this was less than one-tenth the cost projected by industry because this market-based system motivated companies to reduce emissions in more efficient ways.

Emissions trading may also be implemented for particulate emissions and volatile organic compounds, and the combined emissions of SO_2, nitrogen oxides (NO_x), and mercury from coal-burning power plants. Environmentalists and health scientists are particularly opposed to using a cap-and-trade program to control emissions of mercury by coal-burning power plants and industries because it is highly toxic, falls out of the atmosphere adjacent to such facilities, and does not break down in the environment. Coal-burning plants choosing to buy permits instead of sharply reducing their mercury emissions would create hot spots with unacceptably high levels of mercury.

Bad news: In 2002, the EPA reported results from an evaluation of the country's oldest and largest emissions trading program, in effect since 1993 in southern California. The EPA study found that this cap-and-trade model "produced far less emissions reductions than were either projected for the program or could have been expected from" the command-and-control system it replaced. The study also found accounting abuses, including emissions caps set 60% higher than current emissions. Cap-and-trade programs need to be carefully monitored.

X HOW WOULD YOU VOTE? Should emissions trading be used to help control emissions of all major air pollutants? Cast your vote online at http://biology.brookscole.com/miller7.

How Can We Reduce Outdoor Air Pollution? Prevention is Best

There are a number of ways to prevent and control air pollution from coal-burning facilities and motor vehicles.

Figure 12-26 summarizes ways to reduce emissions of sulfur oxides, nitrogen oxides, and particulate matter from stationary sources such as electric power plants and industrial plants that burn coal.

About 20,000 older coal burning plants, industrial plants, and oil refineries in the United States have not been required to meet the air pollution standards required for new facilities under the Clean Air Acts. Environmentalists and officials of states subject to pollution from such plants have been trying to get Congress reverse this situation since 1970. However, they have been not been successful because of strong lobbying efforts by U.S. coal and oil industries.

X *HOW WOULD YOU VOTE?* Should older coal-burning power and industrial plants have to meet the same air pollution standards as new facilities? Cast your vote online at http://biology.brookscole.com/miller7.

Figure 12-27 lists ways to reduce emissions from motor vehicles, the primary culprits in producing photochemical smog.

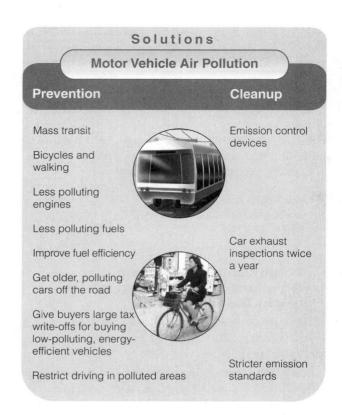

Figure 12-27 Solutions: methods for reducing emissions from motor vehicles. Which two of these solutions do you believe are the most important?

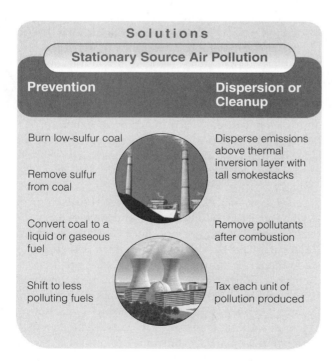

Figure 12-26 Solutions: methods for reducing emissions of sulfur oxides, nitrogen oxides, and particulate matter from stationary sources such as coal-burning electric power plants and industrial plants. Which two of these solutions do you believe are the most important?

Good news: Over the next 10–20 years, air pollution from motor vehicles should decrease from increased use of *partial-zero-emission-vehicles (PZEV)* that emit almost no air pollutants because of improved engine and emission systems, *hybrid-electric vehicles,* and vehicles powered by *fuel cells running on hydrogen.*

Bad news: The growing number of motor vehicles in urban areas of many developing countries is contributing to the already poor air quality there. Many of these vehicles are 10 or more years old, have no pollution control devices, and continue to burn leaded gasoline.

How Can We Reduce Indoor Air Pollution? Emphasize Prevention

Little effort has been spent on reducing indoor air pollution even though it is a much greater threat to human health than outdoor air pollution.

Reducing indoor air pollution does not require setting indoor air quality standards and monitoring the more than 100 million homes and buildings in the United States (or the buildings in any country). Instead, air pollution experts suggest several ways to prevent or reduce indoor air pollution (Figure 12-28). Another

Solutions

Indoor Air Pollution

Prevention	Cleanup or Dilution

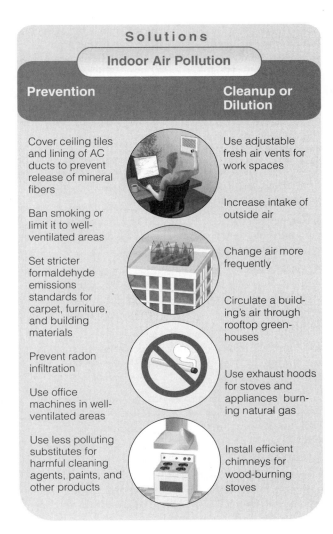

Prevention

Cover ceiling tiles and lining of AC ducts to prevent release of mineral fibers

Ban smoking or limit it to well-ventilated areas

Set stricter formaldehyde emissions standards for carpet, furniture, and building materials

Prevent radon infiltration

Use office machines in well-ventilated areas

Use less polluting substitutes for harmful cleaning agents, paints, and other products

Cleanup or Dilution

Use adjustable fresh air vents for work spaces

Increase intake of outside air

Change air more frequently

Circulate a building's air through rooftop greenhouses

Use exhaust hoods for stoves and appliances burning natural gas

Install efficient chimneys for wood-burning stoves

Figure 12-28 Solutions: ways to prevent and reduce indoor air pollution. Which two of these solutions do you believe are the most important?

possibility for cleaner indoor air in some high-rise buildings is rooftop greenhouses through which building air can be circulated.

In developing countries, indoor air pollution from open fires and leaky and inefficient stoves that burn wood, charcoal, or coal could be reduced if governments gave people inexpensive clay or metal stoves, which burn biofuels more efficiently while venting their exhausts to the outside, or stoves that use solar energy to cook food (solar cookers) in sunny areas. Doing this would also reduce deforestation by using less fuelwood and charcoal.

What Is the Next Step? Individuals Matter

We need to focus on preventing air pollution, with emphasis on sharply reducing indoor air pollution in developing countries.

It is encouraging that since 1970 most of the world's developed countries have enacted laws and regulations that have significantly reduced outdoor air pollution. But without individuals and organized groups putting strong pressure on elected officials in the 1970s and 1980s, these laws and regulations would not have been enacted, funded, and implemented. In turn, these legal requirements spurred companies, scientists, and engineers to come up with better ways to control outdoor pollution.

The current laws are a useful output approach to controlling pollution. To environmentalists, however, the next step is to shift to preventing air pollution. With this approach, the question is not "What can we do about the air pollutants we produce?" but "How can we not produce such pollutants in the first place?"

Figure 12-29 shows ways to prevent outdoor and indoor air pollution over the next 30–40 years. Like the shift to *controlling air pollution* between 1970 and 2000, this new shift to *preventing air pollution* will not take

Solutions

Air Pollution

Outdoor	Indoor

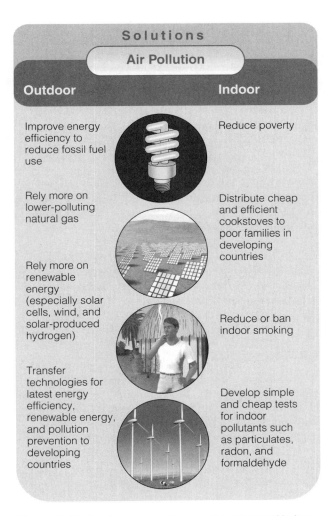

Outdoor

Improve energy efficiency to reduce fossil fuel use

Rely more on lower-polluting natural gas

Rely more on renewable energy (especially solar cells, wind, and solar-produced hydrogen)

Transfer technologies for latest energy efficiency, renewable energy, and pollution prevention to developing countries

Indoor

Reduce poverty

Distribute cheap and efficient cookstoves to poor families in developing countries

Reduce or ban indoor smoking

Develop simple and cheap tests for indoor pollutants such as particulates, radon, and formaldehyde

Figure 12-29 Solutions: ways to prevent outdoor and indoor air pollution over the next 30–40 years.

Indoor Air Pollution

- Test for radon and formaldehyde inside your home and take corrective measures as needed.

- Do not buy furniture and other products containing formaldehyde.

- Remove your shoes before entering your house to reduce inputs of dust, lead, and pesticides.

- Test your house or workplace for asbestos fiber levels and for any crumbling asbestos materials if it was built before 1980.

- Don't live in a pre-1980 house without having its indoor air tested for asbestos and lead.

- Do not store gasoline, solvents, or other volatile hazardous chemicals inside a home or attached garage.

- If you smoke, do it outside or in a closed room vented to the outside.

- Make sure that wood-burning stoves, fireplaces, and kerosene- and gas-burning heaters are properly installed, vented, and maintained.

- Install carbon monoxide detectors in all sleeping areas.

Figure 12-30 What can you do? Ways to reduce your exposure to air pollution.

place without political pressure on elected officials by individual citizens and groups of such citizens. Figure 12-30 lists some ways that you can reduce your exposure to indoor air pollution.

Turning the corner on air pollution requires moving beyond patchwork, end-of-pipe approaches to confront pollution at its sources. This will mean reorienting energy, transportation, and industrial structures toward prevention.

HILARY F. FRENCH

CRITICAL THINKING

1. In preparation for the 1992 UN Conference on the Human Environment in Rio de Janeiro, President George H. W. Bush's top economic adviser gave an address in Williamsburg, Virginia, to representatives of governments from a number of countries. He told his audience not to worry about global warming because the average temperature increases scientists were predicting were much less than the temperature increase he experienced in coming from Washington, D.C., to Williamsburg. What is the fundamental flaw in this reasoning?

2. What changes might occur in **(a)** the global hydrologic cycle (Figure 2-25, p. 39) and **(b)** the global carbon cycle (Figure 2-26, p. 40) if the atmosphere experienced significant warming? Explain.

3. Of the three schools of thought on what should be done about possible global warming (p. 265), which do you favor? Explain.

4. Of the proposals in Figure 12-13 (p. 265) for reducing emissions of greenhouse gases into the troposphere, with which do you disagree? Why?

5. What consumption patterns and other features of your lifestyle directly add greenhouse gases to the atmosphere? Which, if any, of these things would you be willing to give up to slow global warming and reduce other forms of air pollution?

6. Identify climate and topographic factors in your local community that **(a)** intensify air pollution and **(b)** help reduce air pollution.

7. Do you agree or disagree with the possible weaknesses of the U.S. Clean Air Acts listed on p. 282–283? Defend each of your choices. Can you identify other weaknesses?

8. Explain why you agree or disagree with each of the proposals listed in Figure 12-29 (p. 285) for shifting the emphasis to preventing air pollution over the next several decades. Which two of these proposals do you believe are the most important?

9. Congratulations! You are in charge of the world. List your three most important actions for dealing with the problems of **(a)** global warming, **(b)** depletion of ozone in the stratosphere, and **(c)** air pollution.

LEARNING ONLINE

The website for this book contains helpful study aids and many ideas for further reading and research. They include a chapter summary, review questions for the entire chapter, flash cards for key terms and concepts, a multiple-choice practice quiz, interesting Internet sites, references, and a guide for accessing thousands of InfoTrac® College Edition articles. Log on to

http://biology.brookscole.com/miller7

Then click on the Chapter-by-Chapter area, choose Chapter 12, and select a learning resource.

13 SOLID AND HAZARDOUS WASTE

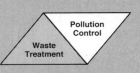

Solid wastes are only raw materials we're too stupid to use.

ARTHUR C. CLARKE

13-1 WASTING RESOURCES

Why Should We Care about Solid Waste? Resource Waste and Pollution

Most solid waste is a symptom of unnecessary waste of resources whose production causes pollution and environmental degradation.

Solid waste is any unwanted or discarded material that is not a liquid or a gas. So what is the big deal? For most people, garbage trucks arrive and whisk away the solid waste they produce—out of sight, out of mind.

In nature, there is essentially no solid waste because the wastes of one organism become nutrients for other organisms. But humans will always produce some solid waste directly and indirectly in almost everything we do. The solid waste we produce directly is called garbage. But most people do not realize that mines, factories, food growers, and businesses that supply people with goods and services produce about 98% of the world's solid waste.

There are two reasons for being concerned about the amount of solid waste we produce directly and indirectly. One is that much of it represents an unnecessary waste of the earth's precious resources. The other is that producing the solid products we use, and often discard, is responsible for huge amounts of air pollution (including greenhouse gases), water pollution, and land degradation.

Some *good news* is that we could reduce our direct and indirect production of solid waste by 75–90%, as you will learn in this chapter.

Case Study: How Much Solid Waste Does the United States Produce? Affluenza in Action

The United States produces about a third of the world's solid waste and buries more than half of it in landfills.

The United States, with only 4.6% of the world's population, produces about one-third of the world's solid waste—a glaring symptom of affluenza (p. 12).

About 98.5% of the solid waste in the United States (and in most developed countries) comes from mining,

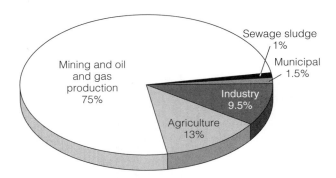

Figure 13-1 Natural capital degradation: Sources of the estimated 11 billion metric tons (12 billion tons) of solid waste produced each year in the United States. Mining, oil and gas production, agricultural, and industrial activities produce 65 times as much solid waste as household activities. (Data from U.S. Environmental Protection Agency and U.S. Bureau of Mines)

oil and natural gas production, agriculture, sewage sludge, and industrial activities (Figure 13-1). This solid waste is produced *indirectly* to provide goods and services to meet the needs and growing wants of consumers.

The remaining 1.5% of solid waste is **municipal solid waste (MSW)**—often called *garbage* or *trash*—generated mostly by homes and workplaces. This small part of the overall solid waste problem is still huge. And between 1960 and 2001, the total amount of MSW in the United States increased 2.6-fold each year and is still rising. Each year, the United States generates enough MSW to fill a bumper-to-bumper convoy of garbage trucks encircling the globe almost eight times! Between 1960 and 1990, the amount of MSW produced per person in the United States increased by 70%.

Canada is the world's second largest per capita producer of MSW. Japan and most developed countries in Europe produce about half as much MSW per person as the United States, and most developing countries produce about one-fourth to one-tenth as much.

According to the U. S. Environmental Protection Agency, about 55% of the MSW produced in the United States is dumped in landfills, 30% is recycled or composted, and 15% is burned in incinerators. Paper makes up about 38% of the trash buried in U.S. landfills, followed by yard waste (12%), food waste (11%), and plastics (10%).

What Does It Mean to Live in a High-Waste Society? A Throwaway Mentality

Most solid waste is a highly visible sign of how a society infected with affluenza wastes valuable resources.

Here are a few of the solid wastes that consumers throw away in the high-waste economy found in the United States:

- Enough aluminum to rebuild the country's entire commercial airline fleet every 3 months

- Enough tires each year to encircle the planet almost three times

- Enough disposable diapers per year that if they were linked end to end, they would reach to the moon and back seven times

- About 2 billion disposable razors, 130 million cell phones, 50 million computers, and 8 million television sets each year. This *electronic waste* or *e-waste*, the fastest growing solid waste problem in the United States and the world, is also a source of toxic and hazardous wastes that include compounds containing lead, mercury, and cadmium that can contaminate the air, surface water, groundwater, and soil.

- Discarded carpet each year that would cover the state of Delaware

- About 2.5 million nonreturnable plastic bottles every hour

- About 670,000 metric tons (1.5 billion pounds) of edible food per year

- Enough office paper each year to build a wall 3.5 meters (11 feet) high across the country from New York City to San Francisco, California

- Some 186 billion pieces of junk mail (an average of 660 pieces per American) each year, about 45% of which are thrown in the trash unopened

13-2   PRODUCING LESS WASTE

What Are Our Options? Management or Prevention

We can try to manage the solid wastes we produce or try to reduce or prevent their production.

We can deal with the solid wastes we create in two ways. One is *waste management*. This is a *high-waste approach* (Figure 4-12, p. 75) that views waste production as a mostly unavoidable product of economic growth. It attempts to manage the wastes that result from economic growth in ways that reduce environmental harm, mostly by mixing and often crushing them to-

gether and then burying them, burning them, or shipping them off to another state or country. In effect, it mixes the wastes we produce together and then transfers them from one part of the environment to another.

The second approach is *waste reduction, a low-waste approach* that recognizes there is no "away." It views most solid waste as potential resources that we should be reusing, recycling, or composting. With this approach, we should be taught to think of trash cans and garbage trucks as *resource containers*. Figure 13-2 lists ways to reduce waste. Study this important figure carefully.

Waste reduction is the preferred solution because it tackles the problem of waste production at the front end—before it occurs—rather than at the back end after wastes have already been produced. It also saves matter and energy resources, reduces pollution (including emissions of greenhouse gases), helps protect biodiversity, and saves money.

Currently, the order of priorities for dealing with solid waste in the United States and in most countries is the reverse of the order suggested by prominent scientists in Figure 13-2. It does not have to be that way. Some scientists and economists estimate that 60–80% of the solid waste we produce can be eliminated by a combination of *reducing waste production, reusing and recycling materials* (including composting), and *redesigning* manufacturing processes and buildings to produce less waste.

Solutions: How Can We Reduce Solid Waste? The Sustainability Six

Reducing consumption and redesigning the products we produce are the best ways to cut waste production and promote sustainability.

Here are six ways to reduce resource use, waste, and pollution—what we might call the *sustainability six*. First, *consume less*. Before buying anything, ask questions such as: Do I really *need* this or do I just *want* it? Can I buy it secondhand (reuse)? Can I borrow or rent it (reuse)?

Second, *redesign manufacturing processes and products to use less material and energy*. A skyscraper built today includes about a third less steel than one the same size built in the 1960s because of the use of lighter-weight but higher-strength steel. The weight of cars has been reduced by about one-fourth by using such steel along with lightweight plastics and composite materials. Plastic milk jugs weigh 40% less than they did in the 1970s, and aluminum drink cans contain one-third less aluminum. All of these changes involve savings in energy use as well as materials.

Third, *redesign manufacturing processes to produce less waste and pollution*. Most toxic organic solvents can be recycled within factories or replaced with water-based

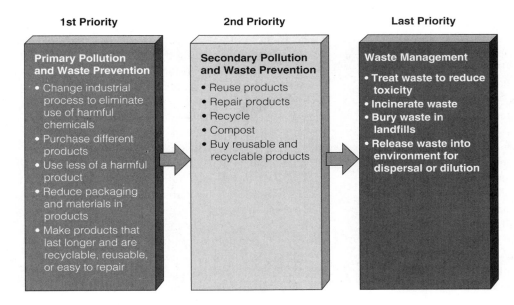

1st Priority	2nd Priority	Last Priority
Primary Pollution and Waste Prevention	**Secondary Pollution and Waste Prevention**	**Waste Management**
• Change industrial process to eliminate use of harmful chemicals • Purchase different products • Use less of a harmful product • Reduce packaging and materials in products • Make products that last longer and are recyclable, reusable, or easy to repair	• Reuse products • Repair products • Recycle • Compost • Buy reusable and recyclable products	• Treat waste to reduce toxicity • Incinerate waste • Bury waste in landfills • Release waste into environment for dispersal or dilution

Figure 13-2 Solutions: priorities suggested by prominent scientists for dealing with material use and solid waste. To date, these waste-reduction priorities have not been followed in the United States (or in most other countries). Instead, most efforts are devoted to waste management (bury it or burn it). (Information from U.S. Environmental Protection Agency and U.S. National Academy of Sciences)

or citrus-based solvents (Individuals Matter, p. 272). Hydrogen peroxide can be used instead of toxic chlorine to bleach paper and other materials.

Fourth, *develop products that are easy to repair, reuse, remanufacture, compost, or recycle.* A new Xerox photocopier with every part reusable or recyclable for easy remanufacturing should eventually save the company $1 billion in manufacturing costs.

Fifth, *design products to last longer.* Today's tires have an average life of 97,000 kilometers (60,000 miles). Researchers believe this use could be extended to at least 160,000 kilometers (100,000 miles).

Sixth, *eliminate or reduce unnecessary packaging.* From an environmental standpoint, the preferred hierarchy for packaging is *no packaging* (nude products), *minimal packaging, reusable packaging,* and *recyclable packaging.* Canada has set a goal of using the first three of these packaging priorities to cut excess packaging in half.

Figure 13-3 lists some ways you can reduce your output of solid waste.

Improvements in resource productivity and environmental design are very important. But we can do much better through a new *resource productivity revolution.* In their 1999 book *Natural Capitalism,* Paul Hawken, Amory Lovins, and Hunter Lovins contend that we have the knowledge and technology to greatly increase resource productivity by getting 75–90% more work or service from each unit of material resources we use. To these analysts, the only major obstacles to such an economic and ecological revolution are laws, policies, taxes, and subsidies that continue to reward inefficient resource use, and fail to reward efficient, resource use. There are many fulfilling career choices for people wanting to become part of the resource productivity revolution.

What Can You Do?
Solid Waste

• Follow the four R's of resource use: Refuse, Reduce, Reuse, and Recycle.

• Ask yourself whether you relly need a particular item.

• Rent, borrow, or barter goods and services when you can.

• Buy things that are reusable, recyclable, or compostable, and be sure to reuse, recycle, and compost them.

• Do not use throwaway paper and plastic plates, cups, and eating utensils, and other disposable items when reusable or refillable versions are available.

• Use e-mail in place of conventional paper mail.

• Read newspapers and magazines online.

• Buy products in concentrated form whenever possible.

Figure 13-3 What can you do? Ways to reduce your output of solid waste.

13-3 THE ECOINDUSTRIAL REVOLUTION AND SELLING SERVICES INSTEAD OF THINGS

What Is the Ecoindustrial Revolution? Reducing Waste Production by Copying Nature

We can make industrial manufacturing processes more sustainable by redesigning them to mimic how nature deals with wastes.

There are growing signs that a new *ecoindustrial revolution* will take place over the next 50 years. The goal is to make industrial manufacturing processes cleaner and more sustainable by redesigning them to mimic how nature deals with wastes. Recall that in nature, the waste outputs of one organism become the nutrient inputs of another organism, so that all of the earth's nutrients are endlessly recycled.

One way we can mimic nature is to recycle and reuse most chemicals used in industries instead of dumping them into the environment. Another is to have industries interact in complex *resource exchange webs* where the wastes of one manufacturer become raw materials for another—similar to food webs in natural ecosystems (Figure 2-17, p. 33).

This is happening in Kalundborg, Denmark, where an electric power plant and a number of nearby industries, farms, and homes are working together to save money and reduce their outputs of waste and pollution. They do this by exchanging waste outputs and thus converting them into resources, as shown in Figure 13-4. Trace the connections in this diagram.

Today there are about 20 ecoindustrial parks similar to the one in Kalundborg in various parts of the world and more are being built or planned. Some are being developed on abandoned industrial sites, called *brownfields,* which are cleaned up and redeveloped.

In Europe, at least one-third of all industrial wastes are sent to waste-material exchanges or clearinghouses where they are sold or given away as raw materials for other industries. About a tenth of the industrial waste in the United States is sent to such clearinghouses, a figure that could be greatly increased.

In addition to eliminating most waste and pollution, these industrial forms of *biomimicry* provide many economic benefits for businesses. They reduce the costs of controlling pollution and complying with pollution regulations. If a company does not add pollutants to the environment, it does not have to worry about government regulations or being sued because the wastes harm someone. The company also improves the health and safety of its workers by reducing their exposure to toxic and hazardous material and thus reduces company health-care insurance costs.

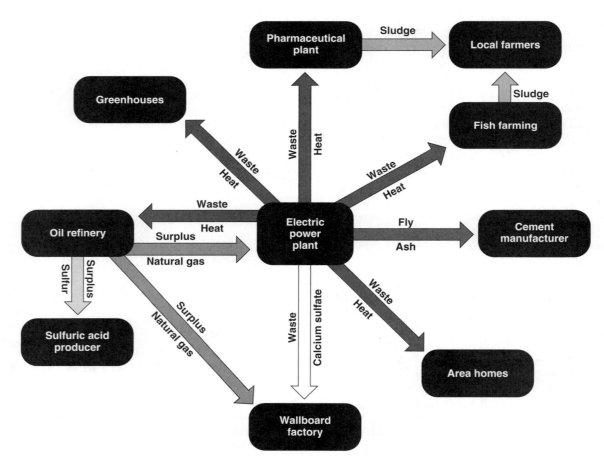

Figure 13-4 Solutions: *industrial ecosystem* in Kalundborg, Denmark, reduces waste production by mimicking a natural food web. The wastes of one business become the raw materials for another. Waste heat in the form of hot air or water can be piped from the power plant to several other sites.

Biomimicry also stimulates companies to come up with new, environmentally beneficial chemicals, processes, and products that can be sold worldwide. Such companies have a better image among consumers based on results rather than public relations campaigns.

In 1975, the Minnesota Mining and Manufacturing Company (3M), which makes 60,000 different products in 100 manufacturing plants, began a Pollution Prevention Pays (3P) program. It redesigned equipment and processes, used fewer hazardous raw materials, identified hazardous chemical outputs (and recycled or sold them as raw materials to other companies), and began making more nonpolluting products.

By 1998, 3M's overall waste production was down by one-third, its air pollutant emissions per unit of production were 70% lower, and the company had saved more than $750 million in waste disposal and material costs. Since 1990, a growing number of companies have adopted similar pollution prevention programs. See the Guest Essay by Peter Montague on cleaner production on the website for this chapter.

What Is a Service-Flow Economy? Selling Services instead of Things

Businesses can greatly decrease their pollution and waste by shifting from selling goods to selling the services the goods provide.

In the mid-1980s, German chemist Michael Braungart and Swiss industry analyst Walter Stahel independently proposed a new economic model that would provide profits while greatly reducing resource use and waste. Their idea for more sustainable economies involves shifting from our current *material flow economy* (Figure 4-12, p. 75) to a *service-flow economy* over the next few decades. Instead of buying most goods outright, customers would use *eco-leasing,* renting the *services* that such goods provide.

In a service-flow economy, a manufacturer makes more money on a product if it uses the minimum amount of materials, lasts as long as possible, and is easy to maintain, repair, remanufacture, reuse, or recycle.

There is evidence that such an economic shift based on eco-leasing is under way. Since 1992, the Xerox Corporation has been leasing most of its copy machines as part of its mission to provide *document services* instead of selling photocopiers. When the service contract expires, Xerox takes the machine back for reuse or remanufacture and has a goal of sending no material to landfills or incinerators. To save money, machines are designed to use recycled paper, have few parts, be energy efficient, and emit as little noise, heat, ozone, and copier chemical waste as possible. Canon in Japan and Fiat in Italy are taking similar measures.

Dow and several other chemical companies are doing a booming business in leasing organic solvents (used mostly to remove grease from surfaces), photographic developing chemicals, and dyes and pigments. In this *chemical service* business, the company delivers the chemicals, helps the client set up a recovery system, takes away the recovered chemicals, and delivers new chemicals as needed.

Finally, Ray Anderson, CEO of a large carpet and tile company, plans to lease rather than sell carpet (Individuals Matter, below). There are many entrepreneurial and career opportunities in the emerging service-flow economy.

Ray Anderson

INDIVIDUALS MATTER

Ray Anderson is CEO of Interface, a company based in Atlanta, Georgia, that makes carpet tiles. The company is the world's largest commercial carpet manufacturer, with 26 factories in six countries, customers in 110 countries, and more than $1 billion in annual sales.

Anderson changed the way he viewed the world and his business after reading Paul Hawken's book *The Ecology of Commerce.* In 1994, he announced plans to develop the nation's first totally sustainable green corporation.

He has implemented hundreds of projects with the goals of zero waste, greatly reduced energy use, and eventually zero use of fossil fuels by relying on renewable solar energy. By 1999, the company had reduced resource waste by almost 30% and reduced energy waste enough to save $100 million. One of Interface's factories in California runs on solar cells to produce the world's first solar-made carpet.

To achieve the goal of zero waste, Anderson plans to stop selling carpet and lease it as a way to encourage recycling. For a monthly fee, the company will install, clean, and inspect the carpet on a monthly basis, repair worn carpet tiles overnight, and recycle worn-out tiles into new carpeting. As Anderson puts it, "We want to harvest yesterday's carpets and recycle them with zero scrap going to the landfill and zero emissions into the ecosystem—and run the whole thing on sunlight."

Anderson is one of a growing number of business leaders committed to finding more economically and ecologically sustainable ways to do business while still making a profit for stockholders. Between 1993 and 1998, the company's revenues doubled and profits tripled, mostly because the company saved $130 million in material costs with an investment of less than $40 million. Andersen says he is having a blast.

13-4 REUSE

What are the Advantages and Disadvantages of Reuse? Improves Environmental Quality for Some, Can Create Hazards for Others

Reusing products is an important way to reduce resource use, waste, and pollution in developed countries, but can create hazards for the poor in developing countries.

Reuse involves cleaning and using materials over and over and thus increasing the typical life span of a product. This form of waste reduction reduces the use of matter and energy resources, cuts pollution and waste, creates local jobs, and saves money. Traditional forms of reuse include salvaging automobile parts from older cars in junkyards and salvaging bricks, doors, fine woodwork, and other items from old houses and buildings.

However, in today's high-throughput societies we have increasingly substituted throwaway tissues for reusable handkerchiefs, disposable paper towels and napkins for reusable cloth ones, throwaway paper plates, cups, and plastic utensils for reusable plates, cups, and silverware, and throwaway beverage containers for refillable ones. We even have disposable cameras.

Reuse is alive and well in most developing countries but can be a health hazard for the poor. About 80% of the e-waste in the United States, including discarded TV sets, computers, and cell phones, is shipped to China, India, Pakistan, and other (mostly Asian) countries where labor is cheap and environmental regulations are weak. Workers there, many of them children, dismantle the products to recover reusable parts and are thus exposed to toxic metals such as lead, mercury, and cadmium. The scrap left over is dumped in waterways and fields, or burned in open fires, which exposes the workers to toxic dioxins.

In cities such as Manila in the Philippines, Mexico City, Mexico, and Cairo, Egypt, large numbers of people—many of them children—eke out a living by scavenging, sorting, and selling materials they get from open city dumps. This exposes them to toxins and infectious diseases.

Should We Use Refillable Containers? Reviving Reuse

Refilling and reusing containers uses less resources and energy, produces less waste, saves money, and creates local jobs.

Two examples of reuse are refillable glass beverage bottles and refillable soft drink bottles made of polyethylene terephthalate (PET) plastic. Typically, such bottles make 15 round-trips before they become too damaged for reuse and then are recycled. Reusing beverage bottles stimulates local economies by creating local jobs related to their collection and refilling. Moreover, studies by Coca-Cola and PepsiCo of Canada show that their soft drinks in 0.5-liter (16-ounce) bottles cost one-third less in refillable bottles than in throwaway bottles.

But big companies make more money by producing and shipping throwaway beverage and food containers at centralized facilities. This shift has put many small local bottling companies, breweries, and canneries out of business.

Denmark and Canada's Prince Edward Island have led the way by banning all beverage containers that cannot be reused. To encourage use of refillable glass bottles, Ecuador has a refundable beverage container deposit fee that is 50% of the cost of the drink. In Finland, 95% of the soft drink, beer, wine, and spirits containers are refillable, and in Germany, about three-fourths are refillable.

X *HOW WOULD YOU VOTE?* Do you support banning all beverage containers that cannot be reused as Denmark has done? Cast your vote online at http://biology.brookscole.com/miller7.

What Are Other Ways to Reuse Things? Reducing Throwaway Items

We can use reusable shopping bags, food containers, and shipping pallets, and borrow tools from tool libraries.

Cloth bags can be used to carry groceries and other items instead of paper or plastic bags. Both paper and plastic bags are environmentally harmful, and the question of which is more damaging has no clear-cut answer. To encourage people to bring reusable bags, stores in the Netherlands and Ireland charge for shopping bags. As a result, the use of plastic shopping bags dropped by 90–95% in both countries. In 2004, supermarkets in Shanghai, China's largest city, began charging shoppers for plastic bags in an attempt to reduce waste.

X *HOW WOULD YOU VOTE?* Should consumers have to pay for plastic or paper bags at grocery and other stores? Cast your vote online at http://biology.brookscole.com/miller7.

Other examples of reusable items are metal or plastic lunchboxes and plastic containers for storing lunchbox items and refrigerator leftovers, instead of using throwaway plastic wrap and aluminum foil.

Manufacturers can use shipping pallets made of recycled plastic waste instead of throwaway wood pallets. In 1991, Toyota shifted entirely to reusable shipping containers. A similar move by the Xerox Corporation saves the company more than $3 million per year.

Another example of reuse involves *tool libraries* (such as those in Berkeley, California, and Takoma Park,

Maryland) where people can check out a variety of power and hand tools.

Figure 13-5 lists several ways for you to reuse some of the items you buy.

13-5   RECYCLING

What Is Recycling? An Environmental Success Story

Recycling is an important way to collect waste materials and turn them into useful products that can be sold in the marketplace.

Recycling involves reprocessing discarded solid materials into new, useful products. Recycling has a number of benefits to people and the environment (Figure 13-6). Households and workplaces produce five major types of materials that can be recycled: *paper products* (including newspaper, magazines, office paper, and cardboard), *glass, aluminum, steel,* and some types of *plastics.*

Materials collected for recycling can be reprocessed in two ways. *Primary* or *closed-loop* recycling occurs

Figure 13-5 What can you do? Ways to reuse some of the items you buy.

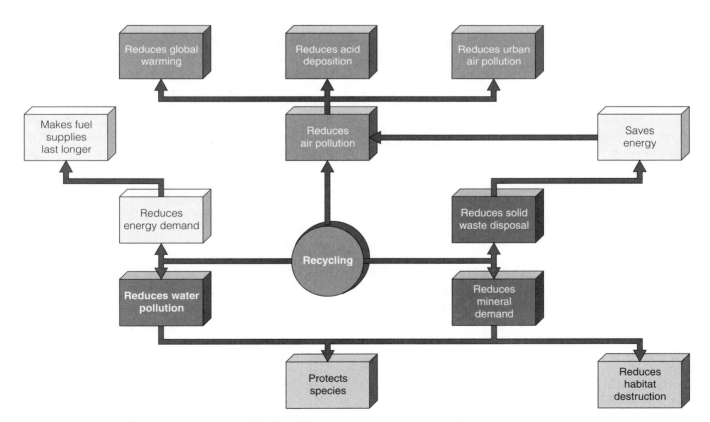

Figure 13-6 Solutions: environmental benefits of recycling. In addition to these environmental benefits, recycling saves more money and creates far more jobs than burning wastes or disposing of them in landfills—assuming that all of these operations receive equal or no government subsidies or tax breaks. Which two of these benefits do you believe are the most important? Despite its many benefits, recycling is still an output approach that deals with wastes after they are produced instead of a way to reduce the overall flow of resources.

when waste is recycled into new products of the same type—turning used newspapers into new newspaper and used aluminum cans into new aluminum cans, for example.

Secondary recycling, *also called* downcycling, *involves converting waste materials into different products.* For example, used tires can be shredded and converted into rubberized road surfacing and newspapers can be converted into cellulose insulation.

Environmentalists distinguish between two types of wastes that can be recycled. One is *preconsumer* or *internal waste.* It consists of waste generated in a manufacturing process and recycled instead of being discarded. The other is *postconsumer* or *external waste* generated by consumer use of products. There is about 25 times more preconsumer than postconsumer waste. It is important to recycle both types.

In theory, just about anything is recyclable, but only two things count. *First,* will the item actually be recycled? Sometimes separated wastes collected for recycling are mixed with other wastes and sent to landfills or incinerated, mostly when prices for recycled raw materials fall sharply.

Second, will businesses and individuals complete the recycling loop by buying products that are made from recycled materials? If we do not buy those products, recycling does not work.

Switzerland and Japan recycle about half of their MSW. The United States recycles about 30% of its MSW—up from 6.4% in 1960. This roughly 5-fold increase in recycling is an impressive achievement. But the country's total amount of solid waste has continued to increase although the MSW per person has leveled off since 1970. Studies indicate that with economic incentives and better design of waste management systems, the United States and other developed countries could recycle 60–80% of their MSW.

How Useful Is Composting? Recycling by Copying Nature

Composting biodegradable organic waste mimics nature by recycling plant nutrients to the soil.

Composting is a simple process in which we copy nature to recycle some of the biodegradable organic wastes we produce. The organic material produced by composting can be added to soil to supply plant nutrients, slow soil erosion, retain water, and improve crop yields.

Some cities in Austria, Belgium, Denmark, Germany, Luxembourg, and Switzerland recover and compost more than 85% of their biodegradable wastes. However, only about 5% of the paper, yard, and vegetable food waste in U.S. MSW is composted, but studies show it could be raised to 35%.

Such wastes can be collected and composted in centralized community facilities, as is done in many European Union countries. The resulting compost can be used as an organic soil fertilizer, topsoil, or landfill cover. It can also be used to help restore eroded soil on hillsides and along highways, and on strip-mined land, overgrazed areas, and eroded cropland.

To be successful, a large-scale composting program must be located carefully and odors must be controlled, because people do not want to live near a giant compost pile or plant. Composting programs must also exclude toxic materials that can contaminate the compost and make it unsafe for fertilizing crops and lawns.

You can easily make your own compost by collecting organic wastes in a backyard bin. For details on composting, see the website for this chapter.

How Should We Recycle Solid Waste? To Separate or Not to Separate?

There is disagreement over whether to send mixed urban wastes to centralized resource recovery plants or have individuals sort recyclables for collection and sale to manufacturers as raw materials.

One way to recycle is to send mixed urban wastes to a centralized *materials-recovery facility (MRF).* There, machines or workers separate the mixed waste to recover valuable materials for sale to manufacturers as raw materials. The remaining paper, plastics, and other combustible wastes are recycled or burned to produce steam or electricity to run the recovery plant or to sell to nearby industries or homes. Ash from the incinerator is buried in a landfill.

Such plants are expensive to build, operate, and maintain. They can emit toxic air pollutants, if not operated properly, and they produce a toxic ash that must be disposed of safely.

MRFs are hungry beasts that must have a large input of garbage to make them financially successful. Thus their owners have a vested interest in increasing *throughput* of matter and energy resources to produce more trash—the reverse of what prominent scientists believe we should be doing (Figure 13-2).

To many experts, it makes more sense economically and environmentally for households and businesses to separate their trash into recyclable categories such as glass, paper, metals, certain types of plastics, and compostable materials. Then these segregated wastes are collected and sold to scrap dealers, compost plants, and manufacturers.

The *source separation* approach has several advantages over the centralized approach. It produces much less air and water pollution and has low start-up costs and moderate operating costs. It also saves more energy, provides more jobs per unit of material, and yields cleaner and usually more valuable recyclables. In addition, it educates people about the need for waste reduction, reuse, and recycling.

To promote separation of wastes for recycling, many communities use a *pay-as-you-throw* (*PAUT*) waste collection system. It charges households and businesses for the amount of mixed waste picked up but does not charge for pickup of materials separated for recycling.

Case Study: Is It Feasible to Recycle Plastics? Some Problems

Recycling many plastics is chemically and economically difficult.

Currently, only about 10% by weight of all plastic wastes in the United States are recycled, for three reasons. *First*, many plastics are difficult to isolate from other wastes because the many different resins used to make them are often difficult to identify, and some plastics are composites of different resins. Most plastics also contain stabilizers and other chemicals that must be removed before recycling.

Second, recovering individual plastic resins does not yield much material because only small amounts of any given resin are used per product. *Third*, the price of oil used to produce petrochemicals for making plastic resins is so low that the cost of virgin plastic resins is much lower than that of recycled resins. An exception is PET (polyethylene terephthalate), used mostly in plastic drink bottles.

Thus, mandating that plastic products contain a certain amount of recycled plastic resins is unlikely to work. It could also hinder the use of recycled plastics in reducing the resource content and weight of many widely used items such as plastic bags and bottles.

Cargill Dow is manufacturing biodegradable and recyclable plastic containers made from a polymer called polyactide (ACT), made from the sugar in corn syrup. Instead of being sent to landfills, containers made from this bio-plastic could be composted to produce a soil conditioner.

Toyota, the world's No. 2 automaker, is investing $38 billion in a process that makes plastics from plants. By 2020, it expects to control two-thirds of the world's supply of such bioplastics.

Does Recycling Make Economic Sense? Yes, for Certain Materials

Recycling materials such as paper and metals has environmental and economic benefits.

Whether recycling makes economic sense depends on how you look at at the economic and environmental benefits and costs of recycling. Critics say recycling does not make sense if it costs more to recycle materials than to send them to a landfill or incinerator. They also point out that recycling is often not needed to save landfill space because many areas are not running out of it.

Critics concede that recycling may make economic sense for valuable and easy-to-recycle materials (such as aluminum, paper, and steel), but not for cheap or plentiful resources such as glass from silica and most plastics that are expensive to recycle.

Critics of recycling also argue that it should pay for itself. But proponents of recycling point out that conventional garbage disposal systems are paid for by charges to households and businesses. So why should recycling be held to a different standard and forced to compete on an uneven playing field?

Proponents also point out that the reducing the use of landfills and incinerators is not as important as the other benefits of recycling (Figure 13-6). And they point to studies showing that the net economic, health, and environmental benefits of recycling far outweigh the costs.

Why Do We Not Have More Reuse and Recycling? Faulty Accounting and an Uneven Economic Playing Field

Prices of goods that do not reflect the ecological truth, too few government subsidies and tax breaks, low landfill dumping costs, and price fluctuations hinder reuse and recycling.

Four factors hinder reuse and recycling. *First* is a faulty accounting system in which the market price of a product does not include the harmful environmental health costs associated with the product during its life cycle (Figure 13-7, p. 296).

Second, there is an uneven economic playing field because in most countries, resource-extracting industries receive more government tax breaks and subsidies than recycling and reuse industries. We get more of what we reward.

Third, charges for depositing wastes in landfills (called tipping fees) in the United States are lower than those in most of Europe. *Fourth*, the demand and thus the price paid for recycled materials fluctuate, mostly because buying goods made with recycled materials is not a priority for most governments, businesses, and individuals.

How can we encourage recycling and reuse? Proponents say that leveling the economic playing field is the best way to start. Governments can *increase* subsidies and tax breaks for reuse and recycling materials (the carrot) and *decrease* subsidies and tax breaks for making items from virgin resources (the stick).

Another way to encourage recycling is to greatly increase use of the *pay-as-you-throw* (*PAUT*) system and encourage or require government purchases of recycled products to help increase demand and lower

LIFE-CYCLE ANALYSIS OF A SHIRT

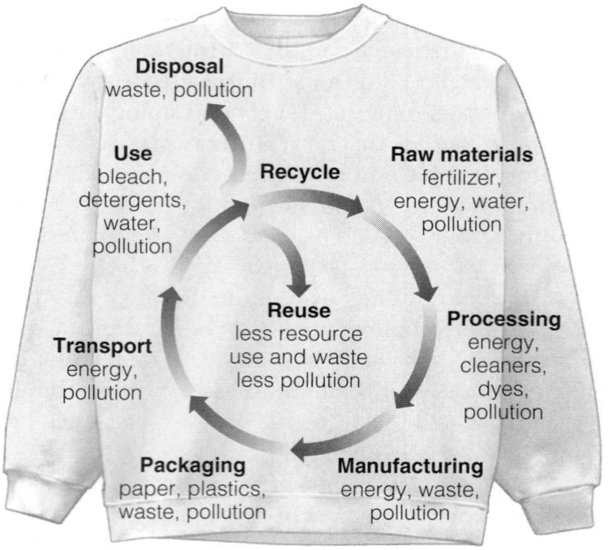

Disposal
waste, pollution

Use
bleach, detergents, water, pollution

Recycle

Raw materials
fertilizer, energy, water, pollution

Reuse
less resource use and waste less pollution

Transport
energy, pollution

Processing
energy, cleaners, dyes, pollution

Packaging
paper, plastics, waste, pollution

Manufacturing
energy, waste, pollution

Figure 13-7 *Life-cycle analysis* of the resources used and pollutants produced over the lifetime of a product. Including all of these costs in the market prices of the products we buy would allow us to compare the harmful environmental costs of different products and thus make more informed choices about products we purchase.

prices. Governments can also pass laws requiring companies to take back and recycle or reuse packaging discarded by consumers. In the Netherlands, all packaging waste is banned from landfills. By 2008, European (EU) countries must recycle 55–80% of all packaging waste. The EU also requires companies to take back electronic products from consumers without charge and bans e-waste in municipal solid waste. These *product stewardship policies* create a strong economic incentive for companies to redesign products for safer and easier recycling, reuse, and remanufacturing.

✗ HOW WOULD YOU VOTE? Should governments pass laws requiring manufacturers to take back and reuse or recycle all packaging waste, appliances, electronic equipment, and motor vehicles at the end of their useful lives? Cast your vote online at http://biology.brookscole.com/miller7.

13-6 BURNING AND BURYING SOLID WASTE

What Are the Advantages and Disadvantages of Burning Solid Waste? A Faded Rose in Some Countries

Japan and a few European countries incinerate most of their municipal waste, but this is done less in the United States and in most European countries.

Globally, municipal solid waste is burned in over 1,000 large *waste-to-energy incinerators*, which boil water to make steam for heating water or space, or for producing electricity. Trace the flow of materials through this process as diagrammed in Figure 13-8. Japan and Switzerland burn more than half of their MSW in incinerators compared to 16% in the United States and about 8% in Canada. Figure 13-9 lists the advantages and disadvantages of using incinerators to burn solid and hazardous waste. Study this figure carefully.

Since 1985, more than 280 new incinerator projects have been delayed or canceled in the United States because of high costs, concern over air pollution, and intense citizen opposition.

✗ HOW WOULD YOU VOTE? Do the advantages of incinerating solid waste outweigh the disadvantages? Cast your vote online at http://biology.brookscole.com/miller7.

What Are the Advantages and Disadvantages of Burying Solid Waste? A Widely Used Last Resort

Most of the world's municipal solid waste is buried in landfills that will eventually leak toxic liquids into the soil and underlying aquifers.

About 54% by weight of the MSW in the United States is buried in sanitary landfills, compared to 90% in the United Kingdom, 80% in Canada, 15% in Japan, and 12% in Switzerland. There are two types of landfills. **Open dumps** are essentially fields or holes in the ground where garbage is deposited and sometimes covered with soil. They are rare in developed countries, but widely used in many developing countries.

In newer landfills, called **sanitary landfills,** solid wastes are spread out in thin layers, compacted, and covered daily with a fresh layer of clay or plastic foam. Modern state-of-the-art landfills on geologically suitable sites and away from lakes, rivers, floodplains, and aquifer recharge zones are lined with clay and plastic before being filled with garbage, as seen in Figure 13-10 (p. 298). Note in the figure that the landfill bottom is covered with a second impermeable liner, usually made of several layers of clay, thick plastic, and sand. This liner collects *leachate* (rainwater contaminated as it percolates through the solid waste) and is intended to prevent its leakage into groundwater. Wells are drilled around the landfill to monitor any leakage.

Collected leachate is pumped from the bottom of the landfill, stored in tanks, and sent to a regular sewage treatment plant or an on-site treatment plant. When full, the landfill is covered with clay, sand, gravel, and topsoil to prevent water from seeping in.

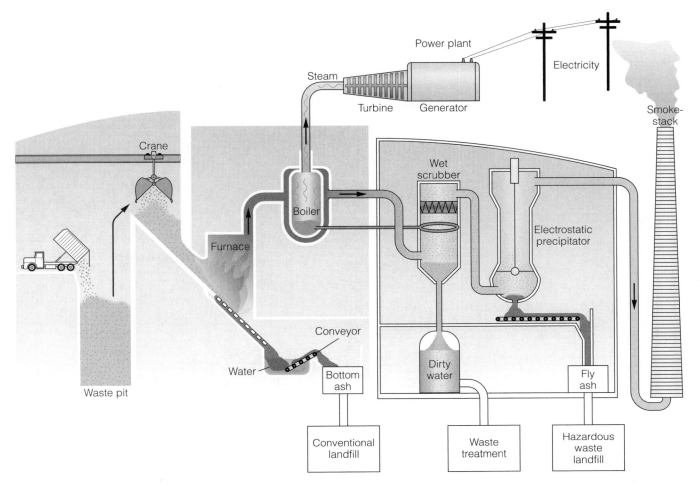

Figure 13-8 Solutions: *waste-to-energy incinerator* with pollution controls that burns mixed solid waste and recovers some of the energy to produce steam used for heating or producing electricity. (Adapted from the EPA, *Let's Reduce and Recycle*)

These new landfills are equipped with a connected network of vent pipes to collect landfill gas (consisting mostly of two greenhouse gases, methane and carbon dioxide) released by the underground decomposition of wastes. The methane is filtered out and burned in small gas turbines to produce steam or electricity for nearby facilities or sold to utilities for use as a fuel. Figure 13-11 (p. 299) lists the advantages and disadvantages of using sanitary landfills to dispose of solid waste. Study this figure carefully. According to the U.S. Environmental Protection Agency (EPA), all landfills eventually leak.

✘ *HOW WOULD YOU VOTE?* Do the advantages of burying solid waste in sanitary landfills outweigh the disadvantages? Cast your vote online at http://biology.brookscole.com /miller7.

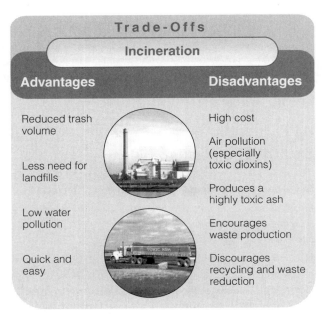

Trade-Offs	
Incineration	
Advantages	**Disadvantages**
Reduced trash volume	High cost
Less need for landfills	Air pollution (especially toxic dioxins)
Low water pollution	Produces a highly toxic ash
	Encourages waste production
Quick and easy	Discourages recycling and waste reduction

Figure 13-9 Trade-offs: advantages and disadvantages of incinerating solid waste. These trade-offs also apply to the incineration of hazardous waste. Pick the single advantage and disadvantage that you think are the most important.

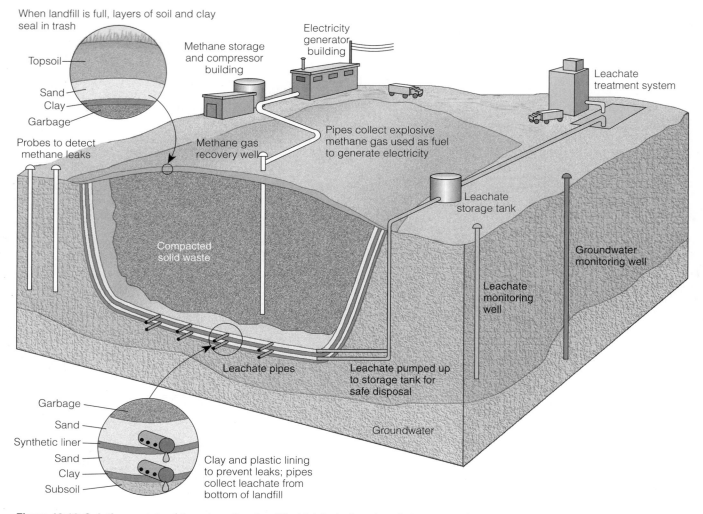

When landfill is full, layers of soil and clay seal in trash

Topsoil
Sand
Clay
Garbage

Methane storage and compressor building

Electricity generator building

Leachate treatment system

Probes to detect methane leaks

Methane gas recovery well

Pipes collect explosive methane gas used as fuel to generate electricity

Compacted solid waste

Leachate storage tank

Groundwater monitoring well

Leachate monitoring well

Leachate pipes

Leachate pumped up to storage tank for safe disposal

Groundwater

Garbage
Sand
Synthetic liner
Sand
Clay
Subsoil

Clay and plastic lining to prevent leaks; pipes collect leachate from bottom of landfill

Figure 13-10 Solutions: state-of-the-art *sanitary landfill*, which is designed to eliminate or minimize environmental problems that plague older landfills. Even such state-of-the-art landfills are expected to leak eventually, passing both the effects of contamination and cleanup costs on to future generations. Since 1997, only this state-of-the-art type of landfill can operate in the United States. As a result, many older and smaller landfills have been closed and replaced with larger local and regional modern landfills.

13-7 HAZARDOUS WASTE

What Is Hazardous Waste? Toxic Threats

Developed countries produce about 80–90% of the world's solid and liquid wastes that can harm people, and most such wastes are not regulated.

Hazardous waste is any discarded solid or liquid material that is *toxic, ignitable, corrosive,* or *reactive* enough to explode or release toxic fumes. According to the UN Environment Programme, developed countries produce 80–90% of these wastes. Figure 13-12 (p. 300) lists some the harmful chemicals found in many homes.

In the United States, about 5% of all hazardous waste is regulated under the Resource Conservation and Recovery Act (RCRA, pronounced "RICK-ra")

and is often referred to as *RCRA hazardous waste.* In other words, about 95% of these wastes are not regulated. In most other countries, especially developing countries, even less, if any, of the hazardous waste is regulated.

In 1980, the U.S. Congress passed the *Comprehensive Environmental Response, Compensation, and Liability Act,* commonly known as the *CERLA* or *Superfund* program. The goals of this law are to identify hazardous waste sites and clean up such sites on a priority basis. The worst sites that represent an immediate and severe threat to human health are put on a *National Priorities List (NPL)* and scheduled for total cleanup using the most cost-effective method.

The U.S. Congress and various state legislatures have also passed laws that encourage the cleanup of

Trade-Offs

Sanitary Landfills

Advantages	Disadvantages
No open burning	Noise and traffic
Little odor	Dust
Low groundwater pollution if sited properly	Air pollution from toxic gases and volatile organic compounds
Can be built quickly	Releases greenhouse gases (methane and CO_2) unless they are collected
Low operating costs	
Can handle large amounts of waste	Groundwater contamination
	Slow decomposition of wastes
Filled land can be used for other purposes	Discourages recycling and waste reduction
No shortage of landfill space in many areas	Eventually leaks and can contaminate groundwater

Figure 13-11 Trade-offs: advantages and disadvantages of using sanitary landfills to dispose of solid waste. Pick the single advantage and disadvantage that you think are the most important.

brownfields—abandoned industrial and commercial sites that in most cases are contaminated with hazardous wastes. Examples include factories, junkyards, older landfills, and gas stations.

Figure 13-13 (p. 300) lists the priorities that prominent scientists believe we should follow in dealing with hazardous waste. Study this important figure carefully. Denmark is following these priorities but most countries are not.

How Can We Remove or Detoxify Hazardous Waste? Science to the Rescue

Chemical and biological methods can be used to remove hazardous wastes or to reduce their toxicity.

In Denmark, all hazardous and toxic waste from industries and households is delivered to 21 transfer stations throughout the country. The waste is then transferred to a large treatment facility. There, about three-fourths of the waste is detoxified by physical, chemical, and biological methods and the rest is buried in a carefully designed and monitored landfill.

Some scientists and engineers consider biological treatment of hazardous waste as the wave of the future for cleaning up some types of toxic and hazardous waste. One approach is *bioremediation,* in which bacteria and enzymes are used to help destroy toxic or hazardous substances or convert them to harmless compounds. See the Guest Essay by John Pichtel on this topic on the website for this chapter.

Another biological way to treat hazardous wastes is *phytoremediation.* It involves using natural or genetically engineered plants to absorb, filter, and remove contaminants from polluted soil and water (Figure 13-14, p. 301). Various plants have been identified as "pollution sponges" to help clean up soil and water contaminated with chemicals such as pesticides, organic solvents, radioactive metals, and toxic metals such as lead, mercury, and arsenic.

Figure 13-15 (p. 302) lists advantages and disadvantages of phytoremediation. Study this figure carefully.

What Are the Advantages and Disadvantages of Burning and Burying Hazardous Waste? Last Resorts

Hazardous waste can be incinerated or disposed of on or underneath the earth's surface, but this can pollute the air and water.

Hazardous waste can be incinerated. This has the same mixture of advantages and disadvantages as burning solid wastes (Figure 13-9). Two major disadvantages of incinerating hazardous waste is that it releases air pollutants such as toxic dioxins and produces a highly toxic ash that must be safely and permanently stored.

Most hazardous waste in the United States is disposed of on land in deep underground wells, surface impoundments such as ponds, pits, or lagoons, and state-of-the-art landfills. In *deep-well disposal,* liquid hazardous wastes are pumped under pressure through a pipe into dry, porous geologic formations or zones of rock far beneath aquifers tapped for drinking and irrigation water. Theoretically, these liquids soak into the porous rock material and are isolated from overlying groundwater by essentially impermeable layers of rock.

Figure 13-16 (p. 302) lists the advantages and disadvantages of deep-well disposal of liquid hazardous wastes. Many scientists believe current regulations for deep-well disposal are inadequate and should be improved.

X *HOW WOULD YOU VOTE?* Do the advantages of deep-well disposal of hazardous waste from soil and water outweigh the disadvantages? Cast your vote online at http://biology .brookscole.com/miller7.

What Harmful Chemicals Are in Your Home?

Cleaning

- Disinfectants
- Drain, toilet, and window cleaners
- Spot removers
- Septic tank cleaners

Paint

- Latex and oil-based paints
- Paint thinners, solvents, and strippers
- Stains, varnishes, and lacquers
- Wood preservatives
- Artist paints and inks

General

- Dry-cell batteries (mercury and cadmium)
- Glues and cements

Gardening

- Pesticides
- Weed killers
- Ant and rodent killers
- Flea powders

Automotive

- Gasoline
- Used motor oil
- Antifreeze
- Battery acid
- Solvents
- Brake and transmission fluid
- Rust inhibitor and rust remover

Figure 13-12 Harmful chemicals found in many homes. Congress has exempted disposal of these household materials from government regulation. Make a survey to see which of these chemicals are in your home.

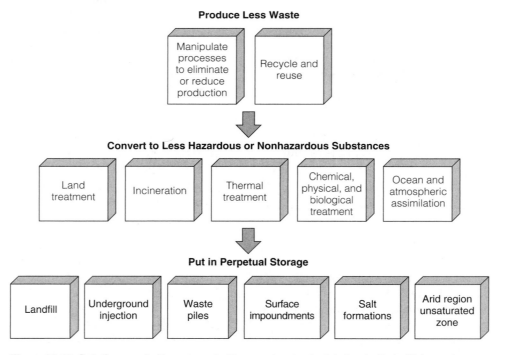

Produce Less Waste

| Manipulate processes to eliminate or reduce production | Recycle and reuse |

Convert to Less Hazardous or Nonhazardous Substances

| Land treatment | Incineration | Thermal treatment | Chemical, physical, and biological treatment | Ocean and atmospheric assimilation |

Put in Perpetual Storage

| Landfill | Underground injection | Waste piles | Surface impoundments | Salt formations | Arid region unsaturated zone |

Figure 13-13 Solutions: priorities suggested by prominent scientists for dealing with hazardous waste. To date, these priorities have not been followed in the United States or most other countries. (Data from U.S. National Academy of Sciences)

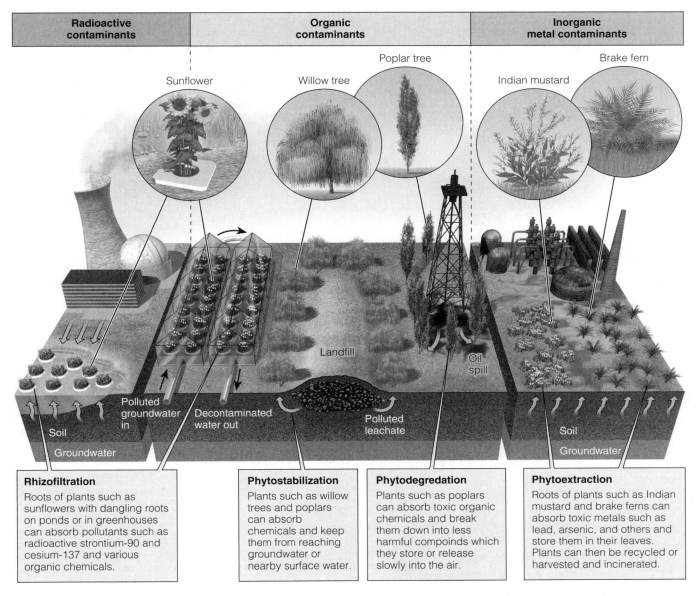

Radioactive contaminants	Organic contaminants	Inorganic metal contaminants

Sunflower

Willow tree

Poplar tree

Indian mustard

Brake fern

Landfill

Oil spill

Polluted groundwater in

Decontaminated water out

Polluted leachate

Soil

Groundwater

Soil

Groundwater

Rhizofiltration

Roots of plants such as sunflowers with dangling roots on ponds or in greenhouses can absorb pollutants such as radioactive strontium-90 and cesium-137 and various organic chemicals.

Phytostabilization

Plants such as willow trees and poplars can absorb chemicals and keep them from reaching groundwater or nearby surface water.

Phytodegredation

Plants such as poplars can absorb toxic organic chemicals and break them down into less harmful compoinds which they store or release slowly into the air.

Phytoextraction

Roots of plants such as Indian mustard and brake ferns can absorb toxic metals such as lead, arsenic, and others and store them in their leaves. Plants can then be recycled or harvested and incinerated.

Figure 13-14 *Phytoremediation.* Ways that various types of plants can be used as pollution sponges to clean up soil and water and radioactive substances (left), organic compounds (center), and toxic metals (right). (Data from American Society of Plant Physiologists, U.S. Environmental Protection Agency, and Endspace)

Surface impoundments are excavated depressions such as ponds, pits, or lagoons into which liquid hazardous wastes are drained and stored (Figure 9-23, p. 191). As water evaporates, the waste settles and becomes more concentrated. Figure 13-17 (p. 302) lists the advantages and disadvantages of this method. Study this figure carefully. EPA studies found that 70% of these storage basins in the United States have no liners, and as many as 90% may threaten groundwater. According to the EPA, all liners are likely to leak eventually and can contaminate groundwater.

Sometimes liquid and solid hazardous waste are put into drums or other containers and buried in carefully designed and monitored *secure hazardous waste landfills* (Figure 13-18, p. 303). Sweden goes further and

buries its concentrated hazardous wastes in underground vaults made of reinforced concrete. By contrast, in the United Kingdom most hazardous wastes are mixed with household garbage and stored in hundreds of conventional landfills throughout the country.

Hazardous wastes can also be stored in carefully designed *aboveground buildings.* This is especially useful in areas where the water table is close to the surface or areas that are above aquifers used for drinking water. These structures are built to withstand storms and to prevent the release of toxic gases. Leaks are monitored and any leakage is collected and treated.

To most environmental scientists, the only solution to the hazardous waste problem is to produce as little as possible in the first place (Figure 13-13)—a prevention

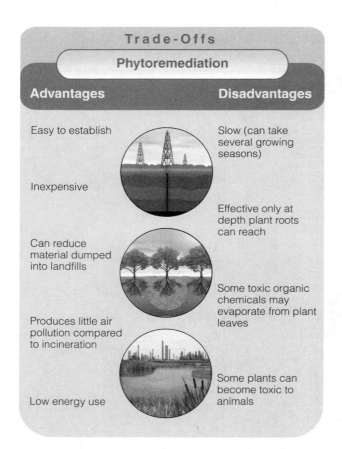

Figure 13-15 Trade-offs: advantages and disadvantages of using *phytoremediation* to remove or detoxify hazardous waste. Pick the single advantage and disadvantage that you think are the most important.

Trade-Offs

Phytoremediation

Advantages	Disadvantages
Easy to establish	Slow (can take several growing seasons)
Inexpensive	
	Effective only at depth plant roots can reach
Can reduce material dumped into landfills	
	Some toxic organic chemicals may evaporate from plant leaves
Produces little air pollution compared to incineration	
Low energy use	Some plants can become toxic to animals

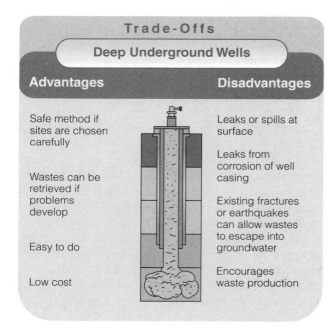

Trade-Offs

Deep Underground Wells

Advantages	Disadvantages
Safe method if sites are chosen carefully	Leaks or spills at surface
	Leaks from corrosion of well casing
Wastes can be retrieved if problems develop	
	Existing fractures or earthquakes can allow wastes to escape into groundwater
Easy to do	
Low cost	Encourages waste production

Figure 13-16 Trade-offs: advantages and disadvantages of injecting liquid hazardous wastes into deep underground wells. Pick the single advantage and disadvantage that you think are the most important.

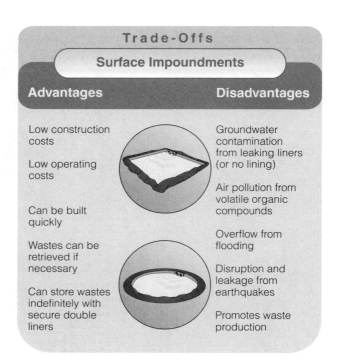

Trade-Offs

Surface Impoundments

Advantages	Disadvantages
Low construction costs	Groundwater contamination from leaking liners (or no lining)
Low operating costs	
	Air pollution from volatile organic compounds
Can be built quickly	
	Overflow from flooding
Wastes can be retrieved if necessary	
	Disruption and leakage from earthquakes
Can store wastes indefinitely with secure double liners	Promotes waste production

Figure 13-17 Trade-offs: advantages and disadvantages of storing liquid hazardous wastes in surface impoundments. Pick the single advantage and disadvantage that you think are the most important.

approach. Figure 13-19 lists some ways you can reduce your output of hazardous waste into the environment.

13-8 CASE STUDY: LEAD

What Is the Threat from Lead? A Toxic Metal

Lead is especially harmful to children and is still used in leaded gasoline and household paints in about 100 countries.

Because it is a chemical element, lead (Pb) does not break down in the environment. Lead is a potent neurotoxin that can harm the nervous system, especially in young children. Each year, 12,000–16,000 American children under age 9 are treated for acute lead poisoning, and about 200 die. About 30% of the survivors suffer from palsy, partial paralysis, blindness, and mental retardation.

Research indicates that children under age 6 and unborn fetuses even with fairly low blood levels of lead are especially vulnerable to nervous system impairment, lowered IQ (by an average of 7.4 points), shortened attention span, hyperactivity, hearing damage, and various behavior disorders.

Good news: Between 1976 and 2000, the percentage of U.S. children ages 1 to 5 with blood lead levels above the safety standard dropped from 85% to 2.2%,

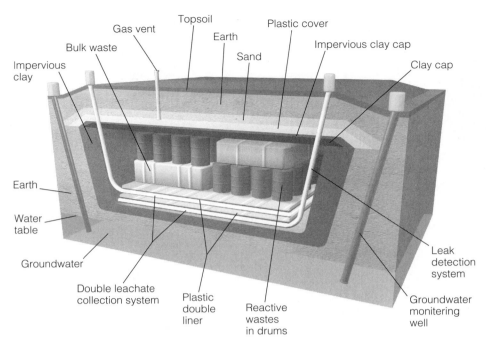

Figure 13-18 Solutions: secure hazardous waste landfill.

What Can You Do?

Hazardous Waste

- Use pesticides in the smallest amount possible.

- Use less harmful substances instead of commercial chemicals for most household cleaners. For example, use liquid ammonia to clean appliances and windows; vinegar to polish metals, clean surfaces, and remove stains and mildew; baking soda to clean household utensils, deodorize, and remove stains; borax to remove stains and mildew.

- Do not dispose of pesticides, paints, solvents, oil, antifreeze, or other products containing hazardous chemicals by flushing them down the toilet, pouring them down the drain, burying them, throwing them into the garbage, or dumping them down storm drains.

Figure 13-19 What can you do? Ways to reduce your input of hazardous waste into the environment.

preventing at least 9 million childhood lead poisonings. The primary reason was that government regulations banned leaded gasoline in 1976 (with complete phaseout by 1986) and lead-based paints in 1970 (but illegal use continued until about 1978)—an excellent example of the power of pollution prevention.

Bad news: Even with the encouraging drop in average blood levels of lead, the U.S. Centers for Disease Control and Prevention estimates that at least 400,000 U.S. children still have unsafe blood levels of lead

caused by exposure from a number of sources. A major source is inhalation or ingestion of lead particles from peeling lead-based paint found in about 38 million houses built before 1960. Lead can also leach from water lines and pipes and faucets containing lead. In addition, a 1993 study by the U.S. National Academy of Sciences and numerous other studies indicate *there is no safe level of lead in children's blood.*

Health scientists have proposed a number of ways to help protect children from lead poisoning, as listed in Figure 13-20 (p. 304).

Although the threat from lead has been reduced in the United States, this is not the case in many developing countries. About 80% of the gasoline sold in the world today is unleaded, but about 100 countries still use leaded gasoline. The World Health Organization (WHO) estimates that 130–200 million children around the world are at risk from lead poisoning, and 15–18 million children in developing countries have permanent brain damage because of lead poisoning—mostly from use of leaded gasoline. *Good news.* China recently phased out leaded gasoline in less than three years.

13-9 ACHIEVING A LOW-WASTE SOCIETY

What Is the Role of Grassroots Action? Making a Difference

In the United States, citizens have kept large numbers of incinerators, landfills, and hazardous waste treatment plants from being built in their local areas.

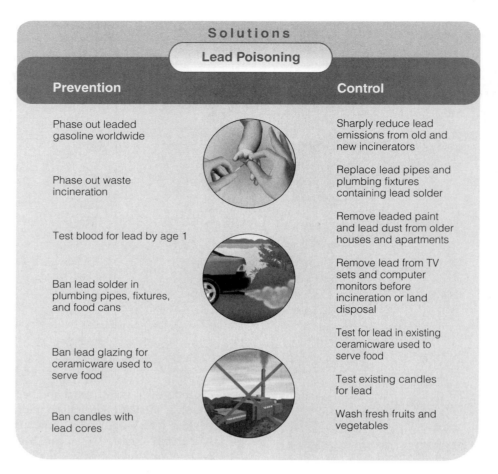

Figure 13-20 Solutions: ways to help protect children from lead poisoning. Which two of these solutions do you believe are the most important?

In the United States, individuals have organized to prevent hundreds of incinerators, landfills, and treatment plants for hazardous and radioactive wastes from being built in or near their communities. Opposition has grown as numerous studies have shown that such facilities have traditionally been located in communities populated mostly by African Americans, Asian Americans, Latinos, and poor whites. This practice has been cited as an example of *environmental injustice.* See the Guest Essay on this subject by Robert Bullard on the website for this chapter.

Health risks from incinerators and landfills, when averaged over the entire country, are quite low, but the risks for people living near these facilities are much higher. They, not the rest of the population, are the ones whose health, lives, and property values are being threatened.

Manufacturers and waste industry officials point out that something must be done with the toxic and hazardous wastes produced to provide people with certain goods and services. They contend that if local citizens adopt a "not in my back yard" (NIMBY) approach, the waste still ends up in someone's back yard.

Many citizens do not accept this argument. To them, the best way to deal with most toxic or hazardous wastes is to produce much less of them, as suggested by the U.S. National Academy of Sciences (Figure 13-13). For such materials, their goal is "not in anyone's back yard" (NIABY) or "not on planet Earth" (NOPE) by emphasizing pollution prevention and use of the precautionary principle.

What Can Be Done at the International Level? The POPs Treaty

An international treaty calls for phasing out the use of harmful persistent organic pollutants (POPs).

Between 1989 and 1994, an international treaty to limit transfer of hazardous waste from one country to another was developed. And in 2000, delegates from 122 countries completed a global treaty to control 12 *persistent organic pollutants (POPs).*

These widely used toxic chemicals are insoluble in water and soluble in fat. This means that in the fatty tissues of humans and other organisms feeding at high trophic levels in food webs, they can become concentrated to levels hundreds of thousand times higher than in the general environment (Figure 7-10, p. 139). These persistent pollutants can also be transported long distances by wind and water.

The list of 12 chemicals, called the *dirty dozen*, includes DDT and 8 other chlorine-containing persistent pesticides, PCBs, dioxins, and furans. The goals of the treaty are to ban or phase out use of these chemicals and detoxify or isolate stockpiles of them. About 25 countries can continue using DDT to combat malaria until safer alternatives are available.

Environmentalists consider the POPs treaty an important milestone in international environmental law because it uses the *precautionary principle* to manage and reduce the risks from toxic chemicals. This list is expected to grow as scientific studies uncover more evidence of toxic and environmental damage from some of the chemicals we use.

In 2000, the Swedish Parliament enacted a law that by 2020 would ban all chemicals that are persistent and can bioaccumulate in living tissue. This law also requires an industry to perform risk assessments on all old and new chemicals and show that these chemicals are safe to use, as opposed to requiring the government to show they are dangerous. In other words, chemicals are assumed guilty until their innocence can be established—the reverse of the current policy in the United States and most countries. There is strong opposition to this approach in the United States, especially by industries producing potentially dangerous chemicals.

How Can We Make the Transition to a Low-Waste Society? A New Vision

A number of the principles and programs discussed in this chapter can be used to make the transition to a low-waste society during this century.

According to physicist Albert Einstein, "A clever person solves a problem, a wise person avoids it." To prevent pollution and reduce waste, many environmental scientists urge us to understand and live by four key principles:

- Everything is connected.

- There is no "away" for the wastes we produce.

- Dilution is not always the solution to pollution.

- The best and cheapest way to deal with waste and pollution is to produce fewer pollutants and to reuse and recycle most of the materials we use.

Good news: There is growing interest in and use of increased *resource productivity, pollution prevention, eco-industrial systems,* and *service-flow* businesses. In addition, at least 24 countries have *eco-labeling programs* that certify a product or service as having met specified environmental standards.

Such changes start off slowly but can accelerate rapidly as their economic, ecological, and health advantages become more apparent to investors, business leaders, elected officials, and citizens.

The key to addressing the challenge of toxics use and wastes rests on a fairly straightforward principle: harness the innovation and technical ingenuity that has characterized the chemicals industry from its beginning and channel these qualities in a new direction that seeks to detoxify our economy.

ANNE PLATT MCGINN

CRITICAL THINKING

1. Collect all of the trash (excluding food waste) that you generate in a typical week. Measure its total weight and volume. Sort it into major categories such as paper, plastic, metal, and glass. Then weigh each category and calculate the percentage by weight in each category. What percentage by weight of this waste consists of materials that could be recycled or reused? What percentage by weight of the items could you have done without? Tally and compare the results for your entire class.

2. Are you for or against bringing about an ecoindustrial revolution in the country and community where you live? Explain. Do you believe it will be possible to phase in such a revolution over the next two to three decades? Explain.

3. Are you for or against shifting to a service-flow economy in the country and community where you live? Explain. Do you believe it will be possible to shift to such an economy over the next two to three decades? Explain. What are the three most important strategies for doing this?

4. Would you oppose having a hazardous waste landfill, waste treatment plant, deep-injection well, or incinerator in your community? Explain. If you oppose these disposal facilities, how do you believe the hazardous waste generated in your community and your state should be managed?

5. Give your reasons for agreeing or disagreeing with each of the following proposals for dealing with hazardous waste:
   a. Reduce the production of hazardous waste and encourage recycling and reuse of hazardous materials by charging producers a tax or fee for each unit of waste generated.

b. Ban all land disposal and incineration of hazardous waste to encourage recycling, reuse, and treatment, and to protect air, water, and soil from contamination.

c. Provide low-interest loans, tax breaks, and other financial incentives to encourage industries producing hazardous waste to reduce, recycle, reuse, treat, and decompose such waste.

6. Congratulations! You are in charge of the world. List the three most important components of your strategy for dealing with **(a)** solid waste and **(b)** hazardous waste.

LEARNING ONLINE

The website for this book contains helpful study aids and many ideas for further reading and research. They include a chapter summary, review questions for the entire chapter, flash cards for key terms and concepts, a multiple-choice practice quiz, interesting Internet sites, references, and a guide for accessing thousands of InfoTrac® College Edition articles. Log on to

http://biology.brookscole.com/miller7

Then click on the Chapter-by-Chapter area, choose Chapter 13, and select a learning resource.

14 ECONOMICS, POLITICS, WORLDVIEWS, AND THE ENVIRONMENT

The main ingredients of an environmental ethic are caring about the planet and all of its inhabitants, allowing unselfishness to control the immediate self-interest that harms others, and living each day so as to leave the lightest possible footprints on the planet.

ROBERT CAHN

14-1 ECONOMIC RESOURCES AND ENVIRONMENTAL PROBLEMS

What Supports and Drives Economies? Three Types of Capital

An economic system produces goods and services by using natural, human, and manufactured resources.

An **economic system** is a social institution through which goods and services are produced, distributed, and consumed to satisfy people's unlimited wants in the most efficient possible way. In a *market-based economic system,* buyers (demanders) and sellers (suppliers) interact in *markets* to make economic decisions about which goods and services are produced, distributed, and consumed.

Three types of resources are used to produce goods and services (Figure 14-1). One is **natural resources,** or **natural capital,** the goods and services produced by the earth's natural processes, which support all economies and all life (top half of back cover). See the Guest Essay on natural capital by Paul Hawken on the website for this chapter.

A second resource type is **human resources** or **human capital,** people's physical and mental talents that provide labor, innovation, culture, and organization. A third type is **manufactured resources,** or **manufactured capital,** items such as machinery, equipment, and factories made from natural resources with the help of human resources.

How Do Economists Differ in Their View of Market-Based Economic Systems? The Importance of Natural Capital

Economists differ on the importance of natural capital and on the long-term sustainability of economic growth.

Neoclassical economists such as Milton Friedman and Robert Samuelson view natural resources as important but not vital because of our ability to find substitutes for scarce resources and ecosystem services. They also contend that economic growth is necessary, desirable, and essentially unlimited.

Ecological economists such as Herman Daly and Robert Costanza view economic systems as subsystems of the environment that depend heavily on the earth's irreplaceable natural resources (Figure 14-2, p. 308). They point out that there are no substitutes for many natural resources, such as air, water, fertile soil, and biodiversity. They also believe that conventional economic growth eventually is unsustainable because it can deplete or degrade the natural resources on which economic systems depend.

In the middle of this debate are *environmental economists.* They generally agree with ecological economists that some forms of economic growth are not

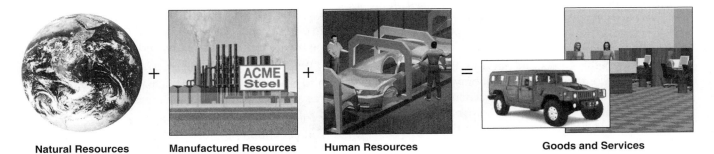

Natural Resources + **Manufactured Resources** + **Human Resources** = **Goods and Services**

Figure 14-1 Three types of resources are used to produce goods and services.

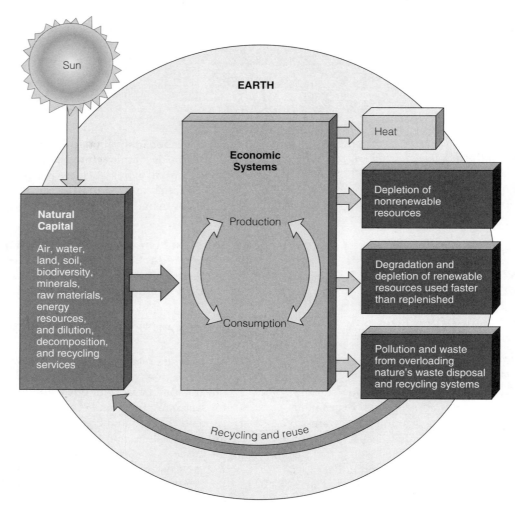

Figure 14-2 *Ecological economists* see all economies as human subsystems that depend on resources and services provided by the sun and the earth's natural resources.

sustainable. But they believe we can modify the principles of neoclassical economics and reform current economic systems, rather than redesigning them to provide more environmentally sustainable economic development.

Ecological and environmental economists call for making a shift from our current economy based on unlimited economic growth to a more *environmentally sustainable economy,* or *eco-economy* over the next several decades. See the Guest Essay on this topic by Herman Daly on the website for this chapter. Figure 14-3 shows some of the goods and services these economists would encourage in an eco-economy.

Ecological and environmental economists have suggested a number of strategies to help make the shift to a more sustainable eco-economy over the next several decades.

■ Use indicators that monitor economic and environmental health.

■ Include in the market prices of goods and services their estimated harmful effects on the environment and human health (full-cost pricing).

■ Phase out environmentally harmful government subsidies and tax breaks while increasing such breaks for environmentally friendly activities, goods and services.

■ Shift taxes by lowering taxes on income and wealth in order to increase *green taxes and fees*—those levied on pollution, resource waste, and environmentally harmful goods and services.

■ Pass laws and regulations to prevent pollution and resource depletion in certain areas.

■ Use tradable permits or rights to pollute or use resources within programs that limit overall pollution and resource use in given areas.

■ Reduce poverty, one of the major ultimate causes of resource depletion and pollution.

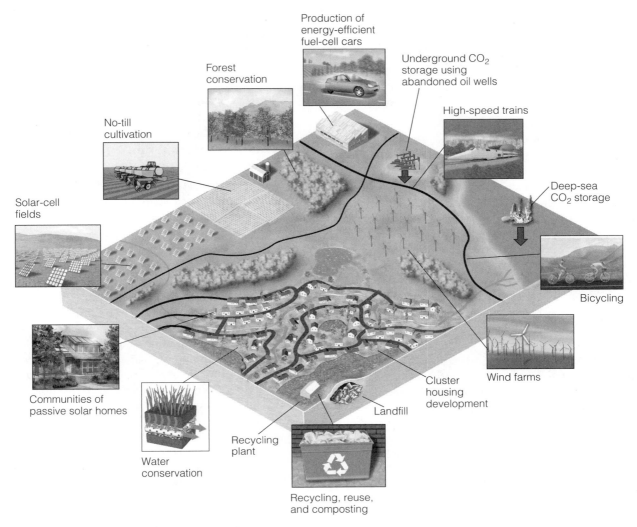

No-till
cultivation

Solar-cell
fields

Forest
conservation

Production of
energy-efficient
fuel-cell cars

Underground CO_2
storage using
abandoned oil wells

High-speed trains

Deep-sea
CO_2 storage

Bicycling

Communities of
passive solar homes

Water
conservation

Recycling
plant

Recycling, reuse,
and composting

Landfill

Cluster
housing
development

Wind farms

Figure 14-3 Solutions: Some components of more environmentally sustainable economic development favored by ecological economists and environmental economists. The goal is to have economic systems that put more emphasis on conserving and sustaining the air, water, soil, biodiversity, and other natural resources that sustain all life and all economies.

Let's look at these proposed solutions in more detail.

14-2 USING ECONOMICS TO IMPROVE ENVIRONMENTAL QUALITY

How Can We Evaluate Environmental Quality and Human Well-Being? Develop and Use New Indicators

We need new indicators that reflect changing levels of environmental quality and human health.

Gross domestic product (GDP), and *per capita GDP* indicators provide a standardized and useful method for measuring and comparing the economic outputs of nations. The GDP is deliberately designed to measure the annual economic value of all goods and services produced within a country without attempting to make any distinction between goods and services that are environmentally or socially beneficial and those that are harmful. Thus, economists who developed the GDP many decades ago never intended it to be used for measuring environmental quality or human well-being.

Environmental and ecological economists and environmental scientists call for the development of new indicators to help monitor environmental quality and human well-being. One approach is to develop indicators that *add* to the GDP things not counted in the marketplace that enhance environmental quality and human well-being. They would also *subtract* from the GDP the costs of things that lead to a lower quality of life and depletion of natural resources.

One such indicator is the *genuine progress indicator (GPI)*, introduced in 1995 by Redefining Progress, a nonprofit organization that develops economics and policy tools to help evaluate and promote sustainability. (This group also developed the concept of ecological footprints, Figure 1-5, p. 9.) Within the GPI, the estimated value of beneficial transactions that meet basic needs, but in which no money changes hands, are added to the GDP. Examples are unpaid volunteer work, healthcare for family members, childcare, and housework. Then the estimated harmful environmental costs (such as pollution and resource depletion and degradation) and social costs (such as crime) are subtracted from the GDP.

Genuine		benefits not		harmful
progress	= GDP +	included in market	−	environmental
indicator		transactions		and social costs

Figure 14-4 compares the per capita GDP and GPI for the United States between 1950 and 2000. Note that while the per capita GDP rose sharply, the per capita GPI stayed nearly flat and declined slightly between 1975 and 2000.

Other environmental and social indicators are in various stages of development. The GPI and other environmental indicators are far from perfect and include many crude estimates. But without such indicators, we do not know much about what is happening to people,

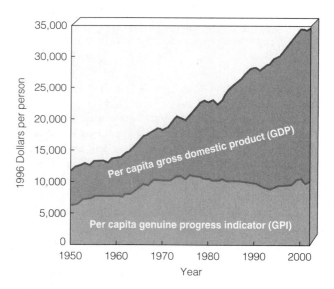

Figure 14-4 Comparison of the per capita gross domestic product (GDP) and per capita genuine progress indicator (GPI) in the United States between 1950 and 2002. (Data from Redefining Progress, 2004)

the environment, and the planet's natural resource base, and we have an effective way to measure what policies work. In effect, according to environmental economists, we are trying to guide national and global economies through treacherous economic and environmental waters at ever-increasing speeds without a good radar system.

What Are Internal and External Costs? Hidden Costs

The direct price you pay for something does not include indirect environmental, health, and other harmful costs associated with its production and use.

All economic goods and services have *internal* or *direct costs* associated with producing them. For example, if you buy a car the direct price you pay includes the costs of raw materials, labor, and shipping, as well as a markup to allow the car company and its dealers some profits. Once you buy the car you must pay additional direct costs for gasoline, maintenance, and repair.

Making, distributing, and using economic goods or services also involve *indirect* or *external costs* not included in their market prices and that affect people other than the buyer and seller. For example, extracting and processing raw materials to make a car use nonrenewable energy and mineral resources, produce solid and hazardous wastes, disturb land, and pollute the air and water. These external costs not included in the price of the car can have short- and long-term harmful effects on other people and on the earth's life-support systems.

Because these harmful costs are not included in the market price, most people do not connect them with car ownership. Still, the car buyer and other people in a society pay these hidden costs sooner or later, in the form of poorer health, higher costs of health care and insurance, higher taxes for pollution control, traffic congestion, and land used for highways and parking.

What Is Full-Cost Pricing? Creating an Environmentally Honest Market

Including external costs in market prices informs consumers about the costs that their purchases have on the earth's life-support systems and human health.

For most economists, creating an *environmentally honest market system* is a way to deal with the harmful costs of goods and services. It requires including the harmful indirect or external costs of goods and services, as much as possible, in the market price of any good or service, such that its price would come as close as possible to its **full cost**—its internal costs plus its external costs. This system would allow consumers to make more informed choices because they would be aware of most or all of the costs involved when they buy something.

With full-cost pricing, some eco-friendly (or green) goods and services that now cost more would eventually cost less because adding external costs to market prices encourages producers to invent more resource-efficient and less-polluting methods of production, thereby cutting their production costs. Jobs would be lost in environmentally harmful businesses as consumers more often choose green products, but jobs would be created in environmentally beneficial businesses. If a shift to full-cost pricing took place over several decades, most environmentally harmful businesses would have time to transform themselves into environmentally beneficial businesses. And consumers would have time to adjust their purchases and buying habits to more environmentally friendly products and services.

Full-cost pricing seems to make a lot of sense. So why is it not used more widely? There are various reasons. One is that many producers of harmful and wasteful goods would have to charge more and some would go out of business. Naturally, they oppose such pricing.

Also, it is difficult to put a price tag on many environmental and health costs. But to ecological and environmental economists, making the best possible estimates is far better than not including such costs in what we pay for most goods and services.

X *How Would You Vote?* Should full-cost pricing be used in setting the market prices of goods and services? Cast your vote online at http://biology.brookscole.com/miller7.

Phasing in such a system requires government action. Few if any companies will volunteer to reduce short-term profits by becoming more environmentally responsible. Governments can use several strategies to encourage or force producers to work toward full-cost pricing. They include phasing out environmentally harmful subsidies, levying taxes on environmentally harmful goods and services, passing laws to regulate pollution and resource depletion, and using tradable permits for pollution or resource use. Let us look at these strategies in more detail.

How Can Ending Certain Subsidies and Tax Breaks Improve Environmental Quality and Reduce Resource Waste? Do Not Reward Environmentally Harmful Activities

We can improve environmental quality and help phase in full-cost pricing by removing environmentally harmful government subsidies and tax breaks.

One way to encourage a shift to full-cost pricing is to *phase out* environmentally harmful subsidies and tax breaks, which cost the world's governments about $1.9 trillion a year, according to studies by Norman Myers and other analysts. Examples include government depletion subsidies and tax breaks for getting minerals and oil out of the ground, cutting timber on public lands, supplying farmers with low-cost water, and encouraging overfishing by providing subsidies and low-cost loans to buy fishing boats.

On paper, phasing out such subsidies may seem like a great idea. But it involves political decisions that often are opposed successfully by powerful interests receiving the subsidies and tax breaks. They want to keep, and if possible increase, these benefits and often oppose subsidies and tax breaks for more environmentally beneficial competitors. Such opposition will only be overcome when enough individuals work together to force elected officials to stop such practices.

Some countries are beginning to reduce environmentally harmful subsidies. Japan, France, and Belgium have phased out all coal subsidies. Germany has cut coal subsidies in half and plans to phase them out completely by 2010. China has cut coal subsidies by about 73% and has imposed a tax on high-sulfur coals.

How Can Green Taxes and Fees and Tax Shifting Improve Environmental Quality and Reduce Resource Waste? Making Polluters and Consumers Pay the Full Price

Taxes and fees on pollution and resource use can take us closer to full-cost pricing, and shifting taxes from wages and profits to pollution and waste helps makes this feasible.

Another way to discourage pollution and resource waste is to *use green taxes* to help include many of the harmful environmental costs of production and consumption in market prices. Taxes can be levied on a per-unit basis on the release of pollution and hazardous or nuclear waste produced, and on the use of fossil fuels, timber, and minerals. Higher fees can also be charged for extracting lumber and minerals from public lands, using water provided by government-financed projects, and using public lands for livestock grazing.

Economists point out two requirements for successful implementation of *green taxes. First,* they should reduce or replace income, payroll, or other taxes. *Second,* the poor and middle class need a safety net to reduce the regressive nature of consumption taxes on essentials such as food, fuel, and housing.

To many analysts, the tax system in most countries is backwards. It can *discourage* what we want more of—jobs, income, and profit-driven innovation—and *encourage* what we want less of—pollution, resource waste, and environmental degradation. A more environmentally sustainable economic system would *lower* taxes on labor, income, and wealth and *raise* taxes on environmentally harmful activities.

With such a shift, for example, a tax on coal would include the increased health costs of breathing polluted air, damages from acid deposition, and estimated costs from climate change. Then taxes on wages and wealth could be reduced by the amount produced by the coal tax.

Some 2,500 economists, including eight Nobel Prize winners, have endorsed the concept of tax shifting. Nine western European countries have begun trial versions of such tax shifting, known as *environmental tax reform*. So far, only a small amount of revenue has been shifted by taxes on emissions of CO_2 and toxic metals, garbage production, and vehicles entering congested cities. But such experience shows that this idea works.

X *How Would You Vote?* Do you favor shifting taxes on wages and profits to pollution and waste? Cast your vote online at http://biology.brookscole.com/miller7.

How Can Environmental Laws and Regulations Improve Environmental Quality and Reduce Resource Waste? Encouraging Innovation

Environmental laws and regulations work best if they motivate companies to find innovative ways to control and prevent pollution and reduce resource waste.

Regulation is a widely used form of government intervention in the marketplace that is used to help control or prevent pollution and reduce resource waste. It involves enacting and enforcing laws and establishing regulations that set pollution standards, regulate harmful activities, ban the release of toxic chemicals into the environment, and require that certain irreplaceable or slowly replenished resources be protected from unsustainable use.

Many environmentalists and business leaders agree that *innovation-friendly regulations* can motivate companies to develop eco-friendly products and industrial processes that can increase profits and competitiveness in national and international markets. But they also agree that some pollution control regulations are too costly and discourage innovation. Examples are regulations that concentrate on cleanup instead of prevention, are too prescriptive (for example, by mandating specific technologies), set compliance deadlines that are too short to allow companies to find innovative solutions, and discourage risk taking and experimentation.

Experience shows that an innovation-friendly regulatory process emphasizes pollution prevention and waste reduction and requires industry and environmental interests to work together in developing realistic standards and timetables. It sets goals, frees industries to meet them in any way that works, and allows enough time for innovation.

For many years, companies mostly resisted environmental regulation and developed an adversarial relationship with government regulators. In recent years, a growing number of companies have realized the economic and competitive advantages of making environmental improvements and recognized that their shareholder value depends in part on having a good environmental record. As a result, a number of firms have begun looking for innovative and profitable ways to reduce resource use, pollution, and waste (Individuals Matter, p. 291). At the same time, many consumers have begun buying green products.

Should We Rely More on Tradable Pollution and Resource-Use Permits? The Marketplace Can Work.

The government can set a limit on pollution emissions or use of a resource, give pollution or resource-use permits to users, and allow them to trade their permits in the marketplace.

A market approach is for the government to *grant tradable pollution and resource-use permits.* The government sets a limit or cap on total emissions of a pollutant or use of a resource such as a fishery. Then it issues or auctions permits that allocate the total among manufacturers or users.

A permit holder not using its entire allocation can use the permit as a credit against future expansion, use it in another part of its operation, or sell it to other companies. In the United States, this approach has been used to reduce the emissions of sulfur dioxide and several other air pollutants, as discussed on p. 283. Tradable rights can also be established among countries to help preserve biodiversity and reduce emissions of greenhouse gases and other pollutants with harmful regional or global effects.

The effectiveness of such programs depends on how high or low the initial cap is set and the rate at which the cap is reduced. However, this approach can allow major polluters and resource wasters to buy their way out instead of making technological improvements. This can create hot spots of pollution and land degradation.

14-3 REDUCING POVERTY TO IMPROVE ENVIRONMENTAL QUALITY AND HUMAN WELL-BEING

How Is the World's Wealth Distributed? Flowing Up to the Rich

Since 1960, most of the financial benefits of global economic growth have flowed up to the rich rather than down to the poor.

Poverty usually is defined as the inability to meet one's basic economic needs. According to a 2000 World Bank study, half of humanity is trying to live on less than $3 (U.S.) a day and one of every five people on the planet is struggling to survive on an income of roughly $1 (U.S.) per day.

Poverty has numerous harmful health and environmental effects (Figure 1-9, p. 12, and Figure 11-5, p. 245) and has been identified as one of the six major causes of the environmental problems we face.

Most neoclassical economists believe a growing economy can help the poor by creating more jobs, enabling more of the wealth created by economic growth to reach workers, and providing greater tax revenues that can be used to help the poor help themselves. Economists call this the *trickle-down* effect.

However, since 1960, most of the benefits of global economic growth as measured by income have flowed up to the rich rather than down to the poor (Figure 14-5). Since 1980, growth of this *wealth gap* has increased. According to Ismail Serageldin, the planet's richest three people have more wealth than the combined GDP of the world's 47 poorest countries. South African President Thabo Mbeki told delegates at the 2003 Johannesburg World Summit on Sustainable Development, "A global human society based on poverty for many and prosperity for a few, characterized by islands of wealth, surrounded by a sea of poverty, is unsustainable."

These trends do not mean that economic growth causes poverty. Instead, they mean that for a variety of reasons rich nations and individuals have not been willing to devote very much of their wealth to helping reduce poverty and its harmful effects on the environment and human well-being. Poverty is also sustained by corruption, the absence of property rights, insufficient legal protection, and the inability of many poor people to borrow money to grow crops or start a small business.

How Can We Reduce Poverty? Help the Poor Help Themselves

We can sharply cut poverty by forgiving the international debts of the poorest countries and greatly increasing international aid and small individual loans to help the poor help themselves.

Analysts point out that reducing poverty requires the governments of most developing countries to make policy changes. One way is to shift more of the national budget to help the rural and urban poor work their way out of poverty. Another is to give villages, villagers, and the urban poor title to common lands and to crops and trees they plant on them.

Analysts suggest that one way to help reduce global poverty is to forgive at least 60% of the $2.4 trillion debt that developing countries owe to developed countries and international lending agencies, and all of the $422 billion debt of the poorest and most heavily indebted countries.

This would be done on the condition that money saved on interest debt is spent on meeting basic human needs. Currently, developing countries pay almost $300 billion per year in interest to developed countries to service this debt.

Developed countries can also:

■ Increase nonmilitary government and private aid with mechanisms to assure that most of it goes directly to the poor to help them become more

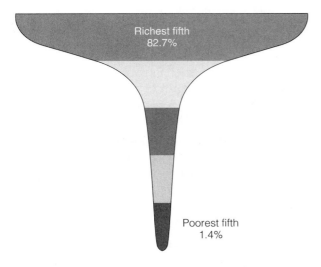

Figure 14-5 Data on the *global distribution of income* show that most of the world's income has flowed up; the richest 20% of the world's population receive more of the world's income than all of the remaining 80%. Each horizontal band in this diagram represents one-fifth of the world's population. This upward flow of global income has accelerated since 1960 and especially since 1980. This trend can increase environmental degradation by increasing average per capita consumption by the richest 20% of the population and causing the poorest 20% of the world's people to survive by using renewable resources faster than they are replenished. (Data from UN Development Programme and Ismail Serageldin, "World Poverty and Hunger—A Challenge for Science," *Science* 296 (2002): 54–58)

Microloans to the Poor

Most of the world's poor desperately want to earn more, become more self-reliant, and have a better life. But they have no credit record. Also, they have few if any assets to use for collateral to secure a loan to buy seeds and fertilizer for farming or tools and materials for a small business.

For almost three decades, an innovative tool called *microlending,* or *microfinance,* has helped deal with this problem. For example, since economist Muhammad Yunus started it in 1976, the Grameen (Village) Bank in Bangladesh has provided more than $4 billion in microloans (varying from $50 to $500) to several million mostly poor, rural, and landless women in 40,000 villages. About 94% of the loans are to women who start their own small

businesses as seamstresses, weavers, bookbinders, or vendors.

To stimulate repayment and provide support, the Grameen Bank organizes microborrowers into five-member "solidarity" groups. If one member of the group misses a weekly payment or defaults on the loan, the other members of the group must make the payments.

The Grameen Bank's experience has shown that microlending is both successful and profitable. For example, less than 3% of microloan repayments to the Grameen Bank are late, and the repayment rate on its loans is an astounding 90–95%, much higher than repayment rate of conventional loans by commercial banks throughout most of the world.

About half of Grameen's borrowers move above the poverty line within 5 years, and domestic violence, divorce, and birth rates are lower among most borrowers.

Microloans to the poor by the Grameen Bank are being used to develop day-care centers, health clinics, reforestation projects, drinking water supply projects, literacy programs, and group insurance programs. Microloans are also being used to bring small-scale solar and wind power systems to rural villages.

Grameen's model has inspired the development of microcredit projects in more than 58 countries that have reached 36 million people (including dependents), and the number is growing rapidly.

Critical Thinking

Why do you think there has been little use of microloans by international development and lending agencies such as the World Bank and the International Monetary Fund? How might this situation be changed?

---

self-reliant and to help provide social safety nets such as welfare, unemployment payments, and pension benefits

- Mount a massive global effort to combat malnutrition and the infectious diseases that kill millions of people prematurely and help perpetuate poverty

- Have lending agencies make small loans to poor people who want to increase their income (Solutions, above)

- Make investments in small-scale infrastructure such as solar-cell power facilities in villages, small scale irrigation projects, and farm-to-market roads

- Make investments in helping sustain and restore the resource bases of fisheries, forests, and small-scale agriculture that provide more than half of the world's jobs

According to the United Nations Development Program (UNDP), it will cost about $50 billion a year to provide universal access to basic services such as education, health, nutrition, family planning, safe water, and sanitation. The UNDP notes that this is less

than 0.1% of the world's annual income and is only a fraction of what the world devotes each year to military spending (Figure 14-6).

How Can We Make a Transition to an Eco-Economy? Copy Nature and Use Full-Cost Pricing

An eco-economy copies nature's four principles of sustainability and uses various economic strategies to help implement full-cost pricing.

An eco-economy mimics the processes that sustain the earth's natural systems (Figure 4-11, p. 74). Figure 14-7 lists principles that Paul Hawken and several other business leaders and economists have suggested for using the sustainability strategies and the various economic tools discussed in this chapter to make the transition to more environmentally sustainable economies over the next several decades. Hawken's simple golden rule for such an economy is this: *"Leave the world better than you found it, take no more than you need, try not to harm life or the environment, and make amends if you do."*

Shifting to eco-economies over the next several decades will require bold leadership by business lead-

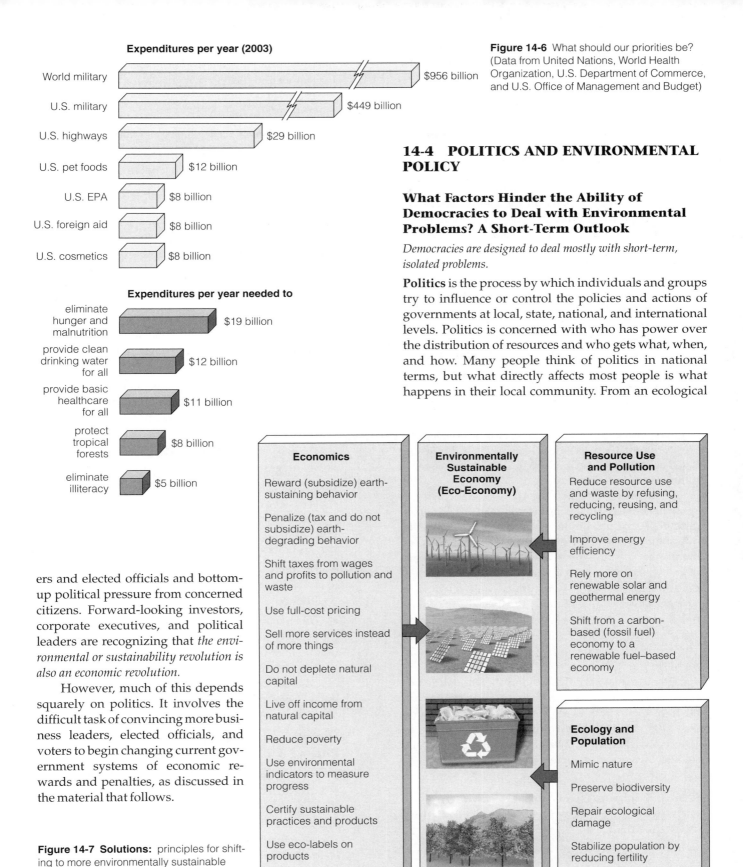

Expenditures per year (2003)

World military	$956 billion
U.S. military	$449 billion
U.S. highways	$29 billion
U.S. pet foods	$12 billion
U.S. EPA	$8 billion
U.S. foreign aid	$8 billion
U.S. cosmetics	$8 billion

Expenditures per year needed to

eliminate hunger and malnutrition	$19 billion
provide clean drinking water for all	$12 billion
provide basic healthcare for all	$11 billion
protect tropical forests	$8 billion
eliminate illiteracy	$5 billion

Figure 14-6 What should our priorities be? (Data from United Nations, World Health Organization, U.S. Department of Commerce, and U.S. Office of Management and Budget)

14-4   POLITICS AND ENVIRONMENTAL POLICY

What Factors Hinder the Ability of Democracies to Deal with Environmental Problems? A Short-Term Outlook

Democracies are designed to deal mostly with short-term, isolated problems.

Politics is the process by which individuals and groups try to influence or control the policies and actions of governments at local, state, national, and international levels. Politics is concerned with who has power over the distribution of resources and who gets what, when, and how. Many people think of politics in national terms, but what directly affects most people is what happens in their local community. From an ecological

ers and elected officials and bottom-up political pressure from concerned citizens. Forward-looking investors, corporate executives, and political leaders are recognizing that *the environmental or sustainability revolution is also an economic revolution.*

However, much of this depends squarely on politics. It involves the difficult task of convincing more business leaders, elected officials, and voters to begin changing current government systems of economic rewards and penalties, as discussed in the material that follows.

Figure 14-7 Solutions: principles for shifting to more environmentally sustainable economies or eco-economies during this century.

Economics

Reward (subsidize) earth-sustaining behavior

Penalize (tax and do not subsidize) earth-degrading behavior

Shift taxes from wages and profits to pollution and waste

Use full-cost pricing

Sell more services instead of more things

Do not deplete natural capital

Live off income from natural capital

Reduce poverty

Use environmental indicators to measure progress

Certify sustainable practices and products

Use eco-labels on products

Environmentally Sustainable Economy (Eco-Economy)

Resource Use and Pollution

Reduce resource use and waste by refusing, reducing, reusing, and recycling

Improve energy efficiency

Rely more on renewable solar and geothermal energy

Shift from a carbon-based (fossil fuel) economy to a renewable fuel–based economy

Ecology and Population

Mimic nature

Preserve biodiversity

Repair ecological damage

Stabilize population by reducing fertility

viewpoint, the ultimate goal of politics is the peaceful achievement of communities that are just and ecologically and economically sustainable.

Democracy is government by the people through elected officials and representatives. In a *constitutional democracy,* a constitution provides the basis of government authority, limits government power by mandating free elections, and guarantees free speech.

Political institutions in constitutional democracies are designed to allow gradual change to ensure economic and political stability. In the United States, for example, rapid and destabilizing change is curbed by a system of checks and balances that distributes power among the three branches of government—*legislative, executive,* and *judicial*—and among federal, state, and local governments.

In passing laws, developing budgets, and formulating regulations, elected and appointed government officials must deal with pressure from many competing *special-interest groups.* Each group advocates passing laws, providing subsidies or tax breaks, or establishing regulations favorable to its cause and weakening or repealing laws and regulations unfavorable to its position.

Some special-interest groups such as corporations are *profit-making organizations,* and others are *nonprofit nongovernmental organizations (NGOs).* Examples of NGOs are labor unions and mainline and grassroots environmental organizations.

The deliberately stable design of democracies is highly desirable. But several related features of democratic governments hinder their ability to deal with environmental problems. Many problems such as climate change, biodiversity loss, and long-lived hazardous waste have long-range effects, are related to one another, and require integrated long-term solutions emphasizing prevention. But because elections are held every few years, most politicians seeking re-election focus on short-term, isolated problems rather than on complex, interrelated, time-consuming, and long-term problems.

Also, too many political leaders have too little understanding of how the earth's natural systems work and how they support all life, economies, and societies. This lack of ecological literacy is dangerous in these times when we are moving through treacherous ecological waters at an increasing speed.

What Principles Can Guide Us in Making Environmental Policy Decisions? Principles Are Important.

Several principles can guide us in making environmental decisions.

An **environmental policy** consists of laws, rules, and regulations related to an environmental problem that are developed, implemented, and enforced by a particular government agency. Analysts have suggested that legislators and individuals evaluating existing or proposed environmental policy should be guided by several principles:

- *The humility principle:* Our understanding of nature and of the consequences of our actions is quite limited.

- *The reversibility principle:* Try not to do something that cannot be reversed later if the decision turns out to be wrong.

- *The precautionary principle:* When substantial evidence indicates an activity threatens to harm human health or the environment, take precautionary measures to prevent or reduce such harm, even if some of the cause-and-effect relationships are not fully established scientifically. In such cases, it is better to be safe than sorry.

- *The prevention principle:* Whenever possible, make decisions that help prevent a problem from occurring or becoming worse.

- *The polluter pays principle:* Develop regulations and use economic tools such as full-cost pricing to insure that polluters bear the cost of the pollutants and the wastes they produce.

- *The integrative principle:* Make decisions that involve integrated solutions to environmental and other problems.

- *The public participation principle:* Citizens should have open access to environmental data and information and the right to participate in developing, criticizing, and modifying environmental policies.

- *The human rights principle:* All people have a right to an environment that does not harm their health and well-being.

- *The environmental justice principle:* Establish environmental policy so that no group of people bears an unfair share of the harms from pollution, degradation, or the execution of environmental laws. See the Guest Essay on this subject by Robert D. Bullard in the website for this chapter.

How Can Individuals Affect Environmental Policy? All Politics Is Local.

Most improvements in environmental quality are the result of millions of citizens putting pressure on elected officials and of individuals developing innovative solutions to environmental problems.

A major theme of this book is that *individuals matter.* History shows that significant change usually comes from the *bottom up* when individuals join with others to bring about change. Without grassroots political action by millions of individual citizens and organized

What Can You Do?

Influencing Environmental Policy

- Become informed on issues
- Run for office (especially at local level)
- Make your views known at public hearings
- Make your views known to elected representatives
- Contribute money and time to candidates for office
- Vote
- Form or join nongovernment organizations (NGOs) seeking change
- Support reform of election campaign financing

Figure 14-8 What can you do? Ways you can influence environmental policy.

groups, the air you breathe and the water you drink today would be much more polluted, and much more of the earth's biodiversity would have disappeared. Figure 14-8 lists ways you can influence and change government policies in constitutional democracies. Which, if any, of these things do you do?

Unfortunately, an increasing number of Americans and citizens of other countries do not actively participate in politics. Many people recycle, buy environmentally friendly products, and take part in other important and responsible activities that help the environment. But in order to influence environmental policy people need to actively work together to improve communities and neighborhoods, as the citizens of Chattanooga, Tennessee and Curitiba, Brazil (Case Study, p. 99) have done. These and other cases have demonstrated the validity of the insight of Aldo Leopold: "All ethics rest upon a single premise: that the individual is a member of a community of interdependent parts."

What Is Environmental Leadership? An Option for Each of Us

Environmental leaders provide vision, focus, resources, and other types of support to people who want to make changes to environmental policies, and each of us can play a leadership role.

Leaders are persons whom other people choose to follow because of their vision, credibility, courage, or charisma.

History judges political leaders by whether and how they respond to the great issues of their day. Today's political leaders face climate change, loss of biodiversity, poverty, and other related environmental problems. One such politician, Prime Minister Tony Blair of the United Kingdom, believes that environmental degradation is the key issue for this generation and that "climate change is unquestionably the most urgent environmental challenge." He calls for the United Kingdom and other nations of the world to reduce carbon dioxide emissions by 60% by 2050 and for a "new international consensus to protect the environment and combat the devastating impacts of climate change."

Solutions to global environmental problems will require leadership from the United States, the world's wealthiest and most powerful society. However, instead of leading us toward solutions, the majority of elected U.S. officials are leading a charge in the opposite direction. They are weakening environmental laws, withdrawing from international efforts to deal with climate change, weakening national and international biodiversity protection, lowering nonmilitary foreign aid for fighting poverty, and encouraging the use of fossil fuels, instead of increasing energy conservation and relying much more on renewable energy resources. The United States has become mired in the politics of confrontation, polarization, and deadlock on environmental and most other major issues. There are many reasons for this situation, but many analysts say that one is a lack of vision and leadership.

The way out of this quagmire, some analysts say, is for individuals to work together to elect ecologically literate officials who will provide environmental leadership for the United States and the world. Thus, leadership at the grassroots level is perhaps the key to solving a leadership crisis at higher levels.

Each of us can also provide leadership in four ways. One is to *lead by example,* using our own lifestyle to show others that change is possible and beneficial.

A second way is to *work within existing economic and political systems to bring about environmental improvement.* We can influence political decisions by campaigning and voting for candidates and by communicating with elected officials. We can also send a message to companies making harmful environment products or policies by *voting with our wallets* and letting them know what we have done. We can also work within the system by choosing environmental careers. Have you considered an environmental career?

A third form of environmental leadership is for you to *run for some sort of local office.* Look in the mirror. Maybe you are one who can make a difference as an office holder.

A fourth form of environmental leadership involves *proposing and working for better solutions to environmental problems.* Leadership is more than being against something. It also involves coming up with better ways to accomplish various goals and getting

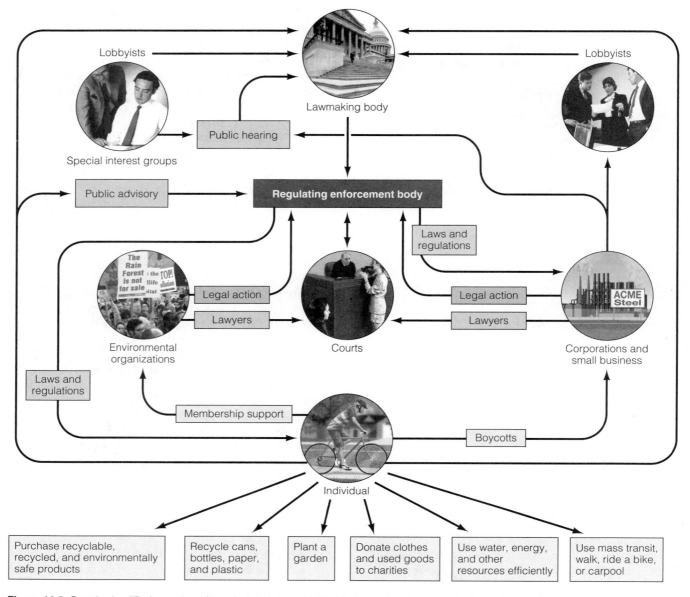

Figure 14-9 Greatly simplified overview of how individuals and lobbyists for and against a particular environmental law interact with the legislative, executive, and judicial branches of government in the United States. The bottom of this diagram also shows some ways in which individuals can bring about environmental change through their own lifestyles. See the website for this book for details on contacting elected representatives.

people to work together to achieve such goals. If we care enough, each of us can make a difference.

Case Study: How Is Environmental Policy Made in the United States? A Complicated and Thorny Process

Formulating, legislating, and executing environmental policy in the United States is a complex, lengthy, and controversial process.

The major function of the federal government in the United States is to develop and implement *policy* for dealing with various issues. Policy is typically composed of various *laws* passed by the legislative branch, *regulations* instituted by the executive branch to put laws into effect, and *funding* to implement and enforce the laws and regulations. Figure 14-9 is a greatly simplified overview of how individuals and lobbyists for and against a particular environmental law interact with the three branches of government in the United States. Trace the flows of information and feedback in this diagram. Figure 14-10 lists some of the major environmental laws passed in the United States since 1969.

Passing an environmental law is not enough to make policy. The next step involves trying to get enough funds appropriated to implement and enforce each law. Indeed, developing and adopting a budget is the most important and controversial activity of the executive and legislative branches.

Once a law has been passed and funded, the appropriate government department or agency must draw up regulations for implementing it. An affected group may take the agency to court for failing to implement and enforce the regulations or for enforcing them too rigidly.

Politics plays an important role in the policies and staffing of environmental regulatory agencies—depending on what political party is in power and the prevailing environmental attitudes. Industries facing environmental regulations often put political pressure on regulatory agencies and lobby to have the president appoint people to high positions in such agencies who come from the industries being regulated. In other words, the regulated try to take over the agencies and become the regulators—described by some as "putting foxes in charge of the henhouse."

In addition, people in regulatory agencies work closely with and often develop friendships with officials in the industries they are regulating. Some industries offer regulatory agency employees high-paying jobs in an attempt to influence their regulatory decisions. This can lead to what is called a *revolving door,* as employees move back and forth between industry and government.

What Are the Roles of Mainline and Grassroots Environmental Groups? Watchdogs and Agents of Change

Environmental groups monitor environmental activities, work to pass and strengthen environmental laws, and work with corporations to find solutions to environmental problems.

The spearhead of the global conservation and environmental movement consists of more than 100,000 nonprofit NGOs working at the international, national, state, and local levels—up from about 2,000 such groups in 1970. The growing influence of these organizations is one of the most important changes influencing environmental decisions and policies.

NGOs range from grassroots groups with just a few members to global organizations like the 5-million-member World Wide Fund for Nature, with offices in 48 countries. Other international groups with large memberships include Greenpeace, the World Wildlife Fund, the Nature Conservancy, Grameen Bank (Solutions, p. 314), and Conservation International.

In the United States, more than 8 million citizens belong to over 30,000 NGOs dealing with environmen-

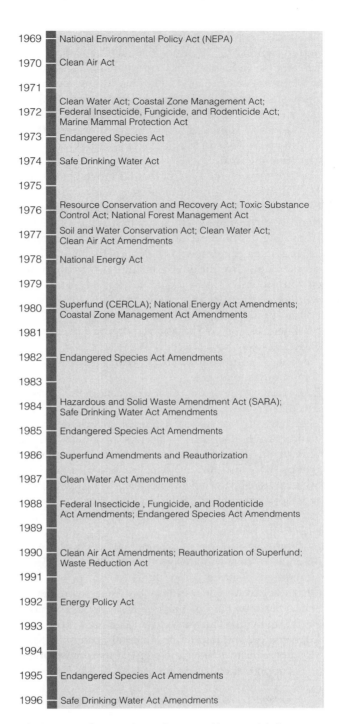

1969	National Environmental Policy Act (NEPA)
1970	Clean Air Act
1971	
1972	Clean Water Act; Coastal Zone Management Act; Federal Insecticide, Fungicide, and Rodenticide Act; Marine Mammal Protection Act
1973	Endangered Species Act
1974	Safe Drinking Water Act
1975	
1976	Resource Conservation and Recovery Act; Toxic Substance Control Act; National Forest Management Act
1977	Soil and Water Conservation Act; Clean Water Act; Clean Air Act Amendments
1978	National Energy Act
1979	
1980	Superfund (CERCLA); National Energy Act Amendments; Coastal Zone Management Act Amendments
1981	
1982	Endangered Species Act Amendments
1983	
1984	Hazardous and Solid Waste Amendment Act (SARA); Safe Drinking Water Act Amendments
1985	Endangered Species Act Amendments
1986	Superfund Amendments and Reauthorization
1987	Clean Water Act Amendments
1988	Federal Insecticide, Fungicide, and Rodenticide Act Amendments; Endangered Species Act Amendments
1989	
1990	Clean Air Act Amendments; Reauthorization of Superfund; Waste Reduction Act
1991	
1992	Energy Policy Act
1993	
1994	
1995	Endangered Species Act Amendments
1996	Safe Drinking Water Act Amendments

Figure 14-10 Some major environmental laws and their amended versions enacted in the United States since 1969. A more detailed list is found on the website for this chapter.

tal issues. They range from small *grassroots* groups to large heavily *funded* groups, staffed by expert lawyers, scientists, and economists.

The large groups have become powerful and important forces within the political system. They have been major forces in persuading the U.S. Congress to pass and strengthen environmental laws (Figure 14-10) and in fighting off attempts to weaken or repeal such laws. Some industries and environmental groups are working together to find solutions to environmental problems. For example, the Environmental Defense has worked with McDonald's to redesign its packaging system to eliminate its plastic hamburger containers, and with General Motors to remove high-pollution cars from the road.

The base of the environmental movement in the United States and throughout the world consists of thousands of grassroots citizens' groups organized to improve environmental quality, often at the local level. According to political analyst Konrad von Moltke, "There isn't a government in the world that would have done anything for the environment if it weren't for the citizen groups." Taken together, a loosely connected worldwide network of grassroots NGOs working for bottom-up political, social, economic, and environmental change can be viewed as an emerging citizen-based *global sustainability movement.*

These groups have worked with individuals and communities to oppose harmful projects such as landfills, waste incinerators, nuclear waste dumps, clearcutting of forests, and various development projects. They have also taken action against environmental injustice. See the Guest Essay on this topic by Robert D. Bullard on the website for this chapter. Grassroots groups have also formed land trusts and other local organizations to save wetlands, forests, farmland, and ranchland from development and they have helped restore forests (Individuals Matter, p. 115), degraded rivers, and wetlands.

Some grassroots environmental groups use the nonviolent and nondestructive tactics of protest marches, tree sitting (Individuals Matter, p. 111), and other devices for generating publicity to help educate and sway members of the public to oppose various environmentally harmful activities. Much more controversial are militant environmental groups that use violent means to achieve their ends. Most environmentalists oppose such tactics.

X *How Would You Vote?* Do you support the use of nonviolent and nondestructive civil disobedience tactics by some environmental groups? Cast your vote online at http://biology .brookscole.com/miller7.

Case Study: Environmental Action by Students in the United States—Making a Difference

Many student environmental groups work to bring about environmental improvements in their schools and local communities.

Since 1988, there has been a boom in environmental awareness on a number of college campuses and public schools across the United States. Most student environmental groups work with members of the faculty and administration to bring about environmental improvements in their schools and local communities.

Many of these groups make environmental audits of their campuses or schools. Then they use the data gathered to propose changes that will make their campus or school more ecologically sustainable, usually saving money in the process. As a result, students have also helped convince almost 80% of universities and colleges in the United States to develop recycling programs.

Students at Oberlin College in Ohio helped design a more sustainable environmental studies building. At Northland College in Wisconsin, students helped design a "green" dorm that features a large wind generator, panels of solar cells, recycled furniture, and waterless (composting) toilets.

At Minnesota's St. Olaf College, students have carried out sustainable agriculture and ecological restoration projects. Students at Brown University studied the impacts of lead and other toxic pollutants in low-income neighborhoods in nearby Providence, Rhode Island.

Such student-spurred environmental activities and research studies are spreading to universities in at least 42 other countries. See Noel Perrin's Guest Essay on environmental activities at U.S. colleges on this chapter's website.

How Successful Have U.S. Environmental Groups and Their Opponents Been? Achievements and Setbacks for Both Groups

Environmental groups have helped educate the public and business and political leaders about environmental issues and pass environmental laws, but an organized movement has undermined many of these efforts.

Since 1970, a variety of environmental groups in the United States and other countries have helped increase public understanding of environmental issues and have gained public support for an array of environmental and resource-use laws in the United States and other countries. In addition, they have helped individuals deal with a number of local environmental problems.

Polls show that about 80% of the U.S. public strongly support environmental laws and regulations and do not want them weakened. But polls also show that less than 10% of the U.S. public views the environment as one of the nation's most pressing problems. As a result, environmental concerns often do not get transferred to the ballot box. As one political scientist put it, "Environmental concerns are like the Florida Everglades, a mile wide but only a few inches deep." And since 1980, a well-organized and well-funded movement has undermined much of the improvement in environmental understanding and support for environmental concerns in the United States.

One problem is that the focus of environmental issues has shifted from easy-to-see dirty smokestacks and burning rivers to more complex and controversial environmental problems such as climate change and biodiversity loss. Explaining such complex issues to the public and mobilizing support for often controversial, long-range solutions to such problems is difficult. See the Guest Essay on environmental reporting by Andrew C. Revkin on the website for this chapter.

Another problem is that many environmentalists have mostly brought bad news. History shows that bearers of bad news are not received well, and opponents of environmentalists have used this to undermine environmental concerns.

History also shows that people are moved to bring about change mostly by an inspiring, positive vision of what the world could be like, one that provides hope for the future. So far, environmentalists with a variety of beliefs and goals have not worked together to develop broad, compelling, and positive visions that can be used as road maps for a more sustainable future for humans and other species.

Instead of using confrontation, some people are working to mediate environmental disputes by getting each side to listen to one another's concerns, to try to find areas of agreement, and to work together to find solutions. This approach is being used successfully in places such as the Netherlands, the Great Lakes (p. 190), the Chesapeake Bay area (p. 192), and Curitiba, Brazil (p. 99).

The biggest challenge to environmentalists is that despite general public approval, there is strong opposition to many environmental proposals, laws, and regulations by three major groups: some corporate leaders and other powerful people who see environmental laws and regulations as threats to their wealth and power; citizens who see environmental laws and regulations as threats to their private property rights and jobs; and state and local government officials who resent having to implement federal environmental laws and regulations without federal funding (unfunded mandates) or who disagree with certain regulations.

Since 1980, businesses, individuals, and some elected officials, often working in large well-funded groups, have mounted a strong campaign to weaken or repeal existing environmental laws and regulations, change the way in which public lands are used, and destroy the reputation and effectiveness of the U.S. environmental movement.

Because of the efforts of their opponents, major environmental groups in the United States have spent most of their time and money since 1980 trying to prevent existing environmental laws and regulations from being weakened or repealed.

14-5 GLOBAL ENVIRONMENTAL POLICY

Should We Expand the Concept of National and Global Security? The Big Three

Many analysts believe that environmental security is as important as military and economic security.

Countries are legitimately concerned with *military security* and *economic security*. However, ecologists point out that all economies are supported by the earth's natural capital.

According to environmental expert Norman Myers,

> If a nation's environmental foundations are degraded or depleted, its economy may well decline, its social fabric deteriorate, and its political structure become destabilized as growing numbers of people seek to sustain themselves from declining resource stocks. Thus, national security is no longer about fighting forces and weaponry alone. It relates increasingly to watersheds, croplands, forests, genetic resources, climate, and other factors that, taken together, are as crucial to a nation's security as are military factors.

Proponents of Myers' view call for all countries to make environmental security a major focus of diplomacy and government policy at all levels. This would be implemented by a council of advisers made up of highly qualified experts in environmental, economic, and military security who integrate all three security concerns in making major decisions.

X *HOW WOULD YOU VOTE?* Is environmental security just as important as economic and military security? Cast your vote online at http://biology.brookscole.com/miller7.

What Is the Role of International Environmental Organizations? Major Players

International environmental organizations gather and evaluate environmental data, help develop environmental treaties, and provide funds and loans for sustainable economic development.

A symphony of international environmental organizations helps shape and set environmental policy. Perhaps the most influential is the United Nations, with a large family of organizations such as the UN Environment Programme (UNEP), the World Health Organization (WHO), United Nations Development Programme (UNDP), and the Food and Agriculture Organization (FAO).

Other organizations that make or influence environmental decisions are the World Bank, the Global Environment Facility (GEF), and the World Conservation Union (IUCN). The website for this chapter has a list of major international environmental organizations.

These organizations have played important roles in

- Expanding understanding of environmental issues

- Gathering and evaluating environmental data

- Helping develop and monitor international environmental treaties

- Providing funds and loans for sustainable economic development and reducing poverty

- Helping more than 100 nations develop environmental laws and institutions

Despite their often limited funding, these diverse organizations have made important contributions to global and national environmental progress since 1970.

What Progress Has Been Made in Developing International Environmental Cooperation and Policy?

Earth summits and international environmental treaties have played important roles in dealing with global environmental problems, but most treaties are not effectively monitored and enforced.

Since the 1972 UN Conference on the Human Environment in Stockholm, Sweden, progress has been made in addressing environmental issues at the global level. Figure 14-11 lists some of the good and bad news about international efforts to deal with global environmental problems such as poverty, climate change, biodiversity loss, and ocean pollution. Carefully study this figure. Overall, do you believe that the good news outweighs the bad news? Explain.

In 2004, environmental leader Gus Speth argued that global environmental problems are getting worse and that international efforts to solve them are inadequate. He proposes the creation of a World Environmental Organization (WEO), on the order of the World

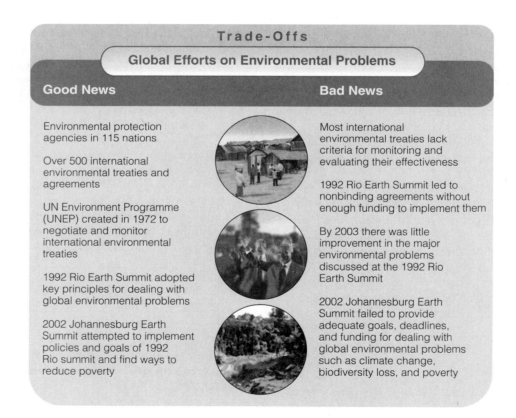

Figure 14-11 Trade-offs: good and bad news about international efforts to deal with global environmental problems. Pick the single piece of good news and bad news that you think are the most important.

Health Organization and the World Trade Organization, to deal with global challenges.

14-6 ENVIRONMENTAL WORLDVIEWS AND ETHICS: CLASHING VALUES AND CULTURES

What Is an Environmental Worldview? Beliefs and Values

Your environmental worldview is how you think the world works, what you believe your role in the world should be, and what you believe is right and wrong environmental behavior.

People disagree about how serious our environmental problems are and what we should do about them. These conflicts arise mostly out of differing **environmental worldviews:** how people think the world works, what they believe their role in the world should be, and what they believe is right and wrong environmental behavior (**environmental ethics**). People with widely differing environmental worldviews can take the same data, be logically consistent, and arrive at quite different conclusions because they start with different assumptions and values.

There are many different types of environmental worldviews. Some are *human centered* (anthropocentric) and others are *life centered* (biocentric, with the primary focus usually on individual species). Other environmental worldviews focus on sustaining the earth's natural life forms (biodiversity), and life-support systems (biosphere) for the benefit of humans and other forms of life.

What Are the Major Human-Centered Environmental Worldviews? We Are In Charge

In the human-centered view, we are the planet's most important species and should manage the earth mostly for our own benefit.

Some people in today's industrial consumer societies have a **planetary management worldview.** According to this anthropocentric worldview, we are the planet's most important and dominant species and we can and should manage the earth mostly for our own benefit. Other species and parts of nature are seen as having only *instrumental value* based on how useful they are to us.

Figure 14-12 (left) summarizes the four major beliefs or assumptions of one version of this worldview.

Environmental Worldviews

Planetary Management

- As the planet's most important species, we are in charge of the earth.

- Because of our ingenuity and technology we will not run out of resources.

- The potential for economic growth is essentially unlimited.

- Our success depends on how well we manage the earth's life-support systems mostly for our benefit.

Stewardship

- We are the planet's most important species but we have an ethical responsibility to care for the rest of nature.

- We will probably not run out of resources, but they should not be wasted.

- We should encourage environmentally beneficial forms of economic growth and discourage environmentally harmful forms.

- Our success depends on how well we manage the earth's life-support systems for our benefit and for the rest of nature.

Environmental Wisdom

- Nature exists for all species and we are not in charge of the earth.

- Resources are limited, should not be wasted, and are not all for us.

- We should encourage earth-sustaining forms of economic growth and discourage earth-degrading forms.

- Our success depends on learning how nature sustains itself and integrating such lessons from nature into the ways we think and act.

Figure 14-12 Comparison of three environmental worldviews. Which of these worldviews comes closer to your own environmental worldview?

Another environmental worldview is the **stewardship worldview**. According to this increasingly popular view, we have an ethical responsibility to be caring and responsible managers, or *stewards*, of the earth. Figure 14-12 (center) summarizes the major beliefs of this worldview.

According to the stewardship view, as we use the earth's natural capital, we are borrowing from the earth and from future generations, and we have an ethical responsibility to pay the debt by leaving the earth in at least as good a condition as we enjoy. In thinking about our responsibility toward future generations, some analysts believe we should consider the wisdom of the 18th century Iroquois Confederation of Native Americans: *In our every deliberation, we must consider the impact of our decisions on the next seven generations.*

Some people believe any human-centered worldview will eventually fail because it wrongly assumes we now have or can gain enough knowledge to become effective managers or stewards of the earth. These critics point out that we do not even know how many species live on the earth, much less what their roles are and how they interact with one another and their nonliving environment. We have only an inkling of what goes on in a handful of soil, a meadow, or any other part of the earth.

Critics believe that human-centered environmental worldviews should be expanded to recognize the *inherent* or *intrinsic value* of all forms of life, regardless of their potential or actual use to us. Most people with a life-centered (biocentric) worldview believe we have an ethical responsibility to avoid causing the premature extinction of species through our activities for three reasons. *First,* each species is a unique storehouse of genetic information that should be respected and protected because it exists (intrinsic value). *Second,* each species is a potential economic good for human use (instrumental value). *Third,* populations of species are capable, through evolution and speciation, of adapting to changing environmental conditions.

Some people believe we must go beyond focusing mostly on species. They believe we have an ethical responsibility not to degrade the earth's ecosystems, biodiversity, and biosphere for this and future generations of humans and other species. In other words, they have an *earth-centered,* or *ecocentric,* environmental worldview, devoted to preserving the earth's biodiversity and the functioning of its life-support systems for all forms of life.

One earth-centered worldview is called the **environmental wisdom worldview.** Figure 14-12 (right) summarizes its major beliefs. In many respects, it is the opposite of the planetary management worldview (Figure 14-12, left). According to this worldview, we are part of, not apart from, the community of life and the ecological processes that sustain all life.

This worldview includes the belief that the earth does not need us managing it in order to go on, whereas we do need the earth in order to survive. According to this view, we cannot save the earth because it does not need saving. What we need to save is the existence of our own species and other species that may become extinct because of our activities. See Lester Milbrath's Guest Essay on this topic on the website for this chapter.

X *HOW WOULD YOU VOTE?* Which one of the following comes closest to your environmental worldview: planetary management, stewardship, or environmental wisdom? Cast your vote online at http://biology.brookscole.com/miller7.

14-7 LIVING MORE SUSTAINABLY

What Are the Main Components of Environmental Literacy? Understanding and Caring about How the Earth Works and Sustains Itself

Environmentally literate citizens and leaders are needed to build more environmentally sustainable and just societies.

Most environmentalists believe that learning how to live more sustainably requires a foundation of environmental education. They cite some of key goals of environmental education or ecological literacy:

- Develop respect or reverence for all life.

- Understand as much as we can about how the earth works and sustains itself, and use such knowledge to guide our lives, communities, and societies.

- Look for connections within the biosphere and between our actions and the biosphere.

- Use critical thinking skills to become seekers of environmental wisdom instead of overfilled vessels of environmental information.

- Understand and evaluate our environmental worldview and see this as a lifelong process.

- Learn how to evaluate the beneficial and harmful consequences to the earth of our choices of lifestyle and profession, today and in the future.

- Foster a desire to make the world a better place and act on this desire.

Specifically, an ecologically literate person should have a basic comprehension of:

- Concepts such as environmental sustainability, natural capital, exponential growth, carrying capacity, and risks and risk analysis

- Environmental history (to help keep us from repeating past mistakes)

- The laws of thermodynamics and the law of conservation of matter
- Basic principles of ecology
- Ways to sustain biodiversity
- Sustainable agriculture and forestry
- Sustainable cities and design
- Sustainable water use
- Nonrenewable and renewable energy resources
- Soil and mineral resources
- Pollution prevention and waste reduction
- Environmentally sustainable economic and political systems
- Environmental ethics

According to environmental educator Mitchell Thomashow, four basic questions should be at the heart of environmental literacy.

First, where do the things I consume come from? *Second*, what do I know about the place where I live? *Third*, how am I connected to the earth and other living things? *Fourth*, what is my purpose and responsibility as a human being?

How we answer these questions determines our *ecological identity*. What are your answers to these four questions?

Figure 14-13 summarizes guidelines and strategies that have been discussed throughout this book for achieving more environmentally sustainable societies.

How Can We Learn from the Earth? Seeking Environmental Wisdom

In addition to formal learning, we need to learn by experiencing nature directly.

Formal environmental education is important, but is it enough? Many analysts say no and urge us to take the time to escape the cultural and technological body armor we use to insulate ourselves from nature and to experience nature directly.

They suggest we kindle a sense of awe, wonder, mystery, and humility by standing under the stars, sitting in a forest, or taking in the majesty and power of an ocean.

We might pick up a handful of soil and try to sense the teeming microscopic life in it that keeps us alive. We might look at a tree, mountain, rock, or bee and try to sense how they are a part of us and we a part of them as interdependent participants in the earth's life-sustaining recycling processes.

Many psychologists believe that consciously or unconsciously, we spend much of our lives in a search for roots: something to anchor us in a bewildering and frightening sea of change. As philosopher Simone Weil

Solutions

Developing Environmentally Sustainable Societies

Guidleines	Strategies
Leave world in as good a shape as—or better than—we found it	Sustain biodiversity
	Eliminate poverty
Do not degrade or deplete the earth's natural capital, and live off the natural income it provides	Develop eco-economies
	Build sustainable communities
Copy nature	Do not use renewable resources faster than nature can replace them
Take no more than we need	Use sustainable agriculture
Do not reduce biodiversity	Depend more on locally available renewable energy from the sun, wind, flowing water, and sustainable biomass
Try not to harm life, air, water, soil	Emphasize pollution prevention and waste reduction
Do not change the world's climate	Do not waste matter and energy resources
Help maintain the earth's capacity for self-repair	Recycle, reuse, and compost 60–80% of matter resources
Do not overshoot the earth's carrying capacity	Maintain a human population size such that needs are met without threatening life-support systems
Repair past ecological damage	Emphasize ecological restoration

Figure 14-13 Solutions: guidelines and strategies for achieving more sustainable societies.

observed, "To be rooted is perhaps the most important and least recognized need of the human soul."

Earth-focused philosophers say that to be rooted, each of us needs to find a *sense of place:* a stream, a mountain, a yard, a neighborhood lot, or any piece of the earth we feel at one with as a place we know, experience emotionally, and love. When we become part of a place, it becomes a part of us. Then we are driven to defend it from harm and to help heal its wounds. This might lead us to recognize that the healing of the earth and the healing of the human spirit are one and the same. We might discover and tap into what Aldo Leopold calls "the green fire that burns in our hearts" and use this as a force for respecting and working with the earth and with one another.

How Can We Live More Simply? Escaping from Affluenza

Some affluent people are voluntarily adopting lifestyles in which they enjoy life more by consuming less.

Many analysts urge us to *learn how to live more simply.* Seeking happiness through the pursuit of material things is considered folly by almost every major religion and philosophy. Yet, it is preached incessantly by modern advertising that encourages us to buy more and more things. As humorist Will Rogers put it, "Too many people spend money they haven't earned to buy things they don't want, to impress people they don't like."

Some affluent people in developed countries are adopting a lifestyle of *voluntary simplicity,* doing and enjoying more with less by learning to live more simply. Voluntary simplicity is based on Mahatma Gandhi's *principle of enoughness:* "The earth provides enough to satisfy every person's need but not every person's greed. . . . When we take more than we need, we are simply taking from each other, borrowing from the future, or destroying the environment and other species."

Most of the world's major religions have similar teachings. "Why do you spend your money for that which is not bread, and your labor for that which does not satisfy?" (Christianity: Old Testament, Isaiah 55:2). "Eat and drink, but waste not by excess" (Islam: Koran 7.31). "One should abstain from acquisitiveness" (Hinduism: Acarangastura 2.119). "He who knows he has enough is rich" (Taoism: Tao Te Ching, Chapter 33).

Implementing this principle means asking oneself, "How much is enough?" This is not easy because people in affluent societies are conditioned to want more and more and to think of wants as basic needs.

How Can We Be More Effective as Environmental Citizens? Avoid Mental Traps, Be Adaptable, and Enjoy Life

We can help make the world a better place by not falling into mental traps that lead to denial and inaction and by keeping our empowering feelings of hope slightly ahead of our immobilizing feelings of despair.

We all make some direct or indirect contributions to the environmental problems we face. However, be-

cause we do not want to feel guilty or bad about the environmental harm our lifestyles may be inflicting, we try not to think about this too much—a path that can lead to denial and inaction.

Analysts suggest that we move beyond fear, denial, apathy, and guilt to more responsible environmental actions in our daily lives by recognizing and avoiding common mental traps that lead to denial, indifference, and inaction. These traps include *gloom-and-doom pessimism* (it is hopeless), *blind technological optimism* (science and technofixes will save us), *fatalism* (we have no control over our actions and the future), *extrapolation to infinity* (if I cannot change the entire world quickly, I will not try to change any of it), *paralysis by analysis* (searching for the perfect worldview, philosophy, solutions, and scientific information before doing anything), and *faith in simple, easy answers.* They also urge us to keep our empowering feelings of hope slightly ahead of our immobilizing feelings of despair.

Recognizing that there is no single correct or best solution to the environmental problems we face is also important. Indeed, one of nature's most important lessons is that preserving diversity—in this case being flexible and adaptable in trying a variety of solutions to our problems—is the best way to adapt to the earth's largely unpredictable, ever-changing conditions.

Finally, we should have fun and take time to enjoy life. Laugh every day and enjoy nature, beauty, friendship, and love. This helps empower us to become good earth citizens who practice Good Earthkeeping.

What Are the Major Components of the Environmental Revolution? A Call for Greatness

The message of environmentalism is not gloom and doom, fear, and catastrophe, but hope, a positive vision of the future, and a call for greatness in dealing with the environmental challenges we face.

The *environmental* or *sustainability revolution* that many environmentalists call for us to bring about during this century would have several components:

- A *biodiversity protection revolution* devoted to protecting and sustaining the genes, species, natural systems, and chemical and biological processes that make up the earth's biodiversity.

- An *efficiency revolution* that minimizes the wasting of matter and energy resources.

- A *solar–hydrogen revolution* based on decreasing our dependence on carbon-based nonrenewable fossil fuels and increasing our dependence on forms of renewable solar energy that can be used to produce hydrogen fuel from water.

- A *pollution prevention revolution* that reduces pollution and environmental degradation from harmful chemicals.

- A *sufficiency revolution* dedicated to meeting the basic needs of all people on the planet while affluent societies learn to live more sustainably but still with quality by living with less.

- A *demographic revolution* based on bringing the size and growth rate of the human population into balance with the earth's ability to support humans and other species sustainably.

- An *economic and political revolution* in which we use economic systems to reward environmentally beneficial behavior and discourage environmentally harmful behavior.

Opponents of such a cultural change like to paint environmentalists as messengers of gloom, doom, and hopelessness. However, *the real message of environmentalism is not gloom and doom, fear, and catastrophe but hope and a positive vision of the future.*

We should rejoice in our environmental accomplishments, but the real question is, Where do we go from here? Can environmental advances be transferred to developing countries? How can humans make the cultural transition to more environmentally sustainable societies based on learning from and working with nature?

As you have read in this book, we have an incredible array of technological and economic solutions to the environmental problems we face. The challenge for environmentalists is to implement such solutions by converting environmental wisdom and beliefs into political action.

This requires understanding that *individuals matter.* Virtually all of the environmental progress we have made during the last few decades occurred because individuals banded together to insist that we can do better.

This journey begins in your own community because in the final analysis *all sustainability is local.* We help make the world more sustainable by working to make our local communities more sustainable. This begins with your own lifestyle. This is the meaning of the motto: "Think globally, act locally."

It is an incredibly exciting time to be alive as we struggle to implement such ideals by entering into a new relationship with the earth that keeps us all alive and supports our economies.

Envision the earth's life-sustaining processes as a beautiful and diverse web of interrelationships—a kaleidoscope of patterns, rhythms, and connections whose very complexity and multitude of possibilities remind us that cooperation, sharing, honesty, humility, compassion, and love should be the guidelines for our behavior toward one another and the earth.

When there is no dream, the people perish.
Proverbs 29:18

CRITICAL THINKING

1. Should we attempt to maximize economic growth by producing and consuming more and more economic goods? Explain. What are the alternatives?

2. Suppose that over the next 20 years the current harmful environmental and health costs of goods and services are internalized so their market prices reflect their total costs. What harmful and beneficial effects might such full-cost pricing have on your lifestyle?

3. List all the goods you use, and then identify those that meet your basic needs and those that satisfy your wants. Identify any economic wants you **(a)** would be willing to give up, **(b)** you believe you should give up but are unwilling to give up, and **(c)** you hope to give up in the future. Relate the results of this analysis to your personal impact on the environment. Compare your results with those of your classmates.

4. What are the greatest strengths and weaknesses of the system of government in your country with respect to **(a)** protecting the environment and **(b)** ensuring environmental justice for all? What three major changes, if any, would you make in this system?

5. This chapter has summarized three major environmental worldviews (Figure 14-12, p. 323). Go through these worldviews and find the beliefs you agree with to describe your own environmental worldview. Which of your beliefs were added or modified as a result of taking this course?

6. Would you or do you use the principle of voluntary simplicity in your life? How?

7. Explain why you agree or disagree with the following ideas: **(a)** everyone has the right to have as many children as they want, **(b)** each member of the human species has a right to use as many resources as they want, **(c)** individuals should have the right to do anything they want with land they own, **(d)** other species exist to be used by humans, and **(e)** all forms of life have an intrinsic value and therefore have a right to exist. Are your answers consistent with the beliefs of your environmental worldview that you described in question 5?

LEARNING ONLINE

The website for this book contains helpful study aids and many ideas for further reading and research. They include a chapter summary, review questions for the entire chapter, flash cards for key terms and concepts, a multiple-choice practice quiz, interesting Internet sites, references, and a guide for accessing thousands of InfoTrac® College Edition articles. Log on to

http://biology.brookscole.com/miller7

Then click on the Chapter-by-Chapter area, choose Chapter 14, and select a learning resource.

APPENDIX 1

UNITS OF MEASUREMENT

LENGTH

Metric

1 kilometer (km) = 1,000 meters (m)
1 meter (m) = 100 centimeters (cm)
1 meter (m) = 1,000 millimeters (mm)
1 centimeter (cm) = 0.01 meter (m)
1 millimeter (mm) = 0.001 meter (m)

English

1 foot (ft) = 12 inches (in)
1 yard (yd) = 3 feet (ft)
1 mile (mi) = 5,280 feet (ft)
1 nautical mile = 1.15 miles

Metric–English

1 kilometer (km) = 0.621 mile (mi)
1 meter (m) = 39.4 inches (in)
1 inch (in) = 2.54 centimeters (cm)
1 foot (ft) = 0.305 meter (m)
1 yard (yd) = 0.914 meter (m)
1 nautical mile = 1.85 kilometers (km)

AREA

Metric

1 square kilometer (km^2) = 1,000,000 square meters (m^2)
1 square meter (m^2) = 1,000,000 square millimeters (mm^2)
1 hectare (ha) = 10,000 square meters (m^2)
1 hectare (ha) = 0.01 square kilometer (km^2)

English

1 square foot (ft^2) = 144 square inches (in^2)
1 square yard (yd^2) = 9 square feet (ft^2)
1 square mile (mi^2) = 27,880,000 square feet (ft^2)
1 acre (ac) = 43,560 square feet (ft^2)

Metric–English

1 hectare (ha) = 2.471 acres (ac)
1 square kilometer (km^2) = 0.386 square mile (mi^2)
1 square meter (m^2) = 1.196 square yards (yd^2)
1 square meter (m^2) = 10.76 square feet (ft^2)
1 square centimeter (cm^2) = 0.155 square inch (in^2)

VOLUME

Metric

1 cubic kilometer (km^3) = 1,000,000,000 cubic meters (m^3)
1 cubic meter (m^3) = 1,000,000 cubic centimeters (cm^3)
1 liter (L) = 1,000 milliliters (mL) = 1,000 cubic centimeters (cm^3)
1 milliliter (mL) = 0.001 liter (L)
1 milliliter (mL) = 1 cubic centimeter (cm^3)

English

1 gallon (gal) = 4 quarts (qt)
1 quart (qt) = 2 pints (pt)

Metric–English

1 liter (L) = 0.265 gallon (gal)
1 liter (L) = 1.06 quarts (qt)
1 liter (L) = 0.0353 cubic foot (ft^3)
1 cubic meter (m^3) = 35.3 cubic feet (ft^3)
1 cubic meter (m^3) = 1.30 cubic yards (yd^3)
1 cubic kilometer (km^3) = 0.24 cubic mile (mi^3)
1 barrel (bbl) = 159 liters (L)
1 barrel (bbl) = 42 U.S. gallons (gal)

MASS

Metric

1 kilogram (kg) = 1,000 grams (g)
1 gram (g) = 1,000 milligrams (mg)
1 gram (g) = 1,000,000 micrograms (μg)
1 milligram (mg) = 0.001 gram (g)
1 microgram (μg) = 0.000001 gram (g)
1 metric ton (mt) = 1,000 kilograms (kg)

English

1 ton (t) = 2,000 pounds (lb)
1 pound (lb) = 16 ounces (oz)

Metric–English

1 metric ton (mt) = 2,200 pounds (lb) = 1.1 tons (t)
1 kilogram (kg) = 2.20 pounds (lb)
1 pound (lb) = 454 grams (g)
1 gram (g) = 0.035 ounce (oz)

ENERGY AND POWER

Metric

1 kilojoule (kJ) = 1,000 joules (J)
1 kilocalorie (kcal) = 1,000 calories (cal)
1 calorie (cal) = 4,184 joules (J)

Metric–English

1 kilojoule (kJ) = 0.949 British thermal unit (Btu)
1 kilojoule (kJ) = 0.000278 kilowatt-hour (kW-h)
1 kilocalorie (kcal) = 3.97 British thermal units (Btu)
1 kilocalorie (kcal) = 0.00116 kilowatt-hour (kW-h)
1 kilowatt-hour (kW-h) = 860 kilocalories (kcal)
1 kilowatt-hour (kW-h) = 3,400 British thermal units (Btu)
1 quad (Q) = 1,050,000,000,000,000 kilojoules (kJ)
1 quad (Q) = 2,930,000,000,000 kilowatt-hours (kW-h)

TEMPERATURE CONVERSIONS

Fahrenheit($°F$) to Celsius($°C$):
$°C = (°F - 32.0) \div 1.80$
Celsius($°C$) to Fahrenheit($°F$):
$°F = (°C \times 1.80) + 32.0$

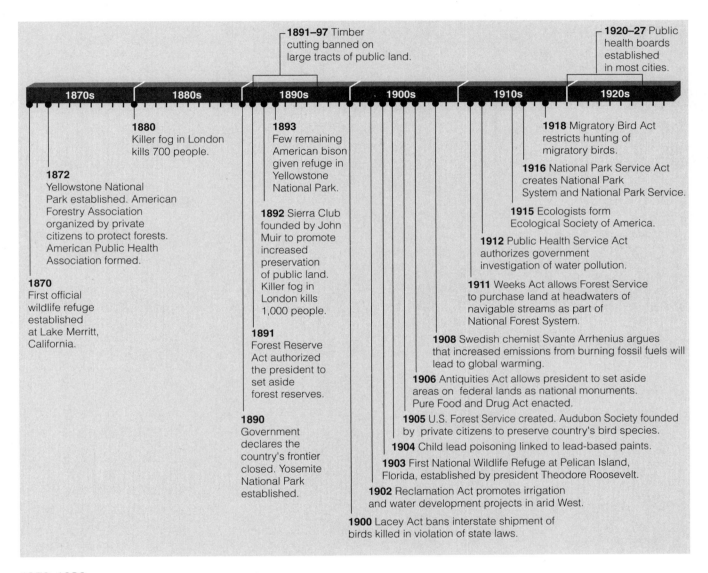

1891–97 Timber cutting banned on large tracts of public land.

1920–27 Public health boards established in most cities.

| 1870s | 1880s | 1890s | 1900s | 1910s | 1920s |

1880 Killer fog in London kills 700 people.

1893 Few remaining American bison given refuge in Yellowstone National Park.

1918 Migratory Bird Act restricts hunting of migratory birds.

1872 Yellowstone National Park established. American Forestry Association organized by private citizens to protect forests. American Public Health Association formed.

1892 Sierra Club founded by John Muir to promote increased preservation of public land. Killer fog in London kills 1,000 people.

1916 National Park Service Act creates National Park System and National Park Service.

1915 Ecologists form Ecological Society of America.

1870 First official wildlife refuge established at Lake Merritt, California.

1891 Forest Reserve Act authorized the president to set aside forest reserves.

1912 Public Health Service Act authorizes government investigation of water pollution.

1911 Weeks Act allows Forest Service to purchase land at headwaters of navigable streams as part of National Forest System.

1908 Swedish chemist Svante Arrhenius argues that increased emissions from burning fossil fuels will lead to global warming.

1890 Government declares the country's frontier closed. Yosemite National Park established.

1906 Antiquities Act allows president to set aside areas on federal lands as national monuments. Pure Food and Drug Act enacted.

1905 U.S. Forest Service created. Audubon Society founded by private citizens to preserve country's bird species.

1904 Child lead poisoning linked to lead-based paints.

1903 First National Wildlife Refuge at Pelican Island, Florida, established by president Theodore Roosevelt.

1902 Reclamation Act promotes irrigation and water development projects in arid West.

1900 Lacey Act bans interstate shipment of birds killed in violation of state laws.

1870–1930

Figure 1 Examples of the increased role of the federal government in resource conservation and public health and establishment of key private environmental groups between 1870 and 1930.

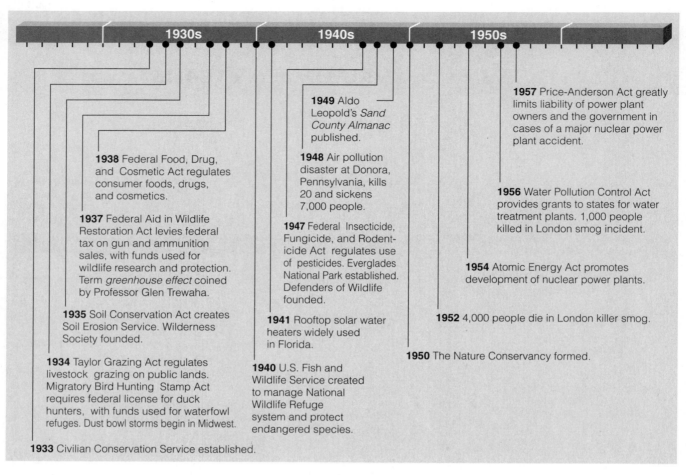

1957 Price-Anderson Act greatly limits liability of power plant owners and the government in cases of a major nuclear power plant accident.

1949 Aldo Leopold's *Sand County Almanac* published.

1938 Federal Food, Drug, and Cosmetic Act regulates consumer foods, drugs, and cosmetics.

1948 Air pollution disaster at Donora, Pennsylvania, kills 20 and sickens 7,000 people.

1956 Water Pollution Control Act provides grants to states for water treatment plants. 1,000 people killed in London smog incident.

1937 Federal Aid in Wildlife Restoration Act levies federal tax on gun and ammunition sales, with funds used for wildlife research and protection. Term *greenhouse effect* coined by Professor Glen Trewaha.

1947 Federal Insecticide, Fungicide, and Rodenticide Act regulates use of pesticides. Everglades National Park established. Defenders of Wildlife founded.

1954 Atomic Energy Act promotes development of nuclear power plants.

1935 Soil Conservation Act creates Soil Erosion Service. Wilderness Society founded.

1941 Rooftop solar water heaters widely used in Florida.

1952 4,000 people die in London killer smog.

1950 The Nature Conservancy formed.

1934 Taylor Grazing Act regulates livestock grazing on public lands. Migratory Bird Hunting Stamp Act requires federal license for duck hunters, with funds used for waterfowl refuges. Dust bowl storms begin in Midwest.

1940 U.S. Fish and Wildlife Service created to manage National Wildlife Refuge system and protect endangered species.

1933 Civilian Conservation Service established.

1930–1960

Figure 2 Some important conservation and environmental events between 1930 and 1960.

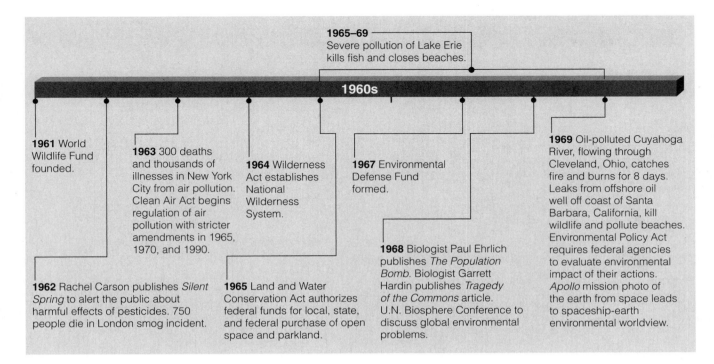

1965–69
Severe pollution of Lake Erie kills fish and closes beaches.

1960s

1961 World Wildlife Fund founded.

1963 300 deaths and thousands of illnesses in New York City from air pollution. Clean Air Act begins regulation of air pollution with stricter amendments in 1965, 1970, and 1990.

1964 Wilderness Act establishes National Wilderness System.

1967 Environmental Defense Fund formed.

1969 Oil-polluted Cuyahoga River, flowing through Cleveland, Ohio, catches fire and burns for 8 days. Leaks from offshore oil well off coast of Santa Barbara, California, kill wildlife and pollute beaches. Environmental Policy Act requires federal agencies to evaluate environmental impact of their actions. *Apollo* mission photo of the earth from space leads to spaceship-earth environmental worldview.

1962 Rachel Carson publishes *Silent Spring* to alert the public about harmful effects of pesticides. 750 people die in London smog incident.

1965 Land and Water Conservation Act authorizes federal funds for local, state, and federal purchase of open space and parkland.

1968 Biologist Paul Ehrlich publishes *The Population Bomb*. Biologist Garrett Hardin publishes *Tragedy of the Commons* article. U.N. Biosphere Conference to discuss global environmental problems.

1960s

Figure 3 Some important environmental events during the 1960s.

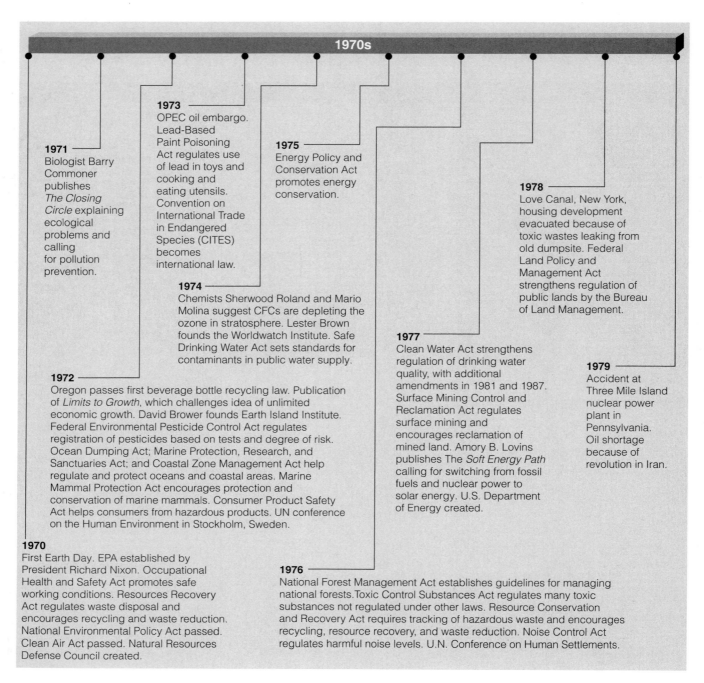

1970s

1971
Biologist Barry Commoner publishes *The Closing Circle* explaining ecological problems and calling for pollution prevention.

1973
OPEC oil embargo. Lead-Based Paint Poisoning Act regulates use of lead in toys and cooking and eating utensils. Convention on International Trade in Endangered Species (CITES) becomes international law.

1975
Energy Policy and Conservation Act promotes energy conservation.

1978
Love Canal, New York, housing development evacuated because of toxic wastes leaking from old dumpsite. Federal Land Policy and Management Act strengthens regulation of public lands by the Bureau of Land Management.

1974
Chemists Sherwood Roland and Mario Molina suggest CFCs are depleting the ozone in stratosphere. Lester Brown founds the Worldwatch Institute. Safe Drinking Water Act sets standards for contaminants in public water supply.

1972
Oregon passes first beverage bottle recycling law. Publication of *Limits to Growth*, which challenges idea of unlimited economic growth. David Brower founds Earth Island Institute. Federal Environmental Pesticide Control Act regulates registration of pesticides based on tests and degree of risk. Ocean Dumping Act; Marine Protection, Research, and Sanctuaries Act; and Coastal Zone Management Act help regulate and protect oceans and coastal areas. Marine Mammal Protection Act encourages protection and conservation of marine mammals. Consumer Product Safety Act helps consumers from hazardous products. UN conference on the Human Environment in Stockholm, Sweden.

1977
Clean Water Act strengthens regulation of drinking water quality, with additional amendments in 1981 and 1987. Surface Mining Control and Reclamation Act regulates surface mining and encourages reclamation of mined land. Amory B. Lovins publishes The *Soft Energy Path* calling for switching from fossil fuels and nuclear power to solar energy. U.S. Department of Energy created.

1979
Accident at Three Mile Island nuclear power plant in Pennsylvania. Oil shortage because of revolution in Iran.

1970
First Earth Day. EPA established by President Richard Nixon. Occupational Health and Safety Act promotes safe working conditions. Resources Recovery Act regulates waste disposal and encourages recycling and waste reduction. National Environmental Policy Act passed. Clean Air Act passed. Natural Resources Defense Council created.

1976
National Forest Management Act establishes guidelines for managing national forests. Toxic Control Substances Act regulates many toxic substances not regulated under other laws. Resource Conservation and Recovery Act requires tracking of hazardous waste and encourages recycling, resource recovery, and waste reduction. Noise Control Act regulates harmful noise levels. U.N. Conference on Human Settlements.

1970s

Figure 4 Some important environmental events during the 1970s, sometimes called the *environmental decade*.

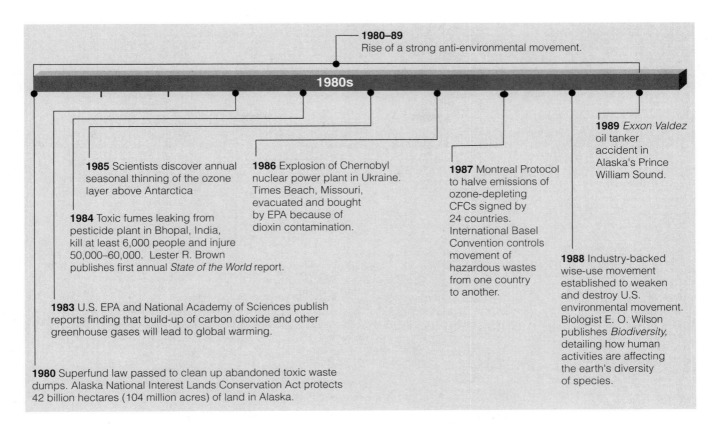

1980–89
Rise of a strong anti-environmental movement.

1980s

1985 Scientists discover annual seasonal thinning of the ozone layer above Antarctica

1986 Explosion of Chernobyl nuclear power plant in Ukraine. Times Beach, Missouri, evacuated and bought by EPA because of dioxin contamination.

1987 Montreal Protocol to halve emissions of ozone-depleting CFCs signed by 24 countries. International Basel Convention controls movement of hazardous wastes from one country to another.

1989 *Exxon Valdez* oil tanker accident in Alaska's Prince William Sound.

1984 Toxic fumes leaking from pesticide plant in Bhopal, India, kill at least 6,000 people and injure 50,000–60,000. Lester R. Brown publishes first annual *State of the World* report.

1988 Industry-backed wise-use movement established to weaken and destroy U.S. environmental movement. Biologist E. O. Wilson publishes *Biodiversity,* detailing how human activities are affecting the earth's diversity of species.

1983 U.S. EPA and National Academy of Sciences publish reports finding that build-up of carbon dioxide and other greenhouse gases will lead to global warming.

1980 Superfund law passed to clean up abandoned toxic waste dumps. Alaska National Interest Lands Conservation Act protects 42 billion hectares (104 million acres) of land in Alaska.

1980s

Figure 5 Some important environmental events during the 1980s.

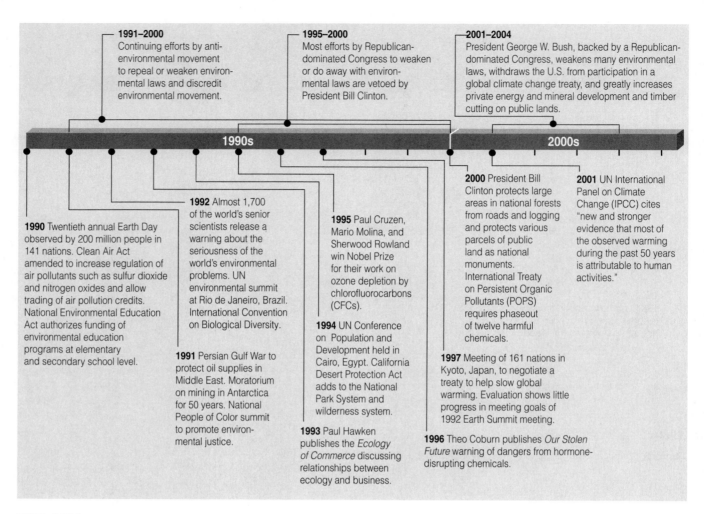

1991–2000
Continuing efforts by anti-environmental movement to repeal or weaken environmental laws and discredit environmental movement.

1995–2000
Most efforts by Republican-dominated Congress to weaken or do away with environmental laws are vetoed by President Bill Clinton.

2001–2004
President George W. Bush, backed by a Republican-dominated Congress, weakens many environmental laws, withdraws the U.S. from participation in a global climate change treaty, and greatly increases private energy and mineral development and timber cutting on public lands.

1990s

2000s

1990 Twentieth annual Earth Day observed by 200 million people in 141 nations. Clean Air Act amended to increase regulation of air pollutants such as sulfur dioxide and nitrogen oxides and allow trading of air pollution credits. National Environmental Education Act authorizes funding of environmental education programs at elementary and secondary school level.

1992 Almost 1,700 of the world's senior scientists release a warning about the seriousness of the world's environmental problems. UN environmental summit at Rio de Janeiro, Brazil. International Convention on Biological Diversity.

1991 Persian Gulf War to protect oil supplies in Middle East. Moratorium on mining in Antarctica for 50 years. National People of Color summit to promote environmental justice.

1995 Paul Cruzen, Mario Molina, and Sherwood Rowland win Nobel Prize for their work on ozone depletion by chlorofluorocarbons (CFCs).

1994 UN Conference on Population and Development held in Cairo, Egypt. California Desert Protection Act adds to the National Park System and wilderness system.

1993 Paul Hawken publishes the *Ecology of Commerce* discussing relationships between ecology and business.

2000 President Bill Clinton protects large areas in national forests from roads and logging and protects various parcels of public land as national monuments. International Treaty on Persistent Organic Pollutants (POPS) requires phaseout of twelve harmful chemicals.

1997 Meeting of 161 nations in Kyoto, Japan, to negotiate a treaty to help slow global warming. Evaluation shows little progress in meeting goals of 1992 Earth Summit meeting.

1996 Theo Coburn publishes *Our Stolen Future* warning of dangers from hormone-disrupting chemicals.

2001 UN International Panel on Climate Change (IPCC) cites "new and stronger evidence that most of the observed warming during the past 50 years is attributable to human activities."

1990–2004

Figure 6 Some important environmental events between 1990 and 2004.

APPENDIX 3

BRIEF HISTORY OF THE AGE OF OIL

Some milestones in the Age of Oil:

- **1905:** Oil supplies 10% of U.S. energy.

- **1925:** United States produces 71% of the world's oil.

- **1930:** Because of an oil glut, oil sells for 10¢ a barrel.

- **1953:** U.S. oil companies account for about half of the world's oil production and the United States is the world's leading oil exporter.

- **1955:** United States has 20% of the world's estimated oil reserves.

- **1960:** OPEC formed so developing countries, with most of the world's known oil and projected oil reserves, can get a higher price for their oil.

- **1973:** United States uses 30% of the world's oil, imports 36% of this oil, and has only 5% of the world's proven oil reserves.

- **1973–1974:** OPEC reduces oil imports to the West and bans oil exports to the U.S. because of its support for Israel in the 18-day Yom Kippur War with Egypt and Syria. World oil prices rise sharply and lead to double-digit inflation in the United States and many other countries and a global economic recession.

- **1975:** Production of estimated U.S. oil reserves peaks.

- **1979:** Iran's Islamic Revolution shuts down most of Iran's oil production and reduces world oil production.

- **1981:** Iran–Iraq war pushes global oil prices to a historic high.

- **1983:** Facing an oil glut, OPEC cuts its oil prices.

- **1985:** U.S. domestic oil production begins to decline and is not expected to increase enough to affect the global price of oil or to reduce U.S. dependence on oil imports.

- **August 1990–June 1991:** United States and its allies fight the Persian Gulf War to oust Iraqi invaders of Kuwait and to protect Western access to Saudi Arabian and Kuwaiti oil supplies.

- **2004:** OPEC has 67% of world oil reserves and produces 40% of the world's oil. U.S. has only 2.9% of oil reserves, uses 26% of the world's oil production, and imports 55% of its oil.

- **2010:** U.S. could be importing at least 61% of the oil it uses as consumption continues to exceed production.

- **2010–2030:** Production of oil from the world's estimated oil reserves is expected to peak. Oil prices expected to increase gradually as the demand for oil increasingly exceeds the supply—unless the world decreases demand by wasting less energy and shifting to other sources of energy.

- **2010–2048:** Domestic U.S. oil reserves projected to be 80% depleted.

- **2042–2083:** Gradual decline in dependence on oil.

GLOSSARY

abiotic   Nonliving. Compare *biotic*.

acid   See *acid solution*.

acid deposition   Falling of acids and acid-forming compounds from the atmosphere to the earth's surface. Acid deposition is commonly known as *acid rain*, a term that refers only to wet deposition of droplets of acids and acid-forming compounds.

acid rain   See *acid deposition*.

acid solution   Any water solution that has more hydrogen ions (H^+) than hydroxide ions (OH^-); any water solution with a pH less than 7. Compare *basic solution, neutral solution*.

active solar heating system   System that uses solar collectors to capture energy from the sun and store it as heat for space heating and water heating. Liquid or air pumped through the collectors transfers the captured heat to a storage system such as an insulated water tank or rock bed. Pumps or fans then distribute the stored heat or hot water throughout a dwelling as needed. Compare *passive solar heating system*.

adaptation   Any genetically controlled structural, physiological, or behavioral characteristic that helps an organism survive and reproduce under a given set of environmental conditions. It usually results from a beneficial mutation. See *biological evolution, differential reproduction, mutation, natural selection*.

adaptive management   Flexible management that views attempts to solve problems as experiments, analyzes failures to see what went wrong, and tries to modify and improve an approach before abandoning it. Because of the inherent unpredictability of complex systems, it often uses the precautionary principle as a management tool. See *precautionary principle*.

adaptive radiation   Process in which numerous new species evolve to fill vacant and new ecological niches in changed environments, usually after a mass extinction. Typically, this takes millions of years.

adaptive trait   See *adaptation*.

aerobic respiration   Complex process that occurs in the cells of most living organisms, in which nutrient organic molecules such as glucose ($C_6H_{12}O_6$) combine with oxygen (O_2) and produce carbon dioxide (CO_2), water (H_2O), and energy. Compare *photosynthesis*.

affluenza   Unsustainable addiction to overconsumption and materialism exhibited in the lifestyles of affluent consumers in the United States and other developed countries.

age structure   Percentage of the population (or number of people of each sex) at each age level in a population.

agroforestry   Planting trees and crops together.

air pollution   One or more chemicals in high enough concentrations in the air to harm humans, other animals, vegetation, or materials. Excess heat and noise can also be considered forms of air pollution. Such chemicals or physical conditions are called air pollutants. See *primary pollutant, secondary pollutant*.

alien species   See *nonnative species*.

alley cropping   Planting of crops in strips with rows of trees or shrubs on each side.

alpha particle   Positively charged matter, consisting of two neutrons and two protons, that is emitted as a form of radioactivity from the nuclei of some radioisotopes. See also *beta particle, gamma rays*.

altitude   Height above sea level. Compare *latitude*.

anaerobic respiration   Form of cellular respiration in which some decomposers get the energy they need through the breakdown of glucose (or other nutrients) in the absence of oxygen. Compare *aerobic respiration*.

ancient forest   See *old-growth forest*.

animal manure   Dung and urine of animals used as a form of organic fertilizer. Compare *green manure*.

anthropocentric   Human-centered. Compare *biocentric*.

aquaculture   Growing and harvesting of fish and shellfish for human use in freshwater ponds, irrigation ditches, and lakes, or in cages or fenced-in areas of coastal lagoons and estuaries. See *fish farming, fish ranching*.

aquatic   Pertaining to water. Compare *terrestrial*.

aquatic life zone   Marine and freshwater portions of the biosphere. Examples include freshwater life zones (such as lakes and streams) and ocean or marine life zones (such as estuaries, coastlines, coral reefs, and the deep ocean).

aquifer   Porous, water-saturated layers of sand, gravel, or bedrock that can yield an economically significant amount of water.

arid   Dry. A desert or other area with an arid climate has little precipitation.

atmosphere   Whole mass of air surrounding the earth. See *stratosphere, troposphere*.

atom   Minute unit made of subatomic particles that is the basic building block of all chemical elements and thus all matter; the smallest unit of an element that can exist and still have the unique characteristics of that element. Compare *ion, molecule*.

atomic number   Number of protons in the nucleus of an atom. Compare *mass number*.

autotroph   See *producer*.

background extinction   Normal extinction of various species as a result of changes in local environmental conditions. Compare *mass depletion, mass extinction*.

basic solution Water solution with more hydroxide ions (OH⁻) than hydrogen ions (H⁺); water solution with a pH greater than 7. Compare *acid solution, neutral solution*.

beta particle Swiftly moving electron emitted by the nucleus of a radioactive isotope. See also *alpha particle, gamma rays*.

bioaccumulation Increase in the concentration of a chemical in specific organs or tissues at a level higher than would normally be expected. Compare *biomagnification*.

biocentric Life centered. Compare *anthropocentric*.

biodegradable Capable of being broken down by decomposers.

biodegradable pollutant Material that can be broken down into simpler substances (elements and compounds) by bacteria or other decomposers. Paper and most organic wastes such as animal manure are biodegradable but can take decades to biodegrade in modern landfills. Compare *degradable pollutant, nondegradable pollutant, slowly degradable pollutant*.

biodiversity Variety of different species (*species diversity*), genetic variability among individuals within each species (*genetic diversity*), variety of ecosystems (*ecological diversity*), and functions such as energy flow and matter cycling needed for the survival of species and biological communities (*functional diversity*).

biofuel Gas or liquid fuel (such as ethyl alcohol) made from plant material (biomass).

biogeochemical cycle Natural processes that recycle nutrients in various chemical forms from the non-living environment to living organisms and then back to the nonliving environment. Examples are the carbon, oxygen, nitrogen, phosphorus, sulfur, and hydrologic cycles.

biological amplification See *biomagnification*.

biological community See *community*.

biological diversity See *biodiversity*.

biological evolution Change in the genetic makeup of a population of a species in successive generations. If continued long enough, it can lead to the formation of a new species. Note that populations—not individuals—evolve. See also *adaptation, differential reproduction, natural selection, theory of evolution*.

biological pest control Control of pest populations by natural predators, parasites, or disease-causing bacteria and viruses (pathogens).

biomagnification Increase in concentration of DDT, PCBs, and other slowly degradable, fat-soluble chemicals in organisms at successively higher trophic levels of a food chain or web. Compare *bioaccumulation*.

biomass Organic matter produced by plants and other photosynthetic producers; total dry weight of all living organisms that can be supported at each trophic level in a food chain or web; dry weight of all organic matter in plants and animals in an ecosystem; plant materials and animal wastes used as fuel.

biome Terrestrial regions inhabited by certain types of life, especially vegetation. Examples are various types of deserts, grasslands, and forests.

biosphere Zone of earth where life is found. It consists of parts of the atmosphere (the troposphere), hydrosphere (mostly surface water and groundwater), and lithosphere (mostly soil and surface rocks and sediments on the bottoms of oceans and other bodies of water) where life is found. Sometimes called the *ecosphere*.

biotic Living organisms. Compare *abiotic*.

biotic potential Maximum rate at which the population of a given species can increase when there are no limits on its rate of growth. See *environmental resistance*.

birth rate See *crude birth rate*.

breeder nuclear fission reactor Nuclear fission reactor that produces more nuclear fuel than it consumes by converting nonfissionable uranium-238 into fissionable plutonium-239.

cancer Group of more than 120 different diseases, one for each type of cell in the human body. Each type of cancer produces a tumor in which cells multiply uncontrollably and invade surrounding tissue.

capitalism See *capitalist market economic system*. Compare *pure command economic system, pure free-market economic system*.

carbon cycle Cyclic movement of carbon in different chemical forms from the environment to organisms and then back to the environment.

carcinogen Chemicals, ionizing radiation, and viruses that cause or promote the development of cancer. See *cancer*. Compare *mutagen, teratogen*.

carnivore Animal that feeds on other animals. Compare *herbivore, omnivore*.

carrying capacity (K) Maximum population of a particular species that a given habitat can support over a given period.

cell Smallest living unit of an organism. Each cell is encased in an outer membrane or wall and contains genetic material (DNA) and other parts to perform its life function. Organisms such as bacteria consist of only one cell, but most of the organisms we are familiar with contain many cells.

CFCs See *chlorofluorocarbons*.

chain reaction Multiple nuclear fissions, taking place within a certain mass of a fissionable isotope, that release an enormous amount of energy in a short time.

chemical One of the millions of different elements and compounds found naturally and synthesized by humans. See *compound, element*.

chemical change Interaction between chemicals in which there is a change in the chemical composition of the elements or compounds involved. Compare *nuclear change, physical change*.

chemical evolution Formation of the earth and its early crust and atmosphere, evolution of the biological molecules necessary for life, and evolution of systems of chemical reactions needed to produce the first living cells. These processes are believed to have occurred about 1 billion years before biological evolution. Compare *biological evolution*.

chemical formula Shorthand way to show the number of atoms (or ions) in the basic structural unit of a compound. Examples are H_2O, $NaCl$, and $C_6H_{12}O_6$.

chemical reaction See *chemical change*.

chemosynthesis Process in which certain organisms (mostly specialized bacteria) extract inorganic compounds from their environment and convert them into organic nutrient compounds without the presence of sunlight. Compare *photosynthesis*.

chlorinated hydrocarbon Organic compound made up of atoms of carbon, hydrogen, and chlorine. Examples are DDT and PCBs.

chlorofluorocarbons (CFCs) Organic compounds made up of atoms of carbon, chlorine, and fluorine. An example is Freon-12 (CCl_2F_2), used as a refrigerant in refrigerators and air conditioners and in making plastics such as Styrofoam. Gaseous CFCs can deplete the ozone layer when they slowly rise into the stratosphere and their chlorine atoms react with ozone molecules. Use of these molecules is being phased out.

chromosome Grouping of various genes and associated proteins in plant and animal cells that carry certain types of genetic information. See *genes*.

clear-cutting Method of timber harvesting in which all trees in a forested area are removed in a single cutting. Compare *seed-tree cutting, selective cutting, shelterwood cutting, strip cutting.*

climate Physical properties of the troposphere of an area based on analysis of its weather records over a long period (at least 30 years). The two main factors determining an area's climate are *temperature,* with its seasonal variations, and the amount and distribution of *precipitation.* Compare *weather.*

climax community See *mature community.*

coal Solid, combustible mixture of organic compounds with 30–98% carbon by weight, mixed with various amounts of water and small amounts of sulfur and nitrogen compounds. It forms in several stages as the remains of plants are subjected to heat and pressure over millions of years.

coal gasification Conversion of solid coal to synthetic natural gas (SNG).

coal liquefaction Conversion of solid coal to a liquid hydrocarbon fuel such as synthetic gasoline or methanol.

coastal wetland Land along a coastline, extending inland from an estuary, that is covered with salt water all or part of the year. Examples are marshes, bays, lagoons, tidal flats, and mangrove swamps. Compare *inland wetland.*

coastal zone Warm, nutrient-rich, shallow part of the ocean that extends from the high-tide mark on land to the edge of a shelflike extension of continental land masses known as the continental shelf. Compare *open sea.*

cogeneration Production of two useful forms of energy, such as high-temperature heat or steam and electricity, from the same fuel source.

commensalism Interaction between organisms of different species in which one type of organism benefits and the other type is neither helped nor harmed to any great degree. Compare *mutualism.*

commercial extinction Depletion of the population of a wild species used as a resource to a level at which it is no longer profitable to harvest the species.

commercial inorganic fertilizer Commercially prepared mixture of plant nutrients such as nitrates, phosphates, and potassium applied to the soil to restore fertility and increase crop yields. Compare *organic fertilizer.*

common-property resource Resource that people normally are free to use; each user can deplete or degrade the available supply. Most are renewable and are owned by no one. Examples are clean air, fish in parts of the ocean not under the control of a coastal country, migratory birds, gases of the lower atmosphere, and the ozone content of the upper atmosphere (stratosphere). See *tragedy of the commons.*

community Populations of all species living and interacting in an area at a particular time.

competition Two or more individual organisms of a single species (*intraspecific competition*) or two or more individuals of different species (*interspecific competition*) attempting to use the same scarce resources in the same ecosystem.

compost Partially decomposed organic plant and animal matter that can be used as a soil conditioner or fertilizer.

compound Combination of atoms, or oppositely charged ions, of two or more different elements held together by attractive forces called chemical bonds. Compare *element.*

concentration Amount of a chemical in a particular volume or weight of air, water, soil, or other medium.

consensus science Scientific data, models, theories, and laws that are widely accepted by scientists considered experts in the area of study. These results of science are very reliable. Compare *frontier science.*

conservation Sensible and careful use of natural resources by humans. People with this view are called *conservationists.*

conservation-tillage farming Crop cultivation in which the soil is disturbed little (minimum-tillage farming) or not at all (no-till farming) to reduce soil erosion, lower labor costs, and save energy. Compare *conventional-tillage farming.*

consumer Organism that cannot synthesize the organic nutrients it needs and gets its organic nutrients by feeding on the tissues of producers or of other consumers; generally divided into *primary consumers* (herbivores), *secondary consumers* (carnivores), *tertiary (higher-level) consumers, omnivores,* and *detritivores* (decomposers and detritus feeders). In economics, one who uses economic goods.

contour farming Plowing and planting across the changing slope of land, rather than in straight lines, to help retain water and reduce soil erosion.

conventional-tillage farming Crop cultivation method in which a planting surface is made by plowing land, breaking up the exposed soil, and then smoothing the surface. Compare *conservation-tillage farming.*

crop rotation Planting a field, or an area of a field, with different crops from year to year to reduce soil nutrient depletion. A plant such as corn, tobacco, or cotton, which removes large amounts of nitrogen from the soil, is planted one year. The next year a legume such as soybeans, which adds nitrogen to the soil, is planted.

crude birth rate Annual number of live births per 1,000 people in the population of a geographic area at the midpoint of a given year. Compare *crude death rate.*

crude death rate Annual number of deaths per 1,000 people in the population of a geographic area at the midpoint of a given year. Compare *crude birth rate.*

crude oil Gooey liquid consisting mostly of hydrocarbon compounds and small amounts of compounds containing oxygen, sulfur, and nitrogen. Extracted from underground accumulations, it is sent to oil refineries, where it is converted to heating oil, diesel fuel, gasoline, tar, and other materials.

cultural eutrophication Overnourishment of aquatic ecosystems with plant nutrients (mostly nitrates and phosphates) because of human activities such as agriculture, urbanization, and discharges from industrial plants and sewage treatment plants. See *eutrophication.*

DDT Dichlorodiphenyltrichloro-ethane, a chlorinated hydrocarbon that has been widely used as a pesticide but is now banned in some countries.

death rate See *crude death rate.*

decomposer Organism that digests parts of dead organisms and cast-off fragments and wastes of living organisms by breaking down the complex organic molecules in those materials into simpler inorganic compounds and then absorbing the soluble nutrients. Producers return most of these chemicals to the soil and water for reuse. Decomposers consist of various bacteria and fungi. Compare *consumer, detritivore, producer.*

deforestation Removal of trees from a forested area without adequate replanting.

degradable pollutant Potentially polluting chemical that is broken down completely or reduced to acceptable levels by natural physical, chemical, and biological processes. Compare *biodegradable pollutant, nondegradable pollutant, slowly degradable pollutant.*

democracy Government by the people through their elected officials and appointed representatives. In a *constitutional democracy,* a constitution provides the basis of government authority and puts restraints on government power through free elections and freely expressed public opinion.

demographic transition Hypothesis that countries, as they become industrialized, have declines in death rates followed by declines in birth rates.

depletion time Time it takes to use a certain fraction, usually 80%, of the known or estimated supply of a nonrenewable resource at an assumed rate of use. Finding and extracting the remaining 20% usually costs more than it is worth.

desalination Purification of salt water or brackish (slightly salty) water by removal of dissolved salts.

desert Biome in which evaporation exceeds precipitation and the average amount of precipitation is less than 25 centimeters (10 inches) a year. Such areas have little vegetation or have widely spaced, mostly low vegetation. Compare *forest, grassland.*

desertification Conversion of rangeland, rain-fed cropland, or irrigated cropland to desertlike land, with a drop in agricultural productivity of 10% or more. It usually is caused by a combi-

nation of overgrazing, soil erosion, prolonged drought, and climate change.

detritivore Consumer organism that feeds on detritus, parts of dead organisms, and cast-off fragments and wastes of living organisms. The two principal types are *detritus feeders* and *decomposers.*

detritus Parts of dead organisms and cast-off fragments and wastes of living organisms.

detritus feeder Organism that extracts nutrients from fragments of dead organisms and their cast-off parts and organic wastes. Examples are earthworms, termites, and crabs. Compare *decomposer.*

developed country Country that is highly industrialized and has a high per capita GNP. Compare *developing country.*

developing country Country that has low to moderate industrialization and low to moderate per capita GNP. Most are located in Africa, Asia, and Latin America. Compare *developed country.*

dieback Sharp reduction in the population of a species when its numbers exceed the carrying capacity of its habitat. See *carrying capacity.*

differential reproduction Phenomenon in which individuals with adaptive genetic traits produce more living offspring than do individuals without such traits. See *natural selection.*

dissolved oxygen (DO) content Amount of oxygen gas (O_2) dissolved in a given volume of water at a particular temperature and pressure, often expressed as a concentration in parts of oxygen per million parts of water.

DNA (deoxyribonucleic acid) Large molecules in the cells of organisms that carry genetic information in living organisms.

dose Amount of a potentially harmful substance an individual ingests, inhales, or absorbs through the skin. Compare *response.* See *dose-response curve, median lethal dose.*

dose-response curve Plot of data showing effects of various dosages of a toxic agent on a group of test organisms. See *dose, median lethal dose response.*

doubling time Time it takes (usually in years) for the quantity of something growing exponentially to double. It can be calculated by dividing the annual percentage growth rate into 70.

drainage basin See *watershed.*

drought Condition in which an area does not get enough water because of lower-than-normal precipitation or higher-than-normal temperatures that increase evaporation.

ecological diversity Variety of forests, deserts, grasslands, oceans, streams, lakes, and other biological communities interacting with one another and with their nonliving environment. See *biodiversity.* Compare *functional diversity, genetic diversity, species diversity.*

ecological efficiency Percentage of energy transferred from one trophic level to another in a food chain or web.

ecological footprint Amount of biologically productive land and water needed to supply each person or population with the renewable resources they use and to absorb or dispose of the wastes from such resource use. It measures the average environmental impact of individuals or populations in different countries and areas.

ecological niche Total way of life or role of a species in an ecosystem. It includes all physical, chemical, and biological conditions a species needs to live and reproduce in an ecosystem.

ecological restoration Deliberate alteration of a degraded habitat or ecosystem to restore as much of its ecological structure and function as possible.

ecological succession Process in which communities of plant and animal species in a particular area are replaced over time by a series of different and often more complex communities. See *primary succession, secondary succession.*

ecology Study of the interactions of living organisms with one another and with their nonliving environment of matter and energy; study of the structure and functions of nature.

economic decision Deciding what goods and services to produce, how to produce them, how much to produce, and how to distribute them to people.

economic depletion Exhaustion of 80% of the estimated supply of a nonrenewable resource. Finding, extracting, and processing the remaining 20% usually costs more than it is worth; may also apply to the depletion of a renewable resource, such as a fish or tree species.

economic development Improvement of living standards by economic growth. Compare *economic growth, environmentally sustainable economic development.*

economic growth   Increase in the capacity to provide people with goods and services produced by an economy; an increase in GDP. Compare *economic development, environmentally sustainable economic development.*

economic resources   Natural resources, human resources, financial resources, and manufactured resources used in an economy to produce material goods and services. See *natural resources.*

economic system   Method that a group of people uses to choose what goods and services to produce, how to produce them, how much to produce, and how to distribute them to people. See *pure free-market economic system.*

economy   System of production, distribution, and consumption of economic goods.

ecosphere   See *biosphere.*

ecosystem   Community of different species interacting with one another and with the chemical and physical factors making up its nonliving environment.

ecosystem services   Natural services or natural resources that support life on the earth and are essential to the quality of human life and the functioning of the world's economies. See *natural resources.*

electromagnetic radiation   Forms of kinetic energy traveling as electromagnetic waves. Examples are radio waves, TV waves, microwaves, infrared radiation, visible light, ultraviolet radiation, X rays, and gamma rays.

electron (e)   Tiny particle moving around outside the nucleus of an atom. Each electron has one unit of negative charge and almost no mass. Compare *neutron, proton.*

element   Chemical, such as hydrogen (H), iron (Fe), sodium (Na), carbon (C), nitrogen (N), or oxygen (O), whose distinctly different atoms serve as the basic building blocks of all matter. Two or more elements combine to form compounds that make up most of the world's matter. Compare *compound.*

endangered species   Wild species with so few individual survivors that the species could soon become extinct in all or most of its natural range. Compare *threatened species.*

endemic species   Species that is found in only one area. Such species are especially vulnerable to extinction.

energy   Capacity to do work by performing mechanical, physical, chemical, or electrical tasks or to cause a heat transfer between two objects at different temperatures.

energy efficiency   Percentage of the total energy input that does useful work and is not converted into low-quality, usually useless heat in an energy conversion system or process. See *energy quality, net energy.* Compare *material efficiency.*

energy productivity   See *energy efficiency.*

energy quality   Ability of a form of energy to do useful work. High-temperature heat and the chemical energy in fossil fuels and nuclear fuels are concentrated high-quality energy. Low-quality energy such as low-temperature heat is dispersed or diluted and cannot do much useful work. See *high-quality energy, low-quality energy.*

environment   All external conditions and factors, living and nonliving (chemicals and energy), that affect an organism or other specified system during its lifetime.

environmental degradation   Depletion or destruction of a potentially renewable resource such as soil, grassland, forest, or wildlife that is used faster than it is naturally replenished. If such use continues, the resource can become nonrenewable (on a human time scale) or nonexistent (extinct). See also *sustainable yield.*

environmental ethics   Human beliefs about what is right or wrong environmental behavior.

environmentalism   A social movement dedicated to protecting the earth's life-support systems for us and other species.

environmentalist   Person concerned about the impact of people on environmental quality who believes some human actions are degrading parts of the earth's life-support systems for humans and many other forms of life.

environmental justice   Fair treatment and meaningful involvement of all people regardless of race, color, sex, national origin, or income with respect to the development, implementation, and enforcement of environmental laws, regulations, and policies.

environmentally sustainable economic development   Development that *encourages* environmentally sustainable forms of economic growth that meet the basic needs of the current

generations of humans and other species without preventing future generations of humans and other species from meeting their basic needs and *discourages* environmentally harmful and unsustainable forms of economic growth. It is the economic component of an *environmentally sustainable society.* Compare *economic development, economic growth.*

environmentally sustainable society   Society that satisfies the basic needs of its people without depleting or degrading its natural resources and thereby preventing current and future generations of humans and other species from meeting their basic needs.

environmental resistance   All the limiting factors that act together to limit the growth of a population. See *biotic potential, limiting factor.*

environmental revolution   Cultural change involving halting population growth and altering lifestyles, political and economic systems, and the way we treat the environment so we can help sustain the earth for ourselves and other species. This involves working with the rest of nature by learning more about how nature sustains itself. See *environmental wisdom worldview, stewardship worldview.*

environmental science   An interdisciplinary study that uses information from the physical sciences and social sciences to learn how the earth works, how we interact with the earth, and how to deal with environmental problems.

environmental scientist   Scientist who uses information from the physical sciences and social sciences to understand how the earth works, learn how humans interact with the earth, and develop solutions to environmental problems.

environmental wisdom worldview   Beliefs that **(1)** nature exists for all the earth's species, not just for us, and we are not in charge of the rest of nature; **(2)** resources are limited, should not be wasted, and are not all for us; **(3)** we should encourage earth-sustaining forms of growth and discourages earth-degrading forms; and **(4)** our success depends on learning to cooperate with one another and with the rest of nature instead of trying to dominate and manage earth's life-support systems primarily for our own use. Compare *planetary management worldview, spaceship-earth worldview, stewardship worldview.*

environmental worldview How people think the world works, what they think their role in the world should be, and what they believe is right and wrong environmental behavior (environmental ethics).

EPA U.S. Environmental Protection Agency; responsible for managing federal efforts to control air and water pollution, radiation and pesticide hazards, environmental research, hazardous waste, and solid waste disposal.

epidemiology Study of the patterns of disease or other harmful effects from toxic exposure within defined groups of people to find out why some people get sick and some do not.

erosion Process or group of processes by which loose or consolidated earth materials are dissolved, loosened, or worn away and removed from one place and deposited in another. See *weathering*.

estuary Partially enclosed coastal area at the mouth of a river where its fresh water, carrying fertile silt and runoff from the land, mixes with salty seawater.

eutrophication Physical, chemical, and biological changes that take place after a lake, estuary, or slow-flowing stream receives inputs of plant nutrients—mostly nitrates and phosphates—from natural erosion and runoff from the surrounding land basin. See *cultural eutrophication*.

even-aged management Method of forest management in which trees, sometimes of a single species in a given stand, are maintained at about the same age and size and are harvested all at once. Compare *uneven-aged management*.

evolution See *biological evolution*.

exhaustible resource See *nonrenewable resource*.

exotic species See *nonnative species*.

exponential growth Growth in which some quantity, such as population size or economic output, increases at a constant rate per unit of time (such as 2% a year); when the increase in quantity over time is plotted, this type of growth yields a curve shaped like the letter J. Compare *linear growth*.

external benefit Beneficial social effect of producing and using an economic good that is not included in the market price of the good. Compare *external cost, full cost*.

external cost Harmful social effect of producing and using an economic good that is not included in the market price of the good. Compare *external benefit, full cost, internal cost*.

externalities Social benefits ("goods") and social costs ("bads") not included in the market price of an economic good. See *external benefit, external cost*. Compare *full cost, internal cost*.

extinction Complete disappearance of a species from the earth. This happens when a species cannot adapt and successfully reproduce under new environmental conditions or when it evolves into one or more new species. Compare *speciation*. See also *endangered species, mass depletion, mass extinction, threatened species*.

family planning Providing information, clinical services, and contraceptives to help people choose the number and spacing of children they want to have.

famine Widespread malnutrition and starvation in a particular area because of a shortage of food, usually caused by drought, war, flood, earthquake, or other catastrophic event that disrupts food production and distribution.

feedlot Confined outdoor or indoor space used to raise hundreds to thousands of domesticated livestock. Compare *rangeland*.

fermentation See *anaerobic respiration*.

fertilizer Substance that adds inorganic or organic plant nutrients to soil and improves its ability to grow crops, trees, or other vegetation. See *commercial inorganic fertilizer, organic fertilizer*.

first law of thermodynamics In any physical or chemical change, no detectable amount of energy is created or destroyed, but in these processes energy can be changed from one form to another; you cannot get more energy out of something than you put in; in terms of energy quantity, you cannot get something for nothing (there is no free lunch). This law does not apply to nuclear changes, in which energy can be produced from small amounts of matter. See also *second law of thermodynamics*.

fishery Concentrations of particular aquatic species suitable for commercial harvesting in a given ocean area or inland body of water.

fish farming Form of aquaculture in which fish are cultivated in a controlled pond or other environment and harvested when they reach the desired size. See also *fish ranching*.

fish ranching Form of aquaculture in which members of a fish species such as salmon are held in captivity for the first few years of their lives, released, and then harvested as adults when they return from the ocean to their freshwater birthplace to spawn. See also *fish farming*.

floodplain Flat valley floor next to a stream channel. For legal purposes the term often is applied to any low area that has the potential for flooding, including certain coastal areas.

flyway Generally fixed route along which waterfowl migrate from one area to another at certain seasons of the year.

food chain Series of organisms in which each eats or decomposes the preceding one. Compare *food web*.

food web Complex network of many interconnected food chains and feeding relationships. Compare *food chain*.

forest Biome with enough average annual precipitation (at least 76 centimeters, or 30 inches) to support growth of various tree species and smaller forms of vegetation. Compare *desert, grassland*.

fossil fuel Products of partial or complete decomposition of plants and animals that occur as crude oil, coal, natural gas, or heavy oils as a result of exposure to heat and pressure in the earth's crust over millions of years. See *coal, crude oil, natural gas*.

fossils Skeletons, bones, skulls, body parts, leaves, seeds, or impressions of such items that provide evidence of organisms that lived long ago.

free-access resources See *common-property resource*.

Freons See *chlorofluorocarbons*.

freshwater life zones Aquatic systems where water with a dissolved salt concentration of less than 1% by volume accumulates on or flows through the surfaces of terrestrial biomes. Examples are *standing* (lentic) bodies of fresh water such as lakes, ponds, and inland wetlands and *flowing* (lotic) systems such as streams and rivers. Compare *biome*.

frontier science Preliminary scientific data, hypotheses, and models that have not been widely tested and accepted. Compare *consensus science*.

full cost Cost of a good when its internal costs and its estimated short-

and long-term external costs are included in its market price. Compare *external cost, internal cost.*

functional diversity Biological and chemical processes or functions such as energy flow and matter cycling needed for the survival of species and biological communities. See *biodiversity, ecological diversity, genetic diversity, species diversity.*

gamma rays Form of electromagnetic radiation with a high energy content emitted by some radioisotopes. They readily penetrate body tissues. See also *alpha particle, beta particle.*

GDP See *gross domestic product.*

gene mutation See *mutation.*

gene pool Sum total of all genes found in the individuals of the population of a particular species.

generalist species Species with a broad ecological niche. They can live in many different places, eat a variety of foods, and tolerate a wide range of environmental conditions. Examples are flies, cockroaches, mice, rats, and human beings. Compare *specialist species.*

genes Coded units of information about specific traits that are passed on from parents to offspring during reproduction. They consist of segments of DNA molecules found in chromosomes.

gene splicing See *genetic engineering.*

genetic adaptation Changes in the genetic makeup of organisms of a species that allow the species to reproduce and gain a competitive advantage under changed environmental conditions. See *differential reproduction, evolution, mutation, natural selection.*

genetically modified organism (GMO) Organism whose genetic makeup has been modified by genetic engineering.

genetic diversity Variability in the genetic makeup among individuals within a single species. See *biodiversity.* Compare *ecological diversity, functional diversity, species diversity.*

genetic engineering Insertion of an alien gene into an organism to give it a beneficial genetic trait. Compare *natural selection.*

geographic isolation Separation of populations of a species for fairly long times into different areas.

geothermal energy Heat transferred from the earth's underground concen-trations of dry steam (steam with no water droplets), wet steam (a mixture of steam and water droplets), or hot water trapped in fractured or porous rock.

global warming Warming of the earth's atmosphere as a result of increases in the concentrations of one or more greenhouse gases primarily as a result of human activities. See *greenhouse effect, greenhouse gases.*

grassland Biome found in regions where moderate annual average precipitation (25–76 centimeters, or 10–30 inches) is enough to support the growth of grass and small plants but not enough to support large stands of trees. Compare *desert, forest.*

greenhouse effect Natural effect that releases heat in the atmosphere (troposphere) near the earth's surface. Water vapor, carbon dioxide, ozone, and several other gases in the lower atmosphere (troposphere) absorb some of the infrared radiation (heat) radiated by the earth's surface. This causes their molecules to vibrate and transform the absorbed energy into longer-wavelength infrared radiation (heat) in the troposphere. If the atmospheric concentrations of these greenhouse gases rise and they are not removed by other natural processes, the average temperature of the lower atmosphere will increase gradually. Compare *global warming.*

greenhouse gases Gases in the earth's lower atmosphere (troposphere) that cause the greenhouse effect. Examples are carbon dioxide, chlorofluorocarbons, ozone, methane, water vapor, and nitrous oxide.

green manure Freshly cut or still-growing green vegetation that is plowed into the soil to increase the organic matter and humus available to support crop growth. Compare *animal manure.*

green revolution Popular term for introduction of scientifically bred or selected varieties of grain (rice, wheat, maize) that, with high enough inputs of fertilizer and water, can greatly increase crop yields.

gross domestic product (GDP) Annual market value of all goods and services produced by all firms and organisms, foreign and domestic, operating within a country.

gross primary productivity (GPP) Rate at which an ecosystem's producers capture and store a given amount of chemical energy as biomass in a given length of time. Compare *net primary productivity.*

groundwater Water that sinks into the soil and is stored in slowly flowing and slowly renewed underground reservoirs called aquifers; underground water in the zone of saturation, below the water table. Compare *runoff, surface water.*

habitat Place or type of place where an organism or population of organisms lives. Compare *ecological niche.*

habitat fragmentation Breakup of a habitat into smaller pieces, usually as a result of human activities.

hazard Something that can cause injury, disease, economic loss, or environmental damage. See also *risk.*

hazardous chemical Chemical that can cause harm because it is flammable or explosive, can irritate or damage the skin or lungs (such as strong acidic or alkaline substances), or can cause allergic reactions of the immune system (allergens). See also *toxic chemical.*

hazardous waste Any solid, liquid, or containerized gas that can catch fire easily, is corrosive to skin tissue or metals, is unstable and can explode or release toxic fumes, or has harmful concentrations of one or more toxic materials that can leach out. See also *toxic waste.*

heat Total kinetic energy of all the randomly moving atoms, ions, or molecules within a given substance, excluding the overall motion of the whole object. Heat always flows spontaneously from a hot sample of matter to a colder sample of matter. This is one way to state the second law of thermodynamics. Compare *temperature.*

herbicide Chemical that kills a plant or inhibits its growth.

herbivore Plant-eating organism. Examples are deer, sheep, grasshoppers, and zooplankton. Compare *carnivore, omnivore.*

heterotroph See *consumer.*

high-input agriculture See *industrialized agriculture.*

high-quality energy Energy that is concentrated and has great ability to perform useful work. Examples are high-temperature heat and the energy in electricity, coal, oil, gasoline, sunlight, and nuclei of uranium-235. Compare *low-quality energy.*

high-quality matter Matter that is concentrated and contains a high concentration of a useful resource. Compare *low-quality matter*.

high-throughput economy Situation in most advanced industrialized countries, in which ever-increasing economic growth is sustained by maximizing the rate at which matter and energy resources are used, with little emphasis on pollution prevention, recycling, reuse, reduction of unnecessary waste, and other forms of resource conservation. Compare *low-throughput economy, matter-recycling economy*.

HIPPO Acronymn for habitat destruction and fragmentation, invasive species, population growth, pollution, and overharvesting.

host Plant or animal on which a parasite feeds.

human resources Physical and mental talents of people used to produce, distribute, and sell an economic good. Compare *manufactured resources, natural resources*.

humus Slightly soluble residue of undigested or partially decomposed organic material in topsoil. This material helps retain water and water-soluble nutrients, which can be taken up by plant roots.

hydrocarbon Organic compound of hydrogen and carbon atoms. The simplest hydrocarbon is methane (CH_4), the major component of natural gas.

hydroelectric power plant Structure in which the energy of falling or flowing water spins a turbine generator to produce electricity.

hydrologic cycle Biogeochemical cycle that collects, purifies, and distributes the earth's fixed supply of water from the environment to living organisms and then back to the environment.

hydropower Electrical energy produced by falling or flowing water. See *hydroelectric power plant*.

hydrosphere The earth's *liquid water* (oceans, lakes, other bodies of surface water, and underground water), *frozen water* (polar ice caps, floating ice caps, and ice in soil, known as permafrost), and *water vapor* in the atmosphere. See also *hydrologic cycle*.

igneous rock Rock formed when molten rock material (magma) wells up from the earth's interior, cools, and solidifies into rock masses. Compare *metamorphic rock, sedimentary rock*. See *rock cycle*.

immigrant species See *nonnative species*.

immigration Migration of people into a country or area to take up permanent residence.

indicator species Species that serve as early warnings that a community or ecosystem is being degraded. Compare *keystone species, native species, nonnative species*.

industrialized agriculture Using large inputs of energy from fossil fuels (especially oil and natural gas), water, fertilizer, and pesticides to produce large quantities of crops and livestock for domestic and foreign sale. Compare *subsistence farming*.

industrial smog Type of air pollution consisting mostly of a mixture of sulfur dioxide, suspended droplets of sulfuric acid formed from some of the sulfur dioxide, and a variety of suspended solid particles. Compare *photochemical smog*.

infant mortality rate Number of babies out of every 1,000 born each year that die before their first birthday.

infiltration Downward movement of water through soil.

inland wetland Land away from the coast, such as a swamp, marsh, or bog, that is covered all or part of the time with fresh water. Compare *coastal wetland*.

inorganic compounds All compounds not classified as organic compounds. See *organic compounds*.

inorganic fertilizer See *commercial inorganic fertilizer*.

input pollution control See *pollution prevention*.

insecticide Chemical that kills insects.

integrated pest management (IPM) Combined use of biological, chemical, and cultivation methods in proper sequence and timing to keep the size of a pest population below the size that causes economically unacceptable loss of a crop or livestock animal.

intercropping Growing two or more different crops at the same time on a plot. For example, a carbohydrate-rich grain that depletes soil nitrogen and a protein-rich legume that adds nitrogen to the soil may be intercropped. Compare *monoculture, polyculture, polyvarietal cultivation*.

internal cost Direct cost paid by the producer and the buyer of an economic good. Compare *external benefit, external cost, full cost*.

interplanting Simultaneously growing a variety of crops on the same plot. See *agroforestry, intercropping, polyculture, polyvarietal cultivation*.

interspecific competition Members of two or more species trying to use the same limited resources in an ecosystem. See *competition*.

intrinsic rate of increase (r) Rate at which a population could grow if it had unlimited resources. Compare *environmental resistance*.

ion Atom or group of atoms with one or more positive (+) or negative (−) electrical charges. Compare *atom, molecule*.

isotopes Two or more forms of a chemical element that have the same number of protons but different mass numbers because they have different numbers of neutrons in their nuclei.

J-shaped curve Curve with a shape similar to that of the letter *J*; can represent prolonged exponential growth.

junk science Scientific results or hypotheses presented as sound science but not having undergone the rigors of the peer review process. Compare *concensus science, frontier science*.

keystone species Species that play roles affecting many other organisms in an ecosystem. Compare *indicator species, native species, nonnative species*.

kinetic energy Energy that matter has because of its mass and speed or velocity. Compare *potential energy*.

lake Large natural body of standing fresh water formed when water from precipitation, land runoff, or groundwater flow fills a depression in the earth.

landfill See *sanitary landfill*.

land-use planning Process for deciding the best present and future use of each parcel of land in an area.

latitude Distance from the equator. Compare *altitude*.

law of conservation of energy See *first law of thermodynamics*.

law of conservation of matter In any physical or chemical change, matter is neither created nor destroyed but merely changed from one form to another; in physical and chemical changes, existing atoms are rearranged into different spatial patterns (physical changes) or different combinations (chemical changes).

law of tolerance Existence, abundance, and distribution of a species in an ecosystem are determined by whether the levels of one or more physical or chemical factors fall within the range tolerated by the species.

LD50 See *median lethal dose*.

LDC See *developing country*.

leaching Process in which various chemicals in upper layers of soil are dissolved and carried to lower layers and, in some cases, to groundwater.

less developed country (LDC) See *developing country*.

life cycle cost Initial cost plus lifetime operating costs of an economic good. Compare *full cost*.

life expectancy Average number of years a newborn infant can be expected to live.

limiting factor Single factor that limits the growth, abundance, or distribution of the population of a species in an ecosystem. See *limiting factor principle*.

limiting factor principle Too much or too little of any abiotic factor can limit or prevent growth of a population of a species in an ecosystem, even if all other factors are at or near the optimum range of tolerance for the species.

linear growth Growth in which a quantity increases by some fixed amount during each unit of time. Compare *exponential growth*.

liquefied natural gas (LNG) Natural gas converted to liquid form by cooling to a very low temperature.

liquefied petroleum gas (LPG) Mixture of liquefied propane (C_3H_8) and butane (C_4H_{10}) gas removed from natural gas and used as a fuel.

lithosphere The earth's crust and upper mantle.

logistic growth Pattern in which exponential population growth occurs when the population is small, and population growth decreases steadily with time as the population approaches the carrying capacity. See *S-shaped curve*.

low-input agriculture See *sustainable agriculture*.

low-quality energy Energy that is dispersed and has little ability to do useful work. An example is low-temperature heat. Compare *high-quality energy*.

low-quality matter Matter that is dilute or dispersed or contains a low concentration of a useful resource. Compare *high-quality matter*.

low-throughput economy Economy based on working with nature by recycling and reusing discarded matter, preventing pollution, conserving matter and energy resources by reducing unnecessary waste and use, not degrading renewable resources, building things that are easy to recycle, reuse, and repair, not allowing population size to exceed the carrying capacity of the environment, and preserving biodiversity. See *environmental worldview*. Compare *high-throughput economy*, *matter-recycling economy*.

low-waste society See *low-throughput economy*.

LPG See *liquefied petroleum gas*.

macroevolution Long-term, large-scale evolutionary changes among groups of species. Compare *microevolution*.

magma Molten rock below the earth's surface.

malnutrition Faulty nutrition, caused by a diet that does not supply an individual with enough protein, essential fats, vitamins, minerals, and other nutrients needed for good health. Compare *overnutrition*, *undernutrition*.

manufactured resources Manufactured items made from natural resources and used to produce and distribute economic goods and services bought by consumers. These include tools, machinery, equipment, factory buildings, and transportation and distribution facilities. Compare *human resources*, *natural resources*.

manure See *animal manure*, *green manure*.

mass Amount of material in an object.

mass depletion Widespread, often global period during which extinction rates are higher than normal but not high enough to classify as a mass extinction. Compare *background extinction*, *mass extinction*.

mass extinction Catastrophic, widespread, often global event in which major groups of species are wiped out over a short time compared with normal (background) extinctions. Compare *background extinction*, *mass depletion*.

mass number Sum of the number of neutrons (n) and the number of protons (p) in the nucleus of an atom. It gives the approximate mass of that atom. Compare *atomic number*.

mass transit Buses, trains, trolleys, and other forms of transportation that carry large numbers of people.

material efficiency Total amount of material needed to produce each unit of goods or services. Also called *resource productivity*. Compare *energy efficiency*.

matter Anything that has mass (the amount of material in an object) and takes up space. On the earth, where gravity is present, we weigh an object to determine its mass.

matter quality Measure of how useful a matter resource is, based on its availability and concentration. See *high-quality matter*, *low-quality matter*.

matter-recycling economy Economy that emphasizes recycling the maximum amount of all resources that can be recycled. The goal is to allow economic growth to continue without depleting matter resources and without producing excessive pollution and environmental degradation. Compare *high-throughput economy*, *low-throughput economy*.

maximum sustainable yield See *sustainable yield*.

MDC See *developed country*.

median lethal dose (LD50) Amount of a toxic material per unit of body weight of test animals that kills half the test population in a certain time.

megacity City with 10 million or more people.

metamorphic rock Rock produced when a preexisting rock is subjected to high temperatures (which may cause it to melt partially), high pressures, chemically active fluids, or a combination of these agents. Compare *igneous rock*, *sedimentary rock*. See *rock cycle*.

metastasis Spread of malignant (cancerous) cells from a tumor to other parts of the body.

microevolution Small genetic changes a population undergoes. Compare *macroevolution*.

microorganisms Organisms such as bacteria that are so small they can be seen only by using a microscope.

mineral Any naturally occurring inorganic substance found in the earth's crust as a crystalline solid. See *mineral resource*.

mineral resource Concentration of naturally occurring solid, liquid, or gaseous material in or on the earth's crust in a form and amount such that extracting and converting it into useful materials or items is currently or potentially profitable. Mineral resources are classified as *metallic* (such as iron and tin ores) or *nonmetallic* (such as fossil fuels, sand, and salt).

minimum-tillage farming See *conservation-tillage farming.*

mixture Combination of one or more elements and compounds.

molecule Combination of two or more atoms of the same chemical element (such as O_2) or different chemical elements (such as H_2O) held together by chemical bonds. Compare *atom, ion.*

monoculture Cultivation of a single crop, usually on a large area of land. Compare *polyculture, polyvarietal cultivation.*

more developed country (MDC) See *developed country.*

municipal solid waste Solid materials discarded by homes and businesses in or near urban areas. See *solid waste.*

mutagen Chemical or form of radiation that causes inheritable changes (mutations) in the DNA molecules in the genes found in chromosomes. See *carcinogen, mutation, teratogen.*

mutation Random change in DNA molecules making up genes that can yield changes in anatomy, physiology, or behavior in offspring. See *mutagen.*

mutualism Type of species interaction in which both participating species generally benefit. Compare *commensalism.*

native species Species that normally live and thrive in a particular ecosystem. Compare *indicator species, keystone species, nonnative species.*

natural capital See *natural resources.*

natural gas Underground deposits of gases consisting of 50–90% by weight methane gas (CH_4) and small amounts of heavier gaseous hydrocarbon compounds such as propane (C_3H_8) and butane (C_4H_{10}).

natural greenhouse effect Heat buildup in the troposphere because of the presence of certain gases, called greenhouse gases. Without this effect, the earth would be nearly as cold as Mars, and life as we know it could not exist. Compare *global warming.*

natural law See *scientific law.*

natural radioactive decay Nuclear change in which unstable nuclei of atoms spontaneously shoot out particles (usually alpha or beta particles) or energy (gamma rays) at a fixed rate.

natural rate of extinction See *background extinction.*

natural resources The earth's natural materials and processes that sustain other species and us. Compare *human resources, manufactured resources.* See diagram on top half of back cover.

natural selection Process by which a particular beneficial gene (or set of genes) is reproduced in succeeding generations more than other genes. The result of natural selection is a population that contains a greater proportion of organisms better adapted to certain environmental conditions. See *adaptation, biological evolution, differential reproduction, mutation.*

net energy Total amount of useful energy available from an energy resource or energy system over its lifetime, minus the amount of energy used (the first law of thermodynamics), automatically wasted (the second law of thermodynamics), and unnecessarily wasted in finding, processing, concentrating, and transporting it to users.

net primary productivity (NPP) Rate at which all the plants in an ecosystem produce net useful chemical energy; equal to the difference between the rate at which the plants in an ecosystem produce useful chemical energy (gross primary productivity) and the rate at which they use some of that energy through cellular respiration. Compare *gross primary productivity.*

neutral solution Water solution containing an equal number of hydrogen ions (H^+) and hydroxide ions (OH^-); water solution with a pH of 7. Compare *acid solution, basic solution.*

neutron (n) Elementary particle in the nuclei of all atoms (except hydrogen-1). It has a relative mass of 1 and no electric charge. Compare *electron, proton.*

niche See *ecological niche.*

nitrogen cycle Cyclic movement of nitrogen in different chemical forms from the environment to organisms and then back to the environment.

noise pollution Any unwanted, disturbing, or harmful sound that impairs or interferes with hearing, causes stress, hampers concentration and work efficiency, or causes accidents.

nondegradable pollutant Material that is not broken down by natural processes. Examples are the toxic elements lead and mercury. Compare *biodegradable pollutant, degradable pollutant, slowly degradable pollutant.*

nonnative species Species that migrate into an ecosystem or are deliberately or accidentally introduced into an ecosystem by humans. Compare *native species.*

nonpersistent pollutant See *degradable pollutant.*

nonpoint source Large or dispersed land areas such as crop fields, streets, and lawns that discharge pollutants into the environment over a large area. Compare *point source.*

nonrenewable resource Resource that exists in a fixed amount (stock) in various places in the earth's crust and has the potential for renewal by geological, physical, and chemical processes taking place over hundreds of millions to billions of years. Examples are copper, aluminum, coal, and oil. We classify these resources as exhaustible because we are extracting and using them at a much faster rate than they were formed. Compare *renewable resource.*

nontransmissible disease Disease that is not caused by living organisms and does not spread from one person to another. Examples are most cancers, diabetes, cardiovascular disease, and malnutrition. Compare *transmissible disease.*

no-till farming See *conservation-tillage farming.*

nuclear change Process in which nuclei of certain isotopes spontaneously change, or are forced to change, into one or more different isotopes. The three principal types of nuclear change are natural radioactivity, nuclear fission, and nuclear fusion. Compare *chemical change, physical change.*

nuclear energy Energy released when atomic nuclei undergo a nuclear reaction such as the spontaneous emission of radioactivity, nuclear fission, or nuclear fusion.

nuclear fission Nuclear change in which the nuclei of certain isotopes with large mass numbers (such as uranium-235 and plutonium-239) are split apart into lighter nuclei when struck by a neutron. This process releases more neutrons and a large amount of energy. Compare *nuclear fusion.*

nuclear fusion Nuclear change in which two nuclei of isotopes of elements with a low mass number (such as hydrogen-2 and hydrogen-3) are forced together at extremely high temperatures until they fuse to form a heavier nucleus (such as helium-4). This process releases a large amount of energy. Compare *nuclear fission*.

nucleus Extremely tiny center of an atom, making up most of the atom's mass. It contains one or more positively charged protons and one or more neutrons with no electrical charge (except for a hydrogen-1 atom, which has one proton and no neutrons in its nucleus).

nutrient Any food, element, or compound an organism must take in to live, grow, or reproduce.

nutrient cycle See *biogeochemical cycle*.

oil See *crude oil*.

old-growth forest Virgin and old, second-growth forests containing trees that are often hundreds, sometimes thousands of years old. Examples include forests of Douglas fir, western hemlock, giant sequoia, and coastal redwoods in the western United States. Compare *second-growth forest, tree plantation*.

omnivore Animal that can use both plants and other animals as food sources. Examples are pigs, rats, cockroaches, and people. Compare *carnivore, herbivore*.

open sea Part of an ocean that is beyond the continental shelf. Compare *coastal zone*.

organic compounds Compounds containing carbon atoms combined with each other and with atoms of one or more other elements such as hydrogen, oxygen, nitrogen, sulfur, phosphorus, chlorine, and fluorine. All other compounds are called *inorganic compounds*.

organic farming Producing crops and livestock naturally by using organic fertilizer (manure, legumes, compost) and natural pest control (bugs that eat harmful bugs, plants that repel bugs, and environmental controls such as crop rotation) instead of using commercial inorganic fertilizers and synthetic pesticides and herbicides. See *sustainable agriculture*.

organic fertilizer Organic material such as animal manure, green manure, and compost, applied to cropland as a source of plant nutrients. Compare *commercial inorganic fertilizer*.

organism Any form of life.

output pollution control See *pollution cleanup*.

overfishing Harvesting so many fish of a species, especially immature fish, that not enough breeding stock is left to replenish the species, such that it is not profitable to harvest them.

overgrazing Destruction of vegetation when too many grazing animals feed too long and exceed the carrying capacity of a rangeland or pasture area.

overnutrition Diet so high in calories, saturated (animal) fats, salt, sugar, and processed foods and so low in vegetables and fruits that the consumer runs high risks of diabetes, hypertension, heart disease, and other health hazards. Compare *malnutrition, undernutrition*.

oxygen-demanding wastes Organic materials that are usually biodegraded by aerobic (oxygen-consuming) bacteria if there is enough dissolved oxygen in the water. See also *dissolved oxygen content*.

ozone depletion Decrease in concentration of ozone (O_3) in the stratosphere. See *ozone layer*.

ozone layer Layer of gaseous ozone (O_3) in the stratosphere that protects life on earth by filtering out most harmful ultraviolet radiation from the sun.

parasite Consumer organism that lives on or in and feeds on a living plant or animal, known as the host, over an extended period of time. The parasite draws nourishment from and gradually weakens its host; it may or may not kill the host. See *parasitism*.

parasitism Interaction between species in which one organism, called the parasite, preys on another organism, called the host, by living on or in the host. See *host, parasite*.

parts per billion (ppb) Number of parts of a chemical found in 1 billion parts of a particular gas, liquid, or solid.

parts per million (ppm) Number of parts of a chemical found in 1 million parts of a particular gas, liquid, or solid.

parts per trillion (ppt) Number of parts of a chemical found in 1 trillion parts of a particular gas, liquid, or solid.

passive solar heating system System that captures sunlight directly within a structure and converts it into low-temperature heat for space heating or for heating water for domestic use without the use of mechanical devices. Compare *active solar heating system*.

pathogen Organism that produces disease. Examples are harmful bacteria, viruses, and protozoa.

per capita GDP Annual gross domestic product (GDP) of a country divided by its total population. See *gross domestic product*.

percolation Passage of a liquid through the spaces of a porous material such as soil.

perpetual resource Essentially inexhaustible resource on a human time scale. Solar energy is an example. See *renewable resource*. Compare *nonrenewable resource, renewable resource*.

persistence How long a pollutant stays in the air, water, soil, or body.

persistent pollutant See *slowly degradable pollutant*.

pest Unwanted organism that directly or indirectly interferes with human activities.

pesticide Any chemical designed to kill or inhibit the growth of an organism that people consider undesirable.

petrochemicals Chemicals obtained by refining (distilling) crude oil. They are used as raw materials in manufacturing most industrial chemicals, fertilizers, pesticides, plastics, synthetic fibers, paints, medicines, and many other products.

petroleum See *crude oil*.

pH Numeric value that indicates the relative acidity or alkalinity of a substance on a scale of 0 to 14, with the neutral point at 7. Acid solutions have pH values lower than 7, and basic or alkaline solutions have pH values greater than 7.

phosphorus cycle Cyclic movement of phosphorus in different chemical forms from the environment to organisms and then back to the environment.

photochemical smog Complex mixture of air pollutants produced in the lower atmosphere by the reaction of hydrocarbons and nitrogen oxides under the influence of sunlight. Especially harmful components include ozone, peroxyacyl nitrates (PANs), and various aldehydes. Compare *industrial smog*.

photosynthesis Complex process that takes place in cells of green plants. Radiant energy from the sun is used to combine carbon dioxide (CO_2) and water (H_2O) to produce oxygen (O_2) and carbohydrates (such as glucose, $C_6H_{12}O_6$) and other nutrient molecules. Compare *aerobic respiration, chemosynthesis.*

photovoltaic cell (solar cell) Device in which radiant (solar) energy is converted directly into electrical energy.

physical change Process that alters one or more physical properties of an element or a compound without altering its chemical composition. Examples are changing the size and shape of a sample of matter (crushing ice and cutting aluminum foil) and changing a sample of matter from one physical state to another (boiling and freezing water). Compare *chemical change, nuclear change.*

phytoplankton Small drifting plants, mostly algae and bacteria, found in aquatic ecosystems. Compare *plankton, zooplankton.*

planetary management worldview Beliefs that **(1)** we are the planet's most important species; **(2)** we will not run out of resources because of our ingenuity in developing and finding new ones; **(3)** the potential for economic growth is essentially limitless; and **(4)** our success depends on how well we can understand, control, and manage the earth's life-support systems for our own benefit. See *spaceship-earth worldview.* Compare *environmental wisdom worldview, stewardship worldview.*

plankton Small plant organisms (phytoplankton) and animal organisms (zooplankton) that float in aquatic ecosystems.

plantation agriculture Growing specialized crops such as bananas, coffee, and cacao in tropical developing countries, primarily for sale to developed countries.

point source Single identifiable source that discharges pollutants into the environment. Examples are the smokestack of a power plant or an industrial plant, drainpipe of a meatpacking plant, chimney of a house, or exhaust pipe of an automobile. Compare *nonpoint source.*

poison A chemical that adversely affects the health of a living human or animal by causing injury, illness, or death.

politics Process through which individuals and groups try to influence or control government policies and actions that affect the local, state, national, and international communities.

pollutant Particular chemical or form of energy that can adversely affect the health, survival, or activities of humans or other living organisms. See *pollution.*

pollution Undesirable change in the physical, chemical, or biological characteristics of air, water, soil, or food that can adversely affect the health, survival, or activities of humans or other living organisms.

pollution cleanup Device or process that removes or reduces the level of a pollutant after it has been produced or has entered the environment. Examples are automobile emission control devices and sewage treatment plants. Compare *pollution prevention.*

pollution prevention Device or process that prevents a potential pollutant from forming or entering the environment or sharply reduces the amount entering the environment. Compare *pollution cleanup.*

polyculture Complex form of intercropping in which a large number of different plants maturing at different times are planted together. See also *intercropping.* Compare *monoculture, polyvarietal cultivation.*

polyvarietal cultivation Planting a plot of land with several varieties of the same crop. Compare *intercropping, monoculture, polyculture.*

population Group of individual organisms of the same species living in a particular area.

population change Increase or decrease in the size of a population. It is equal to (Births + Immigration) − (Deaths + Emigration).

population density Number of organisms in a particular population found in a specified area or volume.

population size Number of individuals making up a population's gene pool.

potential energy Energy stored in an object because of its position or the position of its parts. Compare *kinetic energy.*

poverty Inability to meet basic needs for food, clothing, and shelter.

ppb See *parts per billion.*

ppm See *parts per million.*

ppt See *parts per trillion.*

precautionary principle When there is scientific uncertainty about potentially serious harm from chemicals or technologies, decision makers should act to prevent harm to humans and the environment. See *pollution prevention.*

precipitation Water in the form of rain, sleet, hail, and snow that falls from the atmosphere onto the land and bodies of water.

predation Situation in which an organism of one species (the predator) captures and feeds on parts or all of an organism of another species (the prey).

predator Organism that captures and feeds on parts or all of an organism of another species (the prey).

predator-prey relationship Interaction between two organisms of different species in which one organism, called the *predator,* captures and feeds on parts or all of another organism, called the *prey.*

prey Organism that is captured and serves as a source of food for an organism of another species (the predator).

primary consumer Organism that feeds on all or part of plants (herbivore) or on other producers. Compare *detritivore, omnivore, secondary consumer.*

primary pollutant Chemical that has been added directly to the air by natural events or human activities and occurs in a harmful concentration. Compare *secondary pollutant.*

primary productivity See *gross primary productivity, net primary productivity.*

primary sewage treatment Mechanical sewage treatment in which large solids are filtered out by screens and suspended solids settle out as sludge in a sedimentation tank. Compare *secondary sewage treatment.*

primary succession Sequential development of communities in a bare area that has never been occupied by a community of organisms. Compare *secondary succession.*

probability Mathematical statement about how likely it is that something will happen.

producer Organism that uses solar energy (green plant) or chemical energy (some bacteria) to manufacture the organic compounds it needs as nutrients from simple inorganic compounds obtained from its environment. Compare *consumer, decomposer.*

proton (p) Positively charged particle in the nuclei of all atoms. Each proton has a relative mass of 1 and a single positive charge. Compare *electron, neutron*.

pure free-market economic system System in which all economic decisions are made in the market, where buyers and sellers of economic goods interact freely, with no government or other interference.

pyramid of energy flow Diagram representing the flow of energy through each trophic level in a food chain or food web. With each energy transfer, only a small part (typically 10%) of the usable energy entering one trophic level is transferred to the organisms at the next trophic level.

radiation Fast-moving particles (particulate radiation) or waves of energy (electromagnetic radiation). See *alpha particle, beta particle, gamma rays*.

radioactive decay Change of a radioisotope to a different isotope by the emission of radioactivity.

radioactive isotope See *radioisotope*.

radioactive waste Waste products of nuclear power plants, research, medicine, weapon production, or other processes involving nuclear reactions. See *radioactivity*.

radioactivity Nuclear change in which unstable nuclei of atoms spontaneously shoot out "chunks" of mass, energy, or both at a fixed rate. The three principal types of radioactivity are gamma rays and fast-moving alpha particles and beta particles.

radioisotope Isotope of an atom that spontaneously emits one or more types of radioactivity (alpha particles, beta particles, or gamma rays).

rangeland Land that supplies forage or vegetation (grasses, grasslike plants, and shrubs) for grazing and browsing animals and is not intensively managed. Compare *feedlot*.

range of tolerance Range of chemical and physical conditions that must be maintained for populations of a particular species to stay alive and grow, develop, and function normally. See *law of tolerance*.

recharge area Any area of land allowing water to pass through it and into an aquifer. See *aquifer*.

reconciliation ecology The science of inventing, establishing, and maintaining new habitats to conserve species diversity in places where people live, work, or play.

recycling Collecting and reprocessing a resource so it can be made into new products. An example is collecting aluminum cans, melting them down, and using the aluminum to make new cans or other aluminum products. Compare *reuse*.

reforestation Renewal of trees and other types of vegetation on land where trees have been removed; can be done naturally by seeds from nearby trees or artificially by planting seeds or seedlings.

reliable runoff Surface runoff of water that generally can be counted on as a stable source of water from year to year. See *runoff*.

renewable resource Resource that can be replenished fairly rapidly (hours to several decades) through natural processes. Examples are trees in forests, grasses in grasslands, wild animals, fresh surface water in lakes and streams, most groundwater, fresh air, and fertile soil. If such a resource is used faster than it is replenished, it can be depleted and converted into a nonrenewable resource. Compare *nonrenewable resource* and *perpetual resource*. See also *environmental degradation*.

replacement-level fertility Number of children a couple must have to replace them. The average for a country or the world usually is slightly higher than 2 children per couple (2.1 in the United States and 2.5 in some developing countries) because some children die before reaching their reproductive years. See also *total fertility rate*.

reproductive isolation Long-term geographic separation of members of a particular sexually reproducing species.

reproductive potential See *biotic potential*.

reserves Resources that have been identified and from which a usable mineral can be extracted profitably at present prices with current mining technology.

resource Anything obtained from the living and nonliving environment to meet human needs and wants. It can also be applied to other species.

resource partitioning Process of dividing up resources in an ecosystem so species with similar needs (overlapping ecological niches) use the same scarce resources at different times, in different ways, or in different places. See *ecological niche*.

resource productivity See *material efficiency*.

respiration See *aerobic respiration*.

response Amount of health damage caused by exposure to a certain dose of a harmful substance or form of radiation. See *dose, dose-response curve, median lethal dose*.

reuse Using a product over and over again in the same form. An example is collecting, washing, and refilling glass beverage bottles. Compare *recycling*.

riparian zones Thin strips and patches of vegetation that surround streams. They are very important habitats and resources for wildlife.

risk Probability that something undesirable will result from deliberate or accidental exposure to a hazard. See *risk analysis, risk assessment, risk-benefit analysis, risk management*.

risk analysis Identifying hazards, evaluating the nature and severity of risks (*risk assessment*), using this and other information to determine options and make decisions about reducing or eliminating risks (*risk management*), and communicating information about risks to decision makers and the public (*risk communication*).

risk assessment Process of gathering data and making assumptions to estimate short- and long-term harmful effects on human health or the environment from exposure to hazards associated with the use of a particular product or technology. See *risk, risk-benefit analysis*.

risk-benefit analysis Estimate of the short- and long-term risks and benefits of using a particular product or technology. See *risk assessment*.

risk communication Communicating information about risks to decision makers and the public. See *risk, risk analysis, risk-benefit analysis*.

risk management Using risk assessment and other information to determine options and make decisions about reducing or eliminating risks. See *risk, risk analysis, risk-benefit analysis, risk communication*.

rock Any material that makes up a large, natural, continuous part of earth's crust. See *mineral*.

rock cycle Largest and slowest of the earth's cycles, consisting of geologic, physical, and chemical processes that form and modify rocks and soil in the earth's crust over millions of years.

rule of 70 Doubling time (in years) = 70/percentage growth rate. See *doubling time, exponential growth.*

runoff Fresh water from precipitation and melting ice that flows on the earth's surface into nearby streams, lakes, wetlands, and reservoirs. See *reliable runoff, surface runoff, surface water.* Compare *groundwater.*

salinity Amount of various salts dissolved in a given volume of water.

salinization Accumulation of salts in soil that can eventually make the soil unable to support plant growth.

saltwater intrusion Movement of salt water into freshwater aquifers in coastal and inland areas as groundwater is withdrawn faster than it is recharged by precipitation.

sanitary landfill Waste disposal site on land in which waste is spread in thin layers, compacted, and covered with a fresh layer of clay or plastic foam each day.

scavenger Organism that feeds on dead organisms that were killed by other organisms or died naturally. Examples are vultures, flies, and crows. Compare *detritivore.*

science Attempts to discover order in nature and use that knowledge to make predictions about what should happen in nature. See *consensus science, frontier science, scientific data, scientific hypothesis, scientific law, scientific methods, scientific model, scientific theory.*

scientific data Facts obtained by making observations and measurements. Compare *scientific hypothesis, scientific law, scientific methods, scientific model, scientific theory.*

scientific hypothesis Educated guess that attempts to explain a scientific law or certain scientific observations. Compare *scientific data, scientific law, scientific methods, scientific model, scientific theory.*

scientific law Description of what scientists find happening in nature over and over in the same way, without known exception. See *first law of thermodynamics, law of conservation of matter, second law of thermodynamics.* Compare *scientific data, scientific hypothesis, scientific methods, scientific model, scientific theory.*

scientific methods Ways that scientists gather data and formulate and test scientific hypotheses, models, theories, and laws. See *scientific data, scientific hypothesis, scientific law, scientific model, scientific theory.*

scientific model Simulation of complex processes and systems. Many are mathematical models that are run and tested using computers.

scientific theory Well-tested and widely accepted scientific hypothesis. Compare *scientific data, scientific hypothesis, scientific law, scientific methods, scientific model.*

secondary consumer Organism that feeds only on primary consumers. Compare *detritivore, omnivore, primary consumer.*

secondary pollutant Harmful chemical formed in the atmosphere when a primary air pollutant reacts with normal air components or other air pollutants. Compare *primary pollutant.*

secondary sewage treatment Second step in most waste treatment systems in which aerobic bacteria break down up to 90% of degradable, oxygen-demanding organic wastes in wastewater. This usually is done by bringing sewage and bacteria together in trickling filters or in the activated sludge process. Compare *primary sewage treatment.*

secondary succession Sequential development of communities in an area in which natural vegetation has been removed or destroyed but the soil is not destroyed. Compare *primary succession.*

second-growth forest Stands of trees resulting from secondary ecological succession. Compare *old-growth forest, tree plantation.*

second law of thermodynamics In any conversion of heat energy to useful work, some of the initial energy input is always degraded to a lower-quality, more dispersed, less useful energy, usually low-temperature heat that flows into the environment; you cannot break even in terms of energy quality. See *first law of thermodynamics.*

sedimentary rock Rock that forms from the accumulated products of erosion and in some cases from the compacted shells, skeletons, and other remains of dead organisms. Compare *igneous rock, metamorphic rock.* See *rock cycle.*

selective cutting Cutting of intermediate-aged, mature, or diseased trees in an uneven-aged forest stand, either singly or in small groups. This encourages the growth of younger trees and maintains an uneven-aged stand. Compare *clear-cutting.*

septic tank Underground tank for treating wastewater from a home in rural and suburban areas. Bacteria in the tank decompose organic wastes and the sludge settles to the bottom of the tank. The effluent flows out of the tank into the ground through a field of drainpipes.

shifting cultivation Clearing a plot of ground in a forest, especially in tropical areas, and planting crops on it for a few years (typically 2–5 years) until the soil is depleted of nutrients or the plot has been invaded by a dense growth of vegetation from the surrounding forest. Then a new plot is cleared and the process is repeated. The abandoned plot cannot successfully grow crops for 10–30 years. See also *slash-and-burn cultivation.*

slash-and-burn cultivation Cutting down trees and other vegetation in a patch of forest, leaving the cut vegetation on the ground to dry, and then burning it. The ashes that are left add nutrients to the nutrient-poor soils found in most tropical forest areas. Crops are planted between tree stumps. Plots must be abandoned after a few years (typically 2–5 years) because of loss of soil fertility or invasion of vegetation from the surrounding forest. See also *shifting cultivation.*

slowly degradable pollutant Material that is slowly broken down into simpler chemicals or reduced to acceptable levels by natural physical, chemical, and biological processes. Compare *biodegradable pollutant, degradable pollutant, nondegradable pollutant.*

sludge Gooey mixture of toxic chemicals, infectious agents, and settled solids removed from wastewater at a sewage treatment plant.

smart growth Form of urban planning that recognizes urban growth will occur but uses zoning laws and an array of other tools to prevent sprawl, direct growth to certain areas, protect ecologically sensitive and important lands and waterways, and develop urban areas that are more environmentally sustainable and more enjoyable places to live.

smog Originally a combination of smoke and fog but now used to describe other mixtures of pollutants in the atmosphere. See *industrial smog, photochemical smog.*

soil Complex mixture of inorganic minerals (clay, silt, pebbles, and sand), decaying organic matter, water, air, and living organisms.

soil conservation Methods used to reduce soil erosion, prevent depletion of soil nutrients, and restore nutrients already lost by erosion, leaching, and excessive crop harvesting.

soil erosion Movement of soil components, especially topsoil, from one place to another, usually by wind, flowing water, or both. This natural process can be greatly accelerated by human activities that remove vegetation from soil.

soil horizons Horizontal zones that make up a particular mature soil. Each horizon has a distinct texture and composition that vary with different types of soils. See *soil profile*.

soil profile Cross-sectional view of the horizons in a soil. See *soil horizon*.

solar capital Solar energy from the sun reaching the earth. Compare *natural resources*.

solar cell See *photovoltaic cell*.

solar collector Device for collecting radiant energy from the sun and converting it into heat. See *active solar heating system, passive solar heating system*.

solar energy Direct radiant energy from the sun and a number of indirect forms of energy produced by the direct input. Principal indirect forms of solar energy include wind, falling and flowing water (hydropower), and biomass (solar energy converted into chemical energy stored in the chemical bonds of organic compounds in trees and other plants).

solid waste Any unwanted or discarded material that is not a liquid or a gas. See *municipal solid waste*.

sound sciences See *consensus science*.

spaceship-earth worldview View of the earth as a spaceship: a machine we can understand, control, and change at will by using advanced technology. See *planetary management worldview*. Compare *environmental wisdom worldview, stewardship worldview*.

specialist species Species with a narrow ecological niche. They may be able to live in only one type of habitat, tolerate only a narrow range of climatic and other environmental conditions, or use only one type or a few types of food. Compare *generalist species*.

speciation Formation of two species from one species as a result of divergent natural selection in response to changes in environmental conditions; usually takes thousands of years. Compare *extinction*.

species Group of organisms that resemble one another in appearance, behavior, chemical makeup and processes, and genetic structure. Organisms that reproduce sexually are classified as members of the same species only if they can actually or potentially interbreed with one another and produce fertile offspring.

species diversity Number of different species (species richness) and their relative abundances (species evenness) in a given area or community. See *biodiversity*. Compare *ecological diversity, genetic diversity*.

S-shaped curve Leveling off of an exponential, *J*-shaped curve when a rapidly growing population exceeds the carrying capacity of its environment and ceases to grow.

stewardship worldview **(1)** We are the planet's most important species but we have an ethical responsibility to care for the rest of the planet; **(2)** we will probably not run out of resources but they should not be wasted; **(3)** we should encourage environmentally beneficial forms of economic growth and discourage environmentally harmful forms of economic growth; and **(4)** our success depends on how well we can understand, control, and care for the earth's life support systems for our benefit and for the rest of nature. Compare *environmental wisdom worldview, planetary management worldview*.

stratosphere Second layer of the atmosphere, extending about 17–48 kilometers (11–30 miles) above the earth's surface. It contains small amounts of gaseous ozone (O_3), which filters out about 99% of the incoming harmful ultraviolet (UV) radiation emitted by the sun. Compare *troposphere*.

stream Flowing body of surface water. Examples are creeks and rivers.

strip cropping Planting regular crops and close-growing plants, such as hay or nitrogen-fixing legumes, in alternating rows or bands to help reduce depletion of soil nutrients.

strip mining Form of surface mining in which bulldozers, power shovels, or stripping wheels remove large chunks of the earth's surface in strips. See *surface mining*. Compare *subsurface mining*.

subatomic particles Extremely small particles—electrons, protons, and neutrons—that make up the internal structure of atoms.

subsidence Slow or rapid sinking of part of the earth's crust that is not slope related.

subsurface mining Extraction of a metal ore or fuel resource such as coal from a deep underground deposit. Compare *surface mining*.

succession See *ecological succession, primary succession, secondary succession*.

superinsulated house House that is heavily insulated and extremely airtight. Typically, active or passive solar collectors are used to heat water, and an air-to-air heat exchanger is used to prevent buildup of excessive moisture and indoor air pollutants.

surface mining Removing soil, subsoil, and other strata and then extracting a mineral deposit found fairly close to the earth's surface. Compare *subsurface mining*.

surface runoff Water flowing off the land into bodies of surface water. See *reliable runoff*.

surface water Precipitation that does not infiltrate the ground or return to the atmosphere by evaporation or transpiration. See *runoff*. Compare *groundwater*.

sustainability Ability of a system to survive for some specified (finite) time.

sustainable agriculture Method of growing crops and raising livestock based on organic fertilizers, soil conservation, water conservation, biological pest control, and minimal use of nonrenewable fossil fuel energy.

sustainable development See *environmentally sustainable economic development*.

sustainable living Taking no more potentially renewable resources from the natural world than can be replenished naturally and not overloading the capacity of the environment to cleanse and renew itself by natural processes.

sustainable society Society that manages its economy and population size without doing irreparable environmental harm by overloading the planet's ability to absorb environmental insults, replenish its resources, and sustain human and other forms of life over a specified period, usually hundreds to thousands of years. During this period, it satisfies the needs of its people without depleting natural resources and thereby jeopardizing the prospects of current and future generations of humans and other species.

sustainable yield (sustained yield) Highest rate at which a potentially renewable resource can be used without reducing its available supply throughout the world or in a particular area. See also *environmental degradation*.

synfuels Synthetic gaseous and liquid fuels produced from solid coal or sources other than natural gas or crude oil.

synthetic natural gas (SNG) Gaseous fuel containing mostly methane produced from solid coal.

temperature Measure of the average speed of motion of the atoms, ions, or molecules in a substance or combination of substances at a given moment. Compare *heat*.

temperature inversion Layer of dense, cool air trapped under a layer of less dense, warm air. This prevents upward-flowing air currents from developing. In a prolonged inversion, air pollution in the trapped layer may build up to harmful levels.

teratogen Chemical, radiation, or virus that causes birth defects. See *carcinogen, mutagen*.

terracing Planting crops on a long steep slope that has been converted into a series of broad, nearly level terraces with short vertical drops from one to another that run along the contour of the land to retain water and reduce soil erosion.

terrestrial Pertaining to land. Compare *aquatic*.

tertiary (higher-level) consumers Animals that feed on animal-eating animals. They feed at high trophic levels in food chains and webs. Examples are hawks, lions, bass, and sharks. Compare *detritivore, primary consumer, secondary consumer*.

theory of evolution Widely accepted scientific idea that all life-forms developed from earlier life-forms. Although this theory conflicts with the creation stories of many religions, it is the way biologists explain how life has changed over the past 3.6–3.8 billion years and why it is so diverse today.

thermal inversion See *temperature inversion*.

threatened species Wild species that is still abundant in its natural range but likely to become endangered because of a decline in numbers. Compare *endangered species*.

throughput Rate of flow of matter, energy, or information through a system.

throwaway society See *high-throughput economy*.

tolerance limits Minimum and maximum limits for physical conditions (such as temperature) and concentrations of chemical substances beyond which no members of a particular species can survive. See *law of tolerance*.

total fertility rate (TFR) Estimate of the average number of children who will be born alive to a woman during her lifetime if she passes through all her childbearing years (ages 15–44) conforming to age-specific fertility rates of a given year. In simpler terms, it is an estimate of the average number of children a woman will have during her childbearing years.

toxic chemical See *poison, carcinogen, hazardous chemical, mutagen, teratogen*.

toxicity Measure of how harmful a substance is.

toxicology Study of the adverse effects of chemicals on health.

toxic waste Form of hazardous waste that causes death or serious injury (such as burns, respiratory diseases, cancers, or genetic mutations). See *hazardous waste*.

traditional intensive agriculture Producing enough food for a farm family's survival and perhaps a surplus that can be sold. This type of agriculture uses higher inputs of labor, fertilizer, and water than traditional subsistence agriculture. See *traditional subsistence agriculture*. Compare *industrialized agriculture*.

traditional subsistence agriculture Production of enough crops or livestock for a farm family's survival and, in good years, a surplus to sell or put aside for hard times. Compare *industrialized agriculture, traditional intensive agriculture*.

tragedy of the commons Depletion or degradation of a potentially renewable resource to which people have free and unmanaged access. An example is the depletion of commercially desirable fish species in the open ocean beyond areas controlled by coastal countries. See *common-property resource*.

transmissible disease Disease caused by living organisms (such as bacteria, viruses, and parasitic worms) that can spread from one person to another by air, water, food, or body fluids (or in some cases by insects or other organisms). Compare *nontransmissible disease*.

tree farm See *tree plantation*.

tree plantation Site planted with one or only a few tree species in an even-aged stand. When the stand matures it is usually harvested by clear-cutting and then replanted. These farms normally are used to grow rapidly growing tree species for fuelwood, timber, or pulpwood. See *even-aged management*. Compare *old-growth forest, second-growth forest, uneven-aged management*.

trophic level All organisms that are the same number of energy transfers away from the original source of energy (for example, sunlight) that enters an ecosystem. For example, all producers belong to the first trophic level, and all herbivores belong to the second trophic level in a food chain or a food web.

troposphere Innermost layer of the atmosphere. It contains about 75% of the mass of earth's air and extends about 17 kilometers (11 miles) above sea level. Compare *stratosphere*.

true cost See *full cost*.

undergrazing Reduction of the net primary productivity of grassland vegetation and grass cover from absence of grazing for long periods (at least 5 years). Compare *overgrazing*.

undernutrition Consuming insufficient food to meet one's minimum daily energy needs for a long enough time to cause harmful effects. Compare *malnutrition, overnutrition*.

uneven-aged management Method of forest management in which trees of different species in a given stand are maintained at many ages and sizes to permit continuous natural regeneration. Compare *even-aged management*.

urban sprawl Growth of low-density development on the edges of cities and towns. See *smart growth*.

water cycle See *hydrologic cycle*.

waterlogging Saturation of soil with irrigation water or excessive precipitation so the water table rises close to the surface.

water pollution Any physical or chemical change in surface water or groundwater that can harm living organisms or make water unfit for certain uses.

watershed Land area that delivers water, sediment, and dissolved substances via small streams to a major stream (river).

water table Upper surface of the zone of saturation, in which all available pores in the soil and rock in the earth's crust are filled with water.

weather Short-term changes in the temperature, barometric pressure, humidity, precipitation, sunshine, cloud cover, wind direction and speed, and other conditions in the troposphere at a given place and time. Compare *climate*.

weathering Physical and chemical processes in which solid rock exposed at earth's surface is changed to separate solid particles and dissolved material, which can then be moved to another place as sediment. See *erosion*.

wetland Land covered all or part of the time with salt water or fresh water, excluding streams, lakes, and the open ocean. See *coastal wetland, inland wetland*.

wilderness Area where the earth and its community of life have not been seriously disturbed by humans and where humans are only temporary visitors.

wind farm Cluster of small to medium-sized wind turbines in a windy area to capture wind energy and convert it into electrical energy.

worldview How people think the world works and what they think their role in the world should be. See *environmental wisdom worldview, planetary management worldview, spaceship-earth worldview, stewardship worldview*.

zone of saturation Area where all available pores in soil and rock in the earth's crust are filled by water. See *water table*.

zoning Regulating how various parcels of land can be used.

zooplankton Animal plankton. Small floating herbivores that feed on plant plankton (phytoplankton). Compare *phytoplankton*.

INDEX

Note: Page numbers of **boldface** type indicate definitions of key terms. Page numbers followed by italicized *f* or *t* indicate figures or tables, and *b* indicates boxes.

Abiotic components in ecosystems, **27,** 28*f*, 31*f*
Abortion, birth rate and availability of, 80
Aboveground buildings, hazardous waste storage in, 301
Abyssal zone, ocean, 59, 60*f*
Acid deposition (acid rain), 42, **277**–79
　formation of, 277*f*
　global regions affected by, 278*f*
　harmful effects of, 278–79
　solutions for reducing, 279*f*
　in United States, 278, 279
Active solar heating system, 225*f*, **226**
Acute effects of toxic chemicals, 239
Adaptation, **46**
　ecological niche and, 48
　limits of, 49
　strategy of, to global warming, 269*f*
Adaptive ecosystem management, 118, 119*f*
Adaptive radiations, **46, 51**
Adaptive trait, **46**
Advanced light-water reactor (ALWRs), 218
Aerobic respiration, **30,** 40
Aesthetic value, 102
Affluence, greater environmental quality and, 13
Affluenza, **12**
　escaping from, 326
　solid waste production and, 287–88
Age structure, population, **82**–84
　AIDS epidemic and, 84, 246*f*
　diagrams of, 82*f*
　effect of, on population growth, 82, 83*f*
　fertility rates and, 84
　making population and economic projections based on, 83–84
　U.S. baby-boom generation, 83*f*
Agribusiness, 149
Agriculture. *See also* Crop(s); Food production
　controlling pests through cultivation methods of, 168
　effects of climate change on, 263*f*
　genetically-engineered plants and, 52, 53*f*, 156–58
　government policies on, 163
　Green Revolutions in, 148–49
　growing techniques in traditional, 149–50
　industrialized (high-input), 146, 149
　irrigation and, 151, 152*f*, 159, 173, 183–84
　monoculture, 72, 106, 148–49
　plantation, 146
　polyculture, 150, 159
　soil conservation and methods of, 152–54
　sustainable, 169–70

traditional intensive, 147
traditional subsistence, 147
tree farms, 106
in tropical forests, 113*f*
in United States, 149
water needs for, 173*f*
water pollution caused by, 186
Agroforestry, **149, 153**
AIDS (acquired immune deficiency syndrome), 245, 246
　effects of, on Botswana's age structure, 246*f*
　population declines attributed to, 84
Air circulation, climate and global, 255*f*
Air pollutants
　formaldehyde as, 280
　indoor, 280*f*
　major types of outdoor, 273, 275*t*
　primary, and secondary, 273, 274*f*
　radon gas as, 281
　soot and black carbon aerosols, 262
Air pollution, **273**–87
　acid deposition and, 277–79
　atmospheric ozone depletion and increased, 271*f*
　climate change and effects of, 262
　fossil-fuel burning as source of, 207*f*, 210, 273
　harmful effects of, on human health, 12, 281–82
　industrial smog as, 274–77
　indoor, 12, 279–81
　motor vehicles as cause of, 95, 284
　outdoor, major types and sources of, 273
　photochemical smog as, 94, 273–74, 276–77
　preventing and reducing, 282–86
　radioactive radon gas as, 281
Algae, production of hydrogen from green, 235*b*
Alien (nonnative) species, **63.** *See also* Nonnative species
Alley cropping, **149, 153***f*
Altitude, climate, biomes, and effects of, **55,** 56*f*
American Council on Science and Health (ACSH), 165
American Wind Energy Association, 220
Ammonification, nitrogen fixation and, 42
Amphibians as indicator species, 63–64
Anderson, Ray, development of sustainable green corporation by, 291*b*
Anemia, 156
Animal(s)
　as food source, 146
　gray wolf, 116*b*
　keystone species, 64–65
　range reductions affecting, 133*f*
　speciation of, 50*f*
　in tropical rain forests, 49*f*

use of, in chemical toxicity lab experiments, 240, 241–42
zoos and aquariums for protection of endangered, 143
Animal manure, **154,** 230
Antarctic, food web in, 33*f*
Antibiotics, overuse, and germ resistance to, 244
Aquaculture, 161, **162**–63
　advantages and disadvantages of, 163*f*
　sustainable, 163*f*
Aquariums, protecting endangered species in, 143
Aquatic life zones, **58**–62
　biodiversity in, 102, 122–24
　biological magnification of PCBs in, 139*f*
　components of, 28*f*
　effects of acid deposition on, 278–79
　food web in, 33*f*
　freshwater, 58, 59–62
　marine (saltwater), 58–59, 60*f*
　net primary productivity in, 34*f*
　range of tolerance for temperature in, 29*f*
　zones of, 58
Aqueducts, 179
Aquifers, **172**–73. *See also* Groundwater
　contamination of, 171*f*
　depletion of, 180–82
　preventing depletion of, 182*f*
Aral Sea water diversion project, ecological disaster of, 180
Arboreta, 143
Arctic National Wildlife Refuge (ANWR), oil and gas development in, 206
Argentine fire ant, 135*f*, 136*f*
Arsenic as water pollutant, 191
Artificial selection, **52**
El-Ashry, Mohamed, 182, 237
Asthma, 281
Atmosphere, **25,** 253–56
　carbon dioxide levels in, over time, 257*f*
　chemical makeup of, 253
　chemical makeup of, and greenhouse effect, 255–56
　climate and, 254–56 (*see also* Climate; Climate change)
　layers and temperature of, 253*f*
　ozone in, 254 (*see also* Ozone (O_3); Ozone depletion)
　stratosphere of, 253–54 (*see* Stratosphere)
　temperature of, 253*f*, 256, 257*f*, 258*f*, 261*f* (*see also* Global warming)
　troposphere of, 253 (*see also* Troposphere)
　weather and, 253 (*see also* Weather)
Atom(s), **18**
　chemical formulas and, 18
　mass number, 20
　structure of, 18–20

Cell(s), **24**
  relationship of chromosomes, genes, and, 20*f*
Center for International Earth Science Information Network, 101
Center for International Forestry Research, 113
Center-pivot low-pressure sprinkler, 183*f*, 184
Centers for Disease Control and Prevention, 239
Central Arizona Project, 179*f*
Central receiver system, 226
Chain reaction, nuclear, **211***f*
Channelization of streams, 176*f*, 177
Chateaubriand, François-Auguste-René de, 101
Chemical(s)
  inorganic, as water pollutant, 187*t*
  ozone depleting, 270
Chemical bonds, 18
Chemical change in matter, 21–22
Chemical formula, **18**
Chemical hazards, 238, 242–44. *See also* Hazardous wastes; Toxic chemicals
  arsenic as, 191
  assessing toxicity of, 239–42, 243
  case studies on lead, 302–3
  effects of, on human body, 243
  in homes, 164, 300*f*
  indoor air pollutants as, 281–82
  lack of sufficient knowledge about, 243
  persistent organic pollutants (POPs), 74, 244, 304–5
  pollution prevention as solution to threat of, 243–44
  toxic and hazardous, 242 (*see also* Toxic chemicals)
  toxicology as assessment of, 239–42
Chemical reaction, **21–22**
Chernobyl nuclear power accident, radioactive pollutants from, 214, 216*b*
Chesapeake Bay, pollution in, 192–93, 194*f*
Children
  cost of raising and educating, 79–80
  in labor force, 79
  protecting from lead poisoning, 302–3, 304*f*
  saving from malnutrition and disease, 155
China
  acid deposition in, 278
  biogas production in, 231
  carbon dioxide emissions reduced by, 268*f*
  coal reserves of, 209
  oil use in, 206
  population growth, and population regulation in, 88–89
  water resources in, 183, 189
Chlorinated hydrocarbons, 18
Chlorofluorocarbons (CFCs), 73, 269
  projected concentrations of, in atmosphere, 273*f*
  replacements for, 272*b*
  stratospheric ozone depletion and role of, 270
Chromosomal DNA, 18, 20*f*
Chromosomes, **18,** 20*f*
Chronic effects of toxic chemicals, 240
Chronic undernutrition, 155
Circle of poison, 167*b*
Cities. *See* Urban areas
Clarke, Arthur C., 287
Clean Air Acts, U.S., 273, 282–83

Cleanup approach
  to acid deposition, 279*f*
  to air pollution, 284*f*, 285*f*
  to coastal water pollution, 195*f*
  to global warming, 265*f*
  to groundwater pollution, 192*f*
Clean Water Act of 1977, U.S., 195, 196, 197, 198
Clear-cutting forests, 107, 108*f*
  advantages and disadvantages of, 109*f*
Climate, 253, **254**
  atmosphere, greenhouse effect, and, 26, 27*f*, 255–56
  biomes and, 54, 56*f* (*see also* Biome(s))
  change in (*see* Climate change)
  global air circulation and, 255*f*
  natural processes involving troposphere and, 260*f* (*see also* Troposphere)
  ocean currents and global zones of, 254*f*, 255
  water resources and, 174
  weather and, 254–55
  wind and, 255
Climate change, 256–59
  atmospheric carbon dioxide levels and, 257*f*
  characteristics of global warming, climate change and, 259*t*
  concentrations of greenhouse gases and, 258*f*
  effects of, on forests, 264*f*
  factors affecting earth's average temperature and, 260–62
  human activities leading to, 73
  major processes in, 260*f*
  natural greenhouse effect and past, 255–56
  options for dealing with, 264–69
  past, 256*f*, 257
  possible effects of, 262–64
  scientific consensus on, 258–59
  as species extinction threat, 138–39
  speed of, as cause of concern, 259, 261*f*
  tropospheric warming and, 257–58 (*see also* Global warming)
Climax community, 69
Clone, 52
Cloud cover, climate and changes in, 261–62
Clownfish, mutualism and, 67, 68*f*
Coal, **209–11**
  advantages and disadvantages of, 210*f*
  converting, into gaseous and liquid fuel, 210, 211*f*
  environmental impact of burning, 207*f*, 210
  formation of, 209*f*
  nuclear energy versus, 214, 215*f*
  supplies of, 209–10
Coal-burning facilities
  energy efficiency of, 221*f*
  reducing air pollution from, 284*f*
  trading emissions from, 283
Coal gasification, **210**
Coastal wetlands, 59
Coastal zones, **58**
  development of, and threats to marine biodiversity, 123
  effects of climate change on, 263*f*
  integrated coastal management, 193
  as marine life zone, 59, 60*f*
  pollution in, 192, 193*f*
  protecting, 194, 195*f*
Cockroaches, 48*b*
Cogeneration, **221**
Colinvaux, Paul A., 253

Coloration, 46
Combined heat and power (CHP) systems, 221
Command-and-control approach to regulating air pollution, 283
Commensalism, **67–68**
Commercial energy
  agricultural use of, in U.S., 149*f*
  flow of, in U.S. economy, 220*f*
  fossil fuels as main source of, 201–2
  U.S. sources of, 202*f*
  world sources of, 202*f*
Commercial extinction, 162
Commercial inorganic fertilizer, **153**–54
Commission on Immigration Reform, U.S., 81
Common-property resources, **8**
Community, **25.** *See also* Ecosystem(s)
  competition in, 66–67
  ecological succession in, 68–70
  keystone species in, 64–65
  parasitism, mutualism, and commensalisms in, 67–68
  population dynamics in, 70–72
  predation in, 67
Comparative risk analysis, 248, 249*f*
Competition among species, 66–67
Competitor species, 72
Complex carbohydrates, 18
Compost and composting, **154**
  of biodegradable organic waste, 294
Compounds, **18**
  organic, and inorganic, 18
Comprehensive Environmental Response, Compensation, and Liability Act (Superfund program), U.S., 298
Confined aquifer, 172*f*
Consensus (sound) science, 3, **17**
Conservation of energy, law of, **22**
Conservation of matter, law of, **22**
Conservation-tillage farming, **152**
Constitutional democracy, 316
Consumers (heterotrophs), **29,** 31*f*
Consumption
  in developed and developing countries, 13*f*
  excessive, 12 (*see also* Affluenza)
  reduced, and simple living, 326
Containers, refillable, 292
Contour farming, **152,** 153*f*
Contraception methods, effectiveness of, 80*f*
Controlled experiment, 17
  on chemical toxicity, 240–41
Convention on Biological Diversity (CBD), 139
Convention on International Trade in Endangered Species (CITES), 139
Copenhagen Protocol, 272, 273*f*
Coral reefs, 59, 123
Corporate Average Fuel Economy (CAFE), 222
Costanza, Robert, 307
Costa Rica
  ecological restoration in, 122
  nature reserves in, 118*f*, 119
Costs
  internal, and external, 310
  of nuclear energy, 218
  pricing based on full, 310–11
Cover crops, **153**
Cowie, Dean, 1
Crime, involvement of international, in wildlife trade, 137–38
Critical thinking, developing, 3–4

Heat
high temperature, 204f, 226–27
ocean storage of, 260, 261f
as water pollutant, 187t, 188f
Heating
net energy ratios for, 204f
solar, 225f, 225f
space, 204f
using renewable energy to provide, 225–26
Heliostats, 226
Hepatitis B (HBV), 246
Herbicides, 164
Heterotrophs, 29, 31f
High-input (industrialized) agriculture, **146,** 149
High-level radioactive wastes, 215–17
options for storage/disposal of, 216
security threats and, 215–17
storage of, in U.S., 217
High-quality energy, **21**
one-way flow of, and life on earth, 26f
High-quality matter, **20,** 21f
High-throughput (high-waste) economy, **75**f
mentality of, 288
shift to service flow economy from, 289–91
waste management strategies for, 288, 289f
Hill, Julia Butterfly, 111b
HIPPO (Habitat destruction and fragmentation, Invasive (alien) species, Population growth, Pollution, Overharvesting), 132
Hirschorn, Joel, 252
HIV (human immunodeficiency virus), 245, 246
Homes
indoor air pollution in, 280f
net energy ratios for heating, 204f
pesticides use in, 164
reducing energy use in, 223, 224
reducing water use in, 184–85
solar heated, 225f
Honeybees
effect of pesticides on, 166
wild African, 134, 135f
Hormone(s)
chemical hazards to human, 243
controlling insects by disrupting, 168f
Hot-dry rock zones, 232
Human capital, **307**
Human health. *See also* Human nutrition
biological hazards (disease) to, 244–48
chemical hazards to, 239–44
comparative risk analysis of, 249f
effects of acid deposition on, 278
effects of air pollution on, 281–82
effects of chemical hazards on, 242, 243
effects of climate change on, 263f
effects of ozone depletion and ultraviolet radiation on, 270, 271f, 272f
impact of coal burning on, 210
pesticides and, 167
poisons and (*see* Poison)
potential effects of climate change on, 263f
preventable problems of, 12
risk, probability, and categories of hazards to, 238
risk analysis of hazards to, 248–52
tobacco, smoking, and damage to, 238–39
toxicology and assessment of threats to, 239–42

Human impact on environment, 72–76. *See also* Environmental degradation; Natural capital degradation
air pollution (*see* Air pollution)
on aquatic biodiversity, 123
alterations of natural systems, types of, 72–73
on biodiversity, 51–52, 101–2
on bird species, 134
on carbon cycle, 41
characteristics of natural systems versus human-dominated systems, 73f
climate change, greenhouse gases, and, 73
on deserts, 55, 57f, 58f
developing sustainable economies to reduce, 75–76
earth's net primary productivity and, 35
ecological footprint and, 9f, 72–73
extracting, processing of mineral resources, 10f
on forests, 57, 58f
on grasslands, 57, 58f
learning from ecology to reduce, 73–74
on marine systems, 60f
on nitrogen cycle, 42–43
pesticide use and, 138–39, 166–67
on phosphorus cycle, 43–44
species extinction and, 127–30, 134
unforeseen consequences of, 75b
on water (hydrologic) cycle, 39–40
Human nutrition
food supplies and, 154, 155f
macronutrients, 155
malnutrition, 12, 84, 154, 155
micronutrients and, 156
overnutrition, 156
saving children from malnutrition and disease, 155
undernutrition, 155
Human population, 77–100
age structure of, 82–84
case studies: India and China, 87–89
dieback of, 71
distribution of, and urbanization, 89–92
effects of climate change on, 263f
factors affecting size of, 77–82
growth of (*see* Human population growth)
influencing size of, 84–87
land-use planning for improved habitat for, 97–99
transportation, urban development, and, 94–96
urban resource and environmental problems linked to, 92–94
Human population growth, 6–7, 77–82
age structure and, 82, 83f
birth/death rates and, 77
case studies on slowing, 87–89
exponential growth of, 77–78
factors affecting birth and fertility rates, 79–80
factors affecting death rates, 80–81
fertility rates and, 78–79
immigration and, 81–82
new vision for reducing, 87
projections of global, to year 2050, 78f
poverty and, 12
rate of, 6f, 77–78
sustainable agriculture linked to reduced, 169
in United States, 78–79, 81–82
Human resources, **307**

Human rights principle, environmental policy and, 316
Humility principle, and environmental policy, 316
Humus, **36**
Hybrid gas-electric internal combustion engine, 222, 284
Hydrocarbons, 18
Hydrogen as fuel, 232–33
advantages/disadvantages of, 233f
production of, from green algae, 235b
role of fuel cells, 222, 223f, 233
Hydrological poverty, 174
Hydrologic (water) cycle, 39f, 40
availability of fresh water and, 171
Hydropower, 5, 228–29
Hydrosphere, **25**

Igneous rock, **44**
Immigration and emigration
population growth and, 70
in United States, 81–82
Immune system, chemical hazards to human, 243
Incineration
of biomass, 202, 230–31
of solid wastes, 296, 297f
Income, global distribution of, 313f
India
ecological footprint of, 9f
population growth, and population regulation in, 87–88
Indicator species, **63**
birds as, 134
amphibians as, 63–64
Individuals, influence and role of, 15, 327
author's introduction to environmental problems, 1
creating green sustainable business, 291b
environmental policy and, 316, 317f
protecting biodiversity and ecosystems, 121f, 124–25
reducing CO_2 emissions, 268f
reducing and preventing air pollution, 285–86
reducing and preventing water pollution, 199, 200f
reducing exposure to UV radiation, 272f
reducing solid waste, 289f
reuse of items, 293f
sustainable agriculture, 170f
water use and waste, 186f
Indoor air pollution, 12, 279–81
formaldehyde and, 280
harmful effects of, 281–82
radon gas, 281
strategies for reducing, 284, 285f, 286f
types and sources of, 280f
Industrial ecosystem, 290f
Industrial forestry, 106
Industrial heat, net energy ratios for high-temperature, 204f
Industrialized (high-input) agriculture, **146**
in United States, 149
Industrial smog, 274, **275,** 276
factors influencing formation of, 276–77
Industry
brownfields and abandoned, 290
improving energy efficiency in, 221
reducing water use by, 184–85
water needs for, 173f
water pollution caused by, 186
Infant mortality rate, 80, **81**
Infectious agents. *See* Pathogens

Solid wastes (cont'd)
    burying of, 296–97, 298f
    ecoindustrial revolution and reduction
        of, 289–91
    as indicator of high-waste society, 288
    individual's role in reducing, 289f
    priorities for dealing with, 289f
    producing less, 288–89
    production of, in U.S., 287, 288
    recycling and, 293–96
    reuse and, 292–93
    service flow economy and, 291
    sources of, 287f
Solutions
    acid deposition, 279f
    air pollution, 284f, 285f
    for developing environmentally-
        sustainable societies, 325f
    global warming, 265f
    lead poisoning, 304f
    managing fisheries, 124f
    preventing and reducing water pollution,
        192f, 199f
    preventing and treating infectious
        disease, 248f
    preventing coastal water pollution, 195f
    preventing groundwater depletion, 182f
    principles of sustainability, 74f
    reducing energy waste, 221f
    reducing water waste, 184f, 185f
    smart growth tools in urban areas
        as, 98f
    soil salinization, 152f
    sustainable agriculture, 170f
    sustainable aquaculture, 163f
    sustainable forestry, 109f, 114f
    sustainable water use, 185f
Songbirds, 63
    resource partitioning by warblers, 66f
Soot in atmosphere, 262
Soulé, Michael, 131
Source zone, streams, 62f
Space heating, net energy ratios for, 204f
Special-interest groups, 316
Specialist species, 48–49, 66–67
    resource partitioning by, 66f
    in tropical rain forest, 49f
Speciation, 50
    crisis of, 130
Species, 24, 63–68
    commensalism among, 67–68
    competition among, 66–67
    diversity of, at different latitudes, 56f
    ecological niches of, 48–49
    ecological roles played by, 63
    endemic, 134
    evolution of, 50 (see also Speciation)
    extinction of, 126–30 (see also Extinction
        of species)
    evolution of, 50 (see also Speciation)
    foundation, 65–66
    human impact on, 127–30, 134
    importance of wild, 130–32
    indicator, 63–64
    interactions of, 66–68
    keystone, 64–65
    microbial, 24
    native, 63
    nonnative, 63
    parasitism and mutualism among,
        67–68
    predation among, 67
    preservation of (see Ecosystem approach
        to sustaining biodiversity; Species ap-
        proach to sustaining biodiversity)

Species approach to sustaining biodiversity,
        126–45
    causes of premature extinction of wild
        species, 132–39
    ecosystem approach versus, 103f
    importance of wild species, 130–32
    legal approach to protecting wild species,
        139–42
    reconciliation ecology and, 143–44
    sanctuary approach to protecting wild
        species, 142–43
    species extinction and, 126–30 (see also
        Extinction of species)
Species-area relationship, 127
Species diversity, 30. See also Biodiversity
Speth, Gus, 322
Squamous cell skin cancer, 271
Squatter settlements, 94
Stegner, Wallace, 119
Stewardship environmental worldview,
        323f, 324
Stewart, Richard B., 267
Stratosphere, 25, 253–54
    ozone in, 254
    ozone depletion in, 269–71
Streams. See Rivers and streams
Strip cropping, 153f
Strip cutting, 107, 108f
Strong, 201
Structural adaptations, 46
Student environmental groups, 320
Subatomic particles, 19
Sulfur dioxide (SO$_2$) as air pollutant, 275t
    emissions trading of, 283, 312
Sun, 5. See also Solar energy
Superfund Act, U.S., 298
Superinsulated house, 223, 224f
Surface impoundment of hazardous waste,
        301, 302f
Surface litter layer (O horizon), soil, 35f, 36
Surface runoff, 39, 60–61, 62f, 171
Surface water, 60, 171–72
    preventing and reducing pollution in,
        195–98
Suspended particulate matter (SPM) as air
        pollutant, 275t
Sustainability
    ecocity concept and, 98, 99
    in food production, 169–70
    lack of, in urban areas, 92f
    low-waste society and, 289–91, 303–5
    of present course, 14–15
    principles of, 74f
    in water use and waste, 185, 186f
Sustainable (low-input) agriculture, 169–70
Sustainable energy future, 234–37
    decentralized power system, 234f
    micropower, 234–35
    role of economics in, 235
    strategies for, 236f
Sustainable forestry, 109f, 110
Sustainable living, 5–6, 324–27
    becoming effective environmental citi-
        zens for, 326
    components of environmental revolution
        and, 326–27
    environmental literacy and, 324–25
    environmental wisdom and, 325–26
    guidelines and strategies for developing,
        73–74, 325f
    living more simply for, 326
    in urban areas, 97–99
Sustainable yield, 8
Suzuki, David, 253
Swamps, 61

Synfuels, 210, 211f
Synthetic natural gas (SNG), 210, 211f
Syria, 175

Tapeworms, 67
Tar sand, 207
Taxes, fees, and subsidies
    alternative transportation and shifts in,
        96
    climate change, global warming and
        government, 267
    environmental tax reform, 312
    for food production, 163
    green, 308, 311–12
    improving environmental quality using,
        311–12
    to reduce energy waste, 225
Technological optimists, 14
Technology
    estimating risks of new, 249–50
    population, consumption, and, 13f, 14
    transfer of, to reduce threat of climate
        change, 267
Teenagers, population growth, age structure
        and, 82
Temperate deciduous forests, soils of, 37f
Temperature
    of earth, 26–27, 54, 56f, 254 256–57, 260f,
        261f (see also Global warming)
    organism's range of tolerance for, 29f
Temperature inversions, 276
Teratogens, 242
Terracing, 152, 153f
Terrestrial ecosystems. See also Biome(s)
    components of, 28f
    degradation of, 101, 102f
    ecological succession in, 68–70
    net primary productivity of, 34f
Terrorism
    nuclear energy and potential, 214, 215,
        216
    September 11, 2001 attack in U.S., 214,
        215
Theory of evolution, 46
Thermodynamics, laws of, 16–17, 22–24
Thomashow, Mitchell, 325
Thoreau, Henry David, 15
Threatened (vulnerable) species, 126,
        128–29f
Threshold dose-response model, 241f, 242
Tiger
    illegal poaching of, 137
    reduced range of Indian, 133f
Tiger salamanders, 48
Tigris-Euphrates River basin, 175f
Timber
    certifying sustainably-grown, 110
    harvesting methods, 106–7, 108f
Tobacco, health hazards of, 238–39
Todd, John, 171, 197
Tool libraries, 292
Topsoil
    A horizon, 35f, 36
    erosion of, 150
Total fertility rate (TFR), 78
    in United States, 78, 79f
Toxic chemicals, 239–44
    defined, 242
    determining toxicity of, 239, 240–42
    dose and lethal dose of, 240t
    effects of, on human body, 242, 243
    eliminating, from sewage, 197
    in homes, 164, 300f
    persistent organic pollutants (POPs), 74,
        244, 304–5